Studies in Computational Intelligence

Volume 1088

The series "Studies in Computational Intelligence" (SCI) publishes new developments and advances in the various areas of computational intelligence—quickly and with a high quality. The intent is to cover the theory, applications, and design methods of computational intelligence, as embedded in the fields of engineering, computer science, physics and life sciences, as well as the methodologies behind them. The series contains monographs, lecture notes and edited volumes in computational intelligence spanning the areas of neural networks, connectionist systems, genetic algorithms, evolutionary computation, artificial intelligence, cellular automata, self-organizing systems, soft computing, fuzzy systems, and hybrid intelligent systems. Of particular value to both the contributors and the readership are the short publication timeframe and the world-wide distribution, which enable both wide and rapid dissemination of research output.

Indexed by SCOPUS, DBLP, WTI Frankfurt eG, zbMATH, SCImago.

All books published in the series are submitted for consideration in Web of Science.

Cosimo Ieracitano · Nadia Mammone ·
Marco Di Clemente · Mufti Mahmud ·
Roberto Furfaro · Francesco Carlo Morabito
Editors

The Use of Artificial Intelligence for Space Applications

Workshop at the 2022 International Conference on Applied Intelligence and Informatics

Springer

Editors
Cosimo Ieracitano
Reggio Calabria, Italy

Marco Di Clemente
Rome, Italy

Roberto Furfaro
Tucson, AZ, USA

Nadia Mammone
Reggio Calabria, Italy

Mufti Mahmud
Nottingham, UK

Francesco Carlo Morabito
Reggio Calabria, Italy

ISSN 1860-949X ISSN 1860-9503 (electronic)
Studies in Computational Intelligence
ISBN 978-3-031-25754-4 ISBN 978-3-031-25755-1 (eBook)
https://doi.org/10.1007/978-3-031-25755-1

This Springer imprint is published by the registered company Springer Nature Switzerland AG
The registered company address is: Gewerbestrasse 11, 6330 Cham, Switzerland

Preface

In recent years, there has been a growing impact of Artificial Intelligence (AI) as an interdisciplinary key-enabling technology in several research areas, including the space sector. Indeed, AI-based systems have been contributing to many space operations such as mission planning, big space data collection and processing, autonomous navigation, spacecraft monitoring, identification and clustering of debris, mapping the moon and other planets, forecasting solar flares, and so on. However, the AI application in such critical mission-oriented and on-board in-service tasks raises yet some skepticism.

The 1st International Workshop on *The Use of Artificial Intelligence for Space Applications*, co-located at the 2nd International Conference on *Applied Intelligence and Informatics* (AII2022), provided an innovative international forum to bring together researchers, practitioners, and experts from academia, agencies, and industry. The goal was to debate and share the cutting-edge research results achieved through the application of AI in solving complex problems related to the space domain. The approach was indeed to examine the *dialogue* between AI and space, stimulating the exchange of ideas between researchers of these two fields and providing, at the same time, an up-to-date picture of the space technologies currently employed or that could be employed in future missions.

The workshop was co-organized by the *University Mediterranea of Reggio Calabria* (Italy); the *UA Space Systems Engineering Laboratory*, Arizona (USA); the *Italian Space Agency* (ASI) (Italy); the *European Space Agency's (ESA) Φ-lab* (Italy); and *Thales Alenia Space* (Italy). It also followed, in principle, the main theme of AII2022, i.e., *Fostering the Reproducibility of Research Results* with the aim to promote open science, methods, and data to facilitate the reproduction of the scientific outcomes. After a long and extremely challenging period of restrictions due to Coronavirus Disease (COVID-19), the Workshop has been successfully held in person, thanks also to the warm hospitality of the *University Mediterranea of Reggio Calabria*, Italy, hosting around 100 international participants.

The submitted papers underwent a single-blind review process, soliciting expert comments from experienced reviewers. After the rigorous review process and based on the review reports, 29 contributions were accepted for oral presentation at the workshop, scheduled during the 3 days of the AII2022 Conference. Accordingly, this volume of proceedings contains all of the 29 papers, which were presented at the workshop in Reggio Calabria, Italy.

Reggio Calabria, Italy	Cosimo Ieracitano
Reggio Calabria, Italy	Nadia Mammone
Rome, Italy	Marco Di Clemente
Nottingham, UK	Mufti Mahmud
Tucson, USA	Roberto Furfaro
Reggio Calabria, Italy	Francesco Carlo Morabito
December 2022	

Acknowledgements

We would like to express our gratitude to all AII2022 Chairs and Committee Members for giving us the opportunity to organize the workshop during the AII2022 Conference. The workshop showed a high-quality and interesting program which would not have been possible without the effort of the program committee members in reviewing the submitted contributions.

We are thankful to the University Mediterranea of Reggio Calabria (UNIRC) with the Departments DICEAM and DIIES, for hosting the AII2022 Conference and the Workshop as well, providing all the necessary instrumental and logistic support. A special mention goes to the AI_Lab and Neurolab's staff members for their great effort, generous, valuable, and unwavering support.

We would especially like to express our deepest gratitude to our financial supporters: the University Mediterranea of Reggio Calabria (UNIRC) with the Departments DICEAM and DIIES and the companies Posytron Engineering S.r.l. and Aubay Italia SpA, without which we would not have been able to organize such a successful event.

We would like to express our sincere appreciation to our collaborators, including the Società Italiana Reti Neuroniche (SIREN); the International Neural Network Society (INNS); Thales Alenia Space Italia (TASI); the Italian Space Agency (ASI); the UA Space Systems Engineering Laboratory, Arizona (USA); and the European Space Agency's (ESA) Φ-lab.

We are also grateful to the whole *Studies in Computational Intelligence* team from Springer-Nature for their continuous support in coordinating the publication of this volume.

Last but not the least, we thank all the participants for having made possible the success of the Workshop *The Use of Artificial Intelligence for Space Applications*.

Reggio Calabria, Italy — Cosimo Ieracitano
Reggio Calabria, Italy — Nadia Mammone
Rome, Italy — Marco Di Clemente
Nottingham, UK — Mufti Mahmud
Tucson, USA — Roberto Furfaro
Reggio Calabria, Italy — Francesco Carlo Morabito
December 2022

Contents

SAR Image Formation: Conventional and AI-Based Approaches on Sentinel-1 Raw Products

Gianluca Maria Campagna and Luca Manca

Abstract The Earth Observation (EO) sector, whose data allow to address complex problems such as disasters management and climate change monitoring, is undergoing deep transformation thanks to the fresh paradigms of the New Space Economy. In remote sensing, Synthetic Aperture Radar (SAR) plays a key role, providing high-resolution images regardless of illumination and weather conditions. The great availability of data encourages Artificial Intelligence (AI) solutions for EO, but this promising field is nearly unexplored for SAR data utilization. This work investigates the SAR image formation process, needed to obtain focused images from radar echoes, for Sentinel-1 case study. First, the focusing problem is tackled from a conventional perspective, by developing a simplified version of the Range Doppler Algorithm for Sentinel-1 StripMap and IW products, obtaining satisfactory metrics concerning ESA requirements. Conventional algorithms require users knowledge and high computational capabilities, so the data processing is delegated to ground. Conversely, an AI approach would allow to reduce the computational load, enabling on-board data processing. So, in the second part, several StripMap products are processed and reorganized in a dataset. Then, a fully convolutional network based on U-Net is proposed, with the aim of learning the mapping between range compressed products and focused ones. One of the trained models reaches an accuracy of $\approx$70% and a similarity index (SSIM) of $\approx$0.6 on the training set. Last, this model is tested on unseen samples, resulting in accuracy and SSIM of $\approx$50% and $\approx$0.5. Considering the lack of similar approaches in literature, those results are promising for future works with larger datasets and tailored architectures.

Keywords SAR image formation · Range Doppler algorithm · Deep learning

---

G. M. Campagna (✉)
Politecnico di Milano – Department of Aerospace Science and Technology, via La Masa, 34, 20134 Milano, MI, Italy
e-mail: gianluca@aikospace.com

L. Manca
AIKO S.R.L., via Dei Mille, 22, 10123 Torino, TO, Italy
e-mail: luca.manca@aikospace.com

C. Ieracitano et al. (eds.), *The Use of Artificial Intelligence for Space Applications*, Studies in Computational Intelligence 1088,
https://doi.org/10.1007/978-3-031-25755-1_1

1 Introduction

In the last years, the Earth Observation sector is experiencing a fast growth, primarily thanks to the increasing awareness about the importance of the health of our Planet. Being able to forecast, manage and recover from natural disasters such as earthquakes, tsunamis or volcanic eruptions is a key asset for governments. Moreover, monitor relevant parameters such as biomass, ice concentration and ocean pollution is fundamental to assess the effects of climate change and to foresee mitigation actions. In this context, remote sensing from space plays a crucial role in providing continuous and reliable data. Among all the technologies used for remote sensing, Synthetic Aperture Radars has always provided a special contribution. This because the exploitation of radio frequencies allows those active sensors, which carry their own illumination, to perform continuous acquisitions regardless of day-night cycle or weather conditions. Moreover, radio waves interact with matter differently from optical ones, so it is possible to retrieve different but complementary information with respect to the widely known optical sensors. However, differently from optical data, the direct visualization of raw SAR products (Level-0) is not meaningful. There is the need of applying signal processing techniques to the acquired data to retrieve the high-resolution images. Up to now, the SAR image formation process has been performed through well-established algorithms, either working in time or frequency domain, characterized by a high computational cost. Therefore, the current paradigm delegates the data processing to the ground segment. On the contrary, an AI-based approach would allow to treat the focusing problem as a "black box", reducing the computational load and allowing on-board data processing to enhance the satellite level of autonomy. The on-board availability of processed Level-1 products would pave the way to interesting applications, either in the recognition of observation opportunities, but also for understanding and processing the acquired data. For example, some applications could be aimed at real time monitoring of a Region Of Interest (ROI) to conduct ship detection and localization in a harbour area, or at discarding useless data, thus lightening the downlink procedures, which can easily become the bottleneck for satellite operations. Such applications (and many others) would both reduce costs and increase the performance of the mission.

This paper is structured according to a preliminary overview of all the topics involved, such as SAR (and in particular Sentinel-1 mission) and Deep Learning, followed by a review of the state of the art of SAR focusing. Then, the results of the developed version of the Range Doppler Algorithm are presented. Finally, the proposed Deep Learning pipeline for the SAR image formation process is detailed, together with the main results obtained, some conclusions and suggestions for further works.

2 Fundamentals

2.1 *Synthetic Aperture Radar*

The concept of Synthetic Aperture Radar is historically attributed to Carl Wiley of *Goodyear Aircraft Corporation*, who realized that it was possible to mount coherent radars—with a precise phase relationship between transmitted and received pulses—on moving platforms such as aircrafts [1]. According to a convenient process of the back-scattered echoes, it was possible to retrieve a 2-D reflectivity map of the surface with high resolution in both slant range (line of sight) and azimuth (flight direction) coordinates. The use of SAR technology for remote sensing from space has its roots in the NASA Seasat mission of 1978 [2]. Nowadays, SAR systems are widely used to provide high-resolution images (resolution up to *cm* scale), which are independent from day-night cycle (the active sensor carries its own illumination) and from weather (microwaves are able to penetrate clouds and storms with low deterioration). Moreover, radar frequencies interact with matter in a different way from optical ones, so SAR images provide complementary information with respect to optical sensors [3]. Products obtained through SAR are employed for a variety of applications which embrace geoscience, climate research, 2-D, 3-D and 4-D (space and time) Earth mapping and even planetary exploration [5].

Figure 1 shows a simplified model for the SAR acquisition geometry. The slant range, line of sight distance from the target, is governed by the *range equation*:

$$R^2(\eta) = R_0^2 + v_r^2\eta^2 \tag{1}$$

where R_0 is the slant range of closest approach, η is the azimuth/slow time and v_r is a pseudovelocity which differs from the actual platform velocity (called v_s) in the satellite case. For what concerns the actual acquisition pattern, Linear Frequency Modulated (LFM) pulses are transmitted every Pulse Repetition Interval (PRI) and the back-scattered echoes are registered. The attainable resolution along range is proportional to the inverse of the transmitted pulse bandwidth. Therefore, high reso-

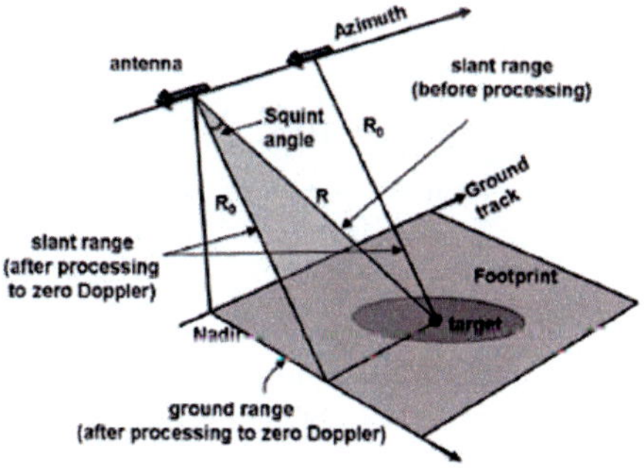

Fig. 1 SAR Geometry [3]

lution can be reached by using very short pulses or, more commonly, through *pulse compression* techniques. The real trick of SAR is hidden in the Doppler effect experienced along azimuth: through convenient processing techniques, it is possible to synthesize a larger antenna and therefore to reach high resolution also along azimuth direction.

2.2 Sentinel-1 Mission

Sentinel-1 is an ESA mission in the context of EU Copernicus Earth Observation Programme [4]. It is composed by a constellation of two satellites, Sentinel-1A and Sentinel-1B, sharing the same Sun-synchronous near-polar orbit with a phasing difference of 180°. Each satellite is equipped with a C-band SAR that can be operated in four exclusive acquisition modes: StripMap (SM), Interferometric Wide Swath (IW), Extra Wide Swath (EW), Wave (WV). The dissemination policy of Copernicus Programme is based on a complete, free and open access to data through its official platform Copernicus Open Access Hub. Sentinel-1 products are distributed at various level of processing, including:

- **Level-0 Products**: raw data compressed using Flexible Dynamic Block Adaptive Quantization (FDBAQ) encoding;
- **Level-1 Products**: focused data that can be either Single-Look Complex (SLC), in slant range-azimuth geometry with complex samples to preserve phase information, or Ground Range Detected (GRD), with pixels detected (only amplitude/intensity information) and projected to ground range—azimuth coordinates.

2.3 Deep Learning

Deep Learning (DL) is a branch of Machine Learning (ML) based on Artificial Neural Networks (ANNs). ANNs are highly parallelized computing systems with simple processors characterized by many interconnections. The fundamental unit of an ANN is a *neuron*, which corresponds to the mathematical formulation of a biological neuron. In fact, this can be seen as a computing unit, composed by a cell body and several branches (*dendrites* and *axon*) linked to other units. Neurons can be organized in a number of connection patterns, leading to different architectures. The most known ones are Fully-Connected Networks, in which a neuron in a layer is connected with all the neurons of the successive layer, Convolutional Neural Networks (CNNs), which exploit convolution operations to efficiently deal with multidimensional arrays, such as images, Recurrent Neural Networks (RNNs), dynamic networks employed to handle sequential inputs such as audio and text. DL architectures are widely used in problems such as Computer Vision (CV) and Natural Language Processing (NLP).

3 SAR Focusing State-of-the-Art Methods

3.1 Conventional SAR Image Formation Techniques

Differently from optical data, the direct visualization of raw SAR products is not meaningful and does not provide information about the imaged region [5]. There is the need of applying signal processing techniques to process the phase history of acquired data and retrieve a 2-D image scene of the mapped area. In principle, SAR processing aims at solving the convolution shown in Eq. 2:

$$S(\tau, \eta) = g(\tau, \eta) * h_{imp}(\tau, \eta) \tag{2}$$

where $S(\tau, \eta)$ is the recorded signal and $h_{imp}(\tau, \eta)$ is the theoretical impulse response of a unitary point target, functions of both fast and slow time. The aim is to determine the ground reflectivity $g(\tau, \eta)$. This is typically done by means of matched filtering techniques, processes used to detect a signal with known features embedded in noise. Dealing with this problem directly in the time domain requires a high computational effort, hence several SAR image formation algorithms working in other domains were developed. Algorithms working in frequency domain exploit fast convolutions based on Fourier transform (FFT) with a noticeable computational gain. Main ones are Range Doppler Algorithm (RDA), Chirp Scaling Algorithm (CSA), Omega-K Algorithm (ωKA) and Spectral Analysis Algorithm (SPECAN). Among those, the RDA is the oldest but still largely used as it represents one of the best trade-offs between accuracy, efficiency and generality. As suggested by its name, most of the processing steps are carried out in the Range time—Azimuth frequency domain. The main steps of the RDA in its basic version are:

- **Range Compression**: the signal replica of the pulse can be defined according to the knowledge of some parameters of the transmitted signal, such as pulse length and bandwidth. The matched filter is the complex conjugate of the FFTed pulse replica and it is applied through a cascade of FFT → multiplication → IFFT to each pulse;
- **Doppler Centroid $\mathbf{f_{DC}}$ estimation**: this frequency is the "center of gravity" of the azimuth spectrum, usually directly estimated from data according either to magnitude-based or phase-based methods;
- **Azimuth FFT**: this step introduces the Range Doppler domain through a FFT performed on each azimuth line. This allows to access the azimuth frequencies f_η;
- **Range Cell Migration Correction (RCMC)**: in the RD domain the range equation can be expressed as

$$R(f_\eta) \approx R_0 + \frac{\lambda^2 R_0 f_\eta^2}{8 v_r^2} \tag{3}$$

From this one can deduce that, due to the movement within the synthetic aperture during the acquisition, each target migrates through several range cells. This is an unwanted effect that has to be corrected: it can be done through a range interpolation or, with the approximation of range-invariant correction, through a convenient phase multiplier;

- **Azimuth Compression**: the azimuth matched filter can be defined directly in frequency domain as

$$H_{az}(f_\eta) = \exp\left(-j\pi \frac{f_\eta^2}{k_a}\right) \tag{4}$$

where $k_a = \frac{2v_r^2}{\lambda R_0}$, and applied to each azimuth line of the data.

The final product is a Single-Look Complex (SLC) matrix whose coordinates are slant range—azimuth, in which the targets are registered to their zero-Doppler positions. Each complex sample contains information about amplitude and phase of the response.

3.2 *Deep Learning for SAR*

Currently, the integration of Deep Learning in remote sensing for Earth Observation is mostly limited to optical data. A comprehensive review of the state of the art related to Deep Learning for SAR is provided in [6]. The key concept to understand in developing AI approaches for SAR-related problems is that SAR products are deeply different from optical ones. Indeed, they are quite unique even in the remote sensing field. The main aspects to take into account are:

- **High dynamic range**: in SAR images, it can extend up to tens of dB. The pixels distribution is extremely asymmetric: most of them lie in the low-value region, while bright scatterers (typical of urban regions) compose a long tail. Normally, networks such as CNNs have problems in handling such ranges;
- **Imaging geometry**: range and azimuth are not arbitrary coordinates, therefore data augmentation is not straightforward. For instance, rotation of SAR products yields not meaningful images;
- **Complex nature**: in processes such as image formation from back-scattered echoes or polarimetric/interferometric data processing, the most precious information of SAR products lies in the phase. Therefore, when developing a tailored architecture, one has to respect the complex nature of SAR data. Nevertheless, for some applications just as semantic segmentation or object detection, the amplitude/intensity information alone is sufficient.

Moreover, SAR products are difficult to collect, process and annotate, therefore there is a lack of high-quality datasets to benchmark the proposed models. Nevertheless, some SAR-related problems have been approached through Deep Learning in the last years. Some of the most relevant ones are Terrain surface classification, Object detection, Despeckling, InSAR and SAR-Optical data fusion. It is crucial to point out that the application of a Deep Learning based approach to the problem of SAR image formation is not investigated in [6]. There, it is stated that conventional algorithms have reached impressive speed and accuracy, and it is assessed that there is no reason for which AI-based approaches would perform better or faster. However, as highlighted in Sect. 1, the proposed DL-based approach would enable on-board data processing capability enhancing satellites level of autonomy. Indeed, all the works present in literature tackle only partially the bigger problem of SAR image formation, without proposing a complete Deep Learning pipeline to skip the traditional steps required for transforming the registered back-scattered echoes in SAR images. Moreover, all the works are pursued with simulated data rather than real SAR products.

4 SAR Image Formation Through Conventional Approach

The *Sentinel-1 Level 1 Detailed Algorithm Definition* document [7] provides a high-level overview of the software implemented at S-1 Instrument Processing Facility (IPF) to obtain Level-1 products starting from raw Level-0 ones. The processor is based on an enhanced version of the Range Doppler Algorithm already used for ENVISAT ASAR data processing. Therefore, two simplified versions of the RDA have been developed to deal with StripMap and Interferometric Wide Swath products.

4.1 StripMap Products

StripMap acquisition mode is mainly used to image coastal zones, little islands and eventually to support emergency management operations. Sentinel-1 mission provides StripMap acquisitions over 6 different sub-beams, named S1 to S6, which differ for some parameters such as acquisition incidence angle, range sampling frequency f_s, PRI and attainable resolution.

The focusing procedure of StripMap products will be described taking as reference an acquisition of Sentinel-1B satellite took on 08/10/2021 at 00:18 UTC, with the S3 sub-beam mode. The acquired region is an area near Houston. The raw product is a huge matrix of complex samples of size 22018×48125, therefore only a slice of 5000 pulses is retained. The Range Compression step can be accomplished by constructing the pulse replica through some parameters of the transmitted pulse such as Length (TXPL), Starting Frequency (TXPSF) and Ramp Rate (TXPRR), that can be read directly from satellite telemetry. As discussed in Sect. 3, the range matched filter is the conjugate of the FFTed pulse replica and can be applied to each range line. After the matched filter throwaway, the range compressed product size is 19072×5000. The Doppler Centroid frequency estimation is carried out according

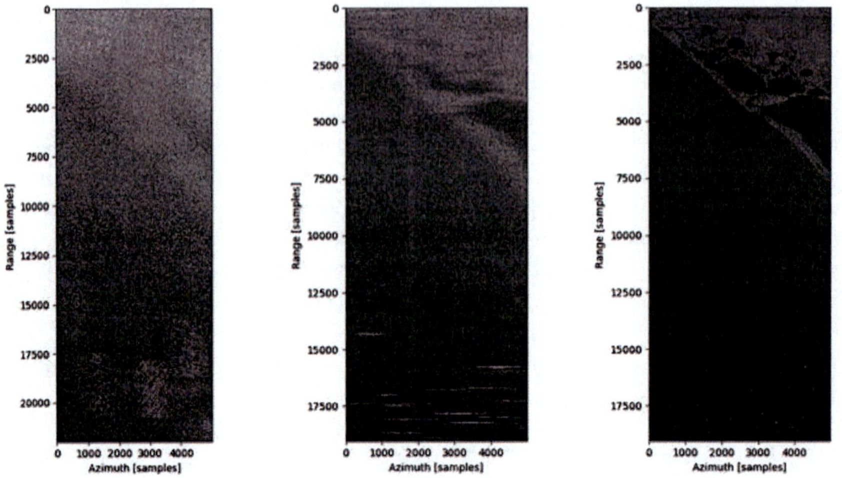

Fig. 2 Houston—StripMap RAW, RangeFoc and SLC data

to the phase-based method suggested in [7], resulting in $f_{DC} = 37.9\,\text{Hz}$. Then, the Range Doppler domain is introduced through a FFT performed on each azimuth line, and the amount of Range Cell Migration to correct is assessed by means of Eq. 3. The RCMC is carried out according to the definition of the phase multiplier

$$G_{RCMC}(f_\tau) = \exp\left(j\frac{4\pi f_\tau \Delta R(f_\eta)}{c}\right) \tag{5}$$

where f_τ are the range frequencies. The phase multiplier is applied to each range line through a cascade of FFT $\rightarrow$ multiplication $\rightarrow$ IFFT. As a last step, the Azimuth Compression is performed through the definition of the azimuth matched filter introduced in Eq. 4, applied to each azimuth line. The azimuth filter is combined with a generalized Hamming window (which is typically introduced only in post-processing phase) used to smooth the results. The obtained Single-Look Complex product is still of size 19072×5000. The raw, range compressed and SLC products are shown in Fig. 2, where the pixel amplitudes are converted in logarithmic scale.

In order to validate the focusing algorithm, some metrics shall be extracted. To this purpose, another acquisition took over the DLR calibration center, near Munich, is processed. In this area, famous to be used for calibration of major SAR missions, more than 30 corner reflectors are implanted. When imaged by a SAR sensor, those artificial objects behave as point-wise targets whose response can be assimilated as an Impulse Response Function (IRF), from which internal evaluation metrics such as Resolution and Peak Sidelobe Ratio (PSLR) can be extracted. From the obtained SLC, it is possible to detect one of the corner reflectors by looking for the brightest target. A neighborhood of 64 pixels near the focused reflector is shown in Fig. 3 in absolute value of pixel intensity.

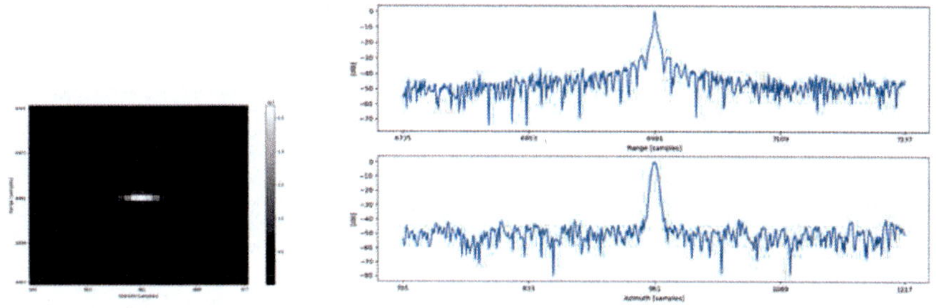

Fig. 3 StripMap response to a corner reflector—Intensity and Principal Axes Cuts

Table 1 RDA performance for StripMap

	ESA PSLR (dB)	Obtained PSLR (dB)	ESA Res (m)	Obtained Res (m)
Range	<−21.2	−28.1	3.3	3.6
Azimuth	<−21.2	−38.2	3.9	15.8

For the evaluation of PSLR and Resolution metrics, it is possible to plot the cuts of the normalized intensity values along the principal axes, i.e. slant range and azimuth. The two cuts, extracted from a neighborhood of 512 pixels around the brightest scatterer, are shown in Fig. 3 in dB scale.

In particular, the PSLR can be measured as the height (in dB) of the first sidelobe, while the Resolution is the width of the response at −3 dB. Obtained results and ESA requirements for StripMap S5 products are shown in Table 1.

As highlighted, PSLR requirements are respected in both range and azimuth directions. Then, for what regards the resolutions, along range direction the developed RDA performs almost perfectly compared to the ESA values. On the contrary, along azimuth direction it is apparent that there is not a perfect focusing. This can be due to the approximations introduced: among the others, there are approximations about constant satellite altitude and velocity, errors in the slant ranges vector definition and implementation of the approximated range-invariant RCMC.

4.2 *Interferometric Wide Swath Products*

Interferometric Wide Swath is the main acquisition mode over land of Sentinel-1 mission. For IW, the acquisition is carried out through the TOPSAR (Terrain Observation with Progressive scans SAR) technique: the sensor acquires bursts (discrete spatial regions) across different sub-swaths at different incidence angles. Moreover, the antenna is steered along azimuth direction from backward to forward. In this way, it is possible to increase the swath width at the expense of spatial resolution.

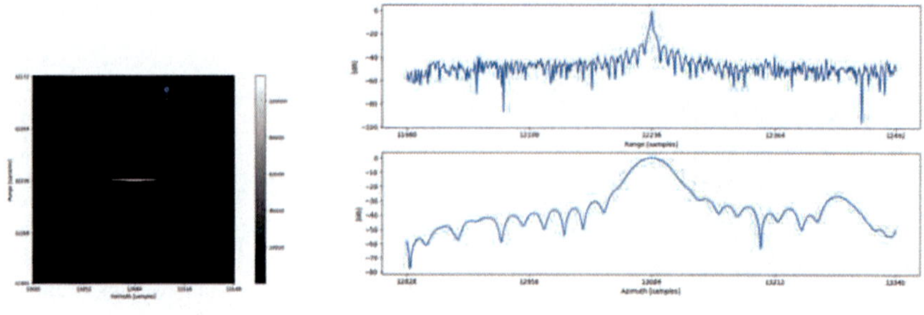

Fig. 4 IW response to a corner reflector—Intensity and Principal Axes Cuts

This acquisition pattern has immediate consequences in the data processing: data are characterized by a much wider bandwidth than the StripMap case, so the Doppler spectrum results aliased. Indeed, when dealing with IW products, one has to solve a double folding of the data, both in time and frequency. This can be done either through the implementation of a tailored approach or by adapting an existing StripMap processor, as discussed in [8]. By following this second path, it is possible to process the 8th burst of the 1st sub-swath of an acquisition performed by Sentinel-1A over the Munich area described in Sect. 4.1.

The raw burst has dimension 23842 × 1409. After the Range Compression step, implemented in the very same way as for StripMap products, the size is 20471 × 1409. Then, it is possible to deal with the frequency folding. This is done by shifting the azimuthal frequencies to zero, according to the linear excursion of the Doppler Centroid due to the antenna sweep, an then performing an azimuth interpolation aimed at oversampling with a factor $N_{folding} = 5$. In this way the product size becomes 20471 × 7045. On the other hand, the time folding can be resolved by correctly representing each burst. Because of the antenna movement, the actual spatial extension of the bursts is larger than the one spanned by the platform during the acquisition time, so a large zero padding is required. The expanding factor values are reported in [8] for each sub-swath. After the zero padding, the support matrix has size 20471 × 32026. It is apparent how much the dimension has increased, therefore to correctly focus even a single burst with this StripMap-based algorithm a high computational load is needed. However, after interpolation and zero padding, the burst can be treated as a StripMap acquisition and so RCMC and Azimuth Compression can be implemented just as described in Sect. 4.1. Then, as done for the StripMap product, it is possible to extract the corner reflector compressed shape, shown in Fig. 4, together with the corresponding cuts along the principal axes.

Along azimuth direction the response seems much coarse with respect to the StripMap case because of the interpolation operation that has reduced the pixel spacing, but also for the less demanding constraints about azimuth resolution for IW products. Results in terms of Resolution and PSLR are summarized in Table 2.

In the IW case, all requirements are met with minor errors. Differently from the StripMap case, the expected resolution is almost reached also along azimuth

Table 2 RDA performance for IW

	ESA PSLR (dB)	Obtained PSLR (dB)	ESA Res (m)	Obtained Res (m)
Range	<−21.2	−28.3	2.7	3.2
Azimuth	<−21.2	−28.2	21.7	23.2

direction. This is due to the fact that the approximations introduced in the developed StripMap-based algorithm are still negligible with respect to the much reduced Doppler history (compared to a StripMap acquisition) of the targets that characterizes the TOPSAR acquisition technique.

5 SAR Image Formation Through AI-Based Approach

After having defined a functional focusing algorithm that allows to pass from Level-0 products to Level-1 ones, it is possible to realize a Deep Learning pipeline for the SAR image formation problem. The main points to be clarified are:

- **Learning task**: the aim is chosen to be the training of a network able to learn the mapping between RangeFoc and SLC products. This because pulse compression is a simple operation that can be applied to a raw product by exploiting a filter always equal to itself, that can be defined a priori by knowing the characteristics of the transmitted signal. Moreover, the real trick of SAR image formation, which coincides with the most demanding computational burden, is represented by the following operations carried out in the Range Doppler domain;
- **Dataset generation**: product slices from several StripMap acquisitions have been extracted, processed, transformed from complex matrices to 2-channels real ones and resized to a common shape of $1024 \times 128 \times 2$. 292 examples of corresponding {RangeFoc, SLC} are considered for the training dataset. Despite this dataset is obviously too small to properly train from scratch the model, the aim is just to assess the feasibility of this approach;
- **Neural network architecture**: a modified version of the U-Net, CNN introduced in 2015 for image segmentation [9], is proposed. Apart from traditional classification tasks, which posed the U-Net in a dominant position with respect to the competitors, in the last years it has been successfully employed also for regression tasks at pixel level. An application to the remote sensing field is shown in [10], where they implement a U-Net like architecture to tackle the pansharpening problem. However, they deal with patches of 33×33 as training examples, so they use a shallower version of the U-Net in order to preserve the localization accuracy rather than the feature abstraction ability. The proposed modifications include Leaky ReLU activation functions for hidden layers and linear activation in the output layer, to adapt the network for the pixel regression task of interest. The

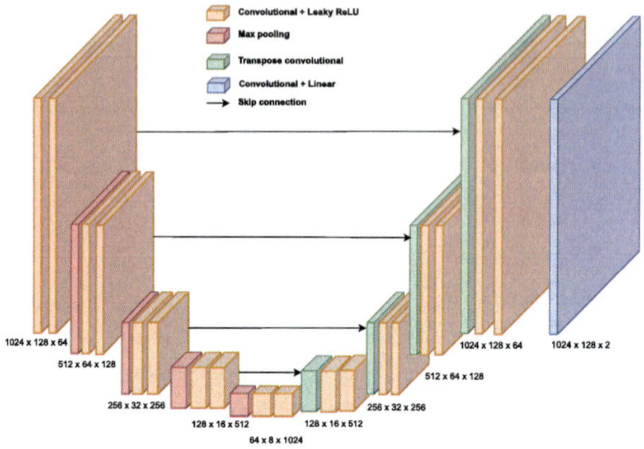

Fig. 5 Proposed network architecture (designed thanks to *diagrams.net*)

Table 3 Parameters distribution for the proposed architecture

Network Block	Output shape	Parameters	Network Block	Output shape	Parameters
Input Layer	1024 × 128 × 2	–	Decoder Block 1	128 × 16 × 512	9,176,576
Encoder Block 1	512 × 64 × 64	38,144	Decoder Block 2	256 × 32 × 256	2,294,528
Encoder Block 2	256 × 32 × 128	221,440	Decoder Block 3	512 × 64 × 128	573,824
Encoder Block 3	128 × 16 × 256	885,248	Decoder Block 4	1024 × 128 × 64	143,552
Encoder Block 4	64 × 8 × 512	3,539,968	Output Layer	1024 × 128 × 2	130
Bridge	64 × 8 × 1024	14,157,824			**31,031,234**

proposed architecture, shown in Fig. 5, contains 31,031,234 trainable parameters, whose distribution across the network is detailed in Table 3.

5.1 *Models Definition and Training*

Classical CNNs are not able to handle the high dynamic range typical of SAR images. This because the loss is dominated by the outliers (high-value pixels), while the multitude of low-value pixels represents a negligible fraction. To partially address this issue, it is thought to define a customized loss function based on the Relative

Mean Squared Error. Three options, each with a different penalty term for the pixel values at the denominator, are proposed:

- **Loss for Model #1**: $\mathcal{L}(y_{pred}, y_{true}) = \frac{1}{n}\sum_n \frac{(y_{pred}-y_{true})^2}{y_{true}^2+\epsilon}$;
- **Loss for Model #2**: $\mathcal{L}(y_{pred}, y_{true}) = \frac{1}{n}\sum_n \frac{(y_{pred}-y_{true})^2}{|y_{true}|+\epsilon}$;
- **Loss for Model #3**: $\mathcal{L}(y_{pred}, y_{true}) = \frac{1}{n}\sum_n \frac{(y_{pred}-y_{true})^2}{\sqrt{|y_{true}|}+\epsilon}$.

where y_{pred} is the network output, y_{true} is the expected output, n is the number of pixels ingested by the optimizer and ϵ is added to the denominator for numerical stability.

First of all, the three models are trained on a single training example, just to assess if they are complex enough to learn the mapping between inputs and outputs. All the training phases are carried out in *TensorFlow* environment with a NVIDIA Tesla P100 GPU. The hyperparameters are Adam optimizer, learning rate $\alpha = 2e-4$ and $\epsilon = 0.1$, while the metrics are accuracy and Structural Similarity Index Measure $(0 \leq SSIM \leq 1)$. After 400 epochs, all the models reach an accuracy $> 84\%$ and a SSIM > 0.92 on the single training example.

Then, the models training is extended to the whole generated dataset, which has been standardized (rescaled to zero mean and unit variance) before being fed to the network. Hyperparameters are quite similar, with the chosen batch size equal to 4 and the introduction of an exponential decay of the learning rate $\alpha = 0.993^{epoch}\alpha_0$ to improve the stability. After 200 epochs, the resulting trends are shown in Fig. 6.

As it can be seen, Model #2 presents intermediate properties among the three. The trained models are used to infer the outcomes starting from a random example chosen from the dataset. The results are shown in Fig. 7.

Interesting insights for models comparison can be obtained by looking at the outliers predictions: Model #1 completely miss them, with percentage errors (in absolute sense) $> 90\%$, while Model #3 provides very good predictions (errors $< 3\%$), but lacks of overall accuracy. Also in this sense, Model #2 provides intermediate performance with errors for outliers in the order of 50%. Given all those considerations, plus the fact that Model #2 outperforms the others in percentage loss decay during

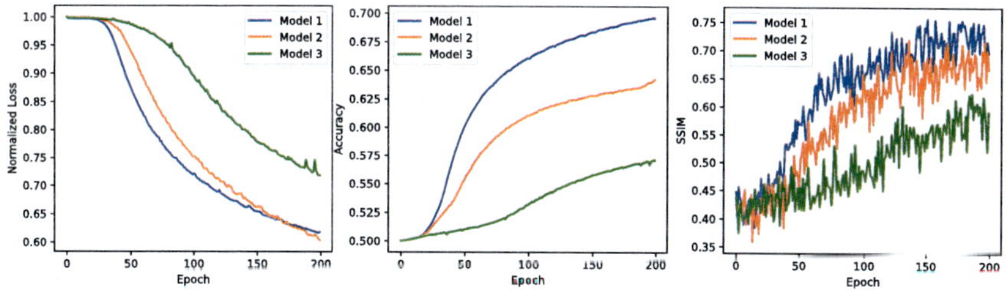

Fig. 6 Training on the dataset—Normalized Loss, Accuracy and Batch SSIM trends

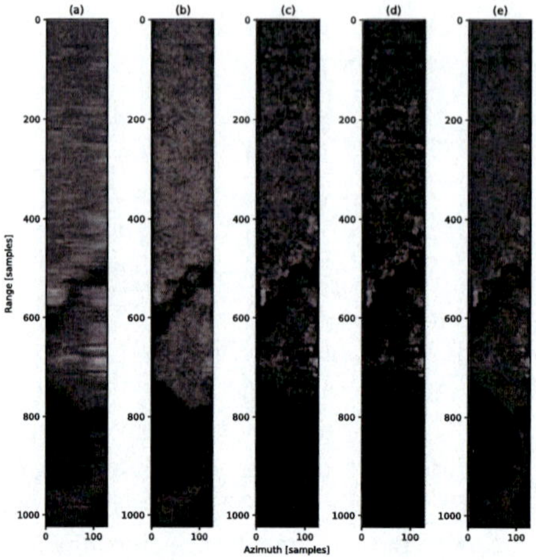

Fig. 7 Training on the dataset—Predictions (**a** RangeFoc, **b** Model #1, **c** Model #2, **d** Model #3, **e** SLC)

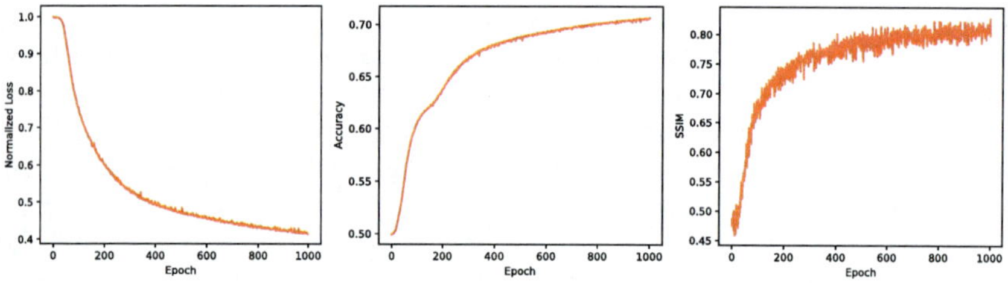

Fig. 8 Training of Model #2—Normalized Loss, Accuracy and Batch SSIM trends

training, it is decided to rely on this one and refine its training to improve the metrics. To this aim, the training is prolongued to 1000 epochs with the introduction of a slightly different decay law for the learning rate. The resulting trends are shown in Fig. 8.

The loss reduction attests to 58.34%, the accuracy is 70.73% while the SSIM averaged on all the training examples is 0.5968. A prediction obtained from a random training example is shown in Fig. 9. In this particular case, the SSIM value is approximately 0.74.

Despite the overall image is reproduced pretty well, the percentage errors for the outliers, both positive and negative in real/imag channels, is still in the order of ≈50%. Nevertheless, with a dynamic range which is basically a half of the initial, one may try to employ the predictions for typical tasks such as landscape classification or object detection in an integrated pipeline. Moreover, the other issue is that there are training examples in the dataset which are not well reproduced (there are some even with $SSIM < 0.1$), in particular those ones with less contrast in the original SLC products.

Fig. 9 Training of Model #2—Prediction (**a** RangeFoc, **b** Model #2, **c** SLC)

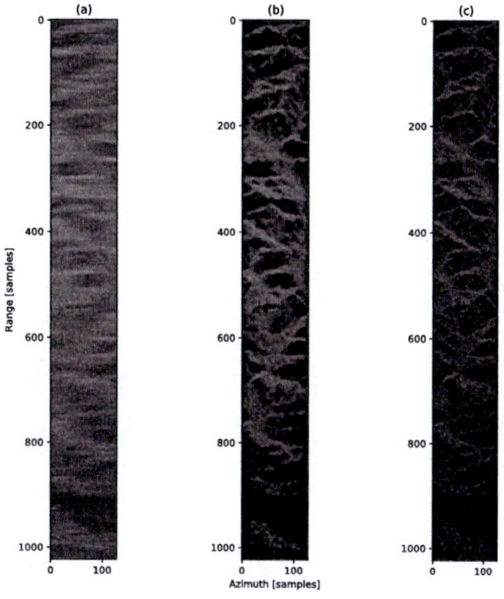

5.2 *Preliminary Testing Phase*

Given the small dataset generated for this preliminary feasibility study, it was decided not to follow the typical 80–20% splitting for training/test subset. Instead, other slices are picked up in a random fashion (with a random azimuth starting index) from the same products used for training, rather than the sequential way adopted for the generation of the training dataset. The obtained product slices result in 50 unseen samples which are further standardized using mean and standard deviation retrieved from the training dataset. The model is evaluated on this dataset, resulting in an accuracy and average SSIM which drop down respectively to 50.05% and 0.5052. This is a sign that the model has overfit to training samples (due to the small dataset and the large number of parameters), but it has learnt some feature extraction capabilities, as it is still able to recognize some patterns in the testing data fed. An example of a prediction made by the model from the test dataset is shown in Fig. 10. In this case, $SSIM \approx 0.85$.

6 Conclusions

This work was aimed at tackling the SAR image formation problem both through conventional techniques and through an innovative AI-based pipeline, on the Sentinel-1 case study. For the first aspect, the main milestone accomplished is the development of a conventional SAR focusing algorithm, obtaining satisfactory metrics for StripMap and IW products with respect to ESA requirements. Then, a labeled dataset, suitable for a supervised learning approach, has been generated and a Deep Learning pipeline

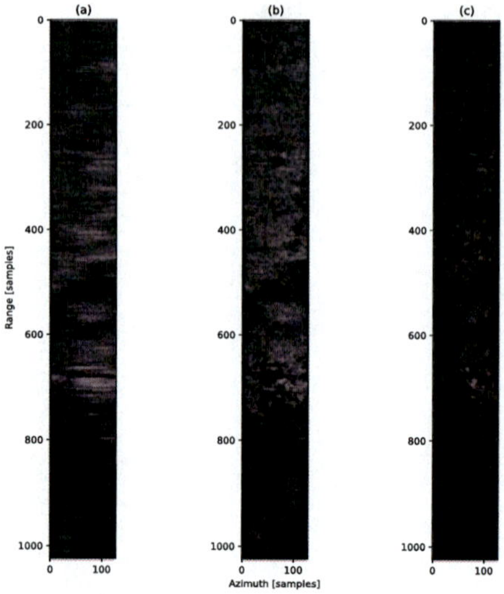

Fig. 10 Testing phase—Prediction (**a** RangeFoc, **b** Model #2, **c** SLC)

based on a modified version of the fully convolutional U-Net has been implemented. The model yields good level of accuracy and SSIM for the training examples and encouraging results also in the preliminary test phase considered. Considering that this approach is the first of its kind in the literature, it is completely open to further improvements. Some directions of future works can be the construction of a bigger dataset, eventually using IW products (the vast majority of distributed products), the definition of a more tailored architecture and the implementation of a use case that proves the advantages of having available on-board a focused Level-1 SAR product, such as ship detection or oil spill monitoring.

References

1. Wiley, C.: Pulsed Doppler Radar Methods and Apparatus, Patent US3196436A (1954)
2. NASA JPL: Seasat. https://www.jpl.nasa.gov/missions/seasat
3. Cumming, I., Wong, F.: Digital Processing of Synthetic Aperture Radar Data. Artech House, Inc. (2005)
4. ESA: Sentinel-1 SAR User Guide. https://sentinels.copernicus.eu/web/sentinel/user-guides/sentinel-1-sar
5. Moreira, A., Prats-Iraola, P., Younis, M., Krieger, G., Hajnsek, I., Papathanassiou, K.: A tutorial on synthetic aperture radar. IEEE Geosci. Remote Sens. Mag. **1**(1), 6–43 (2013)
6. Zhu, X., Montazeri, S., Ali, M., Hua, Y., Wang, Y., Mou, L., Shi, Y., Xu, F., Bamler, R.: Deep learning meets SAR: concepts, models, pitfalls, and perspectives. IEEE Geosci. Remote Sens. Mag. **9**(4), 143–172 (2021)
7. Sentinel-1 Mission Performance Centre: Sentinel-1 Level 1 Detailed Algorithm Definition, 2.2 version (2019)

8. Fusco, A., Pepe, A., Berardino, P., De Luca, C., Buonanno, S., Lanari, R.: A phase-preserving focusing technique for TOPS Mode SAR raw data based on conventional processing methods. Sensors **19**(15) (2019). https://doi.org/10.3390/s19153321
9. Ronneberger, O., Fischer, P., Brox, T.: U-net: convolutional networks for biomedical image segmentation. In: International Conference on Medical Image Computing and Computer-Assisted Intervention, pp. 234–241. Springer (2015)
10. Yao, W., Zeng, Z., Lian, C., Tang, H.: Pixel-wise regression using U-Net and its application on pansharpening. Neurocomputing **312**, 364–371 (2018)

YOLO v4 Based Algorithm for Resident Space Object Detection and Tracking

Marco Mastrofini, Gilberto Goracci, Ivan Agostinelli, and Fabio Curti

Abstract Resident Space Objects (RSOs) detection and tracking are relevant challenges in the framework of Space Situational Awareness (SSA). The growing number of active and inactive platforms and the incoming era of mega constellations is increasing the traffic in the near Earth segment. Recently, more and more research efforts have been focused on this problem. This, combined with the popularity of Artificial Intelligence (AI) applications, has led to interesting solutions. The potential of an AI based approach for image processing, objects detection and tracking oriented to space optical sensors applications has already been proved. In this work, the architecture of a Convolutional Neural Network (CNN) based algorithm has been developed and tested. The image processing and object detection tasks are demanded to Neural Network (NN) modules (U-Net and YOLO v4, respectively) while the tracking of objects inside the sensor's Field Of View (FOV) is formulated as an optimization problem. A performance comparison in terms of detection capabilities has been carried out with respect to a previous version of the algorithm based on YOLO v3. Reported results, based on real and simulated night sky images, show a notable performance improvement from v3 to v4.

Keywords RSOs · Detection · Image processing · Neural network · YOLO · Tracking · Optical sensor · Star tracker

1 Introduction

For SSA and Space Traffic Management (STM), the problems of RSOs detection and tracking to maintain and update ground catalogues are of great concern [2]. RSOs are composed of active platforms and space debris, and the latter constitute the main important slice of RSOs population and are the real threat for operating space-

M. Mastrofini (✉) · G. Goracci · I. Agostinelli · F. Curti
School of Aerospace Engineering, Sapienza University of Rome, Via Salaria 851, 00138 Rome, Italy
e-mail: marco.mastrofini@uniroma1.it

C. Ieracitano et al. (eds.), *The Use of Artificial Intelligence for Space Applications*, Studies in Computational Intelligence 1088,
https://doi.org/10.1007/978-3-031-25755-1_2

crafts and human space missions. Their growth causes an increase of the chance of collision and the exponential growth of debris (Kessler syndrome [6]). Because of this, the importance of detection and tracking of RSOs is increased in recent years. These two operations have always been performed through radar and optical ground stations only, with their consequent limitations: power consumption for radar and weather/daylight conditions for ground optical devices. Nowadays Space-Based Space Surveillance (SBSS) [15] missions are being developed to increase RSOs monitoring coverage, leading to the use of on-board optical sensors which are generally composed of telescopes. Recently, the idea of using star trackers to perform Space Surveillance and Tracking (SST) tasks has been under study [14] and it is in the In-Orbit Validation (IOV) phase [5].

There is a growing effort towards on-board applications where traditional filtering and image processing techniques are still being used [14] for RSOs detection. These techniques have the advantage of a long heritage due to their extensive application. On the contrary, their limitation is the need of complex filtering routines to perform easily human understandable tasks such as objects detection and background removal. For this reason the use of AI appears to be promising for space image data processing and space debris mitigation applications. A particular interest goes towards AI technologies to improve flexibility, autonomy and capability to handle failures and performance degradation while satisfying the reliability standards of a safety space system.

Recent works investigate the application of CNNs for RSOs detection: in [13] event based ground optical sensors are considered while in [3] point-like stars and streak-like stars scenarios are analysed with the consequent training of two different CNNs for object detection. A different approach based on semantic segmentation for dim small targets detection is exploited in [16], achieving satisfying performances in segmentation at the price of a complex network's training and architecture. The advantages shown by CNNs in previous works could be combined with traditional hardcoded modules into a simple architecture for on board RSOs detection applications through star sensors, adding also tracking capabilities to the detection task. Because of this, the development of a dedicated and reliable AI based software to perform on board RSOs detection and tracking could be crucial for space based SSA activities. A preliminary step in this direction has been taken in the present work where segmentation and detection actions are demanded to CNN based modules while the tracking function is made of a hardcoded routine. The proposed algorithm differs from the state of the art because the possibility of performing tracking and because of the detection of targets that is performed on segmented images and not directly on raw ones to improve weak targets detection capability through a suitable segmentation module.

The improvement and test of an AI based module for RSOs detection and tracking through optical sensors are provided. The improved software aims to have a better object detection capability with respect to its predecessor [8] in order to provide a more reliable product. The use of YOLO model for object detection is due to the possibility of its use for near real time or real time applications while keeping a high accuracy [4].

Contributes of this research are:

- Performance comparison of object detection and tracking algorithms based on the YOLO v3 [11] and the YOLO v4 [4] networks.
- Test on real night sky images acquired with a ground based equipment.
- Test on High Fidelity Star Tracker simulator images.

The work is organized as follows: a brief overview of CNN along with the Algorithm structure and modules descriptions and the proposed dataset are provided in the *Methodology* section. In the *Training, Testing and Results* section comparison with the two version of the algorithm are presented. Real images have been considered and processed to validate the improved modules and the capability to perform RSOs detection and tracking over multiple passages. Performance scores and indexes are provided in terms of precision, recall and F1. Future developments are reported in the *Conclusion* section.

2 Methodology

In this section, the algorithm structure for RSOs detection and tracking from star tracker raw images is described. It is based on a combination of Machine Learning (ML) and classical programming techniques:

- The detection function is achieved by stacking a U-Net [12] and a YOLO v4 [1, 4, 10]. In this way the detection operation is helped by the image segmentation process performed by U-Net [9].
- The tracking function is faced through a traditional hardcoded algorithm based on the minimization of a cost function.

Object detection is performed on a mask rather than on the original raw image because with U-Net output, the weakest streaks are easier to be detected by YOLO v4. Indeed, these will result bright enough inside the segmented image to be easily detected by the YOLO v4. For the tracking task, a classical coding approach has been used to reduce the computational effort of the whole algorithm while achieving the desired RSOs tracking inside the FOV.

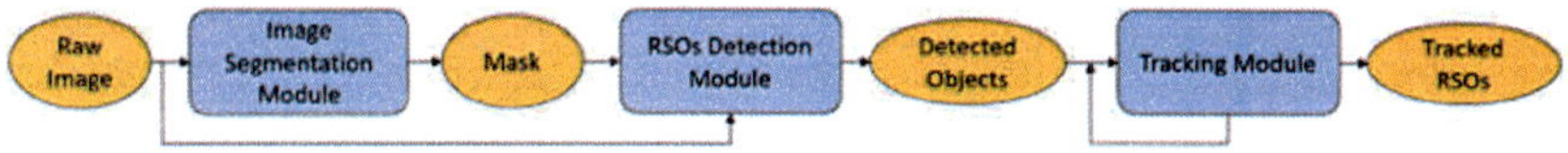

Fig. 1 RSOs Tracking module algorithm structure

2.1 Algorithm Design and Configuration

The algorithm structure is shown in Fig. 1 with the main blocks. The input raw image from the star tracker arrives to the *Image Segmentation Module* where it is binarized, getting what is called *Mask* (a map of foreground and background pixels). The *Mask* is used by the *RSOs Detection Module* to detect the streaks in the night sky image with rectangular boxes. The convolution between the box-delimited regions in the *Mask* and the same sky portions in the *Raw image* is needed for the streaks *centroids* and *extremal points* computation (see next subsections). By this info and the *exposure time* is then possible the computation of the horizontal and vertical components of the streak velocity in the image plane.

These outputs are used by the *Tracking Module* to resolve the objects inside the FOV and provide their time evolution in terms of image coordinates as the output of the algorithm.

Here, RSOs are considered as streaks over a background made of point-like stars under the implicit assumption of low angular velocity of the orbiting platform. This assumption is reasonable since most of the missions involving star trackers require a sidereal pointing.

Image Segmentation Module This module uses the same U-Net reported in [9] (Fig. 3) to segment the image and removes a large part of the noise background, leaving the brightest objects in the FOV. It receives a 960 × 640 *px Raw image* and then just the brightest objects inside the FOV are kept by the U-Net which returns in output a 960 × 640 pixel matrix with the highest pixels' values related to the brightest objects. Then through a suitable binarization level (p_{min}) in the *Thresholding filter*, the foreground and background pixels are selected and a *Mask* is obtained. The U-Net has an initial number of filters equal to 128 while all the other configuration parameters are contained in Table 1 and the p_{min} value was chosen to maximize the similarity between segmentation module predictions and reference *Masks*. The module structure is reported in Fig. 2. The U-Net was trained in Jupyter Notebook environment with Keras and Tensorflow libraries, reaching an accuracy higher than 99% and a loss lower than 0.01. The Dataset for the U-Net training is made of 100 real images taken in a night sky acquisition campaign at Campo Imperatore, Gran Sasso, Italy and 100 reference *Masks*.

RSOs Detection Module In Fig. 4 the RSOs Detection module is shown. The *Mask* is processed by a YOLO v4 model, classifying the streaks and localizing

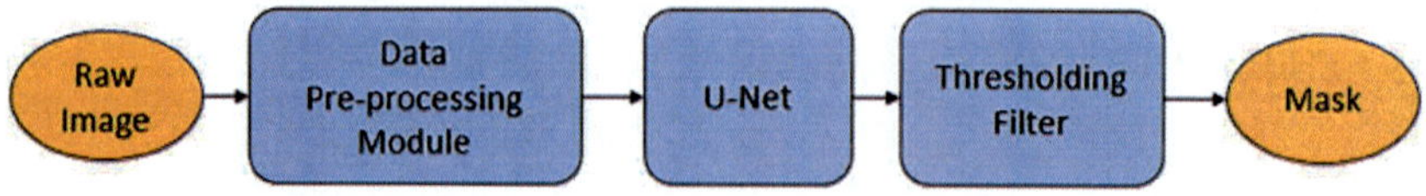

Fig. 2 Image Segmentation Module algorithm structure

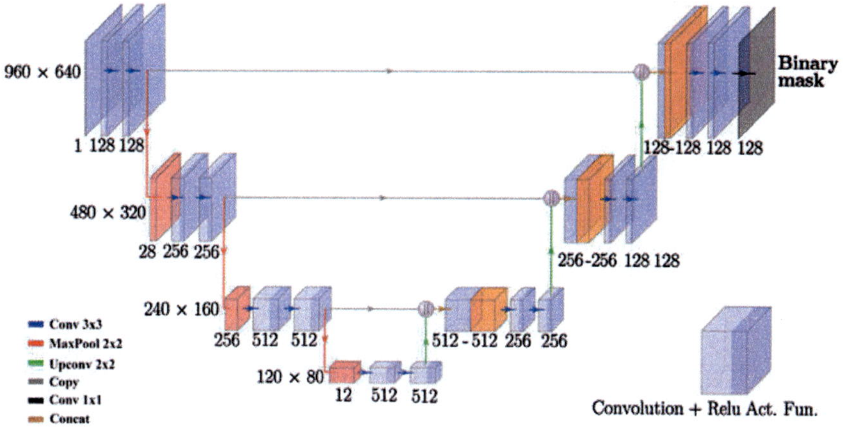

Fig. 3 The implemented U-Net architecture

Table 1 U-Net configuration parameters

Parameter	Value
Image size	960 × 640
Initial filters	128
Learning Rate	10^{-4}
Regularization Factor	10^{-5}
Dropout Rate	0.25
Kernel Size	3
Kernel initializer	$'he\_normal'$

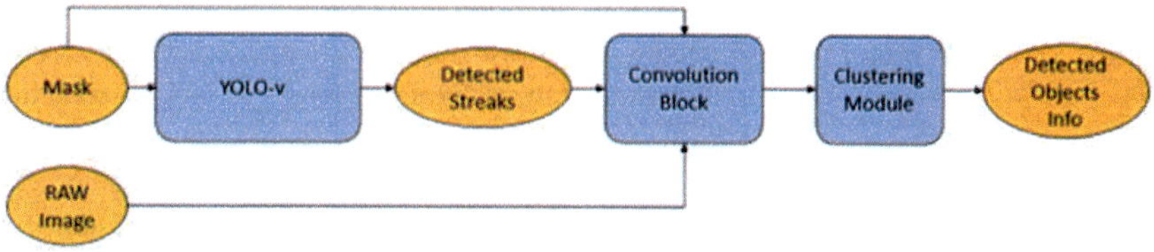

Fig. 4 RSOs Detection Module algorithm structure

them through rectangular boxes. For every detected streak, the box top left pixel coordinates ($\mathbf{P}_{box}$), width and height and class probability are known. This set of information is then used to extract the relative rectangular area both on the *Mask* and the *Raw Image*. Through the convolution of these two arrays, the *Clustering Module* computes the streak *centroid* with a weighted energy based average (Eq. (1)).

$$\mathbf{P}_{cluster,i} = \mathbf{P}_{box} + \sum_{j} \left(\frac{E_{px,j} \cdot \mathbf{p}_{px,j}}{E_{cluster,i}} \right) \tag{1}$$

with

$$E_{cluster,i} = \sum_{j} E_{px,j} \tag{2}$$

where $\mathbf{P}_{cluster,i}$ are the *absolute* coordinates of the ith cluster, $\mathbf{p}_{px,j}$ are the *local* coordinates of the jth pixel in the box and $E_{cluster,i}$ is the sum of the energies $E_{px,j}$ of the ith cluster's pixels.

RSOs Tracking Module The *Tracking Module* algorithm associates every detected object to one that has been previously classified or to a new one. The aim of the algorithm is to distinguish RSOs as they cross the FOV. Hereafter, the number of objects stored in memory will be referred to as n_{obj} while the number of detected objects (boxes) in the current image will be indicated with n_{box}.

The *Tracking Module* is based on the minimization of a Cost Function which is computed for each box-object pair. For each image this Cost Function takes into account the position and orientation of the box and compares them to the corresponding expected values determined from the previous image.

A further and more detailed description of this algorithm is given in [8].

2.2 Algorithm Implementation Methodology

This subsection is devoted to explain how each module was implemented in a software development environment. The U-Net architecture and its training were performed in a Jupyter environment with Keras and Tensorflow libraries. The trained U-Net and *Tresholding filter* were then loaded into a Python script which receives the *RAW input image* and provides the *Mask*. YOLO v4 model is then sequentially runned in Python environment through the Darknet framework to perform the object detection and getting boxes and class probabilities. These data are then used in a Python script which implements the *Tracking Module* where the streaks parameters (*centroids* coordinates, velocity components and angle orientation) are computed and the objects are tracked. All these Python scripts (U-Net inference, YOLO v4 inference and *Tracking Module*) are sequentially runned into a main Python file which processes groups of RAW images into an *Input Folder* and provides output *Masks* and a text file into an *Output Folder*. This text file contains all the tracked objects information.

2.3 Dataset Creation

Datasets of real night sky images have been used as a baseline to create extended datasets via data augmentation to train the two versions of the YOLO Neural Network (NN): v3 and v4. Each image consists of stars and streaks on a dark background and it is associated to a text file containing the information about the streaks in YOLO format. Three datasets have been used: one consisting of 5000 segmented images with artificial streaks; a second one consisting of 5000 segmented images in which some streaks appear to be fragmented for a better resemblance to the real case and finally a dataset of 30000 segmented and fragmented images. These datasets will be used to perform several training strategies that will be discussed in the next section. The aim of these different strategies is to evaluate which is the one that improves more the model accuracy.

3 Training, Testing and Results

Several approaches for YOLO training have been used to increase the robustness to False Negatives (FNs) and False Positives (FPs). The strategy foresees a training on the first dataset (*Original Dataset*), then a Transfer Learning (TL) process over the second dataset and finally a second TL over the third dataset. As Transfer Learning process it is meant here a training of a previously trained Network: a Network has been trained on a specific dataset, the last saved weights are then loaded and the same network is trained again on a different dataset. This technique is used to increase performances of the network against FPs and FNs. A second training from scratch has been performed on the third dataset to verify the effectiveness of the double TL process. Mean Average Precision (mAP) computed with an Intersection Over Union (IOU) equal to 50% and True Positives (TPs), FPs, FNs values are reported in Table 2 for the tested configurations. This performance evaluation is carried out on the test subset of the *Original Dataset*.

As can be seen, the most promising training strategy for YOLO networks is a double TL on different datasets. The percentage of TP increases together with a

Table 2 YOLO v3 and v4 Performance report

Strategy	YOLO v3				YOLO v4			
	mAP [%]	TP [%]	FP [%]	FN [%]	mAP [%]	TP [%]	FP [%]	FN [%]
Original DS	78.4	72.4	8.5	19.1	82.82	74.7	9.5	15.8
First TL	79.8	74.4	8.5	17.1	84.50	75.8	10.1	14.1
Second TL	82.7	77.3	7.7	15.0	**85.95**	**77.1**	**10.3**	**12.6**
Scratch 30k	78.0	71.9	8.3	19.8	82.12	74.2	10.0	15.8

decrease of FN. An interesting fact is that with the YOLO v4 network double TL, the percentage of FP is increased if compared with the achieved one of the double TL with YOLO v3. Moreover it seems that the substitution of the YOLO v4 with the YOLO v3 model does not bring so much improvement in the algorithm performances due to a slight and reduced percentage of TP. Even if the best YOLO v4 model reaches a 77.1% TP percentage, it shows a reduced percentage of FN as the TL process meant to match. The drawback of this double TL process is in the increment of FP percentage which is associated to a multiple and undesired detection of the same streak as shown in the following Fig. 6. However this does not constitute a big deal because this multiple detection case is well handled by the *Tracking module* as next sections will show. The highlighted values in Table 2 are referred to the best YOLO v4 trained model that is used in the next tests.

3.1 Comparison on Real RSOs Passages

In this section, the computation of TPs, FPs and FNs for the best YOLO v4 model has been performed and shown. It is made on a test set of 31 real RSOs passages which were acquired during a real images acquisition campaign. Moreover a comparison with the best YOLO v3 model performances from our previous work is presented [8] in Tables 3 and 4.

Table 3 represents a pseudo confusion matrix due to the impossibility of True Negatives (TNs) computation. Indeed if no streak is present in the image the network is not able to detect the absence of streaks and thus the info of TN is impossible to be provided by us.

By looking at the comparison tables, it is possible to assess that in YOLO v4 predictions there is a perceptible decrement of FN and an increment of FP with respect to YOLO v3 model. In particular, the increased robustness to FN brings YOLO v4 model to detect correctly heavily fragmented streaks, resulting in an increasing probability of dim target detection. On the contrary the drawback is the previous cited

Table 3 YOLO v3 and v4 overall Real RSOs passages performance

YOLO v3			YOLO v4		
FP	FN	TP	FP	FN	TP
13	**37**	**201**	**18**	**26**	**208**

Table 4 YOLO v3 and v4 Real RSOs passages comparison

YOLO v3			YOLO v4		
Precision [%]	Recall [%]	F1 [%]	Precision [%]	Recall [%]	F1 [%]
93.93	84.45	88.94	92.04	88.89	90.44

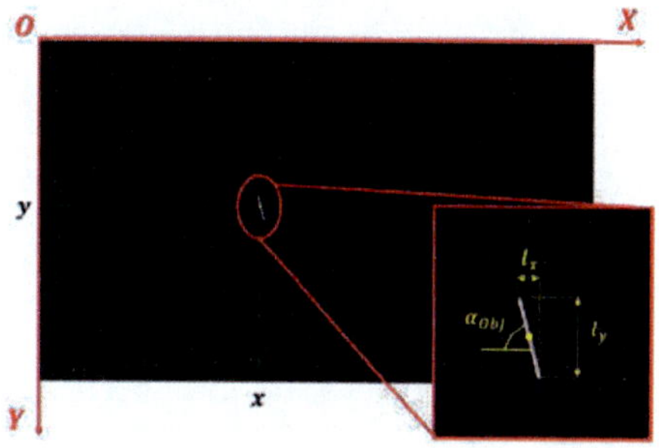

Fig. 5 Image Reference Frame and streak's relevant quantities

higher number of FPs: sometimes the model detects twice clear streaks in the night sky. This was noticed while testing the v4 model on the 31 passages. A case where this occurs is shown in Fig. 6. The same streak is contained in both the boxes and this does not represent a problem for the *centroid*, *extremal points* and velocity components computation. Indeed, the tracking algorithm will assign these redundant info to the same object and then they can be easily filtered out or combined into a single information after the *Tracking module*. Table 5 shows *centroids*' coordinates, orientation angle and velocity components of the detected object as stored by the *Tracking Module*. In particular, in Frame 4 the object detected as double is associated to the same object and tracked during its motion in the FOV through the *Tracking Module*. The double detection info differs mostly in streak orientation due to a different box applied to the object region, resulting in some object pixels on the box contour which do not participate to *centroid* and *extremal point* computation. Anyway, the redundant information can be combined to compensate this inconvenient, providing a better estimate of the RSO coordinates. For our purposes this double detection case does not represent a problem because of the *Tracking Module* existence and the easy implementation of a module which removes the redundant information. In Table 5 double detection and tracking are shown, where x and y are the *centroids*' coordinates (a *centroid* is the center of energy of the streak cluster), α_{Obj} is the orientation angle of the streak with respect to the image horizontal axis, V_x and V_y are the *centroids*' velocity components obtained dividing the streaks length components

Table 5 Double detection and *Tracking Module* reaction

Frame	$x\,[px]$	$y\,[px]$	$\alpha_{Obj1}\,[deg]$	$Vx\,[px/s]$	$Vy\,[px/s]$
1	285.57	278.98	32.01	5.8	3.0
2	–	–	–	–	–
3	325.28	300.18	32.83	5.2	2.8
4	364.02	321.26	39.40	4.8	2.6
4	364.42	321.46	35.54	5.0	2.6
6	399.09	340.43	33.18	4.6	2.6

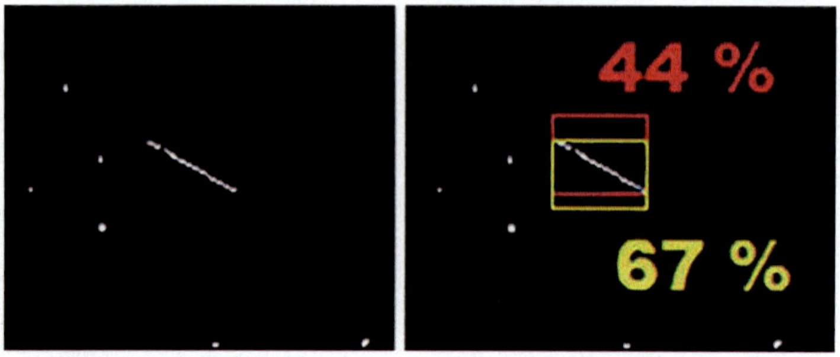

Fig. 6 An example of double detection of the same RSO. The segmented streak (left). The double detection (right)

along each axis (l_x and l_y) by the Exposure Time (Fig. 5). Velocity components are computed without sign.

3.2 *Test of a Real RSO Passage*

In this subsection the ISS (brightest object) and Starlink (weakest object) passage is processed by the algorithm. The whole sequence of 18 images is condensed into a single image (Fig. 7). They were acquired with a reflex camera which exposure time was 5 s and the interval between two consecutive images was set to 2 s. Every image is segmented by U-Net and then RSOs information is extracted by YOLO v4 and differentiated into objects by *Tracking Module*. This passage was already tested in our previous work [8] with YOLO v3 model and here the test is repeated for comparison purposes. Data of the detected RSOs are collected in Table 6. The main differences with respect to the YOLO v3 results (Table 7) are:

- Correctly detection of ISS in frame 3 and Starlink in frame 4 thanks to YOLO v4 upgrade and thus to the improved capability in fragmented streaks detection.
- Slight differences in streak orientation values due to an improved way in *extremal point* computation.
- Differences between computed objects coordinates are generally smaller than 1 px. This is not due to a different U-Net segmentation action but to a better way in detecting the area around the box (less not-RSOs foreground pixels in the box which affect the *centroids* and *extremal points* computation).

Fig. 7 ISS (brightest) and Starlink (weakest) passage over Campo Imperatore

Table 6 ISS and Starlink Tracking Table with YOLO v4

Frame	x_{Obj1} $[px]$	y_{Obj1} $[px]$	α_{Obj1} $[deg]$	x_{Obj2} $[px]$	y_{Obj2} $[px]$	α_{Obj2} $[deg]$
1	–	–	–	–	–	–
2	291.27	187.61	67.75	–	–	–
3	303.95	235.67	66.97	–	–	–
4	316.04	281.70	65.43	61.23	213.97	53.97
5	327.34	325.06	66.37	83.71	253.10	55.89
6	337.46	363.92	65.10	–	–	–
7	346.44	398.65	63.43	–	–	–
8	354.76	431.05	65.38	–	–	–
9	362.32	460.86	64.54	–	–	–
10	369.05	487.48	67.17	–	–	–
11	375.55	512.53	64.65	–	–	–
12	381.51	537.50	64.80	–	–	–
13	387.08	560.15	64.80	–	–	–
14	392.10	580.76	68.20	–	–	–
15	397.09	601.41	68.20	–	–	–
16	401.54	620.10	66.37	–	–	–
17	–	–	–	–	–	–
18	–	–	–	–	–	–

Table 7 ISS and Starlink Tracking Table with YOLO v3

Frame	x_{Obj1} $[px]$	y_{Obj1} $[px]$	α_{Obj1} $[deg]$	x_{Obj2} $[px]$	y_{Obj2} $[px]$	α_{Obj2} $[deg]$
1	–	–	–	–	–	–
2	291.26	187.60	67.38	–	–	–
3	–	–	–	–	–	–
4	316.04	281.70	68.20	–	–	–
5	327.25	324.71	66.57	85.60	256.46	57.99
6	337.46	363.92	69.23	–	–	–
7	346.57	399.18	66.50	–	–	–
8	354.44	429.78	68.63	–	–	–
9	362.37	461.05	66.80	–	–	–
10	369.05	487.48	67.17	–	–	–
11	375.27	512.51	67.62	–	–	–
12	381.61	537.91	70.02	–	–	–
13	386.78	559.23	67.62	–	–	–
14	392.18	581.07	66.37	–	–	–
15	397.00	601.01	66.80	–	–	–
16	401.53	620.09	72.47	–	–	–
17	–	–	–	–	–	–
18	–	–	–	–	–	–

3.3 *Test with High Fidelity Star Tracker Simulator Images*

In this test, 20 High Fidelity Star Tracker Simulator images have been tested with the present algorithm [7]. Five objects have been generated into the FOV. The simulator is part of the Automation Robotics & Control for Aerospace Laboratory (ARCALab) at School of Aerospace Engineering, Sapienza University of Rome. The simulator is a heritage of ARCALab cooperation in several project on Star Trackers and on board debris surveillance mission cooperation. For this set the exposure time has been set to 0.2 s with a negligible time interval between the end of one acquisition and the starting of the successive one, in order to simulate the behaviour of a possible star tracker. The algorithm application with simulated Star Tracker images provides 6 detected objects where for each frame the computed *centroids* coordinates are shown (Table 8). Where coordinates do not appear a blank box or a "-" can be found. The former means that the detected object is not in the camera FOV anymore while the latter means the presence of the object within the image without a successful detection.

By inspection of the Objects coordinates evolution, it can be noticed that the Objects 1 and 2 coordinates are really close at first frame. Indeed a double detection of the same physical RSO is present in the first frame but the reason why this leads to the creation of two different objects comes from the frame number: it is the first frame

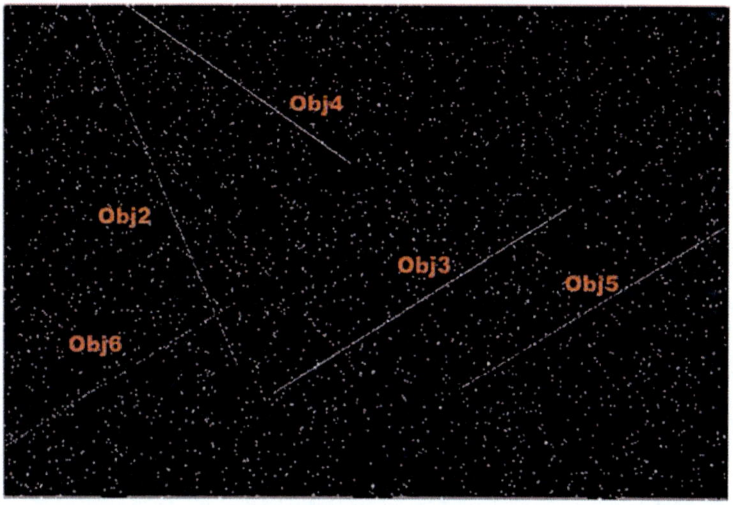

Fig. 8 High Fidelity Star Tracker simulator images passage. The 5 objects are correctly detected

and thus the tracking algorithm assigns every detected box to a specific object. While the double detection does not represent an inconvenient in the successive frames (see frame 8 with double detection for Object 5), from this test it is shown that could lead to a wrong number of detected objects if it occurs during the Object initialization phase. Anyway this is not a problem because one of the two box is recorded just in the first frame and can be easily filtered out. Object 2 disappears after frame 8 where it is not correctly detected. Object 3 is always detected while Object 4 disappears after frame 17 with a missed detection at the second frame. Object 5 is correctly detected till frame 8 when it disappears. Object 6 is detected just in frame 4 because it is heavily fragmented. The last row of Table 8 shows the Percentage of Correct Detection (PoCD) which is the ratio between the correct detections and the number of frames in which the object is actually within the FOV. It is not computed for Object 1 and 6 because they do not correspond to real object or are redundant objects.

The whole passage composed of 20 images with detected objects is shown in Fig. 8.

The YOLO v4 network, despite the FPs cases which do not constitute a problem for the correct working of the algorithm, is able to detect correctly all the 4 of the 5 objects avoiding the undesired FP which occurred with YOLO v3 in the same test. Indeed here a border streak was detected as potential object when it was a clear false object [8].

Table 8 High-Fidelity Star Tracker Simulator tracking results with YOLO v4

	Obj_1		Obj_2		Obj_3		Obj_4		Obj_5		Obj_6	
Frame	x [px]	y [px]	x [px]	y [px]	x [px]	y [px]	x [px]	y [px]	x [px]	y [px]	x [px]	y [px]
1	292.2	428.3	298.8	438.1	366.8	497.6	452.5	200.0	619.6	487.5	–	–
2			273.2	379.9	388.2	484.8	–	–	660.4	462.3	–	–
3			247.0	319.2	407.9	472.1	426.9	182.3	709.8	433.7	–	–
4			221.0	258.9	426.8	460.4	412.3	172.1	755.4	406.1	108.1	506.6
5			193.4	195.2	446.9	448.0	398.2	162.4	801.5	378.8	–	–
6			170.0	140.7	468.5	434.7	383.3	152.1	845.9	352.3	–	–
7			143.6	81.1	487.9	422.9	367.4	141.1	892.4	324.5		
8			–	–	504.1	412.6	351.7	130.0	934.4 934.4	299.8 299.8		
9					524.3	400.2	334.7	118.5				
10					544.6	387.3	318.6	107.0				
11					564.1	375.5	300.9	94.8				
12					585.2	362.2	281.7	81.5				
13					606.6	349.2	262.2	68.1				
14					624.2	338.3	241.6	53.8				
15					642.5	327.0	222.0	40.2				
16					663.9	313.8	200.8	25.5				
17					684.5	301.0	177.2	9.0				
18					703.4	289.3						
19					723.7	276.8						
20					739.6	266.8						
PoCD [%]	–		87.50		100		94.11		100		16.67	

4 Conclusion

This paper shows:

- Comparison of the present algorithm with a previous YOLO v3 based version [8];
- Transition from YOLO v3 to YOLO v4 which improves Algorithm overall accuracy and a better capability in dim weak RSOs detection;
- Capability of *Tracking Module* in resolving objects even in case of multiple detection cases;
- Capability of the algorithm in working well both with real night sky images and simulated Star Tracker images.

An improvement in *Tracking Module* filtering routines to couple with multiple detections in the first frame together with the merging of redundant information related to the same object will be implemented in the future. Moreover, an AI based module is being studied and developed for the *extremal points* computation of the detected streaks.

References

1. Darknet for yolo v3/v4 networks. https://github.com/AlexeyAB/darknet, accessed: 2022-04-01
2. Space debris and human spacecraft. https://www.nasa.gov/mission_pages/station/news/orbital_debris.html, accessed: 2022-01-10
3. Bobrovsky, A., Galeeva, M., Morozov, A., Pavlov, V., Tsytsulin, A.: Automatic detection of objects on star sky images by using the convolutional neural network. In: Journal of Physics: Conference Series, vol. 1236, p. 012066. IOP Publishing (2019)
4. Bochkovskiy, A., Wang, C.Y., Liao, H.Y.M.: Yolov4: optimal speed and accuracy of object detection. arXiv preprint arXiv:2004.10934 (2020)
5. Farissi, M.S., Agostinelli, I., Mastrofini, M., Curti, F., Marzo, C., Facchinetti, C., Ansalone, L.: Hardware implementation of the spot payload for orbiting objects detection using star sensors. In: 72th International Astronautical Congress (IAC), 25–29 Oct. 2021 (2021)
6. Kessler, D.J., Johnson, N.L., Liou, J., Matney, M.: The Kessler syndrome: implications to future space operations. Adv. Astronaut. Sci. **137**(8), 2010 (2010)
7. Mastrofini, M.: Test on 20 image sequences. https://www.youtube.com/watch?v=RGZbLHRshSI, accessed: 2022-06-22
8. Mastrofini, M., Goracci, G., Agostinelli, I., Farissi, M.S., Curti, F.: Resident space objects detection and tracking based on artificial intelligence. In: Astrodynamics Specialist Conference AAS/AIAA, Charlotte, North Carolina, USA, 7–11 Aug. 2022 (2022)
9. Mastrofini, M., Latorre, F., Agostinelli, I., Curti, F.: A convolutional neural network approach to star sensors image processing algorithms
10. Redmon, J., Divvala, S., Girshick, R., Farhadi, A.: You only look once: unified, real-time object detection. In: Proceedings of the IEEE Conference on Computer Vision and Pattern Recognition, pp. 779–788 (2016)
11. Redmon, J., Farhadi, A.: Yolov3: an incremental improvement. arXiv preprint arXiv:1804.02767 (2018)
12. Ronneberger, O., Fischer, P., Brox, T.: U-net: convolutional networks for biomedical image segmentation. In: International Conference on Medical Image Computing and Computer-Assisted Intervention, pp. 234–241. Springer (2015)
13. Salvatore, N., Fletcher, J.: Learned event-based visual perception for improved space object detection. In: Proceedings of the IEEE/CVF Winter Conference on Applications of Computer Vision, pp. 2888–2897 (2022)
14. Spiller, D., Magionami, E., Schiattarella, V., Curti, F., Facchinetti, C., Ansalone, L., Tuozzi, A.: On-orbit recognition of resident space objects by using star trackers. Acta Astronaut. **177**, 478–496 (2020)
15. Utzmann, J., Wagner, A.: SBSS demonstrator: a space-based telescope for space surveillance and tracking
16. Xue, D., Sun, J., Hu, Y., Zheng, Y., Zhu, Y., Zhang, Y.: Dim small target detection based on convolutional neural network in star image. Multimed. Tools Appl. **79**(7), 4681–4698 (2020)

GPU@SAT: A General-Purpose Programmable Accelerator for on Board Data Processing and Satellite Autonomy

Roberto Ciardi, Gianluca Giuffrida, Gionata Benelli, Christian Cardenio, and Riccardo Maderna

Abstract In the last decade, Artificial Intelligence (AI) algorithms are becoming very influential in several sectors thanks to their flexibility and adaptability. Among the several fields of the AI, the most attractive and with the greatest results is Machine Learning (ML). It aims to extract and elaborating characteristics starting from a huge amount of data. Space missions could strongly benefit of these algorithms, allowing the elaboration and processing of data directly where they are acquired. Thanks to the technological advancements in designing low-power hardware accelerators, several enterprises (both private and public) have used Components Off-The-Shelf (COTS) hardware accelerators to run their on-the-edge applications. This has allowed to mitigate the server Graphical Processing Unit (GPU) high power consumption and overheating while maintaining good performance. However, these devices cannot be used in the harsh space environment, due to their low radiation tolerance, that is a fundamental requirement for safely working in space. GPU@SAT aims to revolutionise the current Artificial Intelligence execution and signal processing analysis, exploiting a general-purpose GPU-like IP-core fitted on a radiation-hardened FPGA. The board will be adopted in the framework of a European Space Agency (ESA) project under the Advanced Research in TElecommunications Systems (ARTES) programme. AIKO, the company leader of this activity, is developing AI algorithms

R. Ciardi (✉) · G. Giuffrida · G. Benelli
IngeniArs S.r.l., Via Ponte a Piglieri 8, 56122 Pisa, Italy
e-mail: roberto.ciardi@ingeniars.com
URL: https://www.ingeniars.com/

G. Giuffrida
e-mail: gianluca.giuffrida@ingeniars.com

G. Benelli
e-mail: gionata.benelli@ingeniars.com

C. Cardenio · R. Maderna
AIKO SRL, Via dei Mille 22, 10123 Torino, Italy
e-mail: christian@aikospace.com
URL: https://www.aikospace.com/

R. Maderna
e-mail: riccardo@aikospace.com

C. Ieracitano et al. (eds.), *The Use of Artificial Intelligence for Space Applications*, Studies in Computational Intelligence 1088,
https://doi.org/10.1007/978-3-031-25755-1_3

for Fault Detection Isolation and Recovery systems. The adoption of such algorithms, executed via dedicated and space-qualified systems, directly on-board of satellites, can strongly increase the life of the mission, while reducing the design costs.

Keywords GPU · Satellites · Radiation hardened · Artificial intelligence · Fault detection isolation and recovery · Hardware accelerator · Software framework

1 Introduction

Thanks to technological advancement and the improved techniques adopted to produce sensors and components for space missions, the amount of scientific data produced every year is growing at incredible speed. To exploit the information acquired on-board of satellites, a dedicated processing phase is usually required. It aims to extract, whenever possible, the relevant data transforming them from raw data to valuable information. Actually, this entire process is usually conducted in three steps: (i) raw data acquisition, (ii) data transmission to ground, and (iii) data elaboration and interpretation. A fourth step could be present, in the case the data elaborated on-ground regard telemetries, requiring the retransmission to the satellites of the corresponding telecommands.

This processing chain is usually needed since the computational power on-board of satellites is not enough. However, for micro- and nano- satellite missions also the transmission of the data to and from ground represents a very problematic task [7]. In fact, the bandwidth is strongly limited to maintain both the cost and size of these satellites as cheaper as possible. Finally, the algorithms used to elaborate the data acquired on-board could require a lot of resources, which are very rare in a satellite system. In particular, the most innovative algorithms based on artificial intelligence (AI) and machine learning (ML) can infer additional information starting from the data acquired directly on-board [8, 10]. However they could require long processing time or dedicated hardware for reducing the computational power [5, 9, 19].

These approaches usually benefit from the use of GPUs, that, by executing the same instruction on different portion of data, i.e., Single Instruction Multiple Thread (SIMT) paradigm, strongly optimising the operation flow while reducing the processing time. GPUs, on the other hand, need a large amount of power, limiting their adoption only on data-center or cloud. To address this issue, in the latter years, several big tech companies have developed dedicated hardware solutions, with the purpose of speeding up the inference time while maintaining a low-power consumption. These dedicated accelerators are often exploited for on-the-edge applications [1, 3, 4], but despite their performance, also in this case these devices do not respect the requirements needed for harsh space environment, resulting not suitable for space missions.

To overcome these problems, large integrators in the space industry usually leverage on dedicated family of FPGA boards and dedicated design techniques, developing ad-hoc accelerators for each mission. This developing technique requires a long design time and reduces the reusability of the design developed (tailored on the specific mission requirements), increasing also time to market for those solutions.

To this end, IngeniArs, an Italian company born in 2014 and specialised in soft-core hardware design for space application, has developed a multi-purposes IP-core accelerator. Taking inspiration by the powerful GPU devices, the GPU@SAT IP-core is composed of two main components: the hardware design, developed exploiting only HDL language, hence portable in several FPGA devices [21], focusing mainly on radiation hardened families; and a dedicated compiler that translates high-level programming language algorithms in dedicated operations for the IP core. The possibility to synthesize this design on top of radiation hardened boards is the keystone that will allow the use of complex algorithms and ML directly on-board of satellites. The developed accelerator consists of three distinct elements:

Processing Elements (PEs) components execute the instructions;
Compute Units (CUs) group the PEs, ensuring that all the PEs execute the same instruction on different batch of data;
Dispatcher assigns at each CU the list of operations to execute.

GPU@SAT is completely customisable, adapting the number of CUs directly on the specifications of the final application. In this way, another level of customisation is introduced, allowing a trade-off between power consumption and processing capability. Furthermore, as already mentioned, the IP core is completely described in HDL, hence, the entire design does not depend on a precise vendor, which eliminate the technological dependencies from a specific vendor.

To exploit the potentiality provided by this innovative platform, IngeniArs has also developed a software framework for programming it. The framework allows to generate the instruction through the GPU@SAT compiler, and to schedule the instruction flows. In particular, the GPU@SAT compiler is designed starting from the LLVM 12 compiler suite [12]. LLVM splits the compiling phases in three separated parts: FrontEnd (FE), that aims to encode the high-level programming languages code in a intermediate representation (IR); Optimizer, that aims to optimise the IR generated at previous step, reducing and improving the resources allocation, without referring to a specific architecture or hardware implementation; BackEnd (BE), that aims to adapt the optimised IR directly on the hardware architecture, taking into account all the possible architectural constraints. The advantages brought by a modular compiler are manifold. First and foremost, the interoperability provided by a modular compiler allows to exchange already developed FEs with other BEs, and vice versa.

To validate and demonstrate the potential of this design, IngeniArs is integrating this solution within an ESA project funded by ASI under the ARTES programme. The consortium, composed of AIKO, Tyvak and IngeniArs, aims to develop an automated system for Fault Detection Isolation and Recovery (FDIR). AIKO is developing an intelligent on-board FDIR system able to improve the health monitoring of the satellite through:

1. early identification of anomalies and unexpected behaviour,
2. predictive health monitoring,
3. reduced dependency from ground operations.

This is made possible by leveraging MiRAGE (Mission Replanning through Autonomous Goal gEneration), an on-board satellite operations automation software, developed by AIKO. Overall, the AI-powered software library analyzes telemetry, payload, and external data, to generate and update the operations plan. Satellites become aware, understanding their surroundings, internal health, and mission events. In particular, MiRAGE enables F2 fault management level. This means that satellites can autonomously re-establish nominal mission operations following an on-board failure, by performing complex telemetry analysis on-board, detecting and isolating faults, and by performing reconfiguration to resume nominal operations.

GPU@SAT paves the way for fully autonomous satellite systems, which will execute on-board algorithms that were relegated to terrestrial use, improving their efficiency and operability. Furthermore, the collaboration between AIKO and IngeniArs will open the door to completely autonomous satellites, powered by AI algorithms that will run on top of radiation tolerant hardware.

In the following sections the design and implementation of the IngeniArs' GPU@SAT is presented. Firstly, an overview of the proposed Hardware design is provided. The Hardware presented in Fig. 1 is described, presenting CUs and PEs functionalities. Later on, the Software architecture of GPU@SAT is described. The software framework is presented, describing the LLVM compiler on which it is based and that guarantees the programmability of the IP-core. Finally, the FDIR system, developed by AIKO and accelerated directly on-board satellite via GPU@SAT system, in the scope of an ESA project, part of ARTES programme, is presented. The FDIR system design is outlined, focusing on its modular architecture and explaining its functionalities. This system accelerated by the IP-core will allow to proof the advantages in adopting such technology on-board of satellites.

2 Hardware Architecture

The hardware structure of the GPU@SAT is depicted in Fig. 1.

It contains (I) the central kernel dispatcher, which has the purpose of distributing the instructions to the Computational Units (CUs), (II) the standard AXI communication, which connects the design with the external memory and the other components, if present, available on the FPGA, and (III) the memory controller. Since the GPU@SAT, as suggested by the name, represents a GPU, its IP-core design is compatible with the main operations defined in OpenCL 1.2 [17].

To this end, the internal architecture has been designed to execute algorithms that follow the Single-Instruction Multiple-Thread (SIMT) paradigm. In this type of architecture, each thread executes the same instruction but on a different portion of

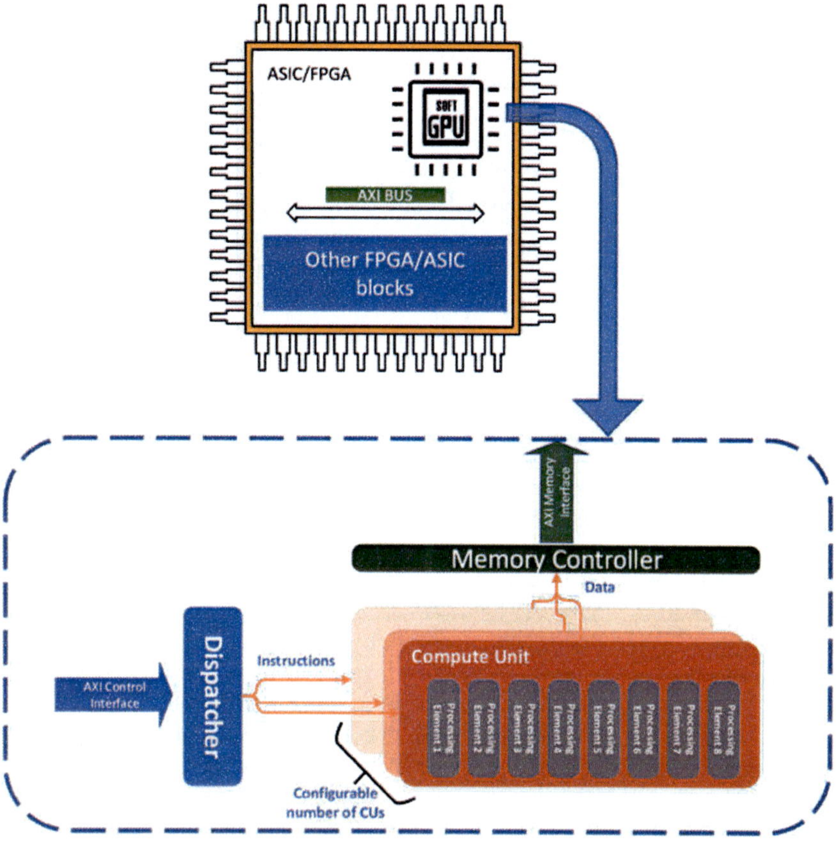

Fig. 1 GPU@SAT hardware Architecture

data. Such execution increases throughput of the application, slightly penalising the latency.

To reflect as much as possible a typical SIMT GPU architecture, GPU@SAT allows to change the number of CUs depending on the application requirements. Each CU contains the Processing Elements (PEs), responsible for executing the general purpose computations, the matrix operations and the logical instructions. The Computation Unit reads the instruction and divides the portion of data to be elaborated by the Processing Elements, while PEs' goal is to execute the instruction assigned to them as fast as possible.

The number of CUs synthesised on the FPGA effectively determines the power consumption. In fact, it can be easily modified directly during the system operation, adapting the computational needs to the available power budget.

The entire architecture is completely programmable using its dedicated software development kit, detailed in Sect. 3, which allows to port high-level algorithms described using OpenCL or TensorFlow.

Within the design, all components are interconnected through an AXI4 Smart interconnect module. Large input/output data-rate for the GPU is ensured thanks to the dedicated Direct Memory Access (DMA) interfaces to other modules on the same FPGA. The rest of the interconnection system is used for command and control purposes. The design of GPU@SAT allows an easy integration with other modules thanks to the AXI interface and with other On-Board Computers, by exploiting standard communication interfaces.

3 Software Architecture

GPU@SAT IP-core can be programmed exploiting its dedicated software development kit. The latter represents an innovative yet essential layer which enables the programmability of the system through general-purpose high-level programming languages, e.g., OpenCL and TensorFlow, offering to end users a way for deploying new applications for GPU@SAT hardware design with minimum effort.

The software framework is composed by a compiler, dedicated to the generation of the executable kernels, and a set of tools, to organise the execution of the kernels and some application examples, as shown in Fig. 2.

The compiler part was developed exploiting the LLVM compiler toolkit, which contains a collection of modular and reusable compiler and toolchain technologies [13]. More in detail, LLVM splits the traditional monolithic compiler in three independent and complementary components: the FrontEnd (FE), the Optimiser (OP) and the BackEnd (BE). FE is responsible to parse the high-level programming language (e.g., C++, OpenCL and TensorFlow) in an Intermediate Representation (IR), a pseudo assembly language that abstracts the generated code from the hardware target. The IR code is then feed back to the OP, which removes the unused/unnecessary functions/operations and generate an optimised version of IR, the so-called "optimised IR". During the optimisation phase some hardware dependent information could be append to the IR, in order to improve the final result of the BE, even if this could reduce the re-usability of the module. Finally, the latter is given as input to the BE, which targets the optimised IR to the selected hardware, generating the runnable code.

The modular architecture of the compiler allows to use multiple FEs for a single BE, and vice versa. In particular, two existing FEs have been integrated in our framework: Clang for OpenCL kernels [11] and XLA for TensorFlow NNs [2]. Since the GPU@SAT represents a new target hardware, IngeniArs has developed its relative BE, which translate the IR in GPU@SAT runnable code. The BE implementation reflects the capability of the designed hardware, allowing the exploitation of the instruction from the high-level programming languages. The new target BE has been implemented to be compliant with LLVM 12.

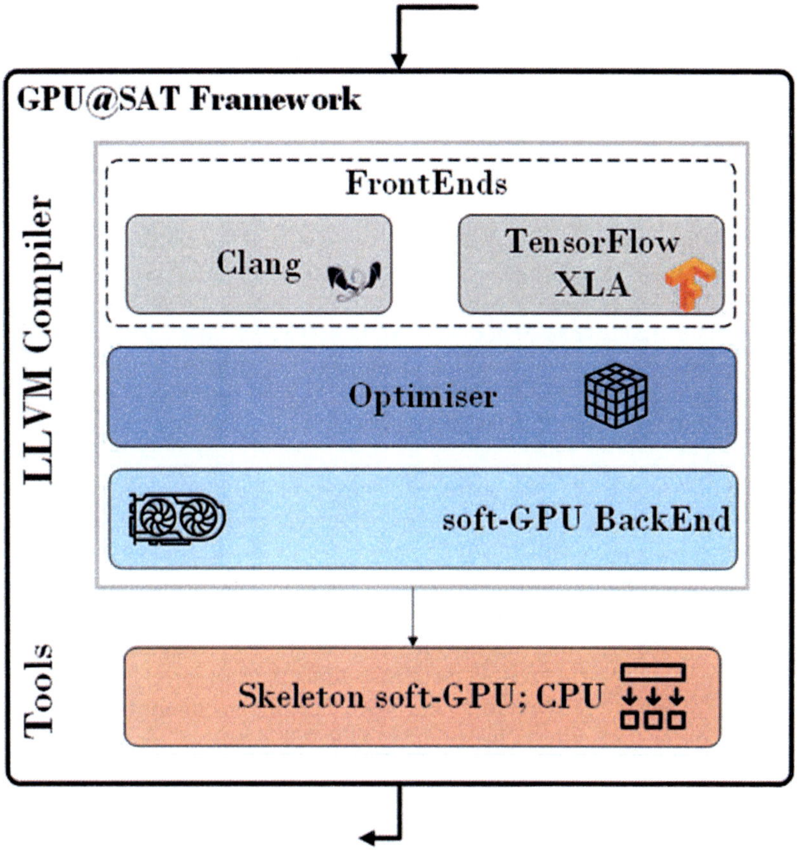

Fig. 2 GPU@SAT software Architecture

In the proposed implementation, the compiler allows to:

- Compile OpenCL code via clang FE compiler
- Convert TensorFlow model exploiting XLA [22] and MLIR [14]
- Notify users of eventual layers incompatibilities
- Generate GPU@SAT debug assembly files

In addition, a simple set of preliminary tools for generating the scheduling & dispatching and the application skeleton has been developed and implemented. In detail, the output of the compiler is used as input for the task scheduler, which divides the RISC-V tasks from the soft-GPU ones. Then, a parser takes the output of the task scheduler tool, and generates a set of files to be used as guideline for the development of the entire application.

In conclusion, the framework, which supports the OpenCL and TensorFlow flows, allows to directly implement AI, CV and signal processing algorithms on the GPU@SAT system, hiding the complex processes of compilation, optimisation, generation of executable code and scheduling of tasks for GPU@SAT IP-core.

4 AI-FDIR Algorithms

To validate the capability of the GPU@SAT system, the presented solution is integrated into an FDIR system, part of an ESA project under the ARTES programme. The FDIR system is being developed by AIKO based on MiRAGE (Mission Replanning through Autonomous Goal gEneration), a proprietary satellite operations automation software powered by AI and Deep Learning (DL). MiRAGE is designed as a modular library of microservices in order to reduce the customisation efforts and promote usability across different platforms and mission concepts (Fig. 3). Thanks to its modularity, MiRAGE can be configured to exploit the hardware available in its host environment - i.e., the spacecraft. Specifically, the execution of DL algorithms on-board can be delegated to purposely developed and dedicated hardware accelerators, such as the GPU@SAT.

While a minimum level of autonomy is required for every spacecraft, past experiments have shown that introducing more sophisticated FDIR systems, powered by AI, can drastically increase the reliability [16] and minimize the performance loss [18] of space-based assets. Overall, the software library analyses telemetry, payload, and external data, to make decisions to improve the health monitoring of the satellite and update the operations plan. In particular, MiRAGE enables F2 fault management level, allowing the satellite to autonomously re-establish nominal mission operations, following an on-board failure.

To do so, MiRAGE offers a set of configurable modules comprising:

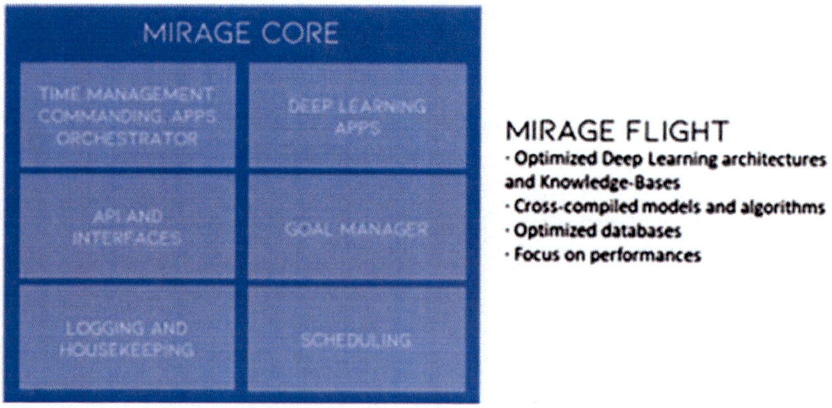

Fig. 3 MiRAGE core building blocks

Orchestrator

The Orchestrator is a software module that manages the interaction among all other modules by providing fundational monitoring, interface, and communication services. In the context of the FDIR system, the Orchestrator enables quick response to the outputs generated by the DL applications. For instance, it generates and distributes notifications related to detected events and packages the high-level information that will then be processed by the Goal Manager.

Deep Learning modules

The Deep Learning modules process payload and telemetry data to extract high-level information. In the context of the FDIR system, DL models will be configured to process on-board telemetries. The output can range from feature extraction (such as patterns and trends in time series) up to anomaly detection (e.g., classification algorithms on a set of known nominal and anomalous behaviors). Different applications can be developed to analyze diverse telemetry streams or extract different types of information in order to address the specific mission and the final user requirements. MiRAGE architecture allows for extended modularity and flexibility in the choice of the employed low-level algorithms, which can range from supervised DL algorithms to unsupervised ones, to prediction modules, and even non-AI-based solutions. The Deep Learning modules constitute the part of MiRAGE that is deployed on the GPU@SAT board.

Goal Manager

The Goal Manager is the automated reasoning unit of MiRAGE, implemented via Knowledge-Based System technology. Its scope is to process all the available information to make accurate decisions related to mission goals, including Failure Identification and Recovery. The "available information" can be anything that is plugged into the Goal Manager, including: the output of DL modules, single spacecraft telemetries, external information such as visibility windows and eclipses, and so on.

Planner

The role of the Planner is to allocate spacecraft resources and plans operations autonomously. In the context of an FDIR system, FDIR actions triggered by the Goal Manager always have the highest priority to ensure the health of the satellite. Therefore, recovery tasks must be executed immediately without planning. Nevertheless, replanning is triggered after recovery to adapt the operation schedule of the spacecraft to the new conditions – e.g., the unavailability of a failed payload or reduced resource availability.

The flow of information among MiRAGE modules, depicted in Fig. 4, originates from state-of-the-art research that indicates that the best use of DL models is to be applied to low-level data to extract higher-level information. Instead, technologies such as Knowledge-Based Systems are best applied to take decisions at the system and mission levels. Specifically, data-driven techniques can easily detect anomalous system behaviour or the occurrence of faults, but often provide little insight into the space system. Instead, a deep understanding of the underlying system, its actions, and limitations is crucial for isolating and recovering from the fault [16]. Likewise, the

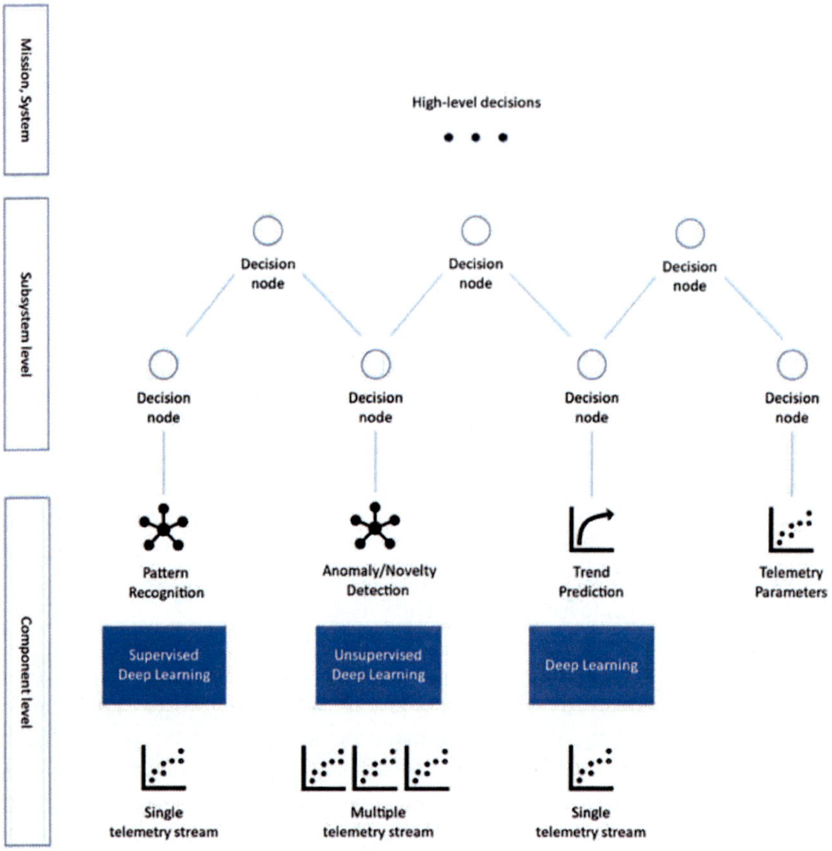

Fig. 4 Hierarchical FDIR design implemented within MiRAGE

ESA ESTEC SAVOIR FDIR Handbook [6] discusses how hierarchical management of failures can minimize the risk of undetected failure modes, with any anomaly not covered at a given level being addressed at a subsequent level.

A comprehensive discussion of the current state of the art in applying machine learning methods to anomaly detection and FDIR can be found in [15, 16, 18], where the latter also presents a comparison among currently available COTS boards for space-based machine learning, and [15] addresses the topic outside the space domain. In addition, [20] identifies Bayesian networks as a suitable computational basis for FDIR, while [15] calls for the creation of hybrid systems, as not one method beats the others in all performance indices.

When compared to the current implementation of onboard FDIR, which mostly relies on monitoring predefined thresholds and taking actions upon threshold violation, the new system is expected to bring several benefits, such as:

Early identification of anomalies and unexpected behaviour
The definition of thresholds is always conservative to prevent false alarms during nominal operations. However, threshold violations are often anticipated by abnormal behavior within the defined range. Moreover, some failures do not result in threshold violations and are difficult to be detected through classical FDIR approaches. Instead, they usually require inspection by human operators or analysis with more sophisticated tools available on-ground. AI-powered algorithms enable early detection of anomalous behavior and, in turn, prompter intervention to prevent or mitigate the failure occurrence. Improved detection capabilities makes it also easier to identify root causes. Thresholds violation can be a delayed symptom of several failures, which can be also not strictly related to the component for which the violation has been detected.

Predictive health monitoring
AI-powered algorithms can be exploited not only for anomaly detection and isolation, but also for predicting the future evolution of the system. Models can learn baseline resource consumption, periodical effects of eclipses and ground contacts, gradual ageing of components.
Information about future states of the spacecraft can be then exploited to better plan mission operations, avoid critical situations, and prevent the emergence of failures.

Reduced dependency from ground operations
Improved anomaly detection and isolation, together with advanced reasoning capabilities provided by MiRAGE, allows the FDIR system to enlarge the set of anomalies that can be managed autonomously onboard by the spacecraft. Furthermore, autonomous replanning capabilities complement anomaly detection and recovery to further increase the autonomy level. This results in a prompter reaction that can prevent the interruption of service and a reduced workload for ground operators.

5 Preliminary Results and Conclusion

This work presents the innovative GPU@SAT on-board accelerator. The proposed IP-core design is a game changer in the field of on-board processing for space missions. It, in fact, aims at enabling the execution of Artificial Intelligence and Computer Vision algorithms on-board thanks to its GPU-like structure.

In particular, IngeniArs has developed the platform on the Zynq ZC706, implementing 4 CU due to the resource limitation of the target device. Exploiting this configuration, we were able to test convolutional layer of different sizes, from 128×128 up to 1024×1024. With such tests we have obtained, for the former (128×128) an inference time of 134 μs, and for the latter (1024×1024) 20136 μs. Future development will aim to reduce the measured time by introducing local memory inside the CU, strongly reducing the amount of data directly exchanged with the central memory. To better characterise the designed solution for space environment, we have evaluated the possibility to synthesis it on different radiation hardened FPGAs. In particular, we tested the KU060 platform from Xilinx and RT4G150 from Microchip. The proposed implementation, using 4 CU, has required the 21,51% of LUT and 15,79% of BRAM on KU060 with an average power consumption of 4,42 W with a frequency of 200 MHz. On the other side, implementing only 1 CU on the RT4G150 has required the 63,63% of LUT and 45,93% of BRAM.

To simplify the final users' effort, a complete software development kit for IP-core programming is provided. This software comprises a compiler, based on the LLVM framework, a task scheduler, to organise the execution of the models, and a set of application examples, that represents the skeleton for future development.

To demonstrate the potentiality of the aforementioned design, the entire system will be adopted as part of an ESA project in the framework of ARTES program, funded by ASI. The consortium composed by AIKO, IngeniArs and Tyvak will exploit Artificial Intelligence algorithms to extend the life of the mission, by executing directly on-board a smart version of a Fault Detection Isolation and Recovery system named MiRAGE, developed by AIKO.

The combination of a dedicated hardware design, together with software that can predict potential satellite faults, represents an enabling technology that can change the way future space missions are conceived.

References

1. Coral. https://coral.ai/
2. Abadi, M., Barham, P., Chen, J., Chen, Z., Davis, A., Dean, J., Devin, M., Ghemawat, S., Irving, G., Isard, M., et al.: Tensorflow: a system for large-scale machine learning. In: 12th USENIX Symposium on Operating Systems Design and Implementation (OSDI 16), pp. 265–283 (2016)
3. Barry, B., Brick, C., Connor, F., Donohoe, D., Moloney, D., Richmond, R., O'Riordan, M., Toma, V.: Always-on vision processing unit for mobile applications. IEEE Micro **35**(2), 56–66 (2015)
4. Cass, S.: Nvidia makes it easy to embed AI: The Jetson nano packs a lot of machine-learning power into DIY projects - [hands on]. IEEE Spectr. **57**(7), 14–16 (2020). https://doi.org/10.1109/MSPEC.2020.9126102
5. Dakir, A., Barramou, F., Alami, O.B.: Opportunities for artificial intelligence in precision agriculture using satellite remote sensing. In: Geospatial Intelligence, pp. 107–117. Springer (2022)
6. ESA/ESTEC: SAVOIR FDIR Handbook 2.0 – SAVOIR-HB-003 (2019)
7. George, A.D., Wilson, C.M.: Onboard processing with hybrid and reconfigurable computing on small satellites. Proc. IEEE **106**(3), 458–470 (2018). https://doi.org/10.1109/JPROC.2018.2802438
8. Giuffrida, G., Diana, L., de Gioia, F., Benelli, G., Meoni, G., Donati, M., Fanucci, L.: Cloudscout: a deep neural network for on-board cloud detection on hyperspectral images. Remote Sens. **12**(14), 2205 (2020)
9. Jacobsen, R., Bernabel, C.A., Hobbs, M., Oishi, N., Puig-Hall, M., Zirbel, S.: Machine learning: paving the way for more efficient disaster relief. In: AIAA SCITECH 2022 Forum, p. 0397 (2022)
10. Kothari, V., Liberis, E., Lane, N.D.: The final frontier: deep learning in space. In: Proceedings of the 21st International Workshop on Mobile Computing Systems and Applications, pp. 45–49 (2020)

11. Lattner, C.: LLVM and Clang: next generation compiler technology. In: The BSD Conference, vol. 5 (2008)
12. Lattner, C., Adve, V.: Llvm: a compilation framework for lifelong program analysis & transformation. In: International Symposium on Code Generation and Optimization, 2004. CGO 2004, pp. 75–86. IEEE (2004)
13. Lattner, C., Adve, V.: LLVM: a compilation framework for lifelong program analysis and transformation, pp. 75–88. San Jose, CA, USA (Mar 2004)
14. Lattner, C., Amini, M., Bondhugula, U., Cohen, A., Davis, A., Pienaar, J., Riddle, R., Shpeisman, T., Vasilache, N., Zinenko, O.: Mlir: a compiler infrastructure for the end of Moore's law. arXiv preprint arXiv:2002.11054 (2020)
15. Liu, R., Yang, B., Zio, E., Chen, X.: Artificial intelligence for fault diagnosis of rotating machinery: a review. Mech. Syst. Signal Process. **108**, 33–47 (2018)
16. Meß, J.G., Dannemann, F., Greif, F.: Techniques of artificial intelligence for space applications-a survey. In: European Workshop on On-Board Data Processing (OBDP2019). European Space Agency (2019)
17. Munshi, A.: The opencl specification. In: 2009 IEEE Hot Chips 21 Symposium (HCS), pp. 1–314. IEEE (2009)
18. Murphy, J., Ward, J.E., Mac Namee, B.: Machine learning in space: a review of machine learning algorithms and hardware for space applications. AICS, pp. 72–83 (2021)
19. Ortiz-Gomez, F.G., Lei, L., Lagunas, E., Martinez, R., Tarchi, D., Querol, J., Salas-Natera, M.A., Chatzinotas, S.: Machine learning for radio resource management in multibeam geo satellite systems. Electronics **11**(7), 992 (2022)
20. Paakko, M., Myllymäki, P., Holsti, N., Tirri, H.: Bayesian networks for advanced FDIR. In: ESA Workshop on On-Board Autonomy, Noordwijk, The Netherlands (2001)
21. Rodriguez-Andina, J.J., Moure, M.J., Valdes, M.D.: Features, design tools, and application domains of FPGAs. IEEE Trans. Ind. Electron. **54**(4), 1810–1823 (2007)
22. Sabne, A.: XLA: compiling machine learning for peak performance (2020)

Overview of Meta-Reinforcement Learning Methods for Autonomous Landing Guidance

Andrea Scorsoglio, Luca Ghilardi, and Roberto Furfaro

Abstract This paper presents a vision-based method for autonomous planetary powered descent and landing using meta-reinforcement learning. The goal is map observations to thrust commands directly using a deep neural network. Two iterations of the method are presented. First the 3-degrees-of-freedom powered descent pinpoint landing is solved using images, altitude and vertical rate as inputs. Then the problem of landing site selection, solved using a secondary neural network that performs semantic segmentation, is introduced. The final model is capable of autonomously select a landing area and guide the spacecraft to the designated landing site using only images as input.

Keywords Meta-reinforcement learning · Spacecraft guidance

1 Introduction

Autonomous planetary landing is becoming increasingly important for solar system exploration. The Artemis mission is bringing focus back to the Moon. The mission, together with SpaceX's and NASA's ongoing Mars exploration programs, will require a new level of autonomy for the future generations of spacecrafts and landers. However, Guidance algorithms have seen relative stagnation over the last decades. Today's methodologies for lander guidance and control are largely based on variations of Apollo-era offline procedures. A reference trajectory is typically calculated on the ground and then provided as input to the lander's guidance system. This trajectory is usually computed by solving an Optimal Control Problem, which can

A. Scorsoglio (✉) · L. Ghilardi · R. Furfaro
University of Arizona, Tucson, AZ 85721, USA
e-mail: andreascorsoglio@arizona.edu

L. Ghilardi
e-mail: lucaghilardi@arizona.edu

R. Furfaro
e-mail: robertof@arizona.edu

C. Ieracitano et al. (eds.), *The Use of Artificial Intelligence for Space Applications*, Studies in Computational Intelligence 1088,
https://doi.org/10.1007/978-3-031-25755-1_4

be done using both indirect and direct methods. Recent advancements in the field include the Theory of Functional Connections (TFC) [15] and Universal Powered Guidance (UPG) [14] for indirect methods, and the convexification technique [1] for direct methods, which has been successful in power descent landing problems. The state-of-the-art methods mentioned above have limitations in that they are not reliable when dealing with unanticipated dynamics and require precise state estimation. This limits their ability to quickly adapt and respond. Additionally, both direct and indirect methods only create open-loop trajectories, which necessitate the use of a separate navigation and control system to follow the trajectory.

Machine Learning (ML) has gained attention for its potential to increase autonomy in autonomous space GN&C. Early examples have shown that deep neural networks can be trained using supervised learning to compute landing trajectories on Mars [26], the Moon [3] and interplanetary trajectories [12]. However, recent studies have revealed that classical neural networks trained via supervised learning are not accurate enough for uncertain environments with complex constraints. Reinforcement Learning (RL) [24] has emerged as a promising alternative for GN&C applications such as planetary landing [4, 7], rendezvous [11, 16], low thrust trajectory design [10, 27] and formation flying [23]. RL is capable of creating feedback guidance laws in complex environments even with non-convex constraints [20, 22] and has the added advantage of having low computational cost at the test phase. This is achieved using state-of-the-art RL algorithms such as Policy Gradient, Deep Deterministic Policy Gradient (DDPG) [13] and Proximal Policy Optimization (PPO) [17].

The examples mentioned above, however, do not address the major shortcomings of classical RL, namely poor performance in uncertain environments and poor sample efficiency. Meta-Reinforcement Learning (MRL), also known as "learning to learn" overcome these shortcomings by training the policy on a distribution of environments or Markov Decision Processes, enabling the system to quickly adapt to new situations. This makes MRL a valuable tool in space exploration, as demonstrated by recent publications on its use in guidance applications and its superiority to classical RL in handling uncertain environments and actuator failures. Examples include its use in a six degrees-of-freedom Mars landing scenario [6, 7], hovering on irregularly shaped asteroids [8, 9], and intercepting hypersonic exoatmospheric targets [5].

In this paper, we describe work they we have done to improve upon previous foundations by incorporating optical observations into a system for planetary landing. The approach is broken down into two steps:

- The first step involves using a deep network to map images and radar altimetry data into thrust commands in a closed-loop fashion [19]. This is done in an adaptive way within defined distribution boundaries, meaning that certain parameters of the environment, such as the gravity acceleration and the initial spacecraft mass, are randomly sampled within certain distributions. This allows the meta-learner to learn over a wide range of instances of a stochastic environment. Additionally, random actuator failures are introduced during training, which makes the algorithm robust to such events. This approach allows for a complete integration of guidance

and navigation in an image-based closed-loop system that is robust to perturbations and un-modeled dynamics.
- The second step is to add hazard detection and avoidance to the loop in order to increase autonomy [18]. We exploit the latest advancements in computer vision and Convolutional Neural Networks (CNN) to implement a hazard detection and landing site selection algorithm. The policy network is trained using PPO in a Meta Reinforcement Learning (MRL) framework using randomized landing sites. The CNN for hazard detection and avoidance is trained to perform semantic segmentation in a supervised manner and is then embedded in the guidance framework at test phase.

A novel integrated environment for photo-realistic rendering [21] has also been developed for fast training and validation. The simulator integrates dynamics and sensor data acquisition seamlessly in a python environment in real-time, and generates accurate images of the lunar surface using digital terrain models (DTM) and a physically-based rendering engine based on Blender[1]. This allows for the efficient training and validation of the system.

The remainder of the paper is organized as follows: Sect. 2 describes the RL methodologies used with a focus on MRL, Sect. 3 describes the first step relative to the image-based landing problem, Sect. 4 describes the second step with the addition of hazard detection, finally Sect. 5 contains some final remarks.

2 Reinforcement Learning and Meta-Reinforcement Learning

Reinforcement Learning (RL) is a method where an agent interacts with an environment repeatedly to learn how to accomplish one or multiple tasks. The environment is typically modeled as a Markov Decision Process (MDP) where the current state only depends on the previous state. However, when the state cannot be directly observed or is influenced by noise, the MDP becomes a Partially Observable MDP (POMDP). In a POMDP, the state $\boldsymbol{x}$ is hidden, and observations o are given through an observation function $\mathcal{O}(x)$. This happens in image or LIDAR based navigation where the observation contains information about the state without making it directly visible to the agent. The agent uses a parametrized policy π_θ to take actions $\boldsymbol{u}_k$, receive rewards r_k and move to the next state based on the current observation o_k. The goal of the reinforcement learning algorithm is to optimize the policy such that it maximizes the sum of rewards collected over an episode, taking into account any path or control constraints specified in the reward function.

RL algorithms are effective in solving time-dependent problems, but they can be computationally inefficient because they often require a large number of trials to learn new tasks. An alternative approach is meta-reinforcement learning (MRL),

[1] https://www.blender.org/.

which is based on how humans learn by "learning how to learn" through experience. MRL is more efficient than RL because it allows for learning over a distribution of tasks and generalizes better to uncertain environments. There are two methods for implementing MRL: Model Agnostic Meta Learning (MAML) [2] and Recurrent Neural Networks (RNN) based on the work of Wang et al. [25]. The paper focuses on the latter approach, where the policy is synthesized as a Gated Recurrent Unit (GRU) RNN to map temporal relationships in input sequences. The hidden states of the policy are updated based on the current task, while the weights are learned across tasks using normal gradient descent. The training method used is Proximal Policy Optimization (PPO) [17], which is a policy gradient algorithm that has shown state-of-the-art performance on many benchmark RL problems.

3 Image-Based Moon Landing with Meta-RL

The goal is to employ RML to create a deep convolutional-recurrent network that maps a sequence of images and radar altimeter data directly to thrust. The task is achieved using a novel integrated platform for modeling and photo-realistic rendering developed specifically for agent training based on Blender. Specifically, the simulator integrates dynamics and sensor data acquisition seamlessly in a python environment, that generates accurate images using lunar digital terrain models (DTM) and a physically-based rendering engine.

3.1 Simulation Environment

This case addresses the 3 Degrees of Freedom (3-DoF) lunar soft landing problem, where the lander's attitude is not considered and its translational motion is described in an inertial reference frame centered on the target landing location. The reference frame, referred to as the *I* frame, has its z-axis pointing upwards. The equations of motion for this problem are as follows:

$$\ddot{\mathbf{r}} = \mathbf{g} + \frac{\mathbf{T}}{m} \tag{1}$$

$$\dot{m} = -\frac{\|\mathbf{T}\|}{I_{sp} g_0} \tag{2}$$

Where the state vector $\mathbf{r} = [r_x, r_y, r_z]^T$ is given in the frame *I*. The gravity vector $\mathbf{g}$ is assumed to be constant, as a flat surface is considered, and the thrust vector $\mathbf{T} = [T_x, T_y, T_z]^T$ is also given in the frame *I*. The lander has the capability to adjust the thrust output in all three directions using its throttable thrusters. The path con-

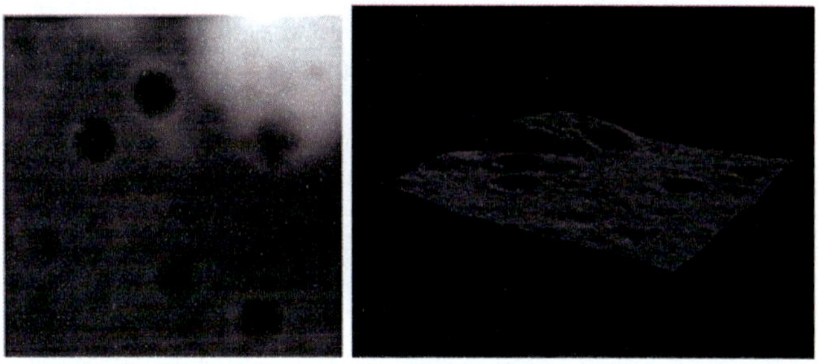

(a) DTM (b) Rendered surface model

Fig. 1 Apollo 16 landing site

straint consists of a fictitious flat surface, located at a certain target altitude, which acts as a virtual ground plane. The initial mass and the gravity vector are varied among the episodes to consider uncertainty in the environment for the reinforcement meta-learning algorithm. Moreover, possible engine failures are considered, which can happen randomly at each time-step. This is done by reducing the computed commanded thrust along the z-axis and along one of the other two axes.

The spacecraft is equipped with radar and optical sensors, which continuously provide a sequence of images and radar altimeter readings. The optical sensor's focus is always pointed in the opposite direction of the z-axis of the I frame. A simulation of the moon's surface is created using real DTM data from the Apollo 16 landing site, obtained by the Lunar Reconnaissance Orbiter (LROC).[2] Images are produced using a software called Cycles, which is a physically-based ray tracer that calculates the interaction of light from all sources in the scene with objects. This allows for the simulation of how parallel light rays from the simulated sun interact with the rough and low-reflective lunar surface. Figure 1 illustrates the DTM and a rendered view of the ground model using the raytracer.

Policy Optimization

Reward Function We used a distributed reward function to address the issue of sparse rewards, which was partially influenced by previous research by the authors [6–9]. The reward at each time step is calculated as the difference between the current velocity vector and a reference velocity determined by a gaze heuristics potential function that stays aligned with the line of sight between the lander and the target. The target's position and velocity are represented as $\boldsymbol{r}_t$ and $\boldsymbol{v}_t$, respectively. This ensures precise landing, not necessarily a soft one. To minimize final velocity, the agent estimates the time-to-go as the ratio of the range and the magnitude of the lander's velocity and reduces the targeted velocity as time-to-go decreases. This is incentivized by giving an additional positive reward at the end of successful trajec-

[2] http://wms.lroc.asu.edu/lroc.

tories. This reward increases in finite increments the closer the final state is to the target state. The reward function has the following form:

$$r = \alpha\|\mathbf{v} - \mathbf{v}_{targ}\| + \eta + \kappa(r_{lim,1} < \|\mathbf{r}(t_f) - \mathbf{r}_t\| < r_{lim,2}) + \xi(v_{lim,1} < \|\mathbf{v}(t_f) - \mathbf{v}_t\| < v_{lim,2}) \tag{3}$$

Where:

$$\begin{aligned}
\mathbf{v}_{targ} &= -v_0\left(\frac{\hat{\mathbf{r}}}{\|\hat{\mathbf{r}}\|}\right)\left[1 - \exp\left(-\frac{t_{go}}{\tau}\right)\right] \\
t_{go} &= \frac{\|\hat{\mathbf{r}}\|}{\|\hat{\mathbf{v}}\|} \\
\hat{\mathbf{r}} &= \begin{cases} \mathbf{r} - \mathbf{r}_{t'} & \text{if} \quad r_z > r_{t',z} \\ \mathbf{r} - \mathbf{r}_t & \text{otherwise} \end{cases} \\
\hat{\mathbf{v}} &= \begin{cases} \mathbf{v} - \mathbf{v}_{t'} & \text{if} \quad r_z > r_{t',z} \\ \mathbf{v} - \mathbf{v}_t & \text{otherwise} \end{cases} \\
\tau &= \begin{cases} \tau_1 = 20 & \text{if} \quad r_z > r_{t',z} \\ \tau_2 = 100 & \text{otherwise} \end{cases}
\end{aligned} \tag{4}$$

with $\mathbf{r}_{t'} = \mathbf{r}_t + [0, 0, \delta z]$ is an intermediate position above the desired final position and:

- $\alpha = -0.01$, a negative term that punishes errors with respect to the target velocity.
- $\eta = 0.01$, a positive term that encourages the agent to take more steps in order to avoid collisions with the constraints.
- κ and ξ, which are bonuses given if the final position and velocity are within specific intervals $[r_{lim,1}, r_{lim,2}]$ and $[v_{lim,1}, v_{lim,2}]$. These intervals and corresponding values can be determined from plots in a specific Fig. 2. These terms are only evaluated at the final time-step.
- v_0, which is the initial velocity of a specific rollout trajectory.

Observations Since the only network being used is the policy network, the critic network can be fed with any information that could help improve learning performance, regardless of whether it can be obtained from sensors. In this case, the critic network is given the velocity error between the true velocity and the reference velocity

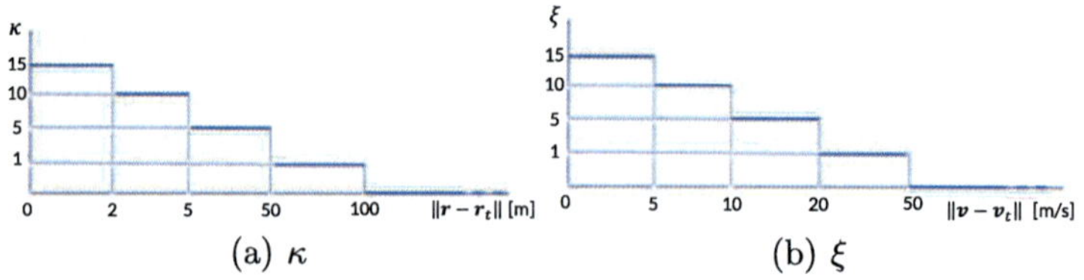

Fig. 2 Parameters κ and ξ vs position and velocity intervals

($\mathbf{v}_{\mathbf{err}} = \|\mathbf{v} - \mathbf{v}_{targ}\|$), the time-to-go ($t_{go}$), the lander state vector ($\mathbf{r}$) and the lander velocity vector ($\mathbf{v}$).

$$o_{VF} = [\mathbf{v}_{\mathbf{err}} \quad t_{go} \quad r_x \quad r_y \quad r_z \quad \dot{r}_x \quad \dot{r}_y \quad \dot{r}_z] \tag{5}$$

Alternatively, the policy only has access to a restricted amount of data. In this project, we utilize images, vertical range, and range rate as measurable data. We denote a 32×32 array of raw grayscale pixel information from the raytracer as $\mathbf{I}$. This array is used as the observation together with the vertical position and velocity.

$$o_{\pi_\theta} = [\mathbf{I} \quad r_z \quad \dot{r}_z] \tag{6}$$

3.2 *Results*

The algorithm was tested on a simulated scenario. The initial position and velocity vectors' components are sampled randomly from uniform distributions. The initial mass for the spacecraft is set at 1500 kg and the final position and velocity are [0, 0, 200] m and [0, 0, -1] m/s, respectively. To improve performance in uncertain conditions, the initial mass and the z component of the gravity vector are randomly sampled in the intervals $[0.9, 1.1]m_0$ and $[0.9, 1.1]g_z$, respectively. Additionally, there is a chance of engine failure which can reduce thrust to 2/3 of the nominal value along the z-axis. The spacecraft's guidance is determined using a RNN with recurrent gated units and a NN that takes in images, position, and velocity as input, using a combination of CNN and fully connected layers, ending with two fully connected layers.

The figures referenced in this passage, Fig. 3a–c, show the results of a simulation. Figure 3a illustrates the relationship between the reward and the number of time steps per trajectory, as well as the decrease in velocity error. This relationship reaches a plateau, but the reward continues to increase due to the decrease in final state error. The final reward value is positive, indicating a successful trajectory. Figure 3b displays the 100 trajectories of the simulation, and Fig. 3c shows the distribution of final position and velocity errors, which indicate that the network can successfully perform pinpoint landing.

4 Adding Hazard Detection and Landing Site Selection

Although the previous results show good performance, the algorithm can only target a specific final state. This can work well in a very controlled and defined environment. The guidance algorithm is robust to uncertainties in the model and failures but it is not flexible to adapt to unknown surface features, and changing the target state would require retraining. For this reason, the work has been expanded to allow for greater

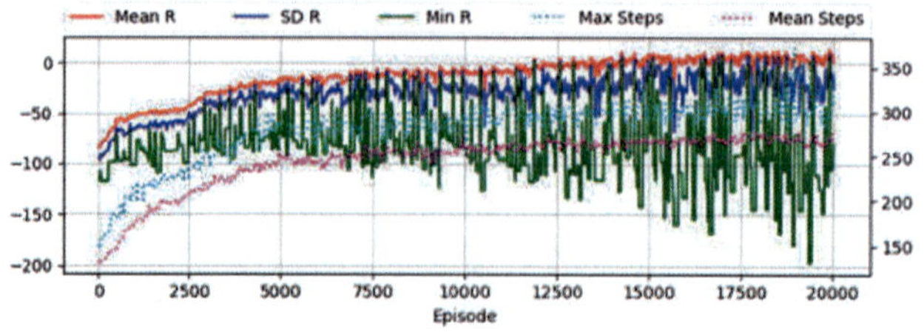

(a) Reward

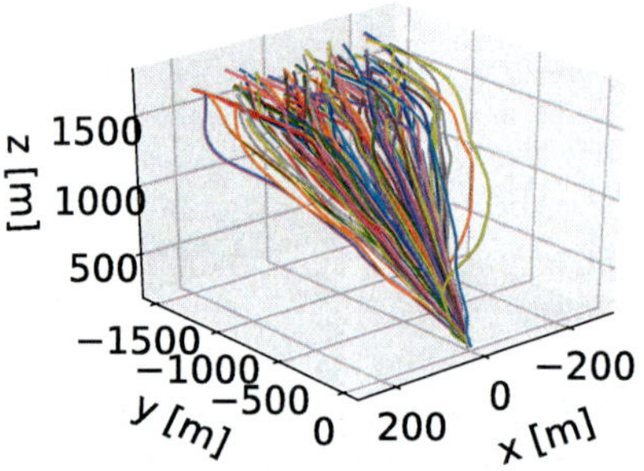

(b) Test trajectories

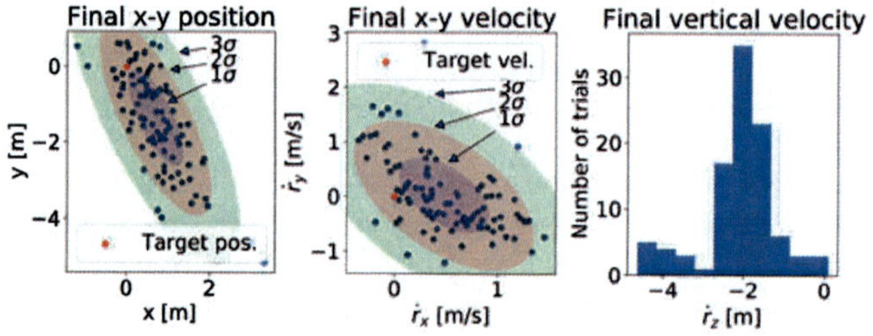

(c) Mean position error. x: 0.70 m(σ=0.66), y: -1.35 m(σ=1.27). Mean velocity error. x: 0.48 m/s(σ=0.43), y: 0.03 m/s(σ=0.67), z: -2.11 m/s(σ=0.94)

Fig. 3 Results

adaptability to varying tasks. Specifically, the system has been augmented with an hazard detection and landing site selection algorithm based on deep-neural networks. The system integrates guidance and navigation, providing an all-in-one solution to the lunar landing problem that integrates image-based navigation to intelligent guidance. We exploit the latest advancements in computer vision and CNN to implement a hazard detection and avoidance algorithm to detect safe landing sites autonomously.

4.1 Hazard Detection and Landing Site Selection Algorithm

To select a safe zone for landing, we developed a hazard detection algorithm. We used a particular kind of neural network that is able to recognize and label different pixels of an image based on a ground truth mask. Specifically, the network is comprised of an encoder and a decoder. The encoder extracts information from the input image by applying a sequence of convolutional layers and reducing the image size. The decoder then upscales that information back to create a labeled image with the same size as the input. The combination of these two elements has a characteristic U shape from which the network takes its name, U-Net. The output of the network is a labeled image in which safe and unsafe pixels are identified with different colors. The algorithm then calculates the minimum distance from the closest hazardous pixel in the image matrix. The safest spot will then be the pixel with the biggest among the minimum computed distances. The position of this pixel in the focal plane is then passed to the guidance algorithm. The details of how this information is used within the control framework will become clearer in the following.

The U-Net is trained in advance via supervised learning. The training dataset is created by sampling the camera position from an equally-spaced 3D grid above the DTM. Both true images and ground truth masks are rendered simultaneously trough the camera with a boresight perpendicular to the ground. The dataset is composed of 18000 grayscale images with a resolution of 64×64 pixels. The chosen parameters are a compromise between resolution and speed. Indeed, the resolution and number of channels influence not only the rendering time, but also the number of learnable parameters in the model, which reflects on the final computational speed. The ground truth is a binary image in which white pixels are considered safe and black pixels are considered unsafe. The safety parameters are computed pixel-wise from the DTM. Specifically, a pixel is considered safe if it respects all these conditions: the local slope with respect to the equipotential surface is smaller than 8 degrees, the roughness is no more than 5%, and the area is not in shadow. The results after training for 50 epochs can be seen in Fig. 4. The algorithm manages to reach above 93% accuracy on the test set. Once the hazard map is computed, we apply a 2D median filter to the output image to reduce the number of small hazardous areas found by the algorithm. Then we adopted an ergodic search algorithm to find, for each pixel, the distance of the nearest hazardous pixel as briefly explained above. However, since the hazard map is computed at each frame, the target pixel may change position considerably. This can cause problems to the tracking algorithm to efficiently close on the target. The solution adopted is a searching window centered in the center of the camera frame with an adaptive size, whose size decreases as the distance from the ground decreases.

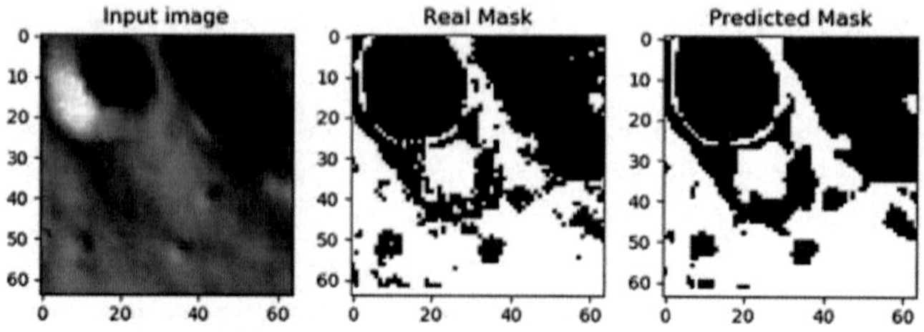

Fig. 4 Hazard detection algorithm performance. Accuracy on 1000 test images—93.93%

4.2 Guidance Architecture

The GNC algorithm integrates the hazard detection and landing site selection algorithm and the RL guidance. It is important to make a distinction between training and testing as some parts of the algorithm are active only in one case of the other.
Training. Training is an iterative process that uses the reinforcement learning theory to optimize the policy. In this case, the observations consist of the landing pixel (LP) coordinates in the camera focal plane and the vertical position (r_z) and velocity ($\dot{r}_z$):

$$obs_{\pi_\theta} = [r_z \quad \dot{r}_z \quad LP_x \quad LP_y] \tag{7}$$

The landing pixel during training is selected randomly at the beginning of each episode within the initial frame. This is then projected on the ground and tracked during the descent. The value function network is fed the same landing pixel coordinates, plus the full state of the lander, so its input is:

$$obs_{VF} = [\mathbf{r} \quad \dot{\mathbf{r}} \quad LP_x \quad LP_y] \tag{8}$$

The action space is continuous and corresponds to $\mathbf{R}^3$ as we are considering the motion to happen in a 3DOF environment.

The value function is implemented as a recurrent network composed of a fully connected layer followed by a single Gated Recurrent Unit (GRU) layer and two fully connected layers. The policy network has a similar structure. The layers specification and activation functions can be seen in Tables 1 and 2.

Table 1 Policy

Layer	Units	Activation function
FC1	60	tanh
GRU	73	–
FC2	30	tanh
FC3	3	linear

Table 2 Critics

Layer	Units	Activation function
FC1	60	tanh
GRU	17	–
FC2	5	tanh
FC3	1	linear

The reward function has been set up similarly to the previous case. One term is related to the error with respect to a gaze heuristic potential function to drive the agent towards the target:

$$r_{track} = \alpha \|\mathbf{v} - \mathbf{v}_{targ}\| + \eta \tag{9}$$

where $\mathbf{v}_{targ}$ has the same form as in the previous case.

This reward is then summed to what will be called target reward r_{target} which ensures that the agent aims to minimize the Euclidean distance between the projection of the landing point on the camera focal plane and the center of that frame:

$$r_{target} = \beta\sqrt{LP_x^2 + LP_y^2} \tag{10}$$

With $\beta = -10$. Maximizing this part of the reward ensures that the agent always tries to align the camera bore-sight with the target on the ground. The overall reward function takes the form:

$$R = r_{track} + r_{target} \tag{11}$$

and is a trade-off between tracking the target velocity to reach the hovering point with zero terminal velocity and having the target in the center of the frame.

The constraint in this case is a simple box out of which the lander should not go. The maximum lateral distance is set to 2000 m to avoid going to close to the borders of the DTM and the lower boundary is set to 200 as we are targeting a hovering altitude above the target rather than a target on the ground.

Testing. We test the policy as it would be deployed on an actual lander. In this case, the random landing point is substituted by the safe landing site selected by the hazard detection and landing site selection subroutine. This is fed with a real image of the ground, and it outputs a feasible landing point in the camera frame. Its coordinates are then fed to the policy as landing pixel $[LP_x, LP_y]$. Figure 5 shows the architecture.

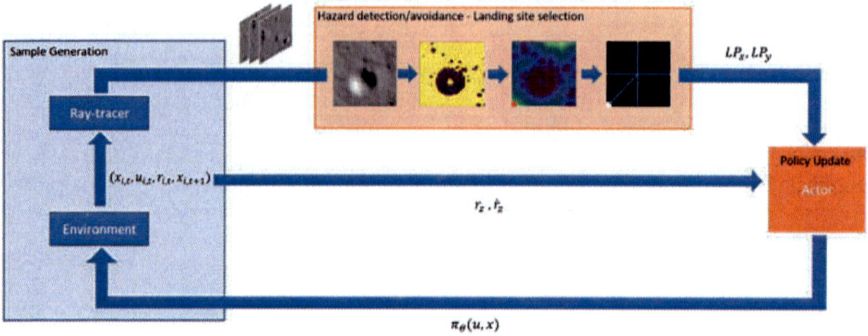

Fig. 5 Deployed guidance architecture

4.3 Results

Training is achieved via repeated interaction with the environment described above, using PPO as an optimizer for the policy. The initial conditions have been sampled using a uniform distribution both in position and velocity spaces. The target point has been selected randomly within the first frame of each trajectory, and is 200 m above the ground level. This altitude offset has been selected to be able to distinguish enough features on the ground in the simulated image. At lower altitudes, the definition of the DTM is insufficient to have enough variation to be able to recognize different features.

Figure 6 shows the training curves after training for approximately 45000 episodes when the training was interrupted because no significant change in the reward was being recorded. Training has been carried out on a machine equipped with an Intel® Core™i7-9700 CPU @ 3.0 GH and an Nvidia GeForce RTX 2060 graphics card and took approximately 14.66 hours. Once trained, the policy has been tested, introducing the landing site selection algorithm in the loop.

In order to reduce chattering of the control system, the landing site is searched within a shrinking area of the image as the lander approaches the ground. The size of

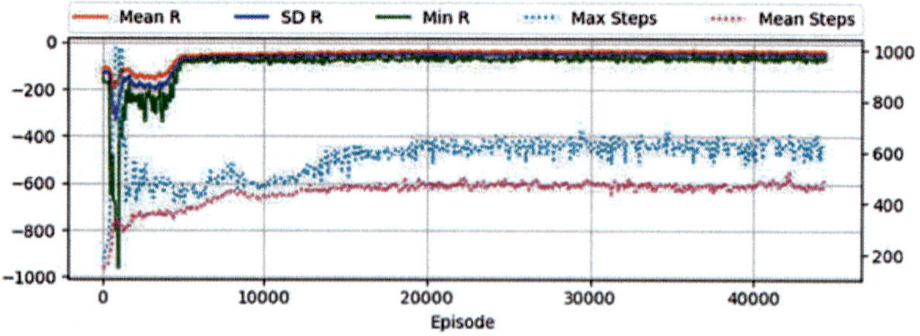

Fig. 6 Training curves

(a) Monte Carlo trajectories

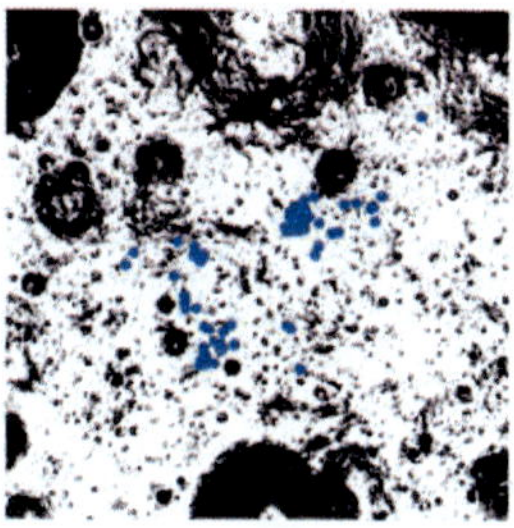

(b) Final position on hazard map

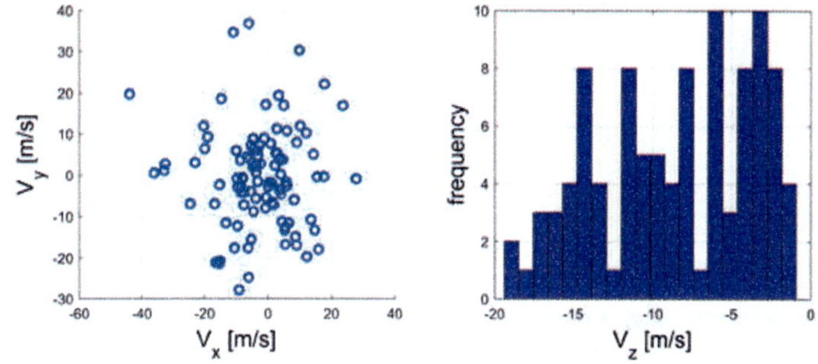

(c) Mean velocity error. x: -2.42 m/s(σ=12.16), y: -0.25 m/s(σ=12.03), z: -8.61 m/s(σ=4.95)

Fig. 7 Monte Carlo simulation

the crop is as big as the whole rendered image (i.e., 64 × 64 pixels) at the beginning of each trajectory and shrinks down linearly towards the center of the frame as the lander approaches the limit altitude of 200 m. The final crop size is 40 × 40. This encourages the agent to look for the landing site in a restricted area towards the end of a trajectory, avoiding making high thrust maneuvers closer to the ground but still retaining enough control authority to steer away from potential hazards. As shown in Fig. 7b, the algorithm is very robust in terms of hazard avoidance, managing to end every trajectory above a safe area. Nevertheless, the erratic movement of the safe spot in the image, with consequential high thrust maneuvers at the end of a trajectory, lead to overall errors in final velocity in all directions. This shows that there is still room for improvements in terms of guidance accuracy, which is what the authors are currently working on.

5 Conclusions and Outlook

In this paper a GNC architecture based on MRL is presented. It is based on two successive work by the authors as part of a broader project on image-based autonomous landing guidance using MRL. The first work solves the problem of creating an image-based landing guidance using CNN-RNN trained using PPO in a MRL framework, capable of landing in a predetermined area using images and altimetry data in uncertain environment and actuator failures. The second work builds upon the fist one by adding a hazard detection and landing site selection algorithm based on CNN trained to perform semantic segmentation. The resulting algorithm is capable of autonomously select a landing site at each time-step and apply the required guidance commands to reach the desired landing site. The algorithm allows for increased autonomy but degrades the soft landing capabilities of the method. Ongoing work by the authors is targeted towards improving the guidance performance of the method while maintaining the same level of autonomy.

References

1. Acikmese, B., Ploen, S.R.: Convex programming approach to powered descent guidance for mars landing. J. Guidance Control Dyn. **30**(5), 1353–1366 (2007)
2. Finn, C., Abbeel, P., Levine, S.: Model-agnostic meta-learning for fast adaptation of deep networks. In: Proceedings of the 34th International Conference on Machine Learning, vol. 70, pp. 1126–1135. JMLR.org (2017)
3. Furfaro, R., Bloise, I., Orlandelli, M., Di Lizia, P., Topputo, F., Linares, R., et al.: Deep learning for autonomous lunar landing. In: 2018 AAS/AIAA Astrodynamics Specialist Conference, pp. 1–22 (2018)
4. Furfaro, R., Scorsoglio, A., Linares, R., Massari, M.: Adaptive generalized ZEM-ZEV feedback guidance for planetary landing via a deep reinforcement learning approach. Acta Astronaut. (2020)
5. Gaudet, B., Furfaro, R., Linares, R., Scorsoglio, A.: Reinforcement metalearning for interception of maneuvering exoatmospheric targets with parasitic attitude loop. J. Spacecr. Rockets **58**(2), 386–399 (2021)
6. Gaudet, B., Linares, R., Furfaro, R.: Adaptive guidance and integrated navigation with reinforcement meta-learning. Acta Astronaut. **169**, 180–190 (2020)
7. Gaudet, B., Linares, R., Furfaro, R.: Deep reinforcement learning for six degree-of-freedom planetary landing. Adv. Space Res. **65**(7), 1723–1741 (2020)
8. Gaudet, B., Linares, R., Furfaro, R.: Six degree-of-freedom body-fixed hovering over unmapped asteroids via lidar altimetry and reinforcement meta-learning. Acta Astronaut. (2020)
9. Gaudet, B., Linares, R., Furfaro, R.: Terminal adaptive guidance via reinforcement meta-learning: applications to autonomous asteroid close-proximity operations. Acta Astronaut. (2020)
10. Holt, H., Armellin, R., Scorsoglio, A., Furfaro, R.: Low-thrust trajectory design using closed-loop feedback-driven control laws and state-dependent parameters. In: AIAA Scitech 2020 Forum, p. 1694 (2020)
11. Hovell, K., Ulrich, S.: On deep reinforcement learning for spacecraft guidance. In: AIAA Scitech 2020 Forum, p. 1600 (2020)

12. Izzo, D., Sprague, C.I., Tailor, D.V.: Machine learning and evolutionary techniques in interplanetary trajectory design. In: Modeling and Optimization in Space Engineering, pp. 191–210. Springer (2019)
13. Lillicrap, T.P., Hunt, J.J., Pritzel, A., Heess, N., Erez, T., Tassa, Y., Silver, D., Wierstra, D.: Continuous control with deep reinforcement learning. arXiv preprint arXiv:1509.02971 (2015)
14. Lu, P.: Propellant-optimal powered descent guidance. J. Guidance Control Dyn. **41**(4), 813–826 (2018)
15. Mortari, D.: The theory of connections: connecting points. Mathematics **5**(4), 57 (2017)
16. Oestreich, C.E., Linares, R., Gondhalekar, R.: Autonomous six-degree-of-freedom spacecraft docking with rotating targets via reinforcement learning. J. Aerosp. Inf. Syst. 1–12 (2021)
17. Schulman, J., Wolski, F., Dhariwal, P., Radford, A., Klimov, O.: Proximal policy optimization algorithms. arXiv preprint arXiv:1707.06347 (2017)
18. Scorsoglio, A., D'Ambrosio, A., Ghilardi, L., Furfaro, R., Gaudet, B., Linares, R., Curti, F.: Safe lunar landing via images: a reinforcement meta-learning application to autonomous hazard avoidance and landing. In: Proceedings of the 2020 AAS/AIAA Astrodynamics Specialist Conference, Virtual, pp. 9–12 (2020)
19. Scorsoglio, A., D'Ambrosio, A., Ghilardi, L., Gaudet, B., Curti, F., Furfaro, R.: Image-based deep reinforcement meta-learning for autonomous lunar landing. J. Spacecr. Rockets **59**(1), 153–165 (2022)
20. Scorsoglio, A., Furfaro, R.: ELM-based actor-critic approach to Lyapunov vector fields relative motion guidance in near-rectilinear orbit. In: 2019 AAS/AIAA Astrodynamics Specialists Conference, pp. 1–20 (2019)
21. Scorsoglio, A., Furfaro, R.: Visualenv: visual gym environments with blender. arXiv preprint arXiv:2111.08096 (2021)
22. Scorsoglio, A., Furfaro, R., Linares, R., Massari, M., et al.: Actor-critic reinforcement learning approach to relative motion guidance in near-rectilinear orbit. In: 29th AAS/AIAA Space Flight Mechanics Meeting, pp. 1–20. American Astronautical Society, San Diego, CA (2019)
23. Silvestrini, S., Lavagna, M.R.: Spacecraft formation relative trajectories identification for collision-free maneuvers using neural-reconstructed dynamics. In: AIAA Scitech 2020 Forum, p. 1918 (2020)
24. Sutton, R.S., Barto, A.G.: Reinforcement Learning: An Introduction. MIT Press (2018)
25. Wang, J.X., Kurth-Nelson, Z., Tirumala, D., Soyer, H., Leibo, J.Z., Munos, R., Blundell, C., Kumaran, D., Botvinick, M.: Learning to reinforcement learn. arXiv preprint arXiv:1611.05763 (2016)
26. You, S., Wan, C., Dai, R., Rea, J.R.: Learning-based onboard guidance for fuel-optimal powered descent. J. Guidance Control Dyn. **44**(3), 601–613 (2021)
27. Zavoli, A., Federici, L.: Reinforcement learning for low-thrust trajectory design of interplanetary missions. arXiv preprint arXiv:2008.08501 (2020)

Satellite IoT for TT&C and Satellite Identification

G. D'Angelo, J. P. Mediano-Alameda, and R. Andreotti

Abstract The new small satellite constellations require real-time and low cost satellite operations for telemetry, tracking and control (TT&C) of all the space assets. Nowadays, these small satellites are launched without truly real-time tracking and ranging capabilities, creating problems of incorrect satellite identification after launch, and delaying the operation and commissioning of the satellites for a significant portion of their lifetimes. Airbus Italia, together with MBI and Inmarsat, has developed a system based on Internet of Things/Machine-to-Machine (IoT/M2M) custom protocol capable to address these well-known issues. Following the activities carried out within European Space Agency (ESA) and with an important satellite operator, Airbus Italia is ready to undertake the industrialization of the system.

Keywords IoT · Mini/Micro satellite · Cubesat · TT&C

1 Introduction

This article describes how a multi-year collaboration between Airbus Italia, MBI and Inmarsat has made possible to modify and optimize a protocol initially targeted to satellite IoT/M2M communications, to remote control and telemetry of satellites, focusing on managing the aspects related to the identification and tracking.

The current operations for low Earth orbit (LEO) satellites involve a terrestrial network of several Earth stations, for tracking, telemetry and control (TT&C), representing a sort of umbilical cord that connects a spacecraft with the operator. However,

G. D'Angelo (✉)
Airbus Italia S.p.A., Via dei Luxardo 22-24, 00156 Roma, Italy
e-mail: giuseppe.dangelo@airbus.com

J. P. Mediano-Alameda
Inmarsat, 99 City Road, London, UK

R. Andreotti
MBI S.r.l., Via F. Squartini 7, 56121 Pisa, Italy

C. Ieracitano et al. (eds.), *The Use of Artificial Intelligence for Space Applications*, Studies in Computational Intelligence 1088,
https://doi.org/10.1007/978-3-031-25755-1_5

such terrestrial network only allows telemetry and telecommands at well-defined times in advance, and only when the satellite is in visibility of the ground station. In order to reduce the latency between satellite and ground interactions, existing Inter Satellite Link (ISL) products have been evaluated, but those as far beyond the target allowed SWaP-C (size, weight, power and cost) of small satellites.

The proposed solution aims to develop a low Size, Weight, Power and Cost (SWaP-C) and fully asynchronous TT&C terminal for each satellite of the constellation. The telemetry is transmitted and the telecommands are received through a geostationary (GEO) satellite acting as transparent data relay linked to the GS. The telemetries are sent asynchronously without any coordination, while the telecommand is implemented with a broadcast multiplexed waveform.

The return link (RL) is based on the adoption of the Enhanced Spread Spectrum Aloha (E-SSA) protocol [1]. This protocol is the evolution of the old Aloha, with two significant features added: the adoption of waveforms based on direct sequence spread spectrum (DS-SS), and the successive interference cancellation (SIC) technique at the receiver. This new adaptation of ESSA allows a distributed and huge population of terminals to transmit simultaneously and asynchronously on the same shared frequency channel. The removal of the coordination among terminals keeps at minimum the need of signaling on the forward link (FL), greatly simplifying the overall system management and cost, not only for the network but also for the terminals. On the one hand, the spreading provided by the DS-SS technique allows to handle the multiple access interference (MAI). On the other hand, it also provides enhanced robustness to the waveform, since it can be demodulated at negative carrier-to-noise power (C/N) values exploiting the processing gain of the spreading factor (SF). In other words, it means that smaller high power amplifier (HPA) could be adopted, since the lower available power can be compensated by a higher spreading factor, although reducing the data rate. On top of this, constant envelope (CE) spreading can be employed, which allows the use of lower class HPAs and hence a further cost reduction at the terminal.

In summary, the telemetry, through the GEO satellite, will be received by the control center which is able to simultaneously manage a very high population of terminals. The terminal will receive the remote controls through a channel shared between all the other terminals. This channel can be used both to provide FL services (e.g. broadcast of application data), and also to manage and/or optimize the terminals' transmission via RL waveform. Such link is an Airbus proprietary solution obtained combining specific techniques derived from different standards like DVB-S2 [2], DVB-SH [3] and 3GPP LTE [4] in order to provide several features. The link is indeed flexible enough to be operated both in GEO and non-GEO scenarios, providing different services (e.g., broadcast, multicast and unicast) in both fixed and mobile environment and supporting different classes of terminals.

Over the past years, several satellite test campaigns have been performed, both in static and dynamic conditions, and with GEO and CubeSat satellites both in transparent mode. The article shows the results obtained, highlights the advantages of this technology and provides a roadmap for the industrialization process.

The paper starts with a brief explanation of the reference scenario and the advantages of the protocol, then moving on to the analysis of the performances obtained from simulations and from field trials. Finally, a roadmap divided into two phases will be drawn up. The first will allow the realization of a prototype based on commercial off-the-shelf (COTS) elements for a validation of the protocol on a real system in a short time, and the second will aim at the realization of a low-cost terminal compatible with CubeSat-type platforms.

2 Methodology

This section describes the main aspects related to the development of the prototypes able to demonstrate the new TT&C concept.

2.1 Reference Scenario

Figure 1 shows the considered reference scenario. The main components of the system are the LEO constellation, the GEO satellite and the Gateway.

The following assumptions are considered:

- Bent-pipe GEO satellite;
- GEO satellite with global coverage footprint;
- LEO satellite with an altitude of 500 km;
- LEO Sun-Synchronous Orbit (SSO);
- LEO/GEO Link in L-Band;

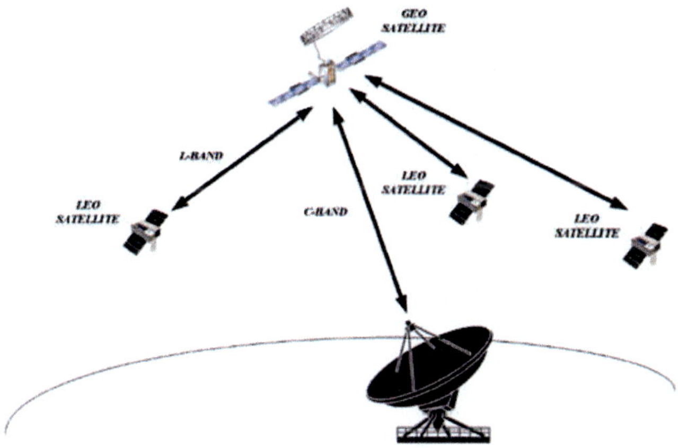

Fig. 1 Reference scenario

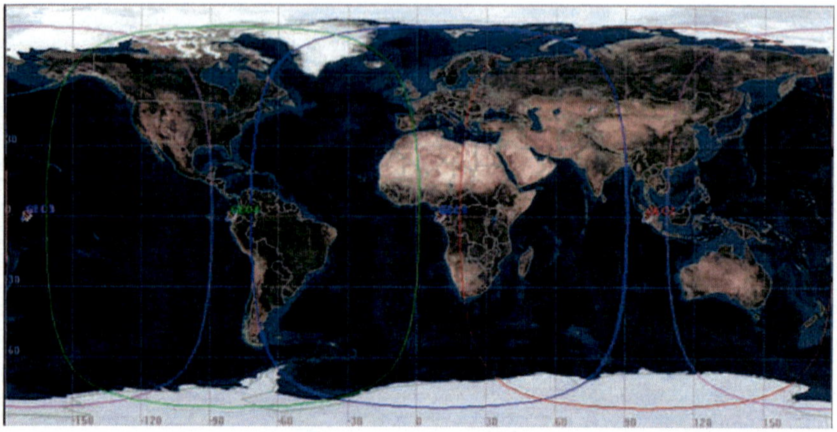

Fig. 2 Example of global coverage

- GEO/Gateway Link in C-Band;
- Omni directional LEO Antenna;
- Forward Link channel width: 100 kHz;
- Return Link channel width: 50 kHz.

Global coverage is achieved with at least three GEO satellites and their respective gateways (Fig. 2).

Concerning the LEO satellite, the main impairments that affect the communication link are:

- Doppler shift: the reception of higher or lower signal frequencies than the original transmitted frequency.
- Doppler rate: the variation over time of Doppler shift which affects the transmission frequencies and the clock of the packets;
- Visibility/transmission window or in-view period: the amount of time during which the LEO satellite is visible by GEO Satellite.
- Variation of the link geometry: the path-loss changes over time during in-view period due to the variation of the relative distance between LEO and GEO satellites.

With the above assumptions, the LEO satellite trajectory within the GEO satellite field of view is illustrated in Fig. 3. The associated Doppler shift and rate curves due to the LEO satellite movement w.r.t. to the GEO satellite, at the carrier frequency of 1.6 GHz, are shown in Fig. 4.

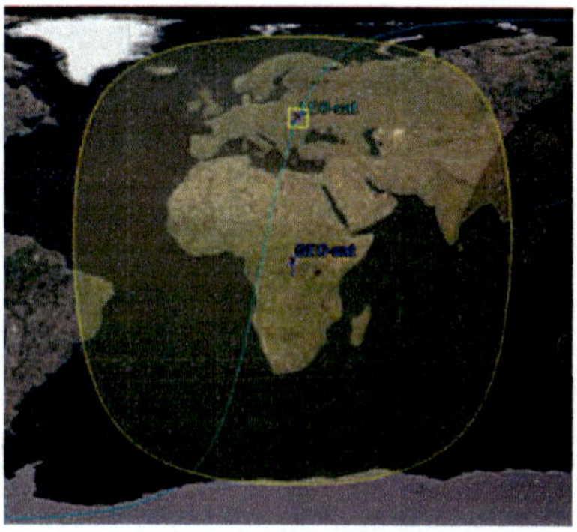

Fig. 3 Example of LEO satellite trajectory within the GEO satellite field of view

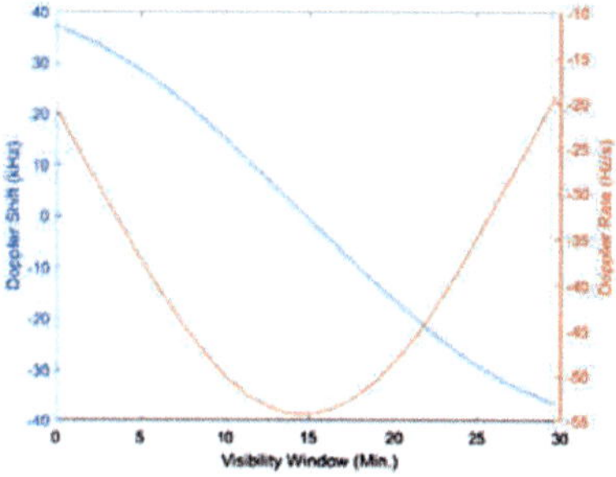

Fig. 4 LEO satellite Doppler shift and rate curves w.r.t. the GEO satellite

2.2 *Used Air Interface Protocols*

This paragraph provides the main specifications used into the definition of the air interface that is able to support the requested scenario.

The air interface design drivers are focused on the following features:

- Efficient support of bursty traffic.
- Support of low terminal data rates.
- Minimization of terminal transmit power.
- Minimization of terminal costs.
- Scalability: from a few terminals to several hundreds of terminals sharing the same bandwidth without degrading the overall system performance.
- Minimization of the satellite spectrum requirements.
- Capability to operate in fully asynchronously mode.
- Minimization of the overhead.
- Support data security and integrity.
- Robustness: communication guaranteed in both directions regardless the channel quality.

Airbus Italia with M.B.I. studied and developed this air interface in the frame of several European Space Agency (ESA) projects [5, 6].

Return Link Air interface.

This section provides the specifications of the RL air interface envisaged to be adequate for uncoordinated access by a massive number of low duty cycle terminals operating in L-band with carrier bandwidths from 50 to 200 kHz.

The RL air interface design drivers are focused on the following features:

- Power saving;
- Low bit rate;
- Low cost terminals;
- Robustness to signal imperfection.

The ideal air interface for such scenario is based on E-SSA [1, 7–9] and, in particular, on the modified E-SSA access scheme proposed in [10] and is dubbed modified E-SSA (ME-SSA).

For the use in the RL, a variant of ME-SSA, named Improved E-SSA, will be defined using the same approach based on a spread spectrum (SS) random access technology. Actually, the baseline of this air interface will be the one already derived in the frame of the MASSIVE project [5, 9], with just few specific modifications required by the target scenario.

In the proposed new air interface, the information word, represented by the bits coming from the upper layer, goes through a 16-bit cyclic redundancy check (CRC), the results of which is appended at the end of the word. The resulting bits are coded by a Turbo code with coding rate 1/3. These coded bits are then mapped into the modulation symbols (either BPSK or QPSK). Pilot symbols are then time division multiplexed with the data symbols, by inserting one pilot every N data symbols. Then, the burst is constructed by concatenating a preamble of N_{pre} symbols to the stream of data and pilot symbols. Preamble and pilot symbols are known at the receiver and are used to ease the acquisition and tracking stages. A field named Transport Frame Information (TFI) may be inserted between the preamble and the data/pilot field, carrying information on the coding, modulation and info word length adopted by the burst, in order for the receiver to correctly tune some demodulation parameters. Finally, the burst symbols are spread by a spreading sequence denoted by a certain SF and the resulting signal is digital-to-analog converted and transmitted at the target carrier frequency.

Figure 5 shows the whole physical layer Tx functional chain also evidencing the presence of FEC, CRC and channel interleaver.

A configuration interface will be used for configuring the burst characteristics, i.e.:

- RF Carrier Frequency
- Chip Rate
- Transmit Power
- Spreading Factor

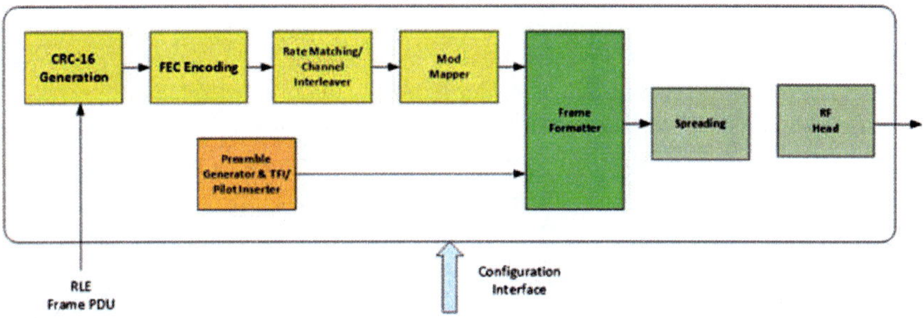

Fig. 5 RL waveform generation

- Data Modulation (BPSK or QPSK)
- Preamble Length
- TFI enable value
- TFI value (if TFI enabled)
- Pilot periodicity
- CRC-16 enable value
- Frame PDU length
- FEC Codeword Length
- FEC information Length.

The organization of the physical layer burst format in the RL is shown in Fig. 6. The Data Part (including CRC) of the burst is assumed divided in slots having a fixed number of channel symbols. Each slot terminates with a pilot symbol. The slot size is 9 symbols (8 data symbols plus one pilot). This slot size was used in most of the simulations performed for validating physical layer performances.

As explained above, the RL is based on the E-SSA protocol that enables several terminals to asynchronously transmit at the same time on the same frequency. The number of simultaneous transmissions N_T that can be successfully demodulated

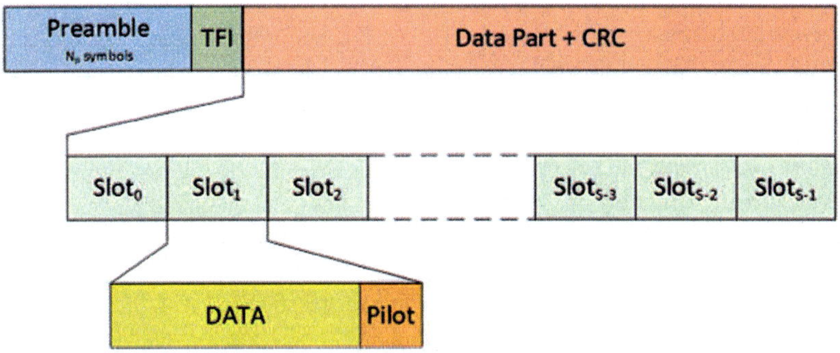

Fig. 6 Return link physical layer burst format

depends on the protocol spectral efficiency (SE)η. Compared to classical Spread-Spectrum Aloha (SSA), E-SSA drastically improves the SE by adopting the SIC methodology at the receiver, hence without increasing complexity and thus costs at the terminal side. The SIC approach consists on locally regenerating each correctly demodulated burst and subtracting it from the current observation window. This cancels the interference it causes to the other bursts, increasing their signal-to-noise-plus-interference ratio and hence their chance to be successfully demodulated.

The SE is usually expressed in bit/chip and is related to N_T by the following formula:

$$\eta = N_T \cdot R_b / R_c \tag{1}$$

where R_b is the per-terminal bit rate and R_c the chip rate. The actual SE depends on the receiver architecture, the algorithms used to demodulate the signal, and on the number of SIC carried out. The higher the number of SIC the higher the SE, but a trade-off with demodulation latency shall be taken into account. Typically, 4 to 6 SIC iterations are enough to reach about the maximum SE. Concerning the latter, in LEO scenario with waveforms of 50 kHz channelization, the SE obtained by a real receiver are about 0.3 bit/chip and 0.5 bit/chip considering SF $=$ 16 and SF $=$ 64, respectively. This means values of N_T in the range between 14 and 96 for BPSK modulation.

Forward Link Air interface.
This section provides the specifications of the FL air interface envisaged to work in the proposed scenario in L-band with channelization from 50 kHz to 1 MHz.

The aim of this FL waveform is to support the management of the terminals.

The goal of the design phase is to envisage a new air interface able to fit in inexpensive terminals and able to efficiently provide bi-directional connectivity for the proposed applications. The main features outcomes of the forward link trade-offs are summarized hereafter:

- Provide a variety of services of different nature: broadcast, multicast and unicast services for both fixed and mobile applications. This is achieved thanks to the powerful channel coding based on Turbo LTE and mobile fading counteraction provided by the adoption of a convolutional interleaver.
- Implement Multi-pipe architecture. Each physical-layer pipe (PLP) can have different modulation (either BPSK or QPSK), coding rate and convolutional interleaver length and a terminal can demodulate only the interested pipe.
- Super-frame structure: such physical layer pips are organized in fixed length frames with fixed header and pilots positions to increase flexibility and ease acquisition and tracking at the receiver.
- Support operation of different classes of terminals, supplied with different quality oscillators and RF components and guaranteeing operation for very low G/T values (in the order of -30 dB/K), even for battery-operated terminals, and of different activity modes to help reduction of terminals power consumption.

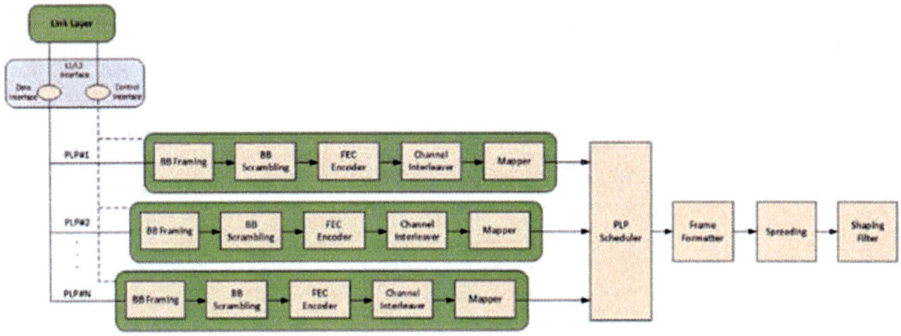

Fig. 7 Functional block scheme of the forward link transmitter

- Support different satellite scenarios: both GEO and non-GEO scenarios are applicable; in case of GEO scenarios, spot beam, regional beam or even global beam coverage is possible. In case of non-GEO constellations, full frequency reuse among the satellites is also an option. This thanks to the use of a limited set of SF values (1, 2 or 4).
- Allow different channelization options.
- Bi-directional link layer procedures for a more efficient network management.

The FL air interface is shown in Fig. 7.

The transmitter has multiple input streams, i.e., the PLPs, which are instantiated by the Link Layer. As discussed above, each PLP has its own physical layer parameters associated and redefines channel coding, channel interleaving and modulation according to such parameters, which are shown in the picture above as N independent modulation blocks from the BB framing to the modulation mapper. Then, the PLP scheduler multiplexes the various pipes and an optional spreading is applied to the multiplexed pipes. Finally, a square-root raised cosine (SRRC) filter is applied to shape the signal spectrum.

Link Budgets.
Table 1 shows the FL link budget (LB), composed of the link between the gateway (GW) and the GEO satellite (uplink), and between the GEO and LEO satellite (downlink).

The GW and GEO satellite are denoted by standard EIRP and G/T values for the case of a global beam coverage and C-band feeder link. Regarding the LEO satellite, a G/T = −25,9 dB/K is considered, obtained considering a minimum antenna gain of 1 dB and an overall system temperature of about 410 K. The LB is evaluated for two cases: the former consists in the LEO satellite seeing the GEO with a low elevation angle (0°), corresponding to the worst-case slant range; the latter to the case in which the elevation angle is maximum, and thus the slant range minimum.

Results show that, employing a FL waveform over a 100 kHz channel, the link margin is positive even in the worst-case scenario.

Table 1 FL link budget

Forward link			
Waveform features			
Channel BW	kHz	100	100
Chip rate	kchip/s	80	80
SF		1	1
Code rate		0,25	0,25
Modulation		BPSK	BPSK
Bit rate	kbit/s	20,00	20,00
Uplink (GW to GEO)			
Up-link Freq	MHz	6375	6375
Gateway EIRP density	dBW/Hz	3	3
Gateway EIRP	dBW	53,00	53,00
Pointing loss	dB	0,9	0,9
Satellite range	km	41800	41800
Free space loss	dB	201,22	201,22
Atmospheric loss	dB	0,5	0,5
Satellite G/T	dB/K	−10,4	−10,4
C/N0	dBHz	69,48	69,48
Es/N0	dB	50,4	50,4
Downlink (GEO to LEO)		**Min. El**	**Max. El**
Carrier power at GEO satellite	dBW	−133,5	−133,5
TXp Gain	dB	168	36300
Satellite TXp gain uncertanty	dB	0,5	0,5
Downlink useful EIRP	dBW	34,00	34,00
NPR	dB	20,00	20,00
Slant range	km	41800	36300
Down-link frequency	MHz	1559	1559
Free spase loss (@ 5° elev)	dB	188,7	187,5
Atmospheric loss	dB	0,2	0,2
LEO G/T	dB/K	−25,9	−25,9
C/N0	dBHz	47,8	49,0
Es/(N0 + IM0)	dB	−1,2	0,0
Total result (uplink + downlink)			
Tot Es/N0	dB	−1,2	0,0
Tot Eb/N0	dB	4,8	6,0
Eb/N0 threshold	dB	1	1
Link margin	**dB**	**3,8**	**5,0**

Table 2 shows the RL, composed of the link between the LEO and GEO satellites (uplink), and between the GEO satellite and the GW (downlink). The GEO satellite and GW are denoted by standard EIRP/transponder gain and G/T values, for the case of a global beam coverage with feeder link in C-band. Concerning the LEO satellite, a useful EIRP = 4 dBW is required to close the LB with an E-SSA based RL waveform with spreading factor 16 over a bandwidth of 50 kHz. With such setting, the link margin is positive even in the worst-case scenario.

3 Results

Airbus Italia have defined a roadmap that will allow, through successive steps, to arrive at a product capable of satisfying market demands.

The roadmap is outlined in Fig. 8 where the steps already completed are shown in green, the steps in progress in yellow, and the activities that have not yet started in white.

As mentioned, a validation procedure is underway, also divided into several steps, described in the block diagram of Fig. 9.

This paragraph shows some results obtained during validation phase.

For the laboratory validation, the following configuration, depicted in Fig. 10, was used. It highlights the presence of a traffic generator, capable of emulating the traffic generated by a population of terminals uniformly distributed in space, thus inserting different impairments for each terminal, for example different power and transmission time, different Doppler shift and rate. Table 3 and Table 4 show the configurations used and the results in the performance validation of the return link, respectively.

The table also shows the Max Packet delay, which is the time at which the last received packet is demodulated (in general the one with the lowest power) after the SIC operations. These SIC operations allow the gateway to "see" the packet that would normally be hidden by packets from other terminals transmitting at higher power at the same time.

Table 5 shows the results and conditions for the FL.

Figures 11 and 12 show the configuration used in the validations in mobile conditions using a geostationary type satellite. During the execution of the tests, over 1000 km were traversed in all environments from rural to urban.

Figure 13 shows, on the map, the position of the received/lost packets measured during the experimentation.

It should be noted that the packets were lost in correspondence with obstacles that prevented the satellite from being visible by the terminal.

Validation of the system using a LEO Satellite is in progress. The current configuration required two antennas with tracking system.

Figure 14 shows the spectrogram at the gateway input. In the figure is clear the effect of the Doppler shift on the signal.

Table 2 RL link budget

Return link			
Waveform features			
Channel BW	kHz	50	50
Chip rate	kchip/s	40	40
SF		16	16
Code rate		0,33	0,33
Modulation		BPSK	BPSK
Bit rate	bit/s	833,33	833,33
Target spectral efficiency	bit/chip	0,30	0,30
Max simultaneous #Carriers (@ Target Spectral Eff)		14,00	14,00
Uplink (LEO to GEO)		**Min. El**	**Max Elev**
Up-link Freq	MHz	1675	1675
LEO satellite useful EIRP	dBW/Hz	4	4
Satellite range	km	41800	36300
Free space loss	dB	189,35	188,12
Atmospheric loss	dB	0,20	0,20
Satellite G/T	dB/K	−9	−3,5
C/N0	dBHz	34,05	40,78
Es/N0	dB	0,1	6,8
Downlink (GEO to GW)			
Aggregated PFD @Sat	dBW/m2	−148,2	−146,9
TXp gain	dB	156	156
Satellite TXp gain uncertanty	dB	0,5	0,5
Aggreg. Down-link EIRP	dBW	−18,10	−16,90
NPR	dB	20,00	20,00
SAS range	km	36535	36535
Down-link frequency	MHz	3675,5	3675,5
Free spase loss (@ 5° elev)	dB	195,0	195,0
Atmospheric loss	dB	0,5	0,5
SAS G/T	dB/K	32,3	32,3
C/N0 (aggregated)	dBHz	47,3	48,5
C/N0 (per carrier)	dBHz	35,9	37,1
Es/N0 (per carrier)	dB	1,9	3,1
Es/IMO (per carrier)	dB	20,6	20,6
Es/(N0 + IM0) (per carrier)	dB	1,8	3,0
Total result (uplink + downlink)			
Tot Es/N0	dB	−2,2	1,5

(continued)

Table 2 (continued)

Return link			
Tot Eb/N0	dB	2,6	6,3
Eb/N0 threshold	dB	1,8	1,8
Link margin	**dB**	**0,8**	**4,5**

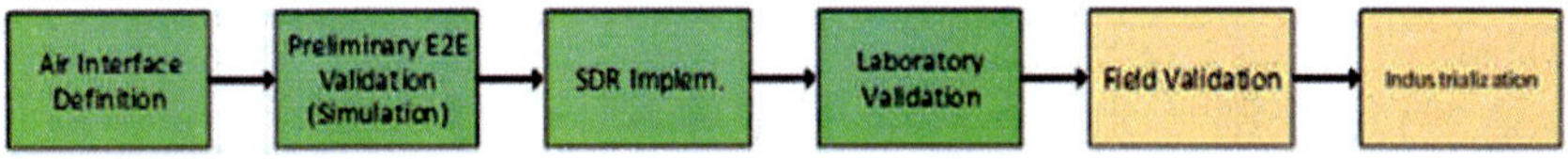

Fig. 8 Roadmap

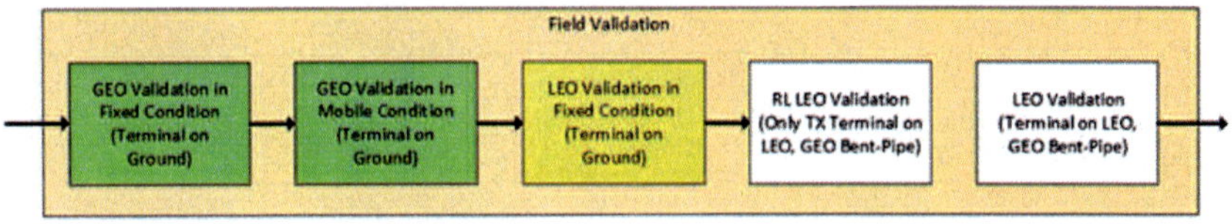

Fig. 9 Field validation activities

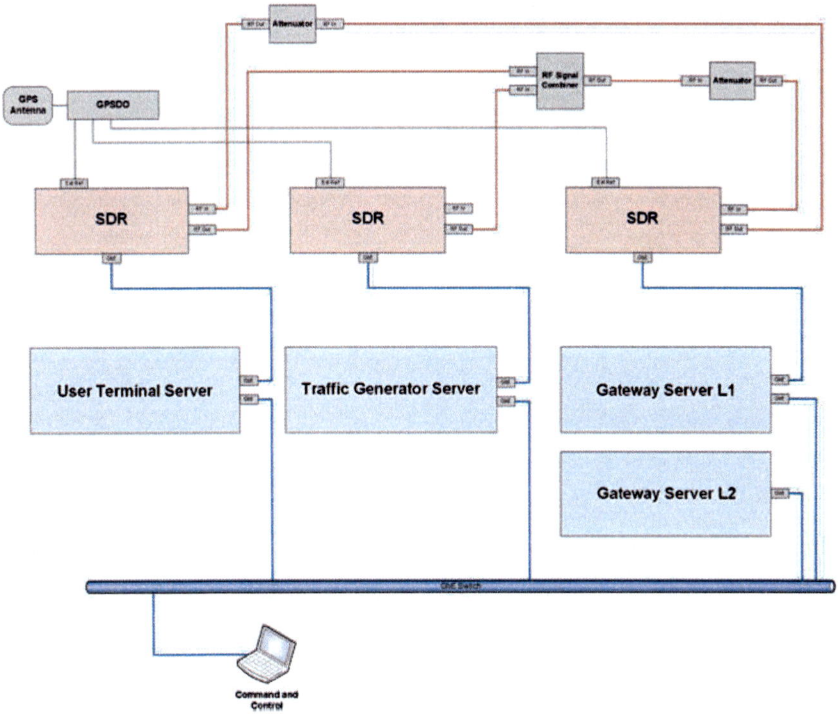

Fig. 10 Laboratory validation setup

Table 3 RL configuration

Traffic generator								
Waveforms	SF	MODCOD	Simulation length [s]	Burst power [dB]	MAC load [pkt/s]	Doppler shift [kHz]	Doppler rate [Hz/s]	Phase noise
TFI #0	16	QPSK 1/3	600	[−42, −35]	26	[−36, 36]	[−430, 0]	ON
TFI #1	16	QPSK 1/3	720	[−42, −35]	15	[−36, 36]	[−430, 0]	ON
TFI #2	64	QPSK 1/3	300	[−42, −35]	35	−36, 36]	[−430, 0]	ON
TFI #3	64	QPSK 1/3	600	[−42, −35]	19	[−36, 36]	[−430, 0]	ON

Table 4 RL results

Waveforms	SF	C/N [dB] (theoretical)	MODCOD	Measured max packet delay [ms]	Measured PLR result
TFI #0	16	−11.5	QPSK 1/3	2600	0.00000
TFI #1	16	−11.5	QPSK 1/3	3200	0.00000
TFI #2	64	−17.5	QPSK 1/3	6500	0.00000
TFI #3	64	−17.5	QPSK 1/3	8200	0.00017

Table 5 FL configuration and result

MODCOD	SF	Doppler shift [Hz]	Doppler rate [Hz/s]	Measured C/N [dB] @10^{-3} PER	Measured PER result
BPSK 1/4	1	±36,000	[−430, 0]	−4.6	0.000000
BPSK 1/3	1	±36,000	[−430, 0]	−3.6	0.000000
BPSK 1/2	1	±36,000	[−430, 0]	−1.4	0.000000
BPSK 1/4	2	±36,000	[−430, 0]	−7.6	0.000000
BPSK 1/3	4	±36,000	[−430, 0]	−9.6	7.77988e-5

4 Conclusion

The completion and excellent results of the early stages of validation with the GEO satellite made possible to start the industrialization process (see Fig. 15) of a reduced Gateway version (RX only). This will allow, in about one year, the installation of the first two equipment, out of 12, to manage a commercial service by a satellite operator: The 12 platforms will be installed within 18 months.

Initially, Airbus Italia focused its efforts on the development of the gateway using an advanced prototype of terminal as a reference for future developments.

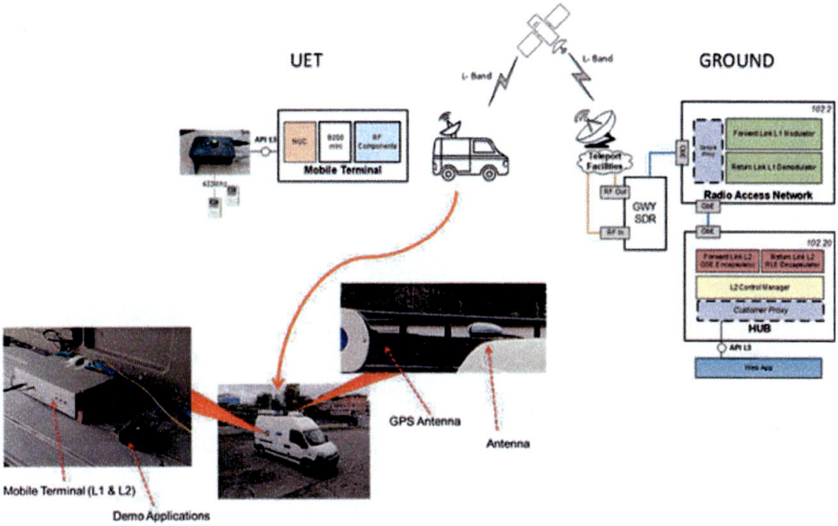

Fig. 11 Setup for validation in mobile condition

Fig. 12 Road route example

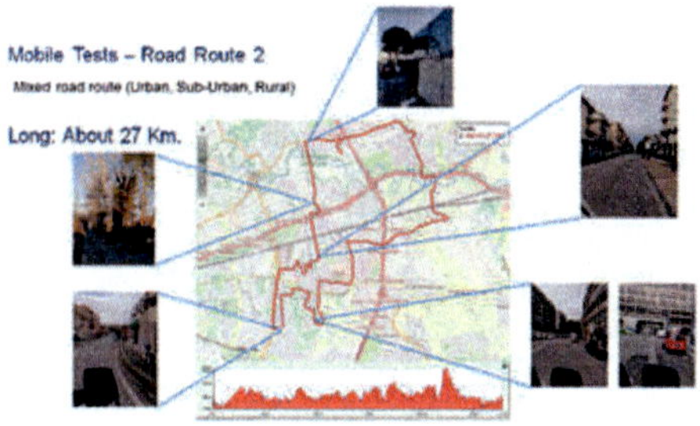

Fig. 13 Result in mobile condition

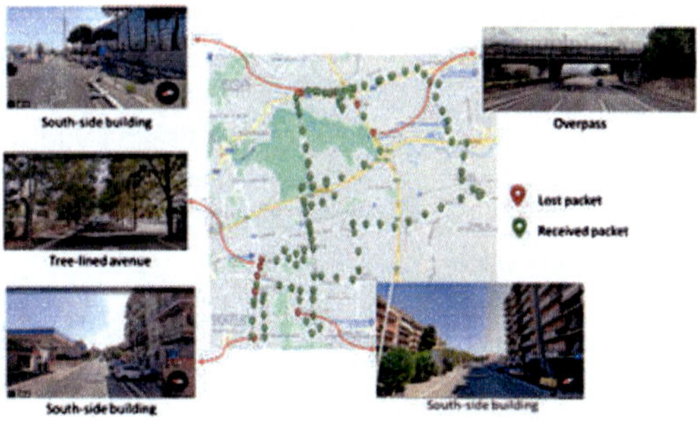

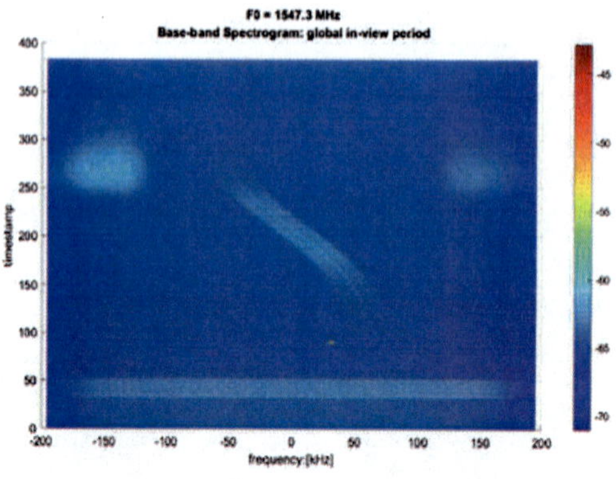

Fig. 14 Spectrogram in LEO environments

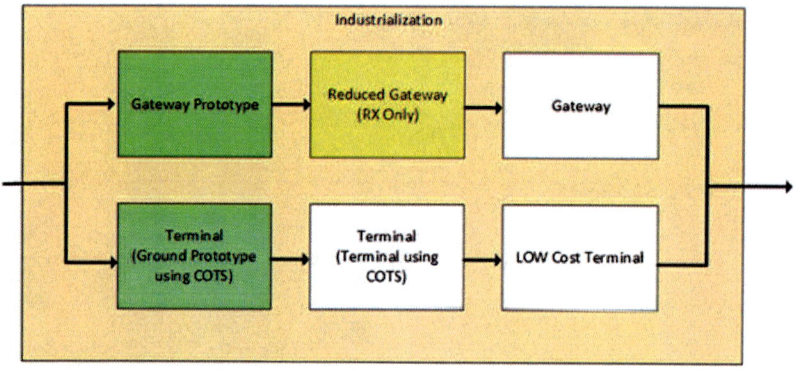

Fig. 15 Industrialization activities

The new concept of the IoT system intended as TT&C service for constellations of small satellites has provided new impetus to the roadmap relating to terminals, thus introducing a new line that is evolving very quickly towards the final product. The main actors of this activity are defining an additional development that will lead to the realization of a prototype of the on-board telemetry system based on COTS SDR capable of being embarked on a CUBESAT to demonstrate the proposed concept by experimentation.

Acknowledgements The authors of this article thank INMARSAT for the support provided in the evolution of these IoT products. INMARSAT, as world leader in global, mobile satellite communications, is able to meet the new needs expressed by its customers and to allow a rapid and precise evolution of the development roadmap.

References

1. Herrero, O.D.R., De Gaudenzi, R.: High efficiency satellite multiple access scheme for machine-to-machine communications. IEEE Trans. Aerosp. Electron. Syst. **48**(4), 2961–2989 (2012)
2. ETSI EN 302 307 (V1.2.1). Digital video broadcasting (DVB); second generation framing structure, channel coding and modulation systems for broadcasting, interactive services, news gathering and other broadband satellite applications (DVB-S2)
3. ESTI EN 302 583 (V1.2.1). Digital video broadcasting (DVB); framing structure, channel coding and modulation for satellite services to handheld devices (SH) below 3 GHz
4. 3GPP TS 36.212. Multiplexing and channel coding (release 8), v.8.8.0 (2009–2012)
5. MASSIVE–ESA ITT AO8871. Satellite gateway development for massive uncoordinated access networks
6. GEMMA–ESA ITT AO8916. Gateway demonstrator for E-SSA-based machine-to-machine application
7. Scalise, S., et al.: S-MIM: a novel radio interface for efficient messaging services over satellite. IEEE Commun. Mag. **51**(3), 119–125 (2013)
8. Arcidiacono, A., et al.: From S-MIM to F-SIM: making satellite interactivity affordable at Ku and Ka-band. Int. J. Satell. Commun. Netw. (2015)
9. Isca, A., Alagha, N., Andreotti, R., Andrenacci, M.: Recent advances in design and implementation of satellite gateways for massive uncoordinated access networks. Sens. (Basel) **22**(2), 565 (2022)
10. Gallinaro, G., et al.: ME-SSA: an advanced random access for the satellite return channel. In: International Conference on Communication 2015 (ICC-2015)

Hardware-in-the-Loop Simulations of Future Autonomous Space Systems Aided by Artificial Intelligence

Andrea Carbone, Dario Spiller, Mohamed Salim Farissi, Sarathchandrakumar T. Sasidharan, Francesco Latorre, and Fabio Curti

Abstract This paper deals with the description of hardware-in-the-loop simulations of future autonomous space systems aided by artificial intelligence and using the MONSTER facility in the ARCALab laboratory at the School of Aerospace Engineering, Sapienza University of Rome. Born as a facility to simulate lunar landing trajectories by means of a Cartesian robotic manipulator, MONSTER is being updated to perform experiments for the verification and validation of new disruptive technologies related to on-board computing with deep learning and reinforcement learning. The facility will be used to test different GNC autonomous systems on a variety of scenarios, e.g., the Moon, the Earth as observed from different remote sensing platforms, other celestial bodies as asteroids. Preliminary results obtained with the proposed experimental facility are discussed, and future planned activities are reported.

Keywords Hardware accelerators · Artificial intelligence · Autonomous satellites · Robotic facility · MONSTER

1 Introduction

Recent trends in academic and industrial research are showing increasing interest in on-board computation (and edge computation as a general methodological approach) supported by methodologies based on artificial intelligence (AI) algorithms [1–4]. This kind of approach is disrupting the way we think to satellites, as many small satellite missions embarking commercial-off-the-shelf (COTS) elements [5] for AI on-board computing are being designed and launched. One remarkable example is the 6U cubesar Phisat-1 which performs cloud detection with an on-board hardware accelerator [6, 7]. Many of the approaches for edge computing are based on the

A. Carbone · D. Spiller · M. S. Farissi · S. T. Sasidharan · F. Latorre · F. Curti (✉)
School of Aerospace Engineering, ARCALab, Sapienza University of Rome, via Salaria 851, Rome 00138, Italy
e-mail: fabio.curti@uniroma1.it

C. Ieracitano et al. (eds.), *The Use of Artificial Intelligence for Space Applications*, Studies in Computational Intelligence 1088,
https://doi.org/10.1007/978-3-031-25755-1_6

usage of convolutional Neural Networks(CNNs) [8, 9], given their ability to reach very good performances in classification and segmentation tasks having a reduced number of parameters with respect to the fully connected networks.

Based on the previous observations, facilities for hardware-in-the-loop (HIL) tests to validate novel edge computing paradigms are of utmost relevance. This work presents MONSTER (Moon Optical Navigation robotic facility on Simulated TERrain), an experimental facility used to simulate guidance, control, and navigation problems on the lunar surface as well as on other different scenarios. The facility consists of a three-dimensional Cartesian manipulator which, once fully operative, will be equipped with a spherical wrist to simulate both attitude and orbital dynamics. All the experiments already performed (and planned for the future) are based on innovative and disruptive approaches using AI algorithms, in line with the current trend to put more autonomy and computing resources in future space missions [1]. Preliminary results related to AI-based control algorithms have been already published in [10], whereas encouraging results of image analysis for Hazard Detection and Avoidance (HDA) with a fully Convolutional Neural Network (CNN) have been reported in [11]. These are only few examples of feasible applications that can be proven hardware-in-the-loop with the MONSTER facility. So far, the NVIDIA Jetson TX2 board has been integrated into the facility to guarantee high computational speed. This device will be interfaced with navigation sensors needed to know position, velocity and attitude of the spacecraft. Finally, the TX2 board will be compared with other hardware accelerators, e.g. the Intel Movidius Myriad and a system-on-chip (SOC) FPGA, on representative machine learning test cases.

This paper is organised as follows. In Sect. 2 the MONSTER facility is described, providing details about the possibility to use different simulated terrains and different AI accelerators. In Sect. 3 a summary of the AI accelerators considered for the current and future research activities is reported, and further details are discussed in Sect. 4 for the lunar landing test case, in Sect. 5 for the remote sensing applications, and in Sect. 6 for preliminary performance comparisons considering different hardware solutions. Final discussions and conclusions are reported in Sect. 7.

2 The MONSTER Robotic Facility

MONSTER is an experimental robotic facility of Sapienza University of Rome located at the School of Aerospace Engineering. It consists of an external computer, which provides the commands to move a Cartesian manipulator, and a hardware-in-the-loop (HIL) experimental platform hosting the on-board computer and mounted on the end-effector of the manipulator. It is composed of a Cartesian robotic manipulator, which can move along the three axes x-y-z (currently only the translational motion can be simulated, but we are updating the facility to include also the attitude motion), and a high-fidelity reproduction of a portion of the lunar terrain. The Cartesian robot is actuated by stepping motors mounted on the principal axes of the robot and providing holding bipolar torque of 820 Ncm. The whole HIL scheme is shown

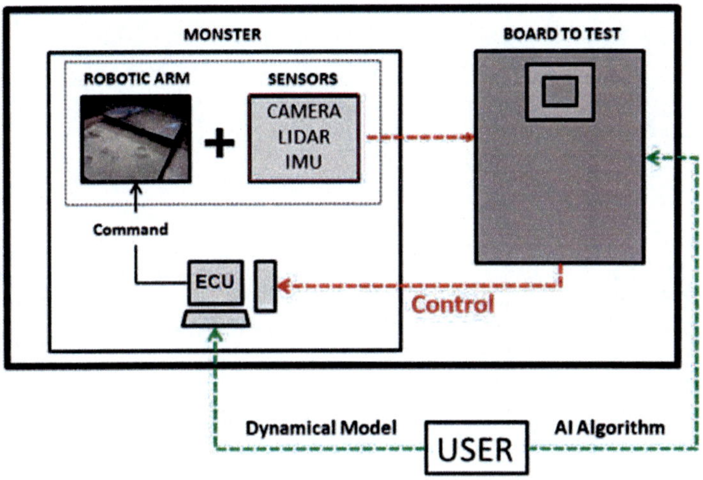

Fig. 1 Hardware-in-the-loop scheme

in Fig. 1. The board to test element can be any hardware accelerators designed to minimize the computational effort of AI algorithms while performing the inference task. The set of sensors installed on MONSTER are required to realistically simulate the flight conditions of the observing platform. Moreover, in addition to the simulation of landing maneuvers on the Moon, MONSTER can also reproduce other operations by overlapping different layers with new *alto-rilievo* or printed scenarios to the current lunar terrain (the dynamical model of the new simulated system must be updated on External Computer Unit (ECU) to get the required motion of the robotic end-effector).

Simulated terrain–So far, MONSTER was used for the simulation of lunar landing trajectories [10, 12]. The current terrain is a high-fidelity reproduction of the lunar equatorial zone known as *Mare Serenitatis* (23° North, 14° East). The simulated terrain has a physical size of 3m × 4m and it corresponds to 6 km × 8 km on the lunar surface. Concerning the possibility to simulate other scenarios, the printed or *alto-rilievo* layers are put on top of a wood support. Accurately playing with the design parameters of the overlapping layers (such as resolution, dimensions, etc.) we can realistically simulate many different kind of observing platforms. For instance, we can change the distance from the observing platform to simulate flights or descent maneuvers at different altitudes, from the aeronautical up to the space domain. For instance, Fig. 2 shows a remote sensing application where the simulated scenario is taken from a PRISMA image (further information on the hyperspectral PRISMA mission can be found in literature, e.g., [13–17], or referring to the PRISMA web portal provided by the Italian Space Agency [18]) showing an active wildfire. In this case, the dimension of the printed terrain is 2m × 3m, with a resolution of 30 m/pixel in the original image. The current dimension and resolution of the printed scenario can be related to flight conditions up to an altitude of approximately 12 km, i.e., we are simulating a fly over the area from an aeronautical platform.

Fig. 2 An example of simulated scenarios installed on the MONSTER facility. The shown setup allows the reader to appreciate a sectional view of the design chosen to reproduce different terrains (lunar terrain and wildfire image)

AI algorithms enabled by on-the-edge computing–MONSTER is designed for the verification and validation of new disruptive guidance, navigation, and control algorithms based on AI methodologies, where edge computation enabled by AI hardware accelerators must be taken into account. Currently, the Nvidia Jetson TX2 GPU module is installed, even though we plan to implement other hardware solutions, such as visual processing units (VPUs) or field programmable gate arrays (FPGAs). For instance, preliminary experiments with the Intel Movidius Myriad have been already carried out, even though it has not yet been installed directly on MONSTER.

3 Enabling Edge Computing with AI Accelerators

Image analysis aided by machine learning usually needs a huge amount of parallel computational. This kind of computational capabilities can be achieved only at the cost of power consumption, which in turn requires devices with optimal performance with respect to power consumption. Moreover, inference times should be as low as possible in order to be compatible with the guidance, navigation, and control system. Using commercial-off-the-shelf (COTS) elements is one of the easiest solution to get the necessary performances from ground-born devices which can turn out to be useful also for space applications [3, 5]. Several technologies can be investigated as on-board AI hardware accelerators [4] to implement *on-the-edge* (or simple *edge*) computing architectures [1, 2]. In the following, the list of the considered solutions is reported.

Nvidia GPU: Jetson TX2 and Nano. GPU technologies have been investigated as radiation tolerant devices for on-board data processing in space [19]. For instance, the NVIDIA Jetson TX2 [20] has been already considered for space applications [21]. This is a high-performance AI on-the-edge devices with multiple GPU and CPU cores, capable of efficiently handling the implementation and the deployemnt of CNNs. The TX2 device includes the Tegra X2 system-on-chip (SoC) processor which incorporates a dual-core 2.0-GHz superscalar ARMv8 Denver processor, a quad-core 2.0-GHz 64-bit ARMv8 A57 processor, and an integrated Pascal GPU. On steady mode, the power consumption of the board is about 0.3 W. Moreover, the integrated GPU lets the GPU share DRAM with the CPU. This allows both the GPU and CPU to run more efficiently on low power, between 5 W at max efficiency and 15 W at max performance.

FPGA system-on-chip. CNNs are extremely demanding and require a considerable amount of computation. Very efficient for this application but power-hungry, GPUs are not necessarily the ideal solution to accelerate the execution of CNNs in embedded systems. FPGAs are particularly efficient as they can provide high-performance execution performance for compute-intensive operations while consuming relatively little power.

So far, PYNQ is used as a development environment to explore the solutions proposed by Xilinx. PYNQ is an open-source project that provides an easy software interface such us Jupyter Notebook for rapid prototyping and development of Zynq-SoC board (ex. PYNQ-Z2) and enabling the possibility to implement CNN applications by the widest possible audience with little FPGA expertise. Furthermore, the PYNQ Framework allows the developers to use higher level languages such as Python while accessing programmable logic (FPGA) overlays to perform the ML acceleration.

Hybrid Neural Networks architecture combining both Formal convolution and Spiking Fully-Connected networks have been already integrated in the European Space Agency's (ESA) Ops-SAT experimental satellite under the designation "Experiment 129–IA4SAT-SpikeFC" [22]. Moreover, binarised neural networks (BNNs) have been already considered for space application to improve inference performances [23], and an FPGA-based approach has been compared with the deployment on the Movidius Myriad-2 board considering the CloudScout mission [24].

Intel Movidius Myriad 2/X. The Intel Movidius Neural Compute Stick is a small fanless deep learning drive, with USB interface, and designed for the deployment of AI models to improve inference performances. The stick is powered by the low-power high performance Movidius Visual Processing Unit (VPU). It contains an Intel Movidius Myriad 2 Vision Processing Unit 4 GB which supports the CNN profiling, prototyping and tuning workflow. This module allows one to achieve real-time on device inference thanks to an energy-efficient CNN processing. Finally, data and power are provided by a single USB type A port, thus simplifying a lot the interface and minimizing the deployment operations. It is worthy to note that this chip has already been tested onboard with the Phisat-1 mission [6, 7] and on the ISS [25].

Google Coral. The Google Coral is an accelerator powered by a quad Cortex-A53 processor supported by a Google edge Tensor Processing Unit (TPU) as a co-processor to provide 4 terabyte operations per second (TOPS, i.e. 1 trillion operations each second) with an average power consumption of 2 W. The Google Coral is available with a large development board model and as a small USB accelerator. The large development board assists software development and debugging before being deployed on the accelerator unit. The Google Coral accelerator has been considered for on-board computing on unmanned aerial vehicles [26] and for future space mission [1].

3.1 Preliminary AI Deployment Analysis and Results

The integration of MONSTER with the different AI accelerator discussed so far is still not fully accomplished. Nonetheless, preliminary deployment of AI models for edge computing have been already carried out and provide some insights and proofs-of-concept of the proposed HIL methodology. In the following, a brief discussion of our results is reported, and some more details will be provided in the next sections.

Deployment of a U-Net Lunar Segmentation Model on the Jetson TX2. The segmentation model used for crater detection and properly discussed in Sect. 4 has been successfully deployed on the Jetson TX2, demonstrating an inference time of about 8 ms on 256×256 images.

Wildfire Detection From Hyperspectral Images Using the Jetson TX2. The methodology reported in [27] has been deployed on the Jetson TX2, and the performances are very promising. More detailed results are reported in Sect. 5.

Volcanic Eruption Detection Using the Movidius Myriad 2. As reported in [28], the Movidius Myriad accelerator is a feasible solution for *small* models. Indeed, what has been shown in [28] demonstrates that the deployed model must be designed in order to be compatible with the maximum performances of the Myriad. This is a limitation that has not be encountered, so far, on the Jetson TX2 GPU module.

Evaluation of FPGA and Jetson TX2 on Example Datasets. As already mentioned, the final goal is to make MONSTER a facility with the possibility of easily changing board or AI algorithm inside the HIL scheme (Fig. 1). To do this, preliminary tests are carried out to verify the performances of difference AI algorithms in two different boards. In particular, two trained BNNs are tested on PYNQ and Jetson TX2 considering the MNIST and CIFAR-10 datasets. The performances of the employed networks are reported in Sect. 6. The preliminary results are very encouraging in term of inference time, making the two boards attractive for real-time applications.

4 The Lunar Landing Test Case

Considering the current lunar landing test case, a hardware-in-the-loop (HIL) test of a GPU board implementing a hazard detection and avoidance module for lunar landing has been already performed, in particular the Nvidia TX2 GPU is exploited.

4.1 *MONSTER Setup*

In this particular test case, the robotic manipulator was used for the simulation of a lunar landing trajectory exploiting a high fidelity reproduction of a portion of the lunar terrain called *Mare Serenitatis*. Since only the last phase of the landing trajectory is simulated, a flat surface is actually considered. Thus, the *Dynamic Model* loaded on the ECU, is composed by the following equations:

$$\dot{\mathbf{r}} = \mathbf{v} \tag{1}$$

$$\dot{\mathbf{v}} = \mathbf{g}_M + \frac{\mathbf{T}}{m} = \mathbf{g}_M + \mathbf{u} \tag{2}$$

where $\mathbf{r} = [x_1, x_2, x_3]$ and $\mathbf{v} = [v_1, v_2, v_3]$ are respectively the position and velocity vector of the lander, $\mathbf{g}_M = [0, 0, -g_M]$ is the Moon's gravitational force ($g_M = 1.62509$ m/s^2), $\mathbf{T}$ and $\mathbf{u}$ are the control thrust and acceleration vectors, m is the lander mass. For what concerns the boundary constraints, they can be described as:

$$\mathbf{r}(t_0) = \mathbf{r}_0 \quad \mathbf{v}(t_0) = \mathbf{v}_0 \tag{3}$$

$$\mathbf{r}(t_f) = \mathbf{r}_f \quad \mathbf{v}(t_f) = \mathbf{v}_f \quad \mathbf{a}(t_f) = \mathbf{a}_f \tag{4}$$

where t_0, t_f, and $\mathbf{a}$ represent respectively the initial and final time, and the acceleration vector of the lander. Moreover, a constraint on the maximum control acceleration (u_{max}) should also be added for guidance purposes:

$$\|\mathbf{u}\| \leq u_{max} = \frac{T_{max}}{m} \tag{5}$$

MONSTER has been equipped with a real-time online HDA algorithm to avoid ending up in craters and to perform a safe landing. For this reason, a camera is mounted on the end-effector of the robotic manipulator. The acquired image is used by the external board (in this case the Jeston TX2) as input of the AI algorithm (CNN) implemented in order to perform the HDA. In particular, the HDA is performed during the approaching phase every Δt_H seconds, as shown in Fig. 3.

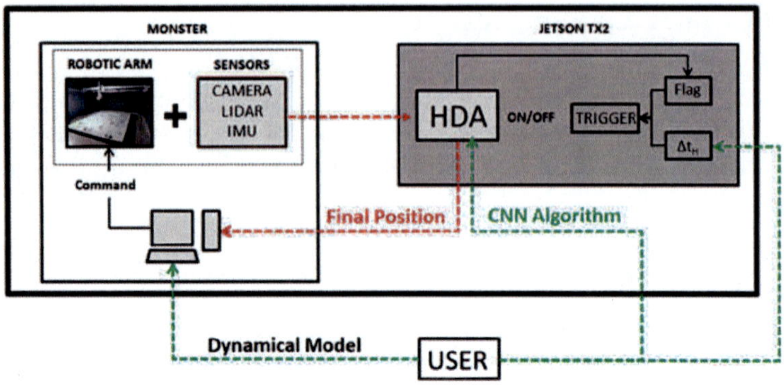

Fig. 3 HIL scheme with HDA implemented on Jetson TX2

4.2 The Autonomous GNC Subsystem

The trajectory is optimized in terms of minimization of the necessary propellant. This optimization is based on the combination of two main techniques/approaches: Particle Swarm Optimization [29] and inverse dynamics, simulated as in Ref. [30]. Moreover, during the entire procedure, the presence of craters on the lunar soil is also considered; therefore, the detection and avoidance of these dangers are also carried out by using a CNN implemented on the Nvidia Jetson TX2 GPU. One should note that more than one PSO can be actually performed in the landing trajectory for re-planning purpose. This happens since, as the lander approaches the surface, new hazards can be detected by the Convolutional Neural Network implemented on the Jetson TX2 and the autonomous system can realize that the desired final position is found to be within a previously non-detected crater. In this case, the ECU receives a new safe landing spot from the Nvidia TX2 and it perform a new optimization by means of PSO. This loop can be repeated each time the Jetson TX2 GPU detects a new hazard.

Current Control Policy with the PSO and Inverse Dynamic: In this paper an Inverse Dynamics approach is used: the external control (**u**) is written as a function of the state (**r**) and its derivatives ($\dot{\mathbf{r}} = \mathbf{v}$, $\ddot{\mathbf{r}} = \mathbf{a}$), by using the Eq. 2. Moreover, the state vector **r** is approximated by means of polynomials. In particular, the following polynomials are chosen to approximate each component of **r**:

$$x_i = a_{i0} + a_{i1}t + a_{i2}t^2 + a_{i3}t^3 + a_{i4}t^4 + a_{i5}t^5, \quad i = 1, 2, 3. \tag{6}$$

Equation 6 must also fulfill the boundary constraints listed in Eqs. 3 and 4. This means that five coefficients of each polynomial are actually fixed, whereas one (for each polynomial) remains a free parameter (in this case a_{12}, a_{22}, a_{32}, respectively) and represents the variable to be optimized in PSO. It is worth noting that, because of the polynomial approximation of the state, a sub-optimal solution is obtained.

As already stated, a fuel efficient problem is taken into account. In fact the total ΔV needed to the control the trajectory represent the basic cost function:

$$J_0 = \int_{t_0}^{t_f} ||\mathbf{u}||dt \tag{7}$$

Moreover, the constraint represented by Eq. 5 has to be satisfy. In order to do that, a penalty function $\bar{J}$ is considered and the total cost function can be written as follows:

$$J = J_0 + \bar{J} \tag{8}$$

where $\bar{J}$ is positive and different from zero if Eq. 5 is not satisfy, otherwise is equal to zero. The entire procedure is explained with more details in [30].

Crater Detection and Avoidance with Deep Learning: The strategy exploited for the HDA is based on a CNNs, which allow the detection of particular image features such as edges, contours, particular shapes and orientation and therefore are very useful when searching for specific shaped objects such as craters. CNNs have been extensively used in a Crater Detection context on Mars [31–33] and Moon [34, 35], reaching very good results in terms of accuracy and crater recognition performances. The CCN used in this work is a U-Net [36], which was first used for segmentation of biological cells and successfully adapted to crater detection on planetary bodies such as Mars and the Moon [37–39]. The network has a U-shaped, symmetric structure, which can be roughly split into two paths: a contracting, *encoding* path distributed in three stages, each of which is made of two 3×3 convolution layers followed by a 2×2 pooling layer. These convolutional layers contain 112, 224 and 448 filters in the three stages, respectively. On the expanding, *decoding* path, the image is up-sampled by means of a 2×2 upconvolution, concatenated with the image on the corresponding downsampling level to better preserve already learned features; it is then passed through a dropout layer to avoid overfitting and two 3×3 convolutional layers (containing 224, 112 and 112 filters, bottom to top) until the image size is restored; at the end, a final 1×1 pixel-wise convolution is applied in order to produce the final mask.

U-Net Training In this paper, a "cold start" approach has been followed, which means that the network has been initialized with random weights at the beginning of the training process. This goes by the name of *training from scratch*. The U-Net is fed with 320×240 visual images taken from the Moon Global Mosaic, labeled using the Head [40] and Povilaitis [41] combined crater catalogue. The output of the network is a segmented image, with pixel values between 0 and 1, in particular the labels equal to 1 are associated with the edges detected, whereas the pixels which don't belong to the edges are reported with labels equal to 0. Later, this output is thresholded to obtain a binary mask upon which detection is performed by means of a circle finding method. The network training is done using 30,000 images for training, 5,000 for

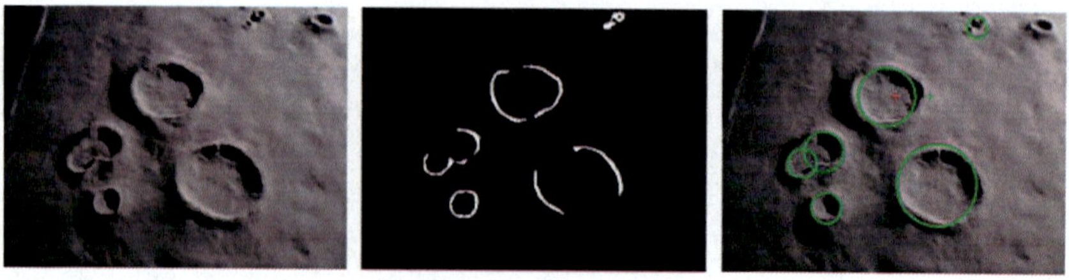

Fig. 4 U-net craters prediction

validation and 5,000 for testing. The loss function to be minimized during the training process is a pixel-wise binary cross entropy function:

$$l_i = x_i - x_i z_i + log(1 + e^{-x_i}) \tag{9}$$

In which x_i is the predicted pixel value and z_i is the ground truth pixel value.

The training was carried on TensorFlow Keras [42], on an NVIDIA GeForce RTX 2060 GPU. The network was compiled using an ADAM optimizer [43].

Transfer Learning In this case, the network trained upon optical images of the Moon is used as a starting point for additional training on 300 annotated images coming from the facility. The reason why this is done is due to the fact that the annotations are handmade and this would result in a huge effort if done for a great amount of images (say, 30,000 images as done in training from scratch); TL allows to use a smaller amount of images while reaching good accuracy and performance. The result of the test on the real time operation of the HDA module by means of U-Net can be seen in the Fig. 4.

4.3 *On-Board Edge-Computing Simulation with the Jetson TX2*

The U-Net model described so far has been already implemented on the Jetson TX2 and put in MONSTER for HIL tests. As far as segmentation performances is concerned, precision, recall and F1 score values computed on 350 images are perfectly compatible with the results obtained with the original model on the PC, being 88.62%, 79.07% and 83.57%, respectively. Moreover, after the proper optimization of the network, the inference time on the TX2 is equal to 8.31 ms, which is completely feasible for real-time applications.

4.4 Transfer Learning Experiments

It is worthy to note that, as already mentioned in [11], the methodology proposed here can be used to perform transfer learning experiments on other celestial bodies, like Mars or Asteroids. The MONSTER facility can be used to this aim by using an *alto-rilievo* layer based on digital elevation models information. With this approach, we will be able to test different transfer learning approaches, from completely freezing the segmentation model to on-board fine-tuning, comparing performance and computational requirements on different AI accelerators.

5 Remote Sensing Test Cases

The MONSTER facility allows our group to test a variety of different scenarios. Specifically, we can perform hardware-in-the-loop tests of different simulated platforms thanks to the opportunity to put different layers over the original lunar terrain. Hence, starting from previous studies from the authors [27, 28], we are now considering remote sensing experiments with RGB composite images from Earth observing satellites. Two test cases are worthy of details, i.e., the wildfire classification example using PRISMA data and the volcanic eruption detection based on optical data. In both cases, AI-accelerators are considered for on-board deployment.

5.1 Wildfire Classification

Referring to Fig. 2, MONSTER has been already updated to run HIL test campaigns with remotely-sensed images, and preliminary tests have been performed considering the original model proposed in [27, 44]. It is worthy to note that in [27, 44] PRISMA hyperspectral images where used, but they cannot be entirely reproduced in MONSTER. Indeed, when printing or reproducing the scenario observed by PRISMA, the input data will no more be the original hyperspectral image, but the RGB composite reported on the simulated scenario. Hence, the spectral content is drastically reduced, and the entire facility can be used to asses operative performances based on the execution of successive tasks, such as image acquisition, image elaboration, evaluation of navigation information and update of the platform guidance.

Referring to the outcome of the research reported in [27, 44], the results of the deployment on the Jetson TX2 are really promising. The classification accuracy is around 98% as in the previous implementations, whereas the computational power is 2.1 W and the inference time is around 3.0 ms on average. This test demonstrates the feasibility of considering on-board computing for disaster management related to wildfire.

5.2 *Volcanic Eruption Detection*

A feasibility study for the detection of volcanic eruptions, with a prototype of an on-board AI model and realistic testing equipment was presented in [28]. Also in that case, the objective was to produce real-time (or near real-time) alerts and allow immediate interventions. The problem was cast as a classification task, where CNNs are tested for on-board deployment considering a computational board composed by the Raspberry PI processing unit and the Movidius Myriad 2.

Uploading the model on the Movidius board requires an optimization process to properly deal with network complexity, inference execution time and number of parameters. As a consequence, two CNNs have been designed and realized from scratch, one big model for optimal inference performances and one small model for optimal inference time. Indeed, the Movidius has limited elaboration power, and machine learning models should usually be optimized or simplified before the deployment. In this case, the original, big reference model was pruned to minimize inference time, because its value represented a strict requirement for on-board execution. Moving from the big to the small model, good inference performances are maintained (i.e., satisfactory results are obtained on the small model even though they are lower than in the big reference one), but the overall inference time passed from 1 to 7 images/second, thus obtaining near real-time operative conditions.

As already discussed for the wildfire test case, the experimental test proposed in [28] basing on a drone platform can be replicated and extended using the MONSTER facility. This will allow us to carry out more experiments in a controlled and safe environment, confirming and improving previous outcomes.

5.3 *Designing Future Missions for Real-Time Extreme Events Management*

The wildfire and volcanic eruption test cases demonstrate the potentialities of new technologies such as hardware accelerators and AI algorithms to help monitoring our planet. The tests performed with MONSTER and with the hardware accelerators are fundamental steps toward understanding that remote sensing missions can be revolutionized to increase their autonomy and lower the response time in case of disasters management. Satellite systems, and specifically distributed systems (e.g., satellite constellations), can significantly help ground-based systems and resources if and only if they can provide useful information in real-time, or maybe even before ground-based information are provided. To this aim, apart from problems related to the design of the space segment (e.g., number of required observing satellites), the feasibility of autonomous on-board computing and elaboration of remotely-sensed imagery must be demonstrated. The results reported in Sects. 5.1 and 5.2 represent significant steps, as they show that even using COTS elements can provide reliable solutions. The demonstration operated by the Phisat-1 about putting and using COTS elements for on-board computing is fundamental to understand that the current technology is mature enough to start thinking about next generation of remote sensing missions. In this scenario, MONSTER offers unique opportunities to perform HIL tests of new disruptive on-board architectures.

6 Preliminary Comparison Between FPGA and Jetson TX2 GPU

The test performed on FPGA and Jetson TX2 GPU is based on a BNN, in particular two different types of BNN are considered for classifying the dataset: the Large Fully Connected network (LFC), inspired by the paper [45], and a convolutional network topology inspired by BinaryNet [46] and VGG-16 [47], and referred with the acronym CNV. The tests were carried out on LARQ environment with TensorFlow Keras [42], regarding the application on the Jetson TX2, meanwhile the PYNQ's test is based on FINN framework ([45]). For each network 1 bit has been allocated for each weight and 1 bit for each activation function. The results of each network are reported in the following sections in terms of power consumption, inference time and accuracy.

6.1 Large Fully Connected Network

The LFC network is a BNN composed of 4 fully connected layers with 1024 neurons each, as deeply reported in [45]. The LFC is used for classifying the MNIST dataset [48]. These networks accept 28×28 binary images and output a 10-bit one-hot vector whose entries indicate the probability of predicting the corresponding class. The LFC network implemented on Jetson TX2 requires an average power consumption of 2.2 W, with a peak power of about 6 W, while the inference time and the accuracy are equal to 0.2 ms and 97.5 % respectively. Meanwhile, the LFC network implemented on PYNQ requires to run a power consumption of 1.6 W, an inference time of 8.4 ms and an accuracy equal to 98.4 %.

6.2 Deep Convolutional Network

The CNV contains a succession of (3×3 convolution, 3×3 convolution, 2×2 max-pool) layers repeated three times with 64-128-256 channels, followed by two fully connected layers of 512 neurons each, as reported on [45]. The CNV is exploited for classifying the CIFAR-10 dataset. Note that the inputs to the first layer and the outputs from the last layer are not binarized; CNV accepts 32×32 images with 24 bits/pixel, and returns a 10-element vector of 16-bit values as the result. The power consumption, the inference time and the accuracy of the CNV implemented on Jeston TX2 are equal to 5 W, 1.8 ms ms and 79.98 %, meanwhile the same parameters result equal to 1.7 W, 1.6 ms and 79.22 % implemented on PYNQ.

7 Discussion and Conclusion

Originally designed as a robotic facility to test lunar landing maneuvers, MONSTER has now been improved to deal with more and diversified scenarios. The facility is currently equipped with the Jetson TX2 board, but it is ready to host other hardware accelerators, realizing a plug-and-play test environment offering the opportunity to install, try and compare different edge computing paradigms. When fully equipped and operative, MONSTER will have sensors and motors to simulate any autonomous flying system such as low-flying drone, Earth-orbiting satellites, and interplanetary probes. The preliminary analysis reported in this work show the potentialities of MONSTER, which can be used to perform hardware-in-the-loop experimental campaigns to test and compare a variety of edge computing approaches, so that specific optimization studies could be carried out for every single addressed problem. A convolutional neural network addressing hazard detection and avoidance for a lunar landing simulation has already been tested in real-time application on the Jetson TX2 GPU module, reporting an inference time of only 8 ms. Preliminary tests related to other types of remotely-sensed disaster events, such as wildfire and volcanic eruption detection, were carried out only with hardware accelerators. The positive results of these simulations in terms of power consumption, accuracy and inference time prove the effectiveness of the proposed approach, where on-board commercial-off-the-shelf accelerators are proposed to run artificial intelligence algorithms for real-time applications.

A current limitation of MONSTER is represented by the fact that only Cartesian motion can be simulated. However, we are currently working for upgrading the facility to include a spherical wrist and thus add the simulation of the attitude dynamics. Having the possibility to simulate both center-of-mass and attitude motions, MONSTER will be able to reproduce 6 degree of freedom maneuvers. In this case, autonomous systems driven by AI algorithms could be tested and validated. For instance, the facility could simulate space platform where artificial intelligence will be responsible for the navigation by means of deep learning methodologies and for the control by means of reinforcement learning algorithms. Also, we could simulate future remote sensing missions with the ability of perform attitude maneuvers to capture specific ground targets based on navigation information evaluated with on-board computing.

References

1. Furano, G., Meoni, G., Dunne, A., Moloney, D., Ferlet-Cavrois, V., Tavoularis, A., Byrne, J., Buckley, L., Psarakis, M., Voss, K.O., Fanucci, L.: Towards the use of artificial intelligence on the edge in space systems: Challenges and opportunities. IEEE Aerosp. Electron. Syst. Mag. **35**(12), 44–56 (2020). https://doi.org/10.1109/MAES.2020.3008468
2. Wang, Y., Yang, J., Guo, X., Qu, Z.: Satellite edge computing for the internet of things in aerospace. Sensors **19**(20) (2019). https://doi.org/10.3390/s19204375
3. Raoofy, A., Dax, G., Serra, V., Ghiglione, M., Werner, M., Trinitis, C.: Benchmarking and feasibility aspects of machine learning in space systems. In: Proceedings of the 19th ACM International Conference on Computing Frontiers, p. 225–226. CF '22, Association for Computing Machinery, New York, NY, USA (2022). https://doi.org/10.1145/3528416.3530986
4. Machine learning application benchmark for in-orbit on-board data processing. In: European Workshop on On-Board Data Processing (2021). https://zenodo.org/record/5520877/files/05.04_OBDP2021_Ghiglione.pdf?download=1
5. Shah, P., Lai, A.: Cots in space: From novelty to necessity. In: 35th Annual Small Satellite Conference (2021)
6. Giuffrida, G., Fanucci, L., Meoni, G., Batič, M., Buckley, L., Dunne, A., van Dijk, C., Esposito, M., Hefele, J., Vercruyssen, N., et al.: The ϕ-sat-1 mission: The first on-board deep neural network demonstrator for satellite earth observation. IEEE Trans. Geosci. Remote Sens. **60**, 1–14 (2021)
7. Esposito, M., Conticello, S.S., Pastena, M., Domínguez, B.C.: In-orbit demonstration of artificial intelligence applied to hyperspectral and thermal sensing from space. In: Pagano, T.S., Norton, C.D., Babu, S.R. (eds.) CubeSats and SmallSats for Remote Sensing III, vol. 11131, pp. 88–96. International Society for Optics and Photonics. SPIE (2019). https://doi.org/10.1117/12.2532262
8. LeCun, Y., et al.: Generalization and network design strategies. Connect. Perspect. **19**(143–155), 18 (1989)
9. Le Cun, Y., Jackel, L.D., Boser, B., Denker, J.S., Graf, H.P., Guyon, I., Henderson, D., Howard, R.E., Hubbard, W.: Handwritten digit recognition: Applications of neural network chips and automatic learning. IEEE Commun. Mag. **27**(11), 41–46 (1989)
10. D'Ambrosio, A., Carbone, A., et al.: *pso*-based soft lunar landing with hazard avoidance: Analysis and experimentation. Aerospace **8**(7) (2021). https://doi.org/10.3390/aerospace8070195
11. Latorre, F., Spiller, D., Curti, F.: Autonomous crater detection on asteroids using a fully-convolutional neural network. In: XXVI International Congress of the Italian Association of Aeronautics and Astronautics. AIDAA (2021). arXiv:2204.42419, https://doi.org/10.48550/arXiv.2204.00477
12. Ansalone, L., Grava, E., Curti, F.: Experimental results of a terrain relative navigation algorithm using a simulated lunar scenario. Acta Astronaut. **116**, 78–92 (2015). https://doi.org/10.1016/j.actaastro.2015.06.022
13. Guarini, R., Loizzo, R., Facchinetti, C., Longo, F., Ponticelli, B., Faraci, M., Dami, M., Cosi, M., Amoruso, L., De Pasquale, V., Taggio, N., Santoro, F., Colandrea, P., Miotti, E., Di Nicolantonio, W.: PRISMA hyperspectral mission products. In: International Geoscience and Remote Sensing Symposium (IGARSS) (2018). https://doi.org/10.1109/IGARSS.2018.8517785
14. Loizzo, R., Guarini, R., Longo, F., Scopa, T., Formaro, R., Facchinetti, C., Varacalli, G.: Prisma: The Italian hyperspectral mission. In: International Geoscience and Remote Sensing Symposium (IGARSS) (2018). https://doi.org/10.1109/IGARSS.2018.8518512
15. Loizzo, R., Daraio, M., Guarini, R., Longo, F., Lorusso, R., DIni, L., Lopinto, E.: Prisma mission status and perspective. In: International Geoscience and Remote Sensing Symposium (IGARSS) (2019). https://doi.org/10.1109/IGARSS.2019.8899272
16. Coppo, P., Brandani, F., Faraci, M., Sarti, F., Cosi, M.: Leonardo spaceborne infrared payloads for earth observation: SLSTRs for copernicus sentinel 3 and PRISMA hyperspectral camera for PRISMA satellite. Proceedings (2019). https://doi.org/10.3390/proceedings2019027001

17. Coppo, P., Brandani, F., Faraci, M., Sarti, F., Dami, M., Chiarantini, L., Ponticelli, B., Giunti, L., Fossati, E., Cosi, M.: Leonardo spaceborne infrared payloads for Earth observation: SLSTRs for copernicus sentinel 3 and PRISMA hyperspectral camera for PRISMA satellite. Appl. Opt. (2020). https://doi.org/10.1364/ao.389485
18. (ASI), I.S.A.: PRISMA Web Portal (2019). https://prisma.asi.it/
19. Bruhn, F.C., Tsog, N., Kunkel, F., Flordal, O., Troxel, I.: Enabling radiation tolerant heterogeneous gpu-based onboard data processing in space. CEAS Space J. **12**(4), 551–564 (2020)
20. Süzen, A.A., Duman, B., Şen, B.: Benchmark analysis of jetson tx2, jetson nano and raspberry pi using deep-cnn. In: 2020 International Congress on Human-Computer Interaction, Optimization and Robotic Applications (HORA), pp. 1–5 (2020). https://doi.org/10.1109/HORA49412.2020.9152915
21. Adams, C., Spain, A., Parker, J., Hevert, M., Roach, J., Cotten, D.: Towards an integrated gpu accelerated soc as a flight computer for small satellites. In: 2019 IEEE Aerospace Conference, pp. 1–7 (2019). https://doi.org/10.1109/AERO.2019.8741765
22. Lemaire, E., Moretti, M., Daniel, L., Miramond, B., Millet, P., Feresin, F., Bilavarn, S.: An fpga-based hybrid neural network accelerator for embedded satellite image classification. In: 2020 IEEE International Symposium on Circuits and Systems (ISCAS), pp. 1–5. IEEE (2020)
23. Wolf, J., Kosmidis, L., Mugarza, I., Agirre, I., Yarza, I., Lussiana, F., Botta, S., Binchi, J., Onaindia, P., Poggi, T., et al.: A taste of binarised neural network inference for on-board fpgas
24. Rapuano, E., Meoni, G., Pacini, T., Dinelli, G., Furano, G., Giuffrida, G., Fanucci, L.: An fpga-based hardware accelerator for cnns inference on board satellites: Benchmarking with myriad 2-based solution for the cloudscout case study. Remote Sens. **13**(8) (2021). https://doi.org/10.3390/rs13081518, https://www.mdpi.com/2072-4292/13/8/1518
25. Dunkel, E., Swope, J., Towfic, Z., Chien, S., Russell, D., Sauvageau, J., Sheldon, D., Romero-Cañas, J., Espinosa-Aranda, J., Buckley, L., et al.: Benchmarking deep learning inference of remote sensing imagery on the qualcomm snapdragon and intel movidius myriad x processors onboard the international space station. In: 2022 IEEE International Geoscience and Remote Sensing Symposium (2022)
26. Kraft, M., Piechocki, M., Ptak, B., Walas, K.: Autonomous, onboard vision-based trash and litter detection in low altitude aerial images collected by an unmanned aerial vehicle. Remote Sens. **13**(5) (2021). https://doi.org/10.3390/rs13050965, https://www.mdpi.com/2072-4292/13/5/965
27. Spiller, D., Ansalone, L., et al.: Analysis and detection of wildfires by using prisma hyperspectral imagery. Int. Arch. Photogram. Remote Sens. Spat. Inf. Sci. **XLIII-B3-2021**, 215–222 (2021). https://doi.org/10.5194/isprs-archives-XLIII-B3-2021-215-2021
28. Del Rosso, M.P., Sebastianelli, A., et al.: On-board volcanic eruption detection through cnns and satellite multispectral imagery. Remote Sens. **13**(17) (2021). https://doi.org/10.3390/rs13173479
29. Kennedy, J., Eberhart, R.: Particle swarm optimization. In: Proceedings of ICNN'95-International Conference on Neural Networks, vol. 4, pp. 1942–1948. IEEE (1995)
30. D'Ambrosio, A., Carbone, A., Spiller, D., Curti, F.: Pso-based soft lunar landing with hazard avoidance: Analysis and experimentation. Aerospace **8**(7), 195 (2021)
31. Palafox, L., Hamilton, C., Scheidt, S., Alvarez, A.: Automated detection of geological landforms on Mars using convolutional neural networks. Comput. Geosci. **101** (2017). https://doi.org/10.1016/j.cageo.2016.12.015
32. Benedix, G.K., Norman, C.J., Bland, P.A., Towner, M.C., Paxman, J., Tan, T.: Automated detection of martian craters using a convolutional neural network. In: Lunar and Planetary Science Conference, p. 2202. Lunar and Planetary Science Conference (2018)
33. Norman, C.J., Paxman, J., Benedix, G.K., Tan, T., Bland, P.A., Towner, M.: Automated detection of craters in martian satellite imagery using convolutional neural networks. In: Planetary Science Informatics and Data Analytics Conference, vol. 2082, p. 6004 (2018)
34. Emami, E., Ahmad, T., Bebis, G., Nefian, A., Fong, T.: Lunar Crater Detection via Region-based Convolutional Neural Networks (2018)

35. Emami, E., Ahmad, T., Bebis, G., Nefian, A., Fong, T.: On Crater Classification using Deep Convolutional Neural Networks (2018)
36. Ronneberger, O., Fischer, P., Brox, T.: U-Net: Convolutional Networks for Biomedical Image Segmentation (2015). https://doi.org/10.1007/978-3-319-24574-4_28
37. Lee, C.: Automated crater detection on Mars using deep learning. Planet. Space Sci. **170**, 16–28 (2019). https://doi.org/10.1016/j.pss.2019.03.008
38. DeLatte, D., Crites, S., Guttenberg, N., Tasker, E., Yairi, T.: Segmentation convolutional neural networks for automatic crater detection on Mars. IEEE J. Sel. Top. Appl. Earth Obs. Remote Sens. 1–14 (2019). https://doi.org/10.1109/JSTARS.2019.2918302
39. Silburt, A., Ali-Dib, M., Zhu, C., Jackson, A., Valencia, D., Kissin, Y., Tamayo, D., Menou, K.: Lunar crater identification via deep learning. Icarus **317** (2018). https://doi.org/10.1016/j.icarus.2018.06.022
40. Head, J., Fassett, C., Kadish, S., Smith, D., Zuber, M., Neumann, G., Mazarico, E.: Global distribution of large lunar craters: Implications for resurfacing and impactor populations. Science **329**, 1504–7 (2010). https://doi.org/10.1126/science.1195050
41. Povilaitis, R., Robinson, M., van der Bogert, C., Hiesinger, H., Meyer, H., Ostrach, L.: Crater density differences: Exploring regional resurfacing, secondary crater populations, and crater saturation equilibrium on the Moon. Planet. Space Sci. **162**, 41–51 (2018). https://doi.org/10.1016/j.pss.2017.05.006
42. Chollet, F.: Deep Learning with Python, 1st edn. Manning Publications Co., USA (2017)
43. Kingma, D., Ba, J.: Adam: A method for stochastic optimization. In: International Conference on Learning Representations (2014). https://doi.org/10.48550/arXiv.1412.6980
44. Spiller, D., Amici, S., Ansalone, L.: Transfer learning analysis for wildfire segmentation using prisma hyperspectral imagery and convolutional neural networks. In: 2022 12th Workshop on Hyperspectral Imaging and Signal Processing: Evolution in Remote Sensing (WHISPERS), pp. 1–5 (2022). https://doi.org/10.1109/WHISPERS56178.2022.9955054
45. Umuroglu, Y., Fraser, N.J., Gambardella, G., Blott, M., Leong, P., Jahre, M., Vissers, K.: Finn: A framework for fast, scalable binarized neural network inference. In: Proceedings of the 2017 ACM/SIGDA International Symposium on Field-Programmable Gate Arrays, pp. 65–74 (2017)
46. Courbariaux, M., Bengio, Y.: Binarynet: Training deep neural networks with weights and activations constrained to +1 or -1. Arxiv 2016. arXiv preprint arXiv:1602.02830
47. Simonyan, K., Zisserman, A.: Very deep convolutional networks for large-scale image recognition. ArXiv preprint arXiv:1409.1556 (2014)
48. LeCun, Y., Bottou, L., Bengio, Y., Haffner, P.: Gradient-based learning applied to document recognition. Proceed. IEEE **86**(11), 2278–2324 (1998)

Deep Reinforcement Learning for Pin-Point Autonomous Lunar Landing: Trajectory Recalculation for Obstacle Avoidance

Giulia Ciabatti, Dario Spiller, Shreyansh Daftry, Roberto Capobianco, and Fabio Curti

Abstract This work aims to present a method to perform autonomous precision landing—pin-point landing—on a planetary environment and perform trajectory recalculation for fault recovery where necessary. In order to achieve this, we choose to implement a Deep Reinforcement Learning—DRL—algorithm, i.e. the Soft Actor-Critic—SAC—architecture. In particular, we select the lunar environment for our experiments, which we perform in a simulated environment, exploiting a real-physics simulator modeled by means of the Bullet/PyBullet physical engine. We show that the SAC algorithm can learn an effective policy for precision landing and trajectory recalculation if fault recovery is made necessary—e.g. for obstacle avoidance.

Keywords Autonomous landing · Reinforcement learning · Pin-point landing · Fault recovery · Trajectory recalculation

1 Introduction

Autonomous systems constitute an important feature for space exploration, especially when it comes to landing and navigation [1, 2]. Despite the fact that Artificial Intelligence (AI)—also system-embedded AI—has made significant progress during the last few years, space exploration still follows a quite traditional approach in real-life

G. Ciabatti (✉) · R. Capobianco
Department of Computer, Control and Management Engineering, Sapienza University of Rome, Via Ariosto 25, 00185 Rome, Italy
e-mail: ciabatti@diag.uniroma1.it

G. Ciabatti · D. Spiller · F. Curti
School of Aerospace Engineering, Sapienza University of Rome, via Salaria 851, 00138 Rome, Italy

S. Daftry
Jet Propulsion Laboratory, California Institute of Technology, Pasadena, CA 91109, USA

R. Capobianco
Sony AI, Rome, Italy

C. Ieracitano et al. (eds.), *The Use of Artificial Intelligence for Space Applications*, Studies in Computational Intelligence 1088,
https://doi.org/10.1007/978-3-031-25755-1_7

missions [3, 4]. Encouraging results have been shown by recent research in the field of autonomous landing. It is the case of the applications of Deep Learning and Meta-Reinforcement [5, 6] or Deep Reinforcement Learning for six-degree-of-freedom Mars landing [7], and the Deep Reinforcement learning application regarding Adaptive Generalized ZEM-ZEV Feedback Guidance [8]. Following up these results, we presented an approach to address the autonomous planetary landing problem in our previous works [9, 10]. We utilized a Deep Reinforcement Learning algorithm to handle high-dimensional and heterogeneous input data, retrieved from the following sensors: RGB camera, LiDAR and pose data. We set up a continuous action space, utilizing the Soft Actor-Critic—SAC—algorithm [11]. In this work, we are going to show that the SAC model can learn an effective landing policy for trajectory recalculation to perform (also) fault recovery, where necessary—e.g. in the case of obstacle avoidance during landing on hazardous terrains. In order to evaluate the performance of SAC on this task, we utilize our simulator, first presented in [9], exploiting the Bullet/PyBullet physical engine and adapting official NASA 3D terrain models—or reconstructing them from SAR imagery retrieved from missions, when none is available. This approach contributes to defining a method for autonomous pin-point landing—with trajectory recalculation and obstacle avoidance—exploiting a physical engine, making the simulation similar to a real-world problem, including also the dynamics. Moreover, the simulator and the implementations utilized in this work, are open source and available for use. The contributions are discussed in detail in the following section. The paper is organized as follows: in Sect. 2 purpose statement and contributions of this work are presented; Sect. 3 contains a detailed description of the simulation environment utilized and the lander design; in Sect. 4 the methodology and SAC algorithm are reported; Sect. 5 showcases the experiment design and results, together with a comparison of other state-of-the-art algorithms and results. Finally, in Sects. 6 and 7 ongoing work and conclusions are presented, respectively.

2 Purpose Statement and Contributions: The Need for Precision

This work is mainly motivated by the need for precision landing, dictated by the most recent advances in space exploration. By achieving an effective pin-point landing, it becomes possible to:

- optimize sample-gathering from sites of scientific interest,
- explore those environments which require the landing precision to be in the order of magnitude of meters—e.g. craters, caves, ridges...
- exploit landers and/or rockets multiple times, similarly to what happens with airplanes. This would guarantee both a consistent reduction of costs for space exploration and a reduction in space debris.

All of these applications are likely to include hazardous terrains with obstacles, that might render the landing task more difficult and require an unplanned shift of a few (tens of) meters from the original target. In order to tackle these issues, we implement an off-policy algorithm, i.e. the SAC algorithm. This approach represents a contribution to the state-of-the-art for the following reasons:

- this algorithm represents an effective way to perform autonomous controlled landing, also in the case of obstacle avoidance and trajectory recalculation.
- The simulator is open source: according to our experience, this can prove very useful, given the lack of open-source simulators;
- specifically, the simulator includes a physical engine [12], it is easily customizable for many diverse proximity operations—e.g. surface navigation and it includes robot models defined by means of the URDF format. Moreover, the software is constantly maintained, adapted and optimized for robotic applications, that are made available for the community.
- The simulator is compatible with ROS.[1]

3 Simulator Overview

The simulator for planetary landing, first presented in [9] is developed in Python. It features the PyBullet/Bullet library [12], which allows to implement a physical engine for vision-based navigation and proximity operations, and lander-environment physical interactions. Our simulator features 3D terrain models and a lander model, which have been developed by means of the classical pipeline to build a visual robot model. The simulator is compatible with ROS and includes OpenAI gym [13] wrappers. The algorithm is trained on the real-physics simulator. Force actions—such as collisions and gravitational field—and heterogeneous terrain have been modeled. In particular, the terrain is shaped to be three-dimensional in order to simulate different heights and roughness. At this purpose, we adapt the 3D mesh models, provided by the official NASA GitHub repository [14] for the Moon terrain—mesh models and model descriptions are also available here [15] and include multiple Lunar and Martian terrains. We utilized a lunar landscape—the “Near Side” of the Moon Fig. 1a. The trajectory recalculation task for this work is performed on this model. Moon’s Far Side presents a denser distribution of craters and visible roughness—Fig. 1b. Mars’ Victoria Crater 3D model is also a good fit to perform autonomous landing and trajectory recalculation Fig. 1c. The 3D meshes have been scaled and textured appropriately. We then utilised them to generate the height map in the simulator workframe and the lander-environment interactions by means of the aforementioned physical engine—Fig. 2a–c.

[1] https://www.ros.org/.

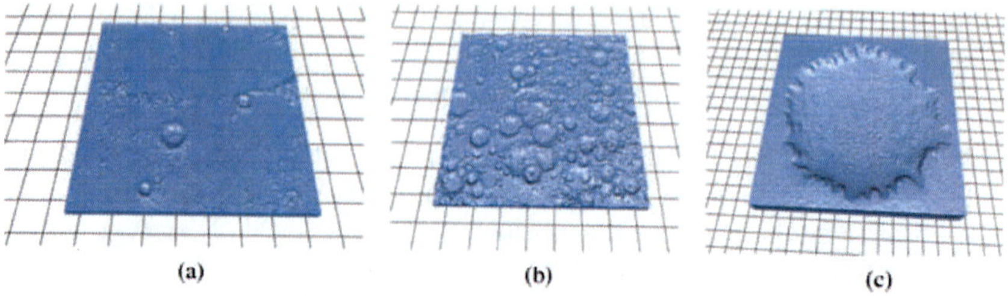

Fig. 1 NASA's 3D Mesh Models of **a**, **b** Lunar terrains **c** Martian terrain

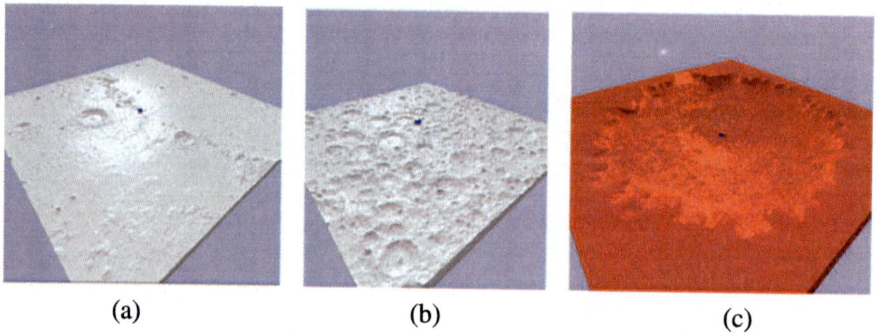

Fig. 2 Terrain 3D rendering of Moon's Near and Far Side **a**, **b** Mars' Victoria Crater **c**

3.1 Reconstructing Terrains: Titan

Because of NASA's interest in future missions to explore Saturn's moon Titan, such as the Dragonfly mission [16], we also choose to reconstruct Titan's terrain. We modeled the Polar Lake district—Fig. 3a—which might represent an important landing site for sample gathering, because of its peculiar morphology, which features an heterogeneous region of small, densely distributed hydrocarbon seas [17].

There was no availability of 3D accurate meshes—such as the ones we use to generate Moon and Mars terrains—for Titan's terrain. Hence, we try to reconstruct a realistic landscape starting from the official Cassini-Huygens mission SAR imagery [18], Fig. 3b, c—provided by NASA [19]. A height map is reconstructed as a 3D mesh. The generated mesh is then rendered to simulate the terrain, similarly to the lunar and martian terrains. In Fig. 4 the final graphical rendering of the Titan terrain is showcased.

3.2 The Lander

The lander model is defined in the Unified Robot Description Format—URDF, according to the classical ROS pipeline to build a visual robot model, i.e. attach-

(a) (b) (c)

Fig. 3 A representation of Titan's Polar Lacus Zone **a** and SAR imagery retrieved by the Cassini-Huygens mission: Titan's Xanadu Annex region and Titan lakes **b, c**

Fig. 4 Titan's terrain graphical rendering in simulation

ing links through joints in a hierarchical fashion, starting from the base. More in detail, the lander model is defined as a central body, composed by a cylinder and a sphere with four links, attached to the base through fixed-type joints and a sensor box. The propulsive forces are designed as four external forces, equidistant from the center of mass.

Two sensors are included in the design of the lander: an RGB camera and a LiDAR. In Fig. 5a, b, it is shown an example of the simulator graphical rendering of the lander model landing on two lunar terrains: the up-left box shows the point of view from the lander's RGB camera, while the "cone" represents the LiDAR detection (green

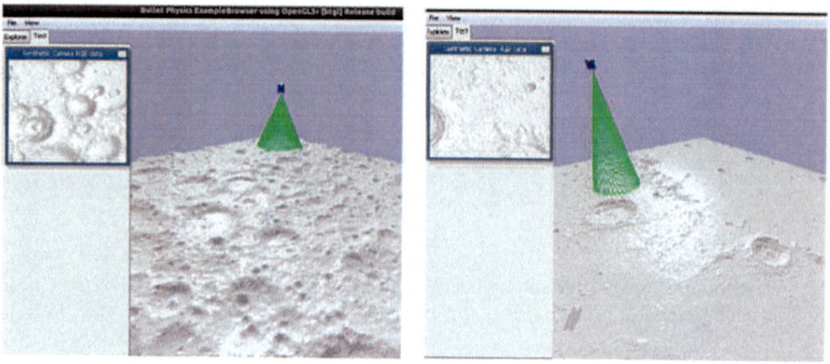

Fig. 5 Simulator snapshots in the case of lunar landing on two different terrains

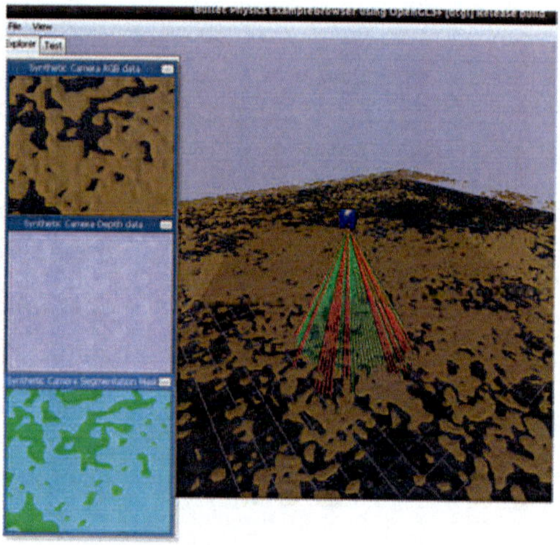

Fig. 6 Simulator snapshot in the case of landing on Titan: Methane Lakes are classified as non-landing zones by the LiDAR

rays mean that the LiDAR is sensing another physical object, i.e. the terrain. The rays are red when no object is detected). In Fig. 6, the landing on Titan's surface is showcased: the dark spots represents the hydrocarbon lakes: they are classified as non-landing zones. It is also possible to detect and visualize the fragmented nature of the terrain combining a Depth Camera and a Segmentation Mask.

4 Methodology: Soft-Actor Critic

The problem is cast as an episodic Deep Reinforcement Learning [20, 21] task, in the form of a Markov Decision Problem—MDP. The lander constitutes the Reinforcement Learning agent: it interacts with the fully-observable environment $\mathcal{E}$. The transition from state $s_t \in \mathcal{S}$—at time t—to the next state s_{t+1} are determined by the action that is chosen $a_t \in \mathcal{A}$, based on the observations. Note: all the actions in this application are real-valued: $\mathcal{A} \subset \mathbb{R}^N$. The agent's behavior is determined by the policy π: it maps states to actions: $\pi : \mathcal{S} \rightarrow \mathcal{A}$. The action-value function describes the expected return obtained after executing an action a_t in state s_t if policy π is followed.

$$Q^{\pi}(s_t, a_t) = \mathbb{E}_{\pi}[\sum_{i=t}^{T} \gamma^{i-t} r_i | s_t, a_t] \tag{1}$$

The Soft Actor-Critic—SAC—architecture [11] combines off-policy updates with a stochastic actor. This algorithm shows to achieve state-of-the-art performance on a range of continuous benchmark tasks. The SAC algorithm is capable of handling also continuous action spaces, proving to be very stable by achieving very similar performance across different random seeds. Beside combining an off-policy actor-critic setup with a stochastic actor the SAC algorithm further aims to maximize the entropy of the actor.

4.1 SAC Algorithm

The SAC Algorithm takes into account a parametrized state value function $\mathcal{V}_{\psi}(s_t)$, soft Q-function $\mathcal{Q}_{\theta}(s_t, a_t)$ and a policy $\pi_{\phi}(a_t|s_t)$. The parameters of these networks are ψ, θ and ϕ. The soft value is approximated by the state value function. The latter aims at minimizing the following:

$$J_V(\psi) = \mathbb{E}_{\mathbf{s}_t \sim \mathcal{D}}[\frac{1}{2}(V_{\psi}(\mathbf{s}_t) - \mathbb{E}_{\mathbf{a}_t \sim \pi_{\phi}}[\mathcal{Q}_{\theta}(\mathbf{s}_t, \mathbf{a}_t) - log\pi_{\phi}(\mathbf{a}_t|\mathbf{s}_t)])^2] \tag{2}$$

which is a squared residual error. Note that $\mathcal{D}$ is the distribution of previously sampled states and actions. The gradient can be estimated with an unbiased estimator:

$$\hat{\nabla}_{\psi} J_V(\psi) = \nabla_{\psi} V_{\psi}(\mathbf{s}_t)(V_{\psi}(\mathbf{s}_t) - \mathcal{Q}_{\theta}(\mathbf{s}_t, \mathbf{a}_t) + log\pi_{\phi}(\mathbf{a}_t|\mathbf{s}_t)) \tag{3}$$

The soft Bellman residual is minimized by means of the Q-function such as:

$$J_Q(\theta) = \mathbb{E}_{(\mathbf{s}_t, \mathbf{a}_t) \sim \mathcal{D}}[\frac{1}{2}(\mathcal{Q}_{\theta}(\mathbf{s}_t, \mathbf{a}_t) - \hat{\mathcal{Q}}(\mathbf{s}_t, \mathbf{a}_t))^2] \tag{4}$$

with

$$\hat{Q}(\mathbf{s}_t, \mathbf{a}_t) = r(\mathbf{s}_t, \mathbf{a}_t) + \gamma \mathbb{E}_{\mathbf{s}_{t+1} \sim p}[V_{\bar{\psi}}(\mathbf{s}_{t+1})] \tag{5}$$

that can be optimized with stochastic gradients:

$$\hat{\nabla}_\theta J_Q(\theta) = \nabla_\theta Q_\theta(\mathbf{s}_t, \mathbf{a}_t)(Q_\theta(\mathbf{s}_t, \mathbf{a}_t) - r\mathbf{s}_t, \mathbf{a}_t) - \gamma V_{\bar{\psi}}(\mathbf{s}_{t+1})) \tag{6}$$

The uses a target value network $V_{\bar{\psi}}$, where $\bar{\psi}$ can be an average of the value network weights that moves exponentially. This approach stabilizes training [22]. Hence, the policy parameters can be learned by directly minimizing the expected Kullback-Leibler divergence as follows:

$$J_\pi(\phi) = \mathbb{E}_{s_t \sim \mathcal{D}}[D_{KL}(\pi_\phi(\cdot|\mathbf{s}_t) \,||\, \frac{exp(Q_\theta(\mathbf{s}_t, \cdot))}{Z_\theta(\mathbf{s}_t})] \tag{7}$$

In order to minimize J_ϕ, the objective is now rewritten as:

$$J_\pi(\phi) = \mathbb{E}_{\mathbf{s}_t \sim \mathcal{D} \sim \mathcal{N}}[log\pi_\phi(f_\phi(\epsilon_t; \mathbf{s}_t)|\mathbf{s}_t) - Q_\theta(\mathbf{s}_t, f_\phi(\epsilon_t; \mathbf{s}_t))] \tag{8}$$

Where the reparameterization trick of the policy has been implemented: $\mathbf{a}_t = f_\phi(\epsilon_t; \mathbf{s}_t)$—by means of a neural network transformation.
Approximating the previously computed gradient as:

$$\hat{\nabla}_\phi J_\pi(\phi) = \nabla_\phi log\pi_\phi(\mathbf{a}_t|\mathbf{s}_t) + \nabla_{\mathbf{a}_t} log\pi_\phi(\mathbf{a}_t|\mathbf{s}_t) - \nabla_{\mathbf{a}_t} Q(\mathbf{s}_t, \mathbf{a}_t))\nabla_\phi f_\phi(\epsilon_t; \mathbf{s}_t) \tag{9}$$

where $\mathbf{a}_t$ is evaluated at $f_\phi(\epsilon_t; \mathbf{s}_t)$. Table 1 showcases the hyperparameters and the model setup for the training sessions implemented in this work. The action value range is set to $\{-15; 15\}$ for $\{T_x, T_y, T_z\}$ along the corresponding three axes—the positive and negative thrust values mean that the thrust can be along both the positive and negative directions, respectively.

Table 1 Hyperparameters and training setup details

Learning rate	Optimizer	Loss fct	Batch size	Replay buffer size	State space size	Action space size
3e-4	Adam	MSE	64	100000	65651	3

5 Experiment Design

In the following subsections we present the environment setup and the experimental results. For these first sets of experiments, we choose to perform the landing task on a lunar environment—in particular, the smoother Moon's near side, previously introduced.

5.1 *Environment Setup*

In order to perform our experiments, we utilise the simulator presented in the previous sections.

Lander Setup We include physical interactions, lander position and attitude. We set up the simulator to visually represent also an ideal landing spot—represented by the red square, invisible to the lander—in Fig. 9. The agent/lander is supposed to steer by means of its thrusters—already represented as a thrusting force along the x–y–z-axes. The lander is equipped with an RGB camera, a LiDAR and an IMU. The landing task ends when contact with the terrain is detected.

Landing Spot Recalculation For this task we take into account not only precision—pin-point—landing, but also the case where trajectory recalculation—e.g. for obstacle avoidance—is necessary. In particular, the initial desired landing spot and lander x-y- positions are randomly initialized. During the descent phase, the x-y- position of the desired landing spot is randomly recalculated, within a maximum of a few tens of meters from the original one: the lander/agent now has to steer to the new landing spot.

Reward Shaping In the first sets of experiments, we choose to model the reward in order to privilege the pin-point landing task. In particular, we design the reward function as a hand-tuned linear combination of several components. In the first place, we take into account the lander attitude: we give a penalty if the lander rolls out of a 45 deg "cone of stability" with respect to the vertical z-axis—this would cause the lander to not only lose stability and prevent it to land correctly without crashing, but also to lose sight of the terrain with both RGB camera and LiDAR. Secondly, we give a penalty if the lander takes too much time to reach the ground: this is aimed at preventing too much fuel to be consumed during the descent phase. Finally, we shape the most significant reward component for this task: we design an "off-course" penalty proportional to the squared distance between the position of the lander and of the landing spot, in the form:

$$R_{oc} = K - \alpha\sqrt{(x_l - x_{ls})^2 + (y_l - y_{ls})^2}$$

where: (x_l, y_l) are the x- y- axis coordinates of the lander and (x_{ls}, y_{ls}) are the x- y- coordinates of the landing spot, K and α are arbitrary constants.

5.2 *Experimental Results*

In the previous tasks—reported in [9] and [10]—a strong constraint was imposed not to allow the lander to regain altitude during the descent phase. For this task that constraint has been loosened to allow for the necessary altitude to be regained to allow the necessary steering to reach the new landing spot in the case hazardous terrains render trajectory recalculation necessary.

When utilizing the SAC algorithm for training a model by means of Deep Reinforcement Learning, it is very important to balance exploration and exploitation phase. In this section we show and comment the outcome of our experiments with different exploration-exploitation setups. In Fig. 7 the average episodic reward for 1000 episodes is showcased. The exploration phase is carried out for 100 episodes, before the exploitation phase. The first part of the curve shows the typical oscillations due to the initial random choice of the actions. Later, the reward scores start to increase, while still oscillating: this the the exploration phase of the training. The second half of the curve becomes more stable and gradually increases the scores: the exploitation phase is being carried out. In Fig. 8 the average episodic reward for 1000 episodes is showcased. This time, the exploration phase is carried out until the 400th episode: this can be seen from the longer oscillation phase, while, at the end of the training, the agent manages to increase the reward scores, with respect to the previous case in Fig. 7. Both the experiments yield a promising outcome, but, in terms of reward scores, the latter seems to perform a little better, meaning that enhancing the exploration phase leads the agent to generalized better during the exploitation phase.

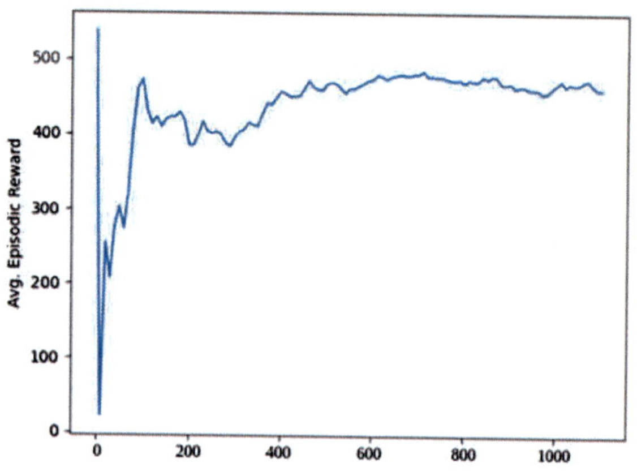

Fig. 7 Average Episode reward for 1000 episodes: Exploration phase until the 100th Episode

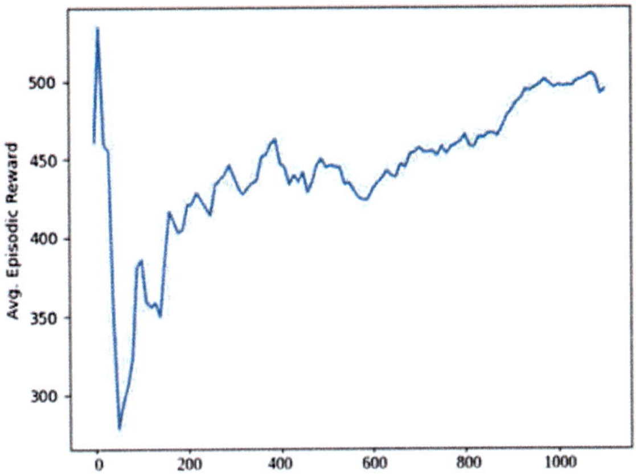

Fig. 8 Average Episode reward for 1000 episodes: Exploration phase until the 400th Episode

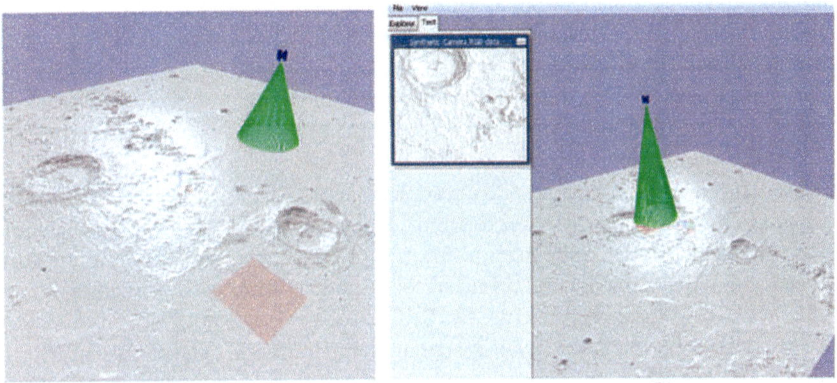

(a) (b)

Fig. 9 Snapshots from the simulator graphical rendering: Fig. 9a (left) shows the agent landing far from the desired landing spot. Figure 9b (right) shows the desired trajectory to land in the desired spot. Note: the red square is not visible to the lander

Figure 9 shows the typical conditions of an untrained agent Fig. 9a (left) and an agent that has managed to steer to the correct landing spot, keeping a good attitude Fig. 9b (right): the first snapshot shows the agent approaching the ground far from the desired landing spot, while the latter shows the agent hovering over the desired landing spot.

5.3 Comparison with Other State-of-the-Art Deep Reinforcement Learning Methods

The performance of a DRL—or, in general, a ML—algorithm is strictly dependent on the environment, the agent and the task to which it is applied. As showcased in the paper, we adapted our own simulator and configured our own agent/lander to perform pin-point landing—we did not utilized one of the typical benchmark simulated environments. The choice to implement the Soft Actor-Critic state-of-the-art approach was based on the results presented in [10] in the case of an autonomous controlled terminal-phase landing task through a deep reinforcement learning approach and transfer learning. In this application, we implemented and compared the Deep Deterministic Policy Gradient—DDPG [23]—and the SAC algorithm. Results in terms of reward are showcased in Figs. 10, 11, 12, 13 and 14. In particular, as shown in Fig. 10, the DDPG algorithm seems to quickly learn a good policy for controlled landing on two lunar environments—namely, Moon Near Side—Fig. 10a, and Moon's Far Side—Fig. 10b, c. When attempting to transfer the policy learnt on another environment, i.e. Mars and Titan, the reward shows visible oscillations. This phenomenon is not visible when exploiting the SAC algorithm: in fact, while the reward plots in Fig. 11 on the two lunar environments show a slower increase at the beginning of the training—during the exploration phase—the sharp increase appears later. When transferring the policy learnt on the Moon on Mars and Titan—Fig. 13, the oscillations in the reward plot do not appear: the SAC algorithm proves to be more stable, also in the case of transfer learning. This behavior is due to the formulation of the stochastic actor and the maximization to the entropy. Figure 14 show a further experiment, where atmospheric disturbances—modeled as lateral wind gusts—have been included in the environments: the SAC model proves stable also in this case. The previous experiments motivated us to implement the SAC algorithm to perform autonomous pin-point landing with trajectory recalculation.

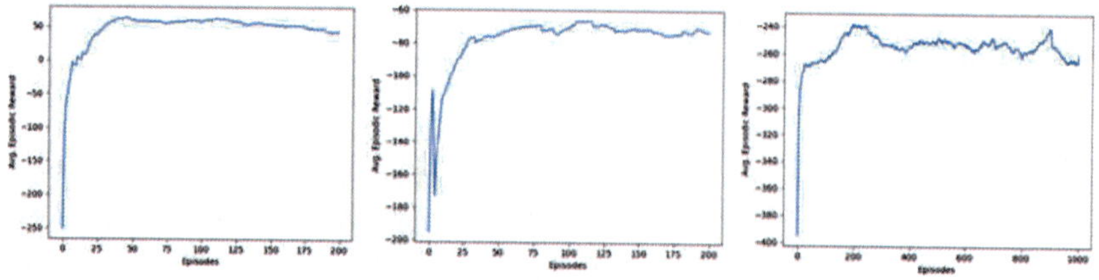

Fig. 10 DDPG: Here the following training results, in terms of average episodic reward, are showcased from left to right: 200 episodes on the Moon's Near Side, 200 episodes on the Moon's Far Side and 1000 episodes on the Moon's Far Side starting from a double altitude

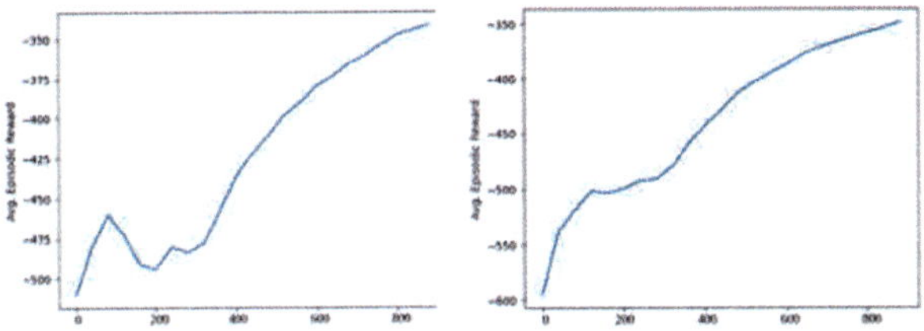

Fig. 11 SAC: Here the following training results, in terms of average episodic reward, are showcased: 1000 episodes on the Moon's Near Side (left), 1000 episodes on the Moon's Far Side (right)

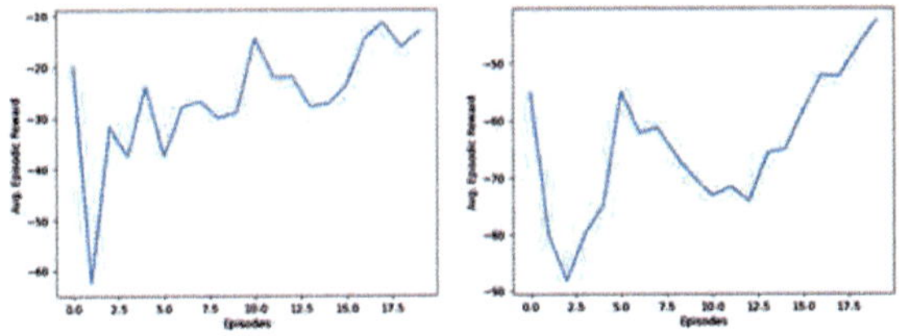

Fig. 12 DDPG: Here the following policy transfer results, in terms of average episodic reward, are showcased: Mars Landing (left) and Titan Landing (right)

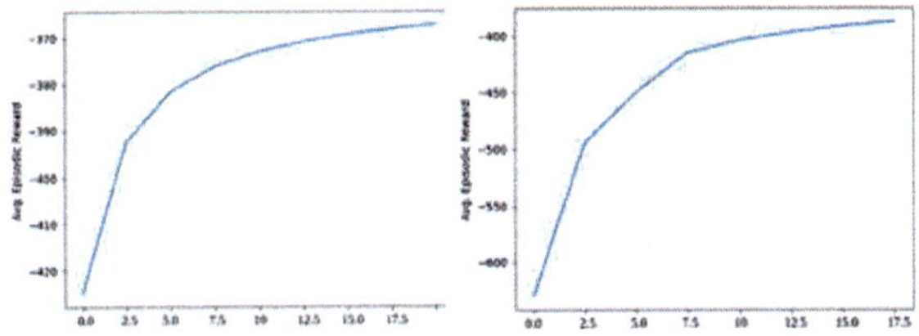

Fig. 13 SAC: Here the following policy transfer results, in terms of average episodic reward, are showcased: Mars Landing (left) and Titan Landing (right)

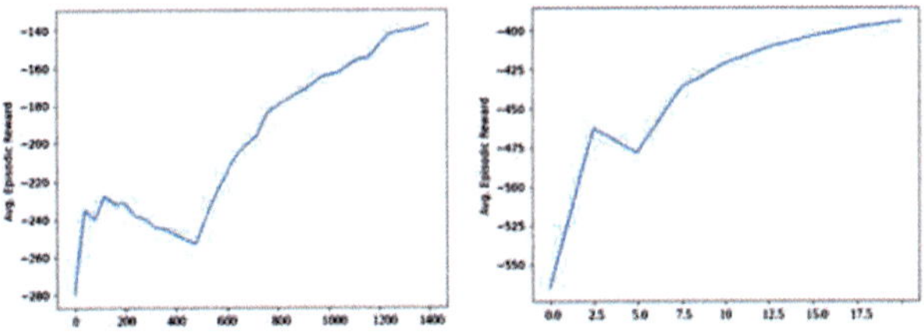

Fig. 14 SAC: Here the following results, in terms of average episodic reward, are showcased: Training session on Martian environment (left). Transfer learning reward scores on Titan environment (right). Both experiments include lateral wind gusts

6 Ongoing Work

We are currently working on the implementation of transfer learning—in particular for what concerns domain transfer. Ideally, a good policy for trajectory recalculation learned on the lunar environment should constitute a good baseline to perform the same task on other, similar domains, like Mars and Titan. Of course, in both these cases, atmosphere and atmospheric disturbances need to be taken into account. In our previous works, we tackled policy transfer in the case of controlled landing during the final Entry-Descent-Landing—EDL—phase. The policy learned by the SAC model proved robust and stable enough to constitute a baseline to perform policy transfer on other domains to shorten the training time and effort in the case of atmospheric disturbances—modeled as lateral (stochastic) wind gusts. Our ongoing work include tackling the pin-point landing task with trajectory recalculation following the same approach, thus also in the case of atmospheric disturbances.

A constraint on fuel consumption can be also added if necessary, by means, of an appropriate reward shaping.

7 Conclusion

In this work, we have presented a new, modified version of our simulator, utilized to model real-physics phenomena by means of the Bullet/PyBullet physical engine and to render realistic terrain 3D models. We adapted 3D models, officially released by NASA. Where such models were not available—like in the case of Titan—we reconstructed a terrain exploiting SAR imagery retrieved during the Cassini-Huygens mission. The lander has been designed by means of the ROS/URDF. We have implemented a Deep Reinforcement Learning, off-policy algorithm—the SAC algorithm—to train our lander/agent to perform a precision—pin-point—landing task. We have also included trajectory recalculation—which can be necessary for obstacle avoidance and/or fault recovery when landing on hazardous terrains. We have tested our approach on a lunar environment with promising results. Given the encouraging results showcased by our approach, we plan to implement a transfer learning approach to execute pin-point landing also on other environments of interest—such as Mars and Titan.

References

1. Johnson, A., Ansar, A., Matthies, L., Trawny, N., Mourikis, A., Roumeliotis, S.: A general approach to terrain relative navigation for planetary landing. In: AIAA Infotech@ Aerospace 2007 Conference and Exhibit
2. Johnson, A.E., Montgomery, J.F.: Overview of terrain relative navigation approaches for precise lunar landing. In: 2008 IEEE Aerospace Conference

3. Ono, M., Rothrock, B., et al.: MAARS: machine learning-based analytics for automated rover systems. In: 2020 IEEE Aerospace Conference
4. Abcouwer, N., Daftry, S., Venkatraman, S., et. al.: Machine Learning Based Path Planning for Improved Rover Navigation (Pre-Print Version)
5. Furfaro, R., Linares, R.: Deep Learning for Autonomous Lunar Landing
6. Scorsoglio, A., Furfaro, R., Linares, R., Gaudet, B.: Image-Based Deep Reinforcement Learning for Autonomous Lunar Landing
7. Gaudet, B., et al.: Deep Reinforcement Learning for Six Degree-of-Freedom Planetary Powered Descent and Landing
8. Furfaro, R., Scorsoglio, A., Linares, R., Massari, M.: Adaptive Generalized ZEM-ZEV Feedback Guidance for Planetary Landing via a Deep Reinforcement Learning Approach
9. Ciabatti, G., Daftry, S., Capobianco, R.: Autonomous planetary landing via deep reinforcement learning and transfer learning. In: Proceedings of the IEEE/CVF Conference on Computer Vision and Pattern Recognition (CVPR) Workshops, pp. 2031–2038
10. Ciabatti, G., Daftry, S., Capobianco, R.: Learning transferable policies for autonomous planetary landing via deep reinforcement learning. In: Proceedings of the Ascend Machine Learning Applications for Autonomous Space Operations Workshop
11. Haarnoja, T., et al.: Soft Actor-Critic: Off-Policy Maximum Entropy Deep Reinforcement Learning with a Stochastic Actor
12. Coumans, E.: Bullet Physics Library
13. Brockman, G., Cheung, V., Pettersson, L.: Openai gym
14. NASA: https://github.com/nasa/NASA-3D-Resources
15. NASA: https://nasa3d.arc.nasa.gov/models
16. Lorenz, R.D., Turtle, E.P., et al.: Dragonfly: a rotorcraft lander concept for scientific exploration at Titan
17. Stofan, E.R., Elachi, C., et al.: The lakes of Titan
18. Paganelli, F., Janssen, M.A., et al.: Titan's surface from Cassini RADAR SAR and high resolution radiometry data of the first five flybys
19. NASA, JPL: https://photojournal.jpl.nasa.gov/target/titan
20. Sutton, R.S., Barto, A.G.: Reinforcement Learning: An Introduction. MIT Press
21. Mnih, V., Kavukcuoglu, K., Silver, D., et al.: Human-level control through deep reinforcement learning. Nature 529–533
22. Mnih, V., et al.: Playing Atari with deep reinforcement learning
23. Lillicrap T.P., et. al.: Continuous Control with Deep Reinforcement Learning

Imbalanced Data Handling for Deep Learning-Based Autonomous Crater Detection Algorithms in Terrain Relative Navigation

Francesco Latorre, Dario Spiller, and Fabio Curti

Abstract This paper investigates the effects of an imbalanced distribution of crater mask labels in datasets used for image segmentation. It can be regarded as an addendum to existing works delving with autonomous Crater Detection tasks neglecting non uniform pixel distribution. In fact, the craters edge pixels represent approximately only 3% of each single mask. The training was performed with the U-Net using the *Focal Tversky Loss*, which is more adaptable to similar real-world data imbalance problems with respect to the *binary cross-entropy* loss. Contextually, the Intersection over Union is computed in the pixel domain for the two cases to assess which approach achieves better segmentation performances. Results of the models trained with Focal Tversky Loss and binary cross entropy are post-processed and compared for performance benchmarking in terms of precision, recall and F_1 in the crater domain. Despite being explicitly targeted at performing Terrain Relative Navigation, these results can prove to be useful both in absolute and relative navigation phases. In fact, these scores give insight about the percentage of image-to-catalogue matched craters and represent the reliability of an artificial intelligence based navigation algorithm.

Keywords Crater detection · Imbalanced data · Terrain relative navigation · Image segmentation · Neural networks · Deep learning · Data mining · Hazard avoidance

1 Introduction

In the last several years, the use of Artificial Intelligence in remote sensing and space exploration missions has become more and more prominent [1–3]. Crater Detection algorithms (CDAs) fall into this batch of applications for space tasks, as they allow real-time autonomous crater cataloguing, hazard avoidance and navigation,

F. Latorre · D. Spiller · F. Curti (✉)
School of Aerospace Engineering, Sapienza University of Rome, via Salaria 851, 00138 Rome, Italy
e-mail: fabio.curti@uniroma1.it

C. Ieracitano et al. (eds.), *The Use of Artificial Intelligence for Space Applications*, Studies in Computational Intelligence 1088,
https://doi.org/10.1007/978-3-031-25755-1_8

eliminating the need for manual catalogue labelling or storage of an absolute map. Robustness is a key feature, as the nature of sensors and acquired data impacts image analysis: the detection results could be different in presence of visual and infrared data, as well as Digital Elevation Models (DEM), the latter being still capable of detecting complex terrain features and less sensitive to actual sunlight variation. For instance, classical machine learning techniques require significant pre-processing [4, 5]. Convolutional neural networks (CNNs) are useful for object detection or segmentation with minimal pre-processing, and therefore faster performance. For image segmentation, Fully-Convolutional Neural Networks (FCNs) take an arbitrary image as input and give the segmented image as output [6]. An instance of FCN is the U-Net, which was initially thought for biomedical image segmentation of biological cells [7] and later adapted to work in other domains such as crater counting, especially when considering deep learning approaches for relative navigation [8]. In the DeepMoon project [9], the U-Net was trained on Lunar DEM images taken from the Lunar Recoinnassance Orbiter (LRO) and Kaguya merged DEM [10]; DeepMars U-Net was trained on both Moon and Mars DEM data [11]; Mars THEMIS Daytime infrared images (NASA Mars Odyssey/THEMIS Team, 2006) were also used during a training phase [12]. Crater Detection with the U-Net for optical Terrain Relative Navigation has been proposed by LunaNet [13], which also proposes spacecraft state improvement using images and altitude measurements. Absolute navigation to get the spacecraft inertial state in a landing maneuver was instead proposed in [14]. All these works simulating on-board implementations with optical (passive) or radar (active) measurements have tackled the crater detection task in a brilliant and successful way, but haven't considered the imbalanced distribution of crater and background pixel values. This is actually one of the main struggles data scientists deal with when delving deep into *data mining* for real-world applications. Data imbalancing can occur in multi-class segmentation but its maximum impact in terms of learning performance degradation is when considering binary segmentation, where the minority class (*hard examples*) is strongly outclassed by the majority class (*easy examples*), a case in which all Crater Detection Algorithms fall into, where the minority class is represented by the crater rim pixels and the majority class instead represents the black background. All these issues must be faced before setting the training phase, and solutions on different levels can be found. *Data sampling* allows oversampling of hard examples or undersampling of easy examples to balance the dataset. The choice of a more efficient *loss function* can also help the training process to weigh the positive class more than the negative one. This consideration has instead become very common and preparatory when facing binary segmentation problems in the medical area [15, 16].

In this work, active measurements, in particular digital elevation data, for image processing, were chosen because of greater robustness to different sensor responses. The data imbalancing aspect will be highlighted, particularly in the post-processing phase. Instead of acting on a data sampling level, the training was set with the aim to minimize a governing loss function for the U-Net training, the *Focal Tversky Loss Function*, while the mean Intersection-Over-Union (IoU) between crater and background pixels has been considered as the network accuracy index. These results will be compared to the ones obtained with a more straightforward, *binary cross-entropy loss* based approach.

2 Related Works

2.1 Autonomous Crater Detection with Deep Learning

While techniques belonging to computer vision and classical machine learning could rely on heavy image pre-processing, deep learning's main focus is on autonomy and very little human effort into image analysis. Convolutional Neural Networks (CNN) do simply what a human eye and brain do: get a glimpse of something and classify it into some category. Standard CNN architectures such as LeNet [17], AlexNet [18], GoogLeNet [19], VGGNet [20] and ResNet [21] are now common knowledge in the AI field and have reached he field of Crater Detection Algorithms as well for applications on the Moon surface [22, 23] or on Mars [24–27]. Fully-Convolutional Neural Networks applications for image segmentation include the already mentioned DeepMoon [9] and DeepMars [11], which adapted the U-Net to feature recognition on celestial terrains. Transfer Learning from a Moon-trained network to an asteroid domain, namely Ceres, was faced in [28].

2.2 Absolute and Relative Navigation

Since autonomy and immediacy are the two main goals of deep learning, Convolutional Neural Network can result to be useful in several space applications, especially in spacecraft navigation, where a high data rate is needed. A navigation algorithm could be referred to as *absolute* or *relative*, depending on the mission phase. For the former, the position of the spacecraft is expressed in an inertial reference frame, while the latter begins when the nominal site of interest is in view. The site is then scanned and the landing trajectory is coarsely corrected [29], whereas fine trajectory corrections are introduced for hazard avoidance. Current Terrain Relative Navigation techniques can be roughly divided into two classes: *passive imaging* and *active ranging*. In the first case, low mass and low power cameras are employed, although a strong dependance on illumination conditions is present. Ranging systems are otherwise more adaptable to every illumination condition but they have limited sensing capabilities with respect to optical sensors. An additional subdivision can be done based on how the sensor measurements are processed on-board. Pattern matching correlates a portion of a sensed image with a stored reference map. Landmark matching similarly works as a correlation algorithm using a reference catalogue, but it acts on a feature level, using easily recognizable terrain features for position estimation. Craters are typically used for landmark matching because they can be effortlessly detected on each scale and illumination appearance, due to the broad dimension distribution in terms of crater diameters. For a more detailed description of TRN approaches in space missions, an exhaustive reference is [30]. Both absolute and relative navigation tasks can tackled with the use of CNNs if optical input data is considered. Therefore, the results obtained in this work can be commented and discussed in a different light, depending on the specific mission phase. Referring to the previous definition of TRN approaches, this work will use a landmark (crater) matching technique for post-processing.

2.3 *Imbalanced Dataset Handling*

Naive approaches in deep learning neglecting the class imbalanced distribution during data pre-processing could lead to very high accuracy but unsatisfying results while performing the chosen task on a test dataset. One important reason is that training a network with standard accuracy as performance index could favorite the learning of the majority class. Moreover, the smaller percentage of the minority class presence with respect to the whole dataset could be read as a noise source by the network, and therefore it may be filtered out during the training process. A survey conducted in [31] tried to review these issues and the corresponding solutions by identifying three main approaches in handling imbalanced data: *data sampling*, in which the positive class is *oversampled* or the negative class is *undersampled* in order to introduce balance into the dataset; the second solution is given by modifying the *algorithmic* approach to training. An algorithm which implements data sampling is SMOTE (Synthetic Minority Oversampling TEchnique [32]); as a a last solution, the training phase can be conducted at a cost function level. For example, *cost-sensitive learning* is guided by a cost matrix that penalizes the misclassified classes.

In the medical area, the dataset of the Multimodal Brain Tumor Image Segmentation (BraTS) Challenge, consisting in a series of 65 MRI images of to-be-segmented brain tumors was used as a test set by [33]. Their approach was based on data sampling and evaluated by comparing the performances of a CNN trained on a random sampled dataset and a balanced dataset, using the Dice score as the governing metric.

The imbalanced dataset issue can additionally be solved by introducing a proper loss function to be minimized during the network training. Surveys of loss functions for semantic segmentation [15, 16] provided a starting place to choose the right loss function. In particular, [15] experiments on that the Tversky loss function and its Focal variant gave the optimal results in terms of Dice coefficient, sensitivity and specificity, if proper parameters are set. For what concerns hazard avoidance in a lunar landing trajectory, data imbalancing was brilliantly addressed in [34], where the task was identifying safe landing trajectories and separate them from hazardous sites. This work used true DEM data from the Lunar Recoinnassance Orbiter (LRO) mission simulating a LIDAR scan during the Hazard Detection and Avoidance (HDA) phase of a lunar landing. The imbalance issue was tackled at a loss function level by minimizing the Jaccard Loss [35] and contextually maximizing the Mean Intersection Over Union (IoU) as accuracy index.

3 Dataset

This section describes the Moon dataset used for Crater Detection with the U-Net. The data is based on a Digital Elevation Model (DEM, built on *elevation* data). Usually, in Crater Detection literature, the terms DEM and DTM are referred to as synonyms, and this work will use them as interchangeable terms as well. For the sake of specificity, there is a formal difference between the two terms: the DEM refers to a smoothed surface model, with all the natural features excluded from it, whereas the DTM includes natural features. It is evident that the simplification is allowable.

The dataset used in this work is made of 256×256 images taken from the Lunar Recoinnassance Orbiter (LRO) and Kaguya merged DEM, spanning a latitude of ± 60 degree and full longitude with a resolution of 512 pixel per degree (or 59 meters per pixel). The DEM was later downsampled to 118 meters per pixel, converted to 16-bit and rescaled to 8-bit PNG. The corresponding binary ground truth masks for each image are created using a combination of Head [36] and Povilaitis [37] catalogues as a source. The dataset, which was used for the work described in [9], can be downloaded at *Zenodo*, contains 30,000 orthographically projected (to prevent crater warping at high latitudes) training images, 5,000 validation images and 5,000 testing images. An example of DEM Moon image with the associated crater mask is shown in Fig. 1. The use of an already existing dataset allows for easy reproducibility of scientific results of other works and for comparison with novel approaches like the one presented here.

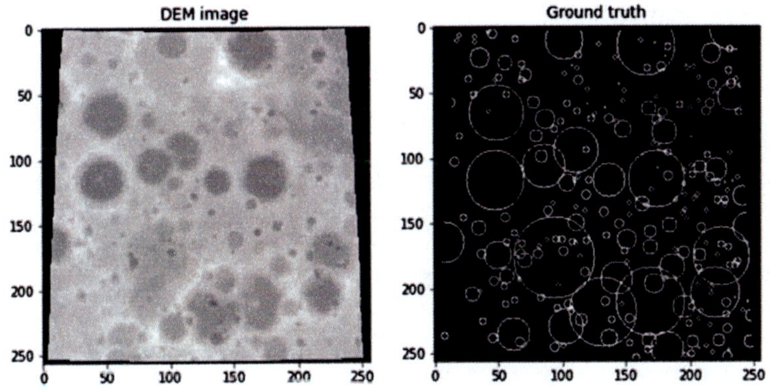

Fig. 1 Sample of Moon DEM image and its corresponding mask

3.1 Imbalanced Dataset

It is evident that a binary image segmentation task introduces the risk that one particular class is more frequent than the other. Crater segmentation makes no exception, leading to an highly imbalanced dataset in the pixel domain for the crater masks. In particular, the black background pixels strongly outclass the white pixels. This aspect is demonstrated through computation of the percentage of pixels belonging to the crater class (crater percentage, CP) and background class (background percentage, BP) with respect to the total number of pixels

$$CP = \frac{w}{t} = 2.89\%, \qquad BP = \frac{b}{t} = 97.11\%, \tag{1}$$

with w denoting the number of white *crater* pixels, b the number of black *background* pixels and t the total number of pixels. These quantities have been counted over 30,000 masks. To verify the high imbalance between the classes, the Imbalance Ratio (IR) has also been computed. It is defined as the ratio of the minority class w with respect to the majority class b:

$$IR = \frac{w}{b} = \frac{CP}{BP} = 0.0297 \tag{2}$$

As a consequence, usual metrics in the pixel domains (such as the pixel accuracy) cannot describe efficiently the result of the training, as well as an accurate choice of the loss function is needed in order to properly consider the classes imbalanced distribution. The next sections will discuss this aspect and will expand the discussion about losses and accuracy metrics outlined in Sect. 2.3.

4 Methodology

4.1 Semantic Segmentation

A particular Fully-Convolutional Neural Network (FCN) [6], the U-Net [7] will be used for the semantic segmentation task. It labels each pixel of the image with a corresponding class, performing what is also called *dense prediction*, with the goal to separate crater rims from the background. In particular, it is an encoder-decoder network. The encoder part is a stack of Relu activated 3×3 convolutional layers (with 112, 224 and 448 filters at each level) and 2×2 pooling layers; the decoder part restores the image dimension by performing 2×2 *upsampling*, followed by additional 3×3 convolutions containing 224, 112 and 112 filters at each level.

In order to better preserve the already learned features during encoding, the downsampled image and the upsampled one at the same depth are merged together; a dropout layer [38] is added before convolutions in the decoding path to avoid overfitting. The final binary mask is produced by means of a 1×1 convolution operation, followed by a sigmoid activation function which produces the final segmented image.

4.2 Segmentation Performance

The segmented images output by the network have to be compared to the ground truth in order to compute the accuracy metrics for an image segmentation problem. No formal way exists to assess segmentation performance, thus its choice depends on the particular problem [39]. Also, a difference between the in-training segmentation performance and the post-processing performance must be pointed out. In accordance to Sect. 3.1, the pixel accuracy is not representative of the task outcome because its definition is generic and equally weighs the contribution of the majority and the minority class, a non negligible aspect in a strongly imbalanced distribution, that could affect the interpretation of results. On the contrary, segmentation metrics, although originally defined at the pixel level, will be applied in the crater domain, i.e., after mapping the results of the U-Net into number and location of the detected craters, where well known metrics can effectively describe the quality of our results.

The group of *precision*, *recall* and F_1 *score* have been used extensively in crater detection:

$$P = \frac{T_P}{T_P + F_P}, \qquad R = \frac{T_P}{T_P + F_N}, \qquad F_1 = \frac{2PR}{P + R} \tag{3}$$

Precision (also called *specificity*) tells how many matches (*True positives*, T_P) are detected with respect to all the craters found by the network. Recall (also called *sensitivity*) tells how many matches are detected with respect to the annotated craters. *False positives* (F_P) are craters found by the U-Net not stored in the ground truth labels, and *false negatives* (F_N) are labelled craters missed by the U-Net. When high precision is reached, recall is slightly penalized and vice versa, and this justifies the introduction of the F1 score (also *Dice score*), as it is the harmonic mean between precision and recall, serving as a balancing index between the two.

Accuracy metrics Although precision, recall and F_1 in the crater domain are the ultimate metrics to be computed for an exhaustive discussion on segmentation results, the in-training results must be monitored and quantified through the accuracy and loss function, which are instead pixel wise. If no data imbalance occurred, the pixel accuracy, defined as below

$$Acc = \frac{T_P + T_N}{T_P + F_N + F_P + T_N} \tag{4}$$

would be a good performance metric during training. T_P, F_P, F_N are true positives, false positives and false negatives, this time defined in the pixel domain. *True neg-*

atives (T_N) are the pixels correctly classified as belonging to the majority class. A high accuracy value is therefore dependent not only on true positives, but also on true negatives. As long as the background pixels are easy and more frequent examples, it is sure that they will be correctly matched to ground truth. This would lead to very high accuracy values, and actually these values are reached in literature works [9, 11, 12]. The main issue with standard accuracy is that in general its final value doesn't represent the actual segmentation performance faithfully, therefore it is not the best performance index to refer to. The *Intersection Over Union (IoU)* metric is instead more representative of imbalanced data behavior during training. It is defined as the ratio between the overlapping pixels (intersection) and the total number of pixels (union) belonging to one particular class:

$$IoU = \frac{T_P}{T_P + F_P + F_N} \tag{5}$$

Since this ratio is specific to a class only, a mean value computed on the crater and background classes is used as a monitoring accuracy score:

$$IoU_m = \frac{1}{2}(IoU_w + IoU_b) \tag{6}$$

The goodness of the segmentation task will be represented better by this metric, because it gives the chance to decide which approach has performed best during training and especially during inference mode, the latter being the phase in which the network sees never before seen data samples.

Both accuracy and mean IoU in training and inference mode will be reported for a model trained with binary cross entropy and focal Tversky loss, which will be described in detail in the next section. It is important to underline that the metrics are defined in the pixel domain, so they are useful only to point out the which training approach is better: the ultimate goal is to compute the precision, recall and F_1 scores in the crater domain.

Loss functions In this paper, a loss function based approach was chosen for imbalanced data handling, as the choice of the loss function also influences the behavior of an imbalanced dataset. This was done because oversampling techniques would create synthetic data hardly representative of a real-world scenario like an HDA phase. Among the segmentation losses, two will be taken into consideration for training and post-processing analysis: the *binary cross-entropy* [40] and the *focal Tversky loss* [41]. The former is defined as

$$BCE = \sum_{i=1}^{N} p_i - p_i g_i + \log(1 + e^{-p_i}) \tag{7}$$

While the latter is built upon the Tversky similarity index TI_c for class c given as

$$TI_c = \frac{\sum_{i=1}^{N} p_{iw} g_{iw} + \epsilon}{\sum_{i=1}^{N} p_{iw} g_{iw} + \alpha \sum_{i=1}^{N} p_{ib} g_{iw} + \beta \sum_{i=1}^{N} p_{iw} g_{ib} + \epsilon}, \tag{8}$$

where p refers to the *predicted* ith pixel value and g refers to the *true* ith pixel value. The subscripts w and b represent the crater and background class, respectively. The values for all these pixel classes vary in the [0, 1] interval. The tunable parameters α and β are introduced to weigh the two classes and are defined such that $\alpha + \beta = 1$. Finally, the Focal Tversky Loss function is defined as

$$FTL_c = \sum_c (1 - TI_c)^{1/\gamma} \tag{9}$$

where γ is the *focal parameter*, which controls the focus on hard examples (the minority class) detected with lower probabilities. It should be noted that all these parameters act as weights, so they can be fixed before training in order to favor or penalize one particular class, depending on the task.

Template matching U-Net predictions must be post-processed before comparison with the ground truth. Global thresholding transforms the raw, grayscale prediction into a binary image. In order to find the pixel coordinates and size of the craters found by the network, the *scikit-image* template matching algorithm is applied. The algorithm works in the following way:

- rings with radius pixel extension from r_{min} to r_{max} are generated and become templates
- the templates are slided through the predictions, and a *correlation map* between the template and the prediction is produced
- if the correlation value exceeds a user defined probability threshold P_m, it is marked as a positive correlation
- the coordinates (x, y) of the crater centroid and its radius r are extracted if a positive correlation is achieved

Crater matching To compute matches between detected craters and ground truth data (true positives), the coordinates extracted with the template matching algorithm are compared to coordinates stored in the catalogue, in the pixel domain. A match between ground truth and prediction is defined if the following two criteria are fulfilled,

$$\frac{(\phi_{ref} - \phi)^2 + (\lambda_{ref} - \lambda)^2}{\min(R_{ref}, R)^2} < D_{x,y}, \tag{10}$$

$$\frac{R_{ref} - R}{\min(R_{ref}, R)} < D_r. \tag{11}$$

The variables $(\phi_{ref}, \lambda_{ref}, R_{ref})$ are the pixel coordinates and size of the ground truth crater which are converted from the geographical longitude and latitude from the catalogue, (ϕ, λ, R) are the pixel coordinates and size of the predicted crater, and $D_{x,y}$, D_r are user-defined threshold parameters that allow to identify a positive match. Crater matching will be applied to predictions obtained with a standard approach and to the ones obtained considering class imbalance, and the two performances will be compared.

5 Results

5.1 *Moon DEM Training*

As pointed out in the rationale of this work, the Moon DEM data has been used for the training of a U-Net using two different approaches:

- a straightforward approach training to minimize the binary cross-entropy, equally weighting the minority and majority class as done in [9, 11, 12]
- an approach considering data imbalancing which can be overcome with the use of focal Tversky loss

The network was trained using TensorFlow *Keras* [40] on a NVIDIA GeForce RTX 2060 GPU. Before the training starts, the network hyperparameters must be tuned. For both approaches, we chose a learning rate of 10^{-4}, a regularization parameter of 10^{-5} and a dropout percentage in the expanding path of 15% as in [9] in order to reduce overfitting. The network was compiled using an ADAM [42] optimizer and trained for 30 epochs with a batch size of 3, a trade off between training time (number of epochs) and intrinsic computational issues of the host machine (batch size). The number of training and validation images are 30,000 and 5,000, respectively.

The best BCE trained model has reached a mean IoU value of 44.02% despite having a 98.60% accuracy in training mode; in inference mode, mean IoU was equal to 44.55% and accuracy was 92.00%. As far as it concerns the FTL training, we chose $\alpha = 0.7$ (and consequently $\beta = 0.3$) and $\gamma = \frac{4}{3}$. The best model reached a 74.03% mean IoU and an accuracy of 92.61% in training mode, while inference mode gave a 66.50% mean IoU and a 89.15% accuracy. These results confirm what we have already pointed out in Sect. 4.2.

5.2 *Crater Post-Processing*

The BCE raw predictions must be globally thresholded for binarization, because their pixel values cover the whole grayscale spectrum, as can be seen in the third image in Fig. 2. Global thresholding constitutes a great disadvantage, as it demands that a wide range of threshold values T must be tested before reaching crater matching optimality in terms of precision, recall and F_1, thus making the predictions poorly

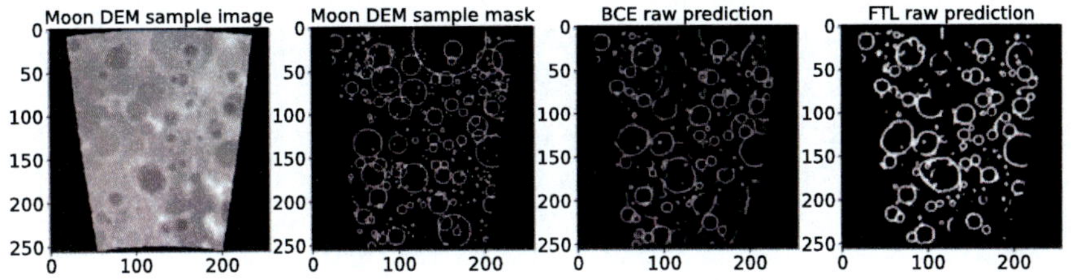

Fig. 2 Raw (non post-processed) predictions after binary cross entropy and focal Tversky loss based U-Net training

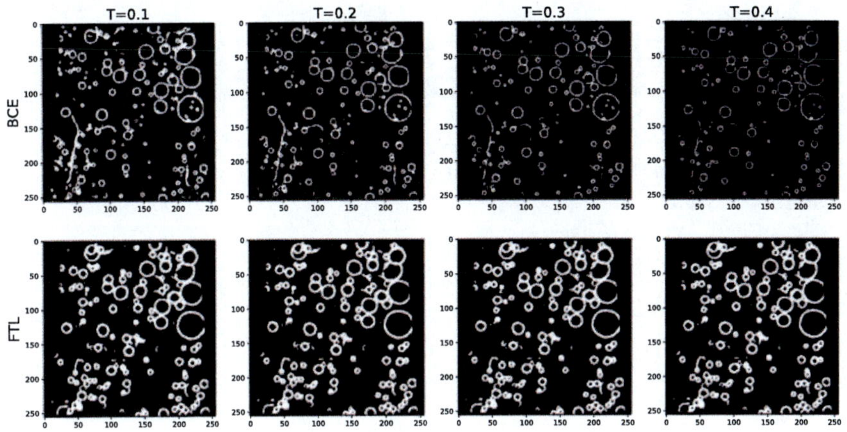

Fig. 3 Globally thresholded predictions after binary cross entropy and focal Tversky loss based U-Net training, with threshold values ranging from 0.1 to 0.4

robust to the user-defined threshold. Instead, the FTL raw prediction, being almost binary, is practically insensitive to any threshold value, producing optimally constant segmentation performances. This aspect is demonstrated in Fig. 3. Post-processing results are presented for different cases considering different threshold values, as shown in Fig. 4. The template matching algorithm spans circle radii from 5 to 120 pixel size; positive matches are the ones whose correlation value exceeds a probability of 0.5. As far as it concerns crater matching, the threshold values are set to 1.8 pixel and 1 pixel for the crater center coordinates and radius thresholds. If these thresholds are not exceeded, the crater found by the network is also a catalogued one. Post-processing was performed on 1,000 test images for threshold values ranging from 0.1 to 0.4 on BCE predictions and FTL predictions. The results are reported in Table 1.

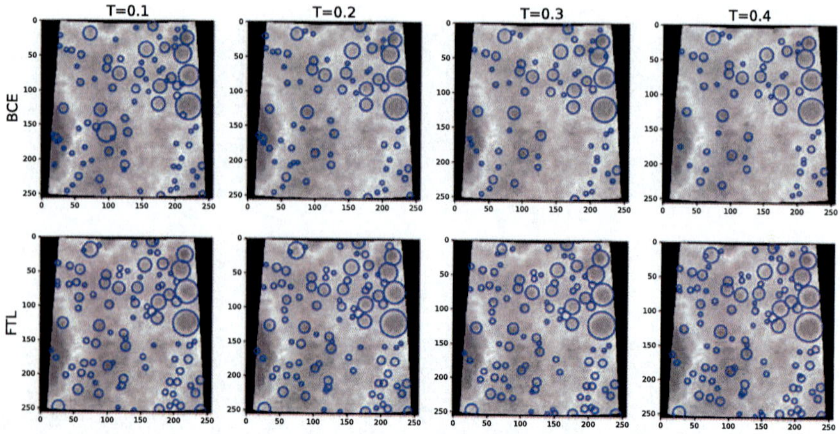

Fig. 4 Post-processed predictions after binary cross entropy and focal Tversky loss based U-Net training, with threshold values ranging from 0.1 to 0.4

Table 1 Crater matching results

(a) Binary cross entropy loss			
T	Precision (%)	Recall (%)	F1 (%)
0.1	65.76	91.20	76.42
0.2	76.55	89.78	82.63
0.3	85.94	85.20	85.56
0.4	91.86	74.60	82.34
(b) Focal Tversky loss			
T	Precision (%)	Recall (%)	F1 (%)
0.1	74.16	91.68	82.00
0.2	74.82	91.52	82.33
0.3	75.20	91.45	82.53
0.4	75.72	91.40	82.82

6 Discussion

6.1 Data sampling Versus Loss Function Choice

The Crater Detection task is strongly affected by imbalance between crater and background pixels in the dataset ground truths. Data imbalance can be addressed with class oversampling/undersampling or with a proper loss function that could mitigate the imbalance effects. The first strategy was discarded, because acting on a sampling level would result in the creation of synthetic data that could be seen by the network as additional noise to filter out, making the oversampling operation useless in the first place. Another idea was acting on ground truth labelling. In this paper, the masks were created by labelling crater rims. The distribution of rim pixels in the 30,000 mask dataset was highly outclassed by black pixels as shown in Sect. 3.1. A different crater labelling strategy representing masked craters as filled circles rather than thin edges, as done in [12] was also initially taken into consideration. Filled circles intuitively increase the percentage of the minority pixel classes. However, this choice would not consider the real crater distribution on the Moon or planetary surface, which contains many overlapped or nested craters (especially smaller examples), as can be seen in Fig. 1. So, even if data imbalancing would surely be limited, the post-processing performance could be damaged, because many nested craters would be missed by the U-Net (Fig. 4).

What we did in our work was handling data imbalance on a loss level, choosing the focal Tversky function as the governing loss during the training phase. The choice was dictated by the adaptability of the function to highly imbalanaced data distribution, as pointed out in [15, 16, 31, 33]. Also, the adaptability of the FTL is demonstrated by Eqs. 7 and 9. In the first case, the binary cross entropy loss is computed as the sum over all pixels, regardless of the class they belong to, whereas in the second case the crater and background pixel values are treated separately, so that the network learning of the minority class learning can be favored after a proper pre-training parameter tuning.

6.2 *BCE-Based Versus FTL-Based Training*

The final objective of a network training is to minimize some user-defined loss function while learning a specific task. What we did was comparing the focal Tversky and binary cross entropy training outcomes. The latter is very popular in U-Net Crater Detection [9, 11–13], but its use is a naive choice which doesn't consider the imbalanced quality of data. The metric referred to the BCE training is the standard accuracy Eq. 4, which is not a useful performance index due to the fact that the number of true negatives is surely very high, as almost the totality of pixel values belongs to the background class (a high value of true negatives).

By switching the training approach to FTL, the performance metric also changes. The Intersection Over Union at the pixel level is very common in the segmentation field, so it was considered as the governing metric during training. To be specific, a mean IoU over the background and crater pixel classes was monitored, reaching 74.03 and 66.50% in training and inference mode, respectively; on the other hand, the mean IoU was much lower for BCE (44.02% training mode versus 44.55% inference mode), despite reaching a very high accuracy with respect to the FTL model.

6.3 Post-Processing Performances

The outcomes of the two training strategies were very different, as already seen in Fig. 2. We also expected the crater matching performances to differ. In fact, the BCE predictions were clearly grayscale, while the FTL outputs were quasi-binary. In both cases, a thresholding value (i.e. putting all values below a certain value equal to zero and values above equal to one) had to be chosen in order to make the raw predictions binary, because the template matching algorithm works with binary images. However, it is not easy to choose the proper thresholding value. This seems to be an issue for BCE predictions: the detected craters become more faint as the threshold value increases, as Fig. 3 is showing. The same figure shows that the FTL predictions are threshold insensitive. This consideration can be confirmed by the values of the precision, recall, F1 metrics group. It should be pointed out that our work doesn't have the goal to prove that a FTL guided training optimizes, in general, crater matching performances. Nonetheless, the results are more stable and robust with respect to BCE predictions and follow a more predictable behavior, as an increasing threshold value corresponds to an increase in the F1 score for the presented test cases, with increasing precision trend and decreasing recall trend. We can't say the same about the BCE predictions, because an increasing threshold value doesn't necessarily mean that the F1 score is higher. Moreover, precision and recall behave in a non predictable way and are case specific. For example, for $T = 0.1$, the post-processed BCE predictions favored recall, while for $T = 0.4$ precision was higher. This means that the FTL predictions are always capable of collecting almost all catalogued craters, with the addition of some false positives that penalize precision: a high recall value is however more important for a navigation algorithm, because almost all real hazard must be predicted as such. The BCE trained network sometimes detects all real craters (in the high recall tests) but sometimes misses them.

7 Conclusion

This work shows how to correctly handle imbalanced data for crater segmentation in Terrain Relative Navigation. It was demonstrated that the focal Tversky loss function adapts more flexibly on learning easy examples (white pixels) with respect to binary cross entropy. This capability also translates into a much better segmentation performance in terms of mean Intersection over Union, but sacrifices accuracy; however, the latter does not provide any useful information on how good the segmentation process was, so any drop in its value is not interpreted as a negative consequence on the overall performance. In general, crater-to-catalogue matching with FTL also reaches greater stability and robustness to threshold decision in the post-processing phase. It is not possible to provide a complete insight on imbalanced data handling in HDA with a single work, but nonetheless we tried to face an issue that to our best knowledge is hardly addressed in Crater Detection, at least at time of writing.

If the loss based approach is carried on, much more can be done on it. For example, more effort can be put in pre-tuning the focal loss parameters α, β and γ in order to provide the best mean Intersection over Union, and the highest precision, recall and F_1, so to optimize performance both in the pixel and crater domain at the same time. This approach can be also paired with (or compared to) a data based method. We mentioned the impossibility of catching overlapping or nested craters using filled circles as crater annotations for a binary pixel-wise segmentation. However, if we extended the problem to a multiclass segmentation task, where each class represent the nesting level of the craters, it would be possible to recover them. In addition to this possibility, the data imbalance would not be an issue anymore, because the class distribution would be uniform. All these suggestions are left for future analysis.

References

1. Girimonte, D., Izzo, D.: Artificial intelligence for space applications. In: Intelligent Computing Everywhere (2007)
2. Esposito, M., et al.: Highly Integration of Hyperspectral, Thermal and Artificial Intelligence for the ESA Phisat-1 Mission (2019)
3. Chien, S., Morris, R.: Space applications of artificial intelligence. AI Mag. **35**, 3–6 (2014)
4. Stepinski, T., Ding, W., Vilalta, R.: Detecting impact craters in planetary images using machine learning. In: Intelligent Data Analysis for Real-Life Applications: Theory and Practice (2012)
5. Di, K., Li, W., Yue, Z., Sun, Y., Liu, Y.: A machine learning approach to crater detection from topographic data. Adv. Space Res. **54** (2014)
6. Long, J., Shelhamer, E., Darrell, T.: Fully convolutional networks for semantic segmentation, pp. 3431–3440 (2015)
7. Ronneberger, O., et al.: U-Net: Convolutional Networks for Biomedical Image Segmentation (2015)
8. Song, J., Rondao, D., Aouf, N.: Deep learning-based spacecraft relative navigation methods: a survey. Acta Astronaut. **191**, 22–40 (2022)
9. Silburt, A., et al.: Lunar crater identification via deep learning. Icarus **317** (2018)
10. Barker, M.K., Mazarico, E., Neumann, G.A., Zuber, M.T., Haruyama, J., Smith, D.E.: A new lunar digital elevation model from the Lunar Orbiter Laser Altimeter and SELENE Terrain Camera. Icarus **273**, 346–355 (2016)
11. Lee, C.: Automated crater detection on Mars using deep learning. Planet. Space Sci. **170**, 16–28 (2019)
12. DeLatte, D., et al.: Segmentation convolutional neural networks for automatic crater detection on Mars. IEEE J. Sel. Top. Appl. Earth Obs. Remote Sens. **PP**, 1–14 (2019)
13. Downes, L., et al.: Lunar Terrain Relative Navigation Using a Convolutional Neural Network for Visual Crater Detection (2020)
14. Silvestrini, S., et al.: Optical navigation for lunar landing based on convolutional neural network crater detector. Aerosp. Sci. Technol. 107503 (2022)
15. Jadon, S.: A survey of loss functions for semantic segmentation. In: 2020 IEEE Conference on Computational Intelligence in Bioinformatics and Computational Biology (CIBCB), pp. 1–7. IEEE (2020)
16. Ma, J.: Segmentation loss odyssey. arXiv preprint arXiv:2005.13449 (2020)
17. Lecun, Y., Bottou, L., Bengio, Y., Haffner, P.: Gradient-based learning applied to document recognition. Proc. IEEE **86**, 2278–2324 (1998)
18. Krizhevsky, A., Sutskever, I., Hinton, G.: ImageNet classification with deep convolutional neural networks. Neural Inf. Process. Syst. **25** (2012)

19. Szegedy, C., Liu, W., Jia, Y., Sermanet, P., Reed, S., Anguelov, D., Erhan, D., Vanhoucke, V., Rabinovich, A.: Going deeper with convolutions, pp. 1–9 (2015)
20. Simonyan, K., Zisserman, A.: Very Deep Convolutional Networks for Large-Scale Image Recognition. arXiv:1409.1556 (2014)
21. He, K., Zhang, X., Ren, S., Sun, J.: Delving deep into rectifiers: surpassing human-level performance on ImageNet classification. In: IEEE International Conference on Computer Vision (ICCV 2015), vol. 1502 (2015)
22. Emami, E., Ahmad, T., Bebis, G., Nefian, A., Fong, T.: On Crater Classification using Deep Convolutional Neural Networks (2018)
23. Emami, E., Ahmad, T., Bebis, G., Nefian, A., Fong, T.: Lunar Crater Detection via Region-Based Convolutional Neural Networks (2018)
24. Cohen, J., Lo, H., Lu, T., Ding, W.: Crater Detection via Convolutional Neural Networks (2016)
25. Palafox, L., Hamilton, C., Scheidt, S., Alvarez, A.: Automated detection of geological landforms on Mars using convolutional neural networks. Comput. Geosci. **101** (2017)
26. Benedix, G.K., Norman, C.J., Bland, P.A., Towner, M.C., Paxman, J., Tan, T.: Automated detection of Martian craters using a convolutional neural network. In: Lunar and Planetary Science Conference, p. 2202. Lunar and Planetary Science Conference (Mar. 2018)
27. Norman, C.J., Paxman, J., Benedix, G.K., Tan, T., Bland, P.A., Towner, M.: Automated detection of craters in Martian satellite imagery using convolutional neural networks. In: Planetary Science Informatics and Data Analytics Conference, vol. 2082, p. 6004 (Apr. 2018)
28. Latorre, F., Spiller, D., Curti, F.: Autonomous crater detection on asteroids using a fully-convolutional neural network. In: Proceedings of XXVI International Congress of the Italian Association of Aeronautics and Astronautics, AIDAA, arXiv preprint arXiv:2204.42419 (2021)
29. D'Ambrosio, A., Carbone, A., Spiller, D., Curti, F.: PSO-based soft lunar landing with hazard avoidance: analysis and experimentation. Aerospace **8**(7) (2021)
30. Johnson, A.E., Montgomery, J.F.: Overview of terrain relative navigation approaches for precise lunar landing. In: 2008 IEEE Aerospace Conference, pp. 1–10. IEEE (2008)
31. López, V., Fernández, A., García, S., Palade, V., Herrera, F.: An insight into classification with imbalanced data: empirical results and current trends on using data intrinsic characteristics. Inf. Sci. **250**, 113–141 (2013)
32. Chawla, N.V., Bowyer, K.W., Hall, L.O., Kegelmeyer, W.P.: Smote: synthetic minority over-sampling technique. J. Artif. Intell. Res. **16**, 321–357 (2002)
33. Small, H., Ventura, J.: Handling unbalanced data in deep image segmentation. University of Colorado (2017)
34. Moghe, R., Zanetti, R.: A deep learning approach to hazard detection for autonomous lunar landing. J. Astronaut. Sci. **67**(4), 1811–1830 (2020)
35. Jaccard, P.: The distribution of the flora in the alpine zone. 1. New Phytol. **11**(2), 37–50 (1912)
36. Head, J., et al.: Global distribution of large lunar craters: implications for resurfacing and impactor populations. Science (New York, N.Y.) **329**, 1504–7 (2010)
37. Povilaitis, R., et al.: Crater density differences: exploring regional resurfacing, secondary crater populations, and crater saturation equilibrium on the Moon. Planet. Space Sci. **162**, 41–51 (2018)
38. Srivastava, N., Hinton, G., Krizhevsky, A., Sutskever, I., Salakhutdinov, R.: Dropout: a simple way to prevent neural networks from overfitting. J. Mach. Learn. Res. **15**, 1929–1958 (2014)
39. Taha, A.A., Hanbury, A., Jimenez-del Toro, O.: A Formal Method For Selecting Evaluation Metrics for Image Segmentation (2014)
40. Chollet, F.: Deep Learning with Python, 1st edn. Manning Publications Co., USA (2017)
41. Abraham, N., Khan, N.M.: A novel focal Tversky loss function with improved attention U-Net for lesion segmentation. In: 2019 IEEE 16th International Symposium on Biomedical Imaging (ISBI 2019), pp. 683–687. IEEE (2019)
42. Kingma, D., Ba, J.: Adam: a method for stochastic optimization. In: International Conference on Learning Representations (2014)

Comparative Analysis of Reinforcement Learning Algorithms for Robust Interplanetary Trajectory Design

Lorenzo Federici, Alessandro Zavoli, and Roberto Furfaro

Abstract This paper focuses on the application of reinforcement learning to the robust design of low-thrust interplanetary trajectories in presence of severe dynamical uncertainties modeled as Gaussian additive process noise. A closed-loop control policy is used to steer the spacecraft to a final target state despite the perturbations. The control policy is approximated by a deep neural network, trained by reinforcement learning to output the optimal control thrust given as input the current spacecraft state. The effectiveness of three different model-free reinforcement learning algorithms is assessed and compared on a three-dimensional low-thrust transfer between Earth and Mars elected as study case.

Keywords Reinforcement learning · Robust trajectory design · Space trajectory optimization · Spacecraft guidance

1 Introduction

Traditional optimal control methods, such as indirect methods based on Pontryagin maximum principle [6] or direct methods based on collocation [11], represent consolidated tools to plan optimal space trajectories in a deterministic, reference scenario. However, in real-world space missions, the spacecraft motion is inevitably affected by several sources of uncertainty, which can cause the actual trajectory to deviate significantly from the deterministic one. Uncertainties may for example arise due to

L. Federici (✉) · A. Zavoli
Sapienza University of Rome, Via Eudossiana 18, 00184 Rome, Italy
e-mail: lorenzo.federici@uniroma1.it

A. Zavoli
e-mail: alessandro.zavoli@uniroma1.it

R. Furfaro
University of Arizona, 1127 E. James E. Rogers Way, Tucson, AZ 85721, USA
e-mail: robertof@arizona.edu

C. Ieracitano et al. (eds.), *The Use of Artificial Intelligence for Space Applications*, Studies in Computational Intelligence 1088,
https://doi.org/10.1007/978-3-031-25755-1_9

unmodeled dynamics, inaccuracies in the orbital determination process, or possible control execution errors.

During mission design, engineers usually check the robustness of the reference trajectory a posteriori, performing a navigation analysis with different uncertainty realizations. The robustness of this nominal design is then improved through an iterative and time-consuming procedure, which may lead to over-conservative margins.

Several stochastic optimization procedures have been proposed by the scientific community to integrate the robustness to uncertainties into the trajectory optimization process. A classical approach is dynamic programming (DP) [2], or its approximate variant differential dynamic programming (DDP) [16], which make use of Bellman equation to find an optimal control policy for every possible state. In practical cases of interest, only the approximate version is viable, due to the curse of dimensionality, but, in turn, DDP struggles with large perturbations due to its local nature. Model predictive control (MPC) has recently emerged as one of the most promising computational guidance methods [3]. MPC consists in repeatedly solving an optimal control problem (OCP) with initial conditions updated at each step on the basis of real-time measurements. In case of general nonlinear dynamics and constraints, the effectiveness of this method decreases as increased computational costs will reduce the guidance update frequency. An alternative solution strategy consists steering the system from an initial probability distribution, or covariance, to a desired value, by optimizing both an open-loop control signal and a state-feedback policy. This approach, referred to as covariance control, has been recently applied to low-thrust interplanetary trajectory design [4]. Nevertheless, using this methodology to tackle more general stochastic problems with bounded uncertainties or non-additive disturbances can be quite challenging.

Artificial intelligence (AI), and in particular deep learning (DL), proved to be an effective and time-efficient approach to the solution of control problems in different research areas, first of all robotics. In space guidance applications, a deep neural network (DNN) can be used as parametric model to map observations of the spacecraft state to corresponding control actions, which define the magnitude and direction of the thrust. Reinforcement learning (RL) represents a way to teach a DNN how to solve an optimal control problem. In RL, training data are directly collected by the network itself (the "agent"), through repeated interactions with the environment, which consists of simulations of the considered mission scenario. The only external feedback is a numerical reward provided at each simulation step, which represents a measure of the current network performance. The control policy is progressively refined by maximizing the average cumulative reward over a single trajectory. Several research papers have already dealt with the use of RL for the robust closed-loop guidance of spacecraft during planetary landing maneuvers [9] and proximity operations [8], as well as for the design of cislunar [12] and interplanetary trajectories [14]. The robust design of an interplanetary mission has been also already addressed via RL by considering the moderate presence of safe mode events [17] or navigation uncertainties [5].

As opposed to traditional solution methods, deep-RL algorithms can be straightforwardly applied to stochastic control problems with arbitrary transition probabilities and observation models, even provided in the form of black-box functions, since an explicit mathematical formulation is not required. Also, the exploratory behavior typical of RL algorithms provides inherent robustness against model uncertainty. Furthermore, the computational effort is almost completely concentrated in the network *training phase*, which is performed pre-flight on high-performing computing hardware. Conversely, the on-board computational time is one forward-pass of the network per guidance step, which is extremely fast.

In this respect, this study aims to investigate and compare several model-free RL algorithms for the robust design of an interplanetary trajectory under dynamical uncertainties, which refer to the presence of unmodeled accelerations and/or badly estimated parameters in the dynamical model.

2 Problem Statement

Let us consider a spacecraft of initial mass m_0, equipped with a low-thrust engine with maximum thrust $T_{\max}$ and effective exhaust velocity c. The spacecraft starts its interplanetary mission leaving a planetary body p_1 at a pre-determined departure epoch $t_0 = 0$, with zero excess of hyperbolic velocity. Its goal is to reach a second planetary body p_2 with null relative velocity at a prescribed arrival epoch t_f with minimum propellant expenditure, and regardless of possible uncertainties on the spacecraft dynamics.

When tackling OCPs with RL, the problem is first reformulated as an equivalent discrete-time problem defined as Markov decision process (MDP). In an MDP, at each time step $h, h = 0, 1, \ldots$, a decision maker (referred to as the agent) chooses an action $\boldsymbol{u}_h$ among the admissible ones, on the basis of the knowledge of the current system (or environment) state $\boldsymbol{x}_h$, according to a closed-loop control policy π: $\boldsymbol{u}_h = \pi(\boldsymbol{x}_h)$. As a consequence of this action, the environment transitions to a new state $\boldsymbol{x}_{h+1}$ and returns a scalar reward $R_h = R(\boldsymbol{x}_h, \boldsymbol{u}_h, \boldsymbol{x}_{h+1})$, which can be intended as a measure of the "goodness" of the last decision maker's choice. The goal in an MDP is to find the control policy $\pi^\star$ that maximizes the expected value of the return received along a trajectory $\tau = \{(\boldsymbol{x}_0, \boldsymbol{u}_0), (\boldsymbol{x}_1, \boldsymbol{u}_1), \ldots\}$

$$J(\pi) = \mathop{\mathrm{E}}_{\tau \sim \pi} [G(\tau)] \tag{1}$$

with

$$G(\tau) = \sum_{h=0,1,\ldots} R_h \tag{2}$$

A possible way to formulate the stochastic trajectory optimization problem as an (almost) equivalent MDP is by approximating the low-thrust trajectory as a series of H ballistic arcs connected by impulsive Δvs, in a similar way as in the well-known Sims-Flanagan model [20].

The spacecraft state $\boldsymbol{x}_h$ at any time t_h, $h = 0, \ldots, H$, before the impulse can be identified by its inertial position $\boldsymbol{r}_h = [x_h \; y_h \; z_h]^T$ and velocity $\boldsymbol{v}_h^- = [v_{x,h}^- \; v_{y,h}^- \; v_{z,h}^-]^T$ with respect to Sun, by its mass m_h^-, and by the current time t_h itself:

$$\boldsymbol{x}_h^- = [\boldsymbol{r}_h^T \; (\boldsymbol{v}_h^-)^T \; m_h^- \; t_h]^T \in \mathbb{R}^7 \times [0, t_f] \tag{3}$$

From now on, the superscripts "−" and "+" will be used to indicate the value of a quantity immediately before and after the impulses, respectively.

The magnitude of the impulse $\Delta\boldsymbol{v}_h$ performed at time t_h is limited by the amount of Δv that could be accumulated over the corresponding trajectory segment (from time t_{h-1} to time t_h) by operating the spacecraft engine at maximum thrust $T_{\max}$. So, its maximum value is

$$\Delta v_{\max,h} = \frac{T_{\max}}{m_h^-} \Delta t \tag{4}$$

where $\Delta t = t_f / H$ denotes the time-length of each trajectory segment, being $t_H = t_f$.

The control $\boldsymbol{u}_h \in [-1, 1]^4$ returned by the control policy π at step h, $h = 0, \ldots, H-1$, defines the impulsive $\Delta\boldsymbol{v}$ maneuver. In particular

$$\Delta\boldsymbol{v}_h = \Gamma(\boldsymbol{u}_h) = \frac{\Delta v_{\max,h}}{\left\| \left[u_h^{(2)} \; u_h^{(3)} \; u_h^{(4)} \right] \right\|} \frac{u_h^{(1)} + 1}{2} \left[u_h^{(2)} \; u_h^{(3)} \; u_h^{(4)} \right]^T \tag{5}$$

This definition of the control has been selected as it inherently meets the constraint on the maximum value of the impulsive velocity variation (Eq. (4)). Furthermore, RL performs better when the actions are sampled from intervals centered about zero.

The terminal $\Delta\boldsymbol{v}$, which would occur at the final time $t_f = t_H$, is not returned by policy π, but it is computed algebraically in order to match the destination planet's velocity, unless this would lead to a violation of the constraint in Eq. (4), that is

$$\Delta\boldsymbol{v}_H = \min\left(\left\| \boldsymbol{v}_{P_2,f} - \boldsymbol{v}_H^- \right\|, \Delta v_{\max,H} \right) \frac{\boldsymbol{v}_{P_2,f} - \boldsymbol{v}_H^-}{\left\| \boldsymbol{v}_{P_2,f} - \boldsymbol{v}_H^- \right\|} \tag{6}$$

Since the spacecraft moves under Keplerian dynamics between any two impulses, in a deterministic scenario the spacecraft state can be propagated analytically with a closed-form transition function ϕ:

$$\boldsymbol{x}_{h+1}^- = \begin{bmatrix} \boldsymbol{r}_{h+1} \\ \boldsymbol{v}_{h+1}^- \\ m_{h+1}^- \\ t_{h+1} \end{bmatrix} = \phi(\boldsymbol{x}_h^-, \Delta\boldsymbol{v}_h) = \begin{bmatrix} \tilde{f}_h \boldsymbol{r}_h + \tilde{g}_h (\boldsymbol{v}_h^- + \Delta\boldsymbol{v}_h) \\ \dot{\tilde{f}}_h \boldsymbol{r}_h + \dot{\tilde{g}}_h (\boldsymbol{v}_h^- + \Delta\boldsymbol{v}_h) \\ m_h^- \exp\left(-\|\Delta\boldsymbol{v}_h\|/c\right) \\ t_h + \Delta t \end{bmatrix} \tag{7}$$

where $\tilde{f}_h$ and $\tilde{g}_h$ are the Lagrange's coefficients at h-th step, defined as in Ref. [1], and the mass update is obtained through the Tsiolkovsky's equation.

However, because of uncertainties on the spacecraft dynamics, the actual spacecraft state at the next time step will be slightly different from the predicted one. In this manuscript, uncertainties are modeled as an additive Gaussian noise on position, velocity, and mass at time t_h, $h = 0, \dots, H-1$, that is

$$\boldsymbol{w}_{\text{st},h} = \left[\delta \boldsymbol{r}_h^{-T} \ (\delta \boldsymbol{v}_h^-)^T \ \delta m_h^- \ \delta t_h\right]^T \sim \mathrm{N}(\boldsymbol{0}_{8,1}, \boldsymbol{\Sigma}) \in \mathbb{R}^8 \tag{8}$$

where $\boldsymbol{\Sigma} = \operatorname{diag}\left(\sigma_r^2 \boldsymbol{I}_{3,3}, \sigma_v^2 \boldsymbol{I}_{3,3}, 0, 0\right)$ is the covariance matrix, with σ_r and σ_v the standard deviations on position and velocity. Thus, in presence of dynamical uncertainties, the stochastic dynamical model of the spacecraft is

$$\boldsymbol{x}_{h+1}^- = \boldsymbol{\phi}(\boldsymbol{x}_h^- + \boldsymbol{w}_{\text{st},h}, \Delta \boldsymbol{v}_h) \tag{9}$$

The goal of the optimization procedure is to maximize the final mass of the spacecraft, while ensuring the compliance, up to a given accuracy ε, with terminal rendezvous constraints on position and velocity. So, the reward R_h obtained at the end of step h has been defined as

$$R_h = \begin{cases} \Delta m_h & h < H \\ \Delta m_h - 50 \max\left(0, e_{x,h+1} - \varepsilon\right) & h = H \end{cases} \tag{10}$$

where Δm_h is the (non-dimensional) consumed propellant mass at time t_h and $e_{x,H}$ is the maximum value between the final position error $e_r = \left\|\boldsymbol{r}_f - \boldsymbol{r}_{P_2,f}\right\| / \left\|\boldsymbol{r}_{P_2,f}\right\|$ and velocity error $e_v = \left\|\boldsymbol{v}_f - \boldsymbol{v}_{P_2,f}\right\| / \left\|\boldsymbol{v}_{P_2,f}\right\|$. The value selected for the tolerance ε on terminal constraint violation is discussed in Sect. 4.1.

So, the MDP formulation is

$$\mathsf{M}: \begin{cases} \max\limits_{\pi} \ \mathop{\mathrm{E}}\limits_{\tau \sim \pi} \left[\sum_{h=0}^{H-1} R_h\right] & \\ \text{s.t. } \ \boldsymbol{x}_{h+1}^- = \boldsymbol{\phi}(\boldsymbol{x}_h^- + \boldsymbol{w}_{\text{st},h}, \boldsymbol{\Gamma}(\boldsymbol{u}_h)), & h = 0, \dots, H-1 \\ \quad \boldsymbol{w}_{\text{st},h} \sim \mathrm{N}(\boldsymbol{0}_{8,1}, \boldsymbol{\Sigma}_{\text{st}}), & h = 0, \dots, H-1 \\ \quad \boldsymbol{u}_h = \pi(\boldsymbol{x}_h^-), & h = 0, \dots, H-1 \\ \quad \boldsymbol{x}_0^- = [\boldsymbol{r}_{P_1,0}^T \ \boldsymbol{v}_{P_1,0}^T \ m_0 \ t_0]^T, & \\ \quad m_H^+ = m_H^- \exp\left(-\|\Delta \boldsymbol{v}_H\|/c\right), & \\ \quad t_H = t_f & \end{cases} \tag{11}$$

3 Reinforcement Learning

Reinforcement Learning is the area of artificial intelligence concerned with how an agent (i.e., a computer program) should behave, that is, map states to actions through a policy π, to maximize the expected trajectory return. Learning occurs through repeated interactions with the environment, modeled as a MDP, and by using the collected state-action-reward tuples to progressively refine the control policy π by gradient-based optimization.

Model-free RL algorithms can be categorized in two different families: Q-learning methods and policy gradient methods. In Q-learning methods, the agent learns an approximation $\hat{Q}$ of the optimal value $Q^\star$ of the action-value function (or Q-function) $Q^\pi(\boldsymbol{x}, \boldsymbol{u})$, defined as:

$$Q^\pi(\boldsymbol{x}, \boldsymbol{u}) = \mathop{\mathrm{E}}_{\tau\sim\pi}\left[G(\tau)|\boldsymbol{x}_0 = \boldsymbol{x},\ \boldsymbol{u}_0 = \boldsymbol{u}\right] \tag{12}$$

This approximation is progressively improved by maximizing an objective function which exploits the Bellmann equation [2]:

$$Q^\star(\boldsymbol{x}, \boldsymbol{u}) = \mathop{\mathrm{E}}_{\boldsymbol{x}'}\left[R(\boldsymbol{x}, \boldsymbol{u}, \boldsymbol{x}') + \max_{\boldsymbol{u}'} Q^\star(\boldsymbol{x}', \boldsymbol{u}')\right] \tag{13}$$

with $\boldsymbol{x}'$ the state resulting from applying the action $\boldsymbol{u}$ in state $\boldsymbol{x}$. At the end of the Q-learning process, the control policy is implicitly found, as it is given by the action that produces the highest $\hat{Q}$ value in the current state.

In policy gradient methods, instead, the agent directly learns an approximation $\hat{\pi}$ of the optimal control policy $\pi^\star$. The approximation is updated by directly maximizing the trajectory return J. A stochastic policy (typically Gaussian) is used during training to enhance the exploration of the solution space.

A number of algorithms lies between these two prototype algorithms, by concurrently approximating both the Q-function and the control policy. When the approximation $\hat{\pi}$ (or, equivalently, $\hat{Q}$) is provided by a deep neural network (DNN) π_θ (or Q_θ) with parameters (i.e., weights and biases) θ, one speaks of deep reinforcement learning (deep-RL), which is the framework adopted in this study.

3.1 Proximal Policy Optimization

Proximal policy optimization (PPO) is a policy gradient method conceived in 2017 by OpenAI [18] as a first-order and unconstrained alternative to trust region methods. The objective in PPO is to take the biggest policy improvement step at each iteration without causing a performance collapse, by limiting the distance between the new and the last policy at each update. Specifically, a surrogate, or clipped, objective function is used in place of the expected trajectory return to force the policy π_θ to stay in a small range $\epsilon \in (0, 1)$ around the value at the previous training iteration. The presence of the clip range ϵ, that is a fundamental hyperparameter of the method, serve to remove incentives for the policy to change dramatically.

PPO is also an actor-critic method, as, in order to compute the gradient of the objective function, it needs both an approximate control policy (the "actor"), and an estimate of the value function V^{π}, provided by a second and independent neural network V_{ϕ} (the "critic"). Being an on-policy algorithm, PPO exploits at each policy update only the data collected by using the most recent version of the policy. Specifically, the actor parameters are updated through n_{opt} iterations of stochastic gradient ascent (SGA), by using an estimate of the gradient of the objective function computed with just the S time steps collected during the last policy rollout. The trajectories are processed through SGA in mini-batches of n_{sb} steps each. The critic parameters, instead, are updated by minimizing an estimate of the mean-squared-error between the value function and the reward-to-go, that is, the cumulative reward obtained from a given step onwards. The training procedure is stopped after K training iterations.

3.2 *Twin Delayed Deep Deterministic Policy Gradient*

Deep determinisitic policy gradient (DDPG) has been first formulated by DeepMind researchers in 2014 [19] as a hybrid method to concurrently learn an action-value function and a deterministic policy.

A good approximation Q_{ϕ} of the optimal Q-function $Q^{\star}$ is learned by minimizing through stochastic gradient descent (SGD) a mean-squared Bellman error (MSBE), which quantifies by how much the approximate action-value function is far from satisfying the Bellman equation. Differently from PPO, the optimization is almost always performed off-policy, that is, the update rule for Q_{ϕ} use data contained in a replay memory, which store a maximum of M transitions collected at any time during the whole training process.

In the MSBE, the network's parameters ϕ appear both in the actual action-value function estimate and in the target value to approximate; this is usually the cause of an unstable behavior during training. To avoid that issue, a second network, named target Q-network, with the same architecture but different parameters ϕ_{targ}, is used and updated once per training iteration using a polyak averaging rule

$$\phi_{\text{targ}} \leftarrow \rho\phi_{\text{targ}} + (1-\rho)\phi \tag{14}$$

with $\rho \in (0, 1)$ (usually close to 1) a hyperparameter called polyak parameter.

An approximate deterministic policy $\boldsymbol{\mu}_{\theta}(\boldsymbol{x})$ is concurrently computed by performing SGA on the expected value of the action-value estimate $Q_{\phi}(\boldsymbol{x}, \boldsymbol{\mu}_{\theta}(\boldsymbol{x}))$. Since a deterministic policy is used in DDPG, in order to enhance exploration, a zero-mean clipped Gaussian noise is added to the deterministic actions at training time. Eventually, a target policy network with parameters θ_{targ} is used to compute the action which approximately maximizes Q_{ϕ} inside the target term in the MSBE, and again updated by polyak averaging the policy parameters.

A variant of the basc algorithm, named twin delayed DDPG (TD3), has been developed to reduce the effect of an overestimation of the action-value function on the policy behavior. It introduces three critical modifications to standard DDPG. The first modification is the clipped double-Q learning: TD3 concurrently learns two

Q-functions, Q_{ϕ_1} and Q_{ϕ_2}, by MSBE minimization, using the smaller of the two Q-values returned by the target networks $\phi_{1,\text{targ}}$, $\phi_{2,\text{targ}}$ to compute the target term in the loss functions.

The second modification is the delayed policy update: the policy network is updated once for every two Q-network updates. The third and last modification is the target policy smoothing: TD3 adds clipped white Gaussian noise also to the target action, in order to make it harder for the policy to exploit errors in the action-value function estimation.

3.3 Soft Actor-Critic

Soft Actor-Critic was created at the University of California Berkeley in 2017 [10]. SAC is an entropy-regularized RL method, that is, the policy is trained to maximize a weighted sum between the expected trajectory return and the entropy, a measure of how random the policy is. The entropy is added to ensure sufficient exploration and prevent premature convergence to a sub-optimal policy. More precisely, the entropy Z of the stochastic policy $\pi(\cdot|\boldsymbol{x})$ is defined as

$$Z(\pi(\cdot|\boldsymbol{x})) = \underset{\boldsymbol{u}\sim\pi}{\mathrm{E}}\left[-\log \pi(\boldsymbol{u}|\boldsymbol{x})\right] \tag{15}$$

In entropy-regularized RL, the objective function is modified as follows

$$J(\pi) = \underset{\tau\sim\pi}{\mathrm{E}}\left[\sum_{h=0}^{H-1}\left(R_h + c_s Z\left(\pi(\cdot|\boldsymbol{x}_h)\right)\right)\right] \tag{16}$$

with c_s the entropy coefficient, one of the main hyperparameters of the method.

Similarly to TD3, SAC learns two action-value functions Q_{ϕ_1}, Q_{ϕ_2}, by minimizing entropy-regularized MSBE losses. Again, two separate target Q-networks are used to compute the target term in the MSBE function and updated once per training iteration using a polyak averaging rule. A stochastic policy π_θ is concurrently learned by maximizing an entropy-corrected value of the action-value function.

4 Numerical Results

A three-dimensional, time-fixed Earth-Mars rendezvous mission, already analyzed in the literature [13, 16], is considered as case study. All problem data are reported in Table 1. In all simulations, the physical quantities have been made non-dimensional by using as reference values the Earth-Sun mean distance $\bar{r} = 149.6 \times 10^6$ km, the corresponding circular velocity $\bar{v} = \sqrt{\mu_\odot/\bar{r}}$, and the initial spacecraft mass $\bar{m} = m_0$.

Table 1 Problem data

Variable	Value
t_f, days	348.79
m_0, kg	1000
$T_{\max}$, N	0.50
c, km/s	19.6133
$\mu_\odot$, km^3/s^2	132712440018
$\boldsymbol{r}_{P1,0}$, km	$[-140699693,\ -51614428,\ 980]^T$
$\boldsymbol{v}_{P1,0}$, km/s	$[9.7746,\ -28.0783,\ 4.3377 \times 10^{-4}]^T$
$\boldsymbol{r}_{P2,f}$, km	$[-172682023,\ 176959469,\ 7948912]^T$
$\boldsymbol{v}_{P2,f}$, km/s	$[-16.4274,\ -14.8605,\ 9.2149 \times 10^{-2}]^T$
σ_r, km	1.0
σ_v, km/s	0.05

The total number of time-steps per episode H has been set equal to 40, as in the original paper by Lantoine and Russell [13].

4.1 Implementation Details

The numerical results presented in this section have been obtained by using pyRL-prob,[1] an in-house library, based on Ray RLlib [15], for easy training, evaluation, and post-processing of RL agents in OpenAI-Gym-based environments.

In order to improve RL effectiveness in solving the constrained problem in Eq. (11), a constraint relaxation technique already used in metaheuristic optimization, named ε-constraint [7], has been adopted in this paper. When using the ε-constraint approach, the terminal constraints are gradually enforced during optimization by decreasing the value of the constraint satisfaction tolerance ε, in order to explore the solution space to a greater extent. A piecewise constant trend with just two values has been used in this application:

$$\varepsilon_{(k)} = \begin{cases} 1 \times 10^{-2} & k \leq K/2 \\ 1 \times 10^{-3} & k > K/2 \end{cases} \tag{17}$$

where k is the current training iteration, and $K = 39000$ the total number of training iterations.

A 3-layer fully-connected architecture has been used for all the networks (actor, critic and Q-networks) in the considered algorithms. Specifically, the three hidden layers have size $h_1 = 10n_i$, $h_2 = \sqrt{h_1 h_3}$, and $h_3 = 10n_o$, respectively, where n_i is

[1] https://github.com/LorenzoFederici/pyrlprob.

Table 2 Hyperparameters

Algorithm	Hyperparameter	Symbol	Value
PPO	Learning rate	α	1.0×10^{-4}
	Clip range	ϵ	0.1
	SGA epochs	n_{opt}	40
	Steps per rollout	S	2880
	Steps per mini-batch	n_{sb}	400
TD3	Policy learning rate	α	5.0×10^{-4}
	Q-function learning rate	β	5.0×10^{-4}
	Polyak parameter	ρ	0.99
	Replay memory size	M	128000
SAC	Policy learning rate	α	2.5×10^{-4}
	Q-function learning rate	β	2.5×10^{-4}
	Entropy coefficient	c_s	0.1
	Polyak parameter	ρ	0.99
	Replay memory size	M	128000

the number of input neurons, and n_o is the number of output neurons of the considered network. Indeed, such architecture has already proven capable of reconstructing the optimal solution of several space trajectory optimization problems with satisfying accuracy [8, 9].

Tuned values for the main hyperparameters of the three algorithms are reported in Table 2. The training with each algorithm lasted in total roughly 110 million steps.

4.2 *Learning Curves*

Figure 1 shows the evolution along training of the mean value (solid line) and interquartile range (shaded region) of the trajectory return (Fig. 1a) of the terminal error on the state (Fig. 1b) and of the final spacecraft mass (Fig. 1c) obtained with the considered RL algorithms.

It is apparent that PPO outperforms the other methods both in terms of final spacecraft mass and terminal constraint violation. Also, PPO improves on average almost monotonically in all test metrics during the training, featuring just a few oscillations related to the stochastic nature of the algorithm. In particular, the average error on terminal position converges to the desired value ($e_{x,H} \leq 10^{-3}$), while the average final spacecraft mass (586.49 kg) is fairly close to the optimal deterministic solution of the problem (603.91 kg). As a further remark, the learning curve features a sudden (small) drop in performance exactly at half the total training time. This is due to the instantaneous reduction in the value of the constraint satisfaction tolerance ε in Eq. (17).

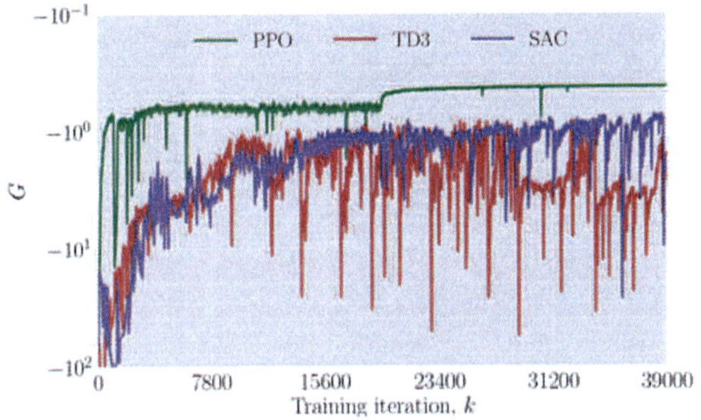

(a) Trajectory return.

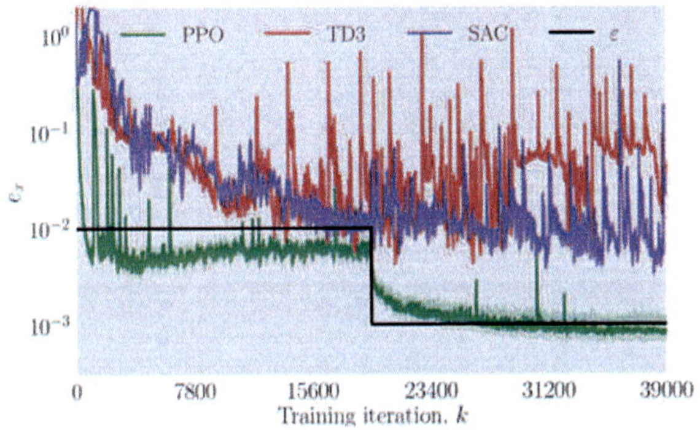

(b) Final error on the state.

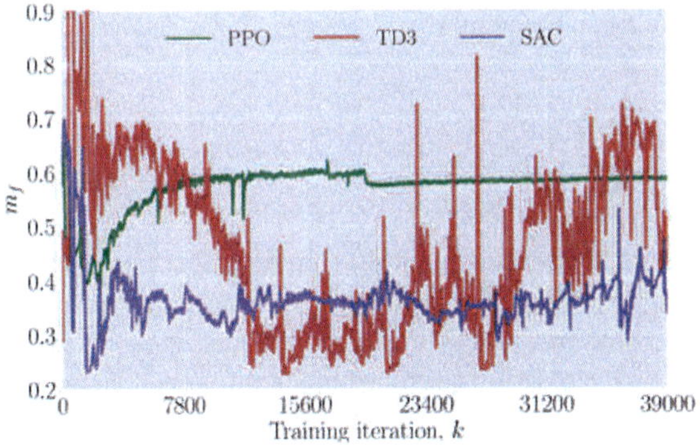

(c) Final spacecraft mass.

Fig. 1 Performance trend of the three algorithms during training

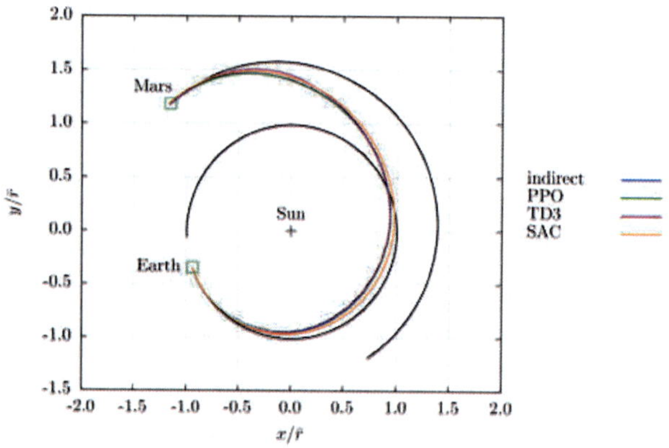

Fig. 2 Robust trajectories obtained with the three RL algorithms

Conversely, the learning curves obtained with both TD3 and SAC feature several oscillations and tend to prematurely converge to a sub-optimal solution ($e_{x,H} \approx 10^{-2}$), quite far from the optimum. This behavior is typical of Q-learning-based methods, which, although tend to be more sample efficient, can fail more easily than policy gradient techniques, as the Q values that they estimate can be very inaccurate [21]. As a consequence, the learning process tends to be less stable.

Specifically, TD3 tends to fail as a result of a dramatic overestimation of the action-value function by the Q-network. When this does happen, the policy optimization process exploits the errors and artificial peaks in the Q-network approximator, thus generating incorrect behavior. Conversely, the large oscillations in SAC are probably a direct effect of the entropy term, which, under some circumstances, may enhance the exploratory behavior of the policy too much, compromising the chance of convergence to a good-quality solution.

4.3 Robust Trajectories

At the end of the training procedure, the reference robust trajectories are obtained by running in an unperturbed environment a deterministic version of the policies trained with each of the three algorithms presented in Sects. 3.1–3.3.

Figure 2 shows the robust trajectory obtained for each trained policy. The deterministic optimal trajectory (in blue), computed with an indirect optimization method, is also reported for the sake of comparison. One can notice that the robust trajectories tend to approach Mars orbit slightly in advance with respect to the optimal deterministic solution, in order to improve their capability of meeting the terminal constraints even in presence of uncertainties in the latter part of the mission.

Table 3 Robust trajectories overview

Algorithm	m_f, kg	e_r, 10^{-3}	e_v, 10^{-3}	G
PPO	591.43	0.92	0.00	−0.41
TD3	308.44	8.96	0.00	−1.09
SAC	385.62	11.98	0.00	−1.16

Table 3 summarizes the main features of these trajectories, that are, the final spacecraft mass m_f, the final position error e_r and velocity error e_v, and the trajectory return $G(\tau)$. The solution corresponding to the PPO policy satisfies the terminal constraints within the prescribed tolerance (10^{-3}). Robustness is obtained by sacrificing just 2% of the final spacecraft mass with respect to the open-loop solution. Conversely, neither TD3 nor SAC manage to reach the desired tolerance, thus resulting in a significantly lower trajectory return.

In all presented cases, the error on the final velocity is zero. This result should not surprise the reader. Indeed, the last Δv is computed algebraically as a difference between the final spacecraft velocity and Mars velocity (see Eq. (6)). Thus, the terminal velocity constraint is automatically satisfied once the computed Δv has a magnitude lower than the maximum admissible for the last trajectory segment, according to the Sims-Flanagan model here adopted.

4.4 Closed-Loop Performance Analysis

Besides returning a reference robust trajectory, the trained network can be also used to provide the spacecraft with a computationally-inexpensive, yet robust, closed-loop guidance. The closed-loop performance of the policies in the uncertain mission scenario is investigated by a Monte Carlo campaign, which consisted in running each policy over a set of 1000 randomly-generated environment realizations, or test episodes.

The spacecraft trajectories resulting from the Monte Carlo campaigns are shown in Fig. 3. Specifically, in each figure, the dark-blue line represents the robust reference trajectory, light-blue arrows indicates the nominal Δvs, and the gray lines represent the randomly-generated test episodes. The differences between each Monte Carlo sample trajectory and the corresponding reference trajectory are up-scaled by a factor 4 for illustration purposes. One can notice that, especially for the policy trained by PPO (Fig. 3a), the Monte Carlo generated trajectories have a greater dispersion in the central part of the mission that progressively reduces, and almost entirely disappears, while approaching Mars, as a direct consequence of the terminal constraints enforcement. Conversely, with the policies trained by TD3 (Fig. 3b) and SAC (Fig. 3c), a substantial part of the trajectories (16% with TD3, 88% with SAC) miss the target

Fig. 3 Monte Carlo trajectories in the uncertain mission scenario

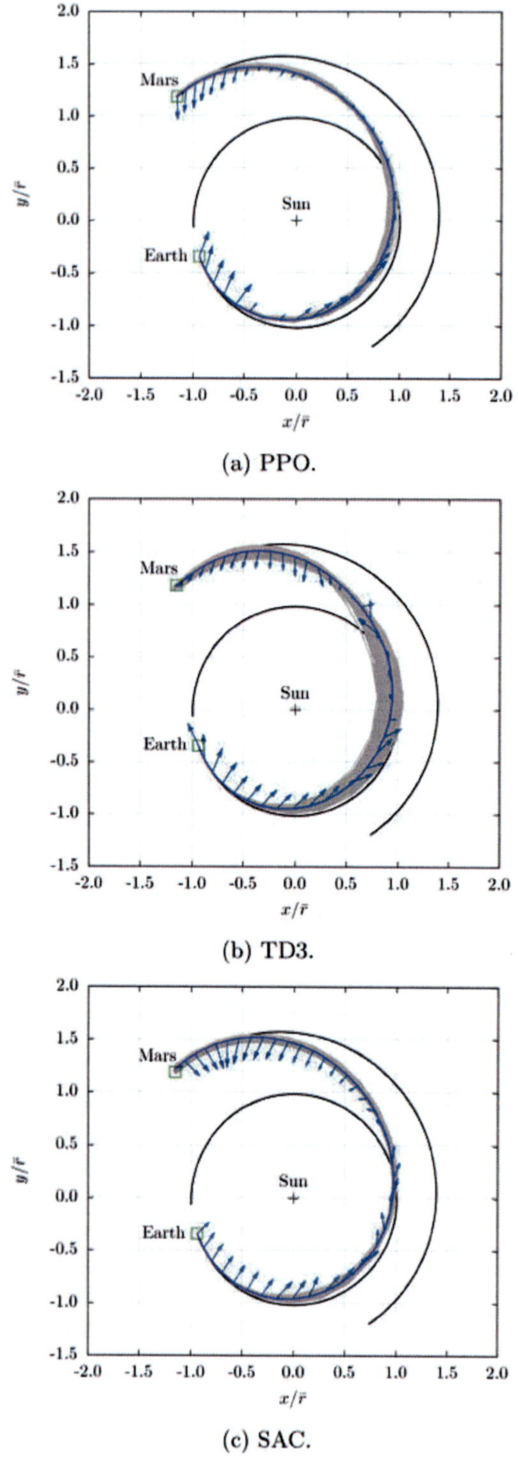

Table 4 Results of the Monte Carlo simulations

Algorithm	m_f, kg			e_r, 10^{-3}			e_v, 10^{-3}			$\varepsilon_{\mathcal{B}}$, 10^{-3}		
	min	mean	max	min	mean	max	min	mean	max	1σ	2σ	3σ
PPO	550.89	586.49	617.91	0.058	1.07	4.37	0.00	0.044	10.89	1.28	2.22	4.26
TD3	298.57	310.11	327.11	3.05	8.22	16.72	0.00	0.65	16.13	8.96	11.75	14.20
SAC	375.84	384.44	392.71	7.41	11.86	16.61	0.00	0.00	0.00	12.89	14.29	15.88

with an error on final position greater than 1% (i.e., more than 10 times the desired tolerance).

The results of the Monte Carlo analysis are summarized in Table 4, which reports, for each environment, the minimum, mean and maximum value of the final spacecraft mass m_f, terminal position error e_r and velocity error e_v, and the tolerance $\varepsilon_{\mathcal{B}}$ that would be required to obtain a success rate equal to 68.3% (1σ), 95.5% (2σ), and 99.7% (3σ), respectively.

According to the obtained results, PPO seems able to cope with the proposed stochastic scenario quite effectively. Indeed, despite the severity of the considered uncertainties, the 95.5% of the trajectories meet the final rendezvous constraint with a precision lower than 2.2×10^{-3}. These results were obtained with a uniform control grid, with maneuvering points spaced by roughly 9 days from each other. Further improvements in constraint handling are expected by introducing additional control points immediately before Mars arrival as done in more traditional guidance schemes. The final spacecraft mass is also, on average, just 3% lower than the optimal one in the nominal scenario (i.e., without uncertainties), thus confirming also the optimality of the solutions. In this respect, an almost bang-off-bang pattern can be clearly recognized in the control law obtained along the reference robust trajectory, as expected in the solution of an optimal control problem with an affine dynamics in the controls.

Conversely, the results obtained with TD3 and SAC do not show the same level of robustness as PPO. In these cases, the average error on terminal position is roughly 10 times higher on average for SAC, while the final spacecraft mass is 36% lower than the nominal value for SAC and 49% lower for TD3. The higher propellant consumption is confirmed by the continuous-thrusting robust trajectories found by the two algorithms (Fig. 3b, c).

5 Conclusion

This paper presented a deep reinforcement learning (RL) framework to deal with the robust design of low-thrust interplanetary trajectories in presence of dynamical uncertainties. Three state-of-the-art model-free RL algorithm, namely proximal policy optimization (PPO), twin-delayed deep deterministic policy gradient (TD3) and soft actor-critic (SAC), have been adopted for the design of a three-dimensional time-fixed Earth-Mars mission. These results suggest that PPO outperforms the other methods in the considered stochastic environment. The PPO solution is effective in terms of both disturbance rejection and propellant consumption. Terminal constraint are correctly enforced, leveraging on the ε-constraint relaxation strategy. Also, PPO shows a stable and monotonic learning curve, and depends on a smaller number of hyperparameters, which have a known effect on its behavior and are rather easy to tune. Conversely, both TD3 and SAC exhibit a strongly oscillating convergence behavior, which leads to a sub-optimal solution that satisfies the terminal constraints with unsatisfactory accuracy. Moreover, their performance is strongly dependent on the value of several hyperparameters (such as the replay memory size, polyak parameter, entropy coefficient, and so on) hard to tune, as their effects on the learning procedure is not completely understood. All these features make PPO a solid candidate for application in complex space guidance problems under uncertainty.

Finally, it is worthwhile remarking that the RL-based optimization methodology proposed in this paper is quite general and can be implemented, with the appropriate changes, to cope with a variety of spacecraft missions and state/control constraints. As an example, time-optimal and/or multi-revolution orbital transfers could be investigated, too. Extensions to arbitrary stochastic dynamical models (e.g., with possibly complex non-Gaussian perturbations) is also straightforward. This is a major advantage with respect to other stochastic optimization techniques presented in the literature, which are based on ad-hoc extensions of traditional optimal control methods. Future works will be addressed to applying a more advanced reinforcement learning methodology, making use of recurrent neural networks (meta-reinforcement learning), to concurrently train the agent on multiple environments with different sources of uncertainty, such as dynamical uncertainties, navigation uncertainties, and control actuation errors, each described by a different probability distribution.

References

1. Bate, R.R., Mueller, D.D., White, J.E.: Fundamentals of Astrodynamics. Dover, NY (1971)
2. Bellman, R.: Dynamic programming. Science **153**(3731), 34–37 (1966). https://doi.org/10.1126/science.153.3731.34
3. Benedikter, B., Zavoli, A., Colasurdo, G., Pizzurro, S., Cavallini, E.: Autonomous upper stage guidance using convex optimization and model predictive control. In: AIAA ASCEND (2020). https://doi.org/10.2514/6.2020-4268
4. Benedikter, B., Zavoli, A., Wang, Z., Pizzurro, S., Cavallini, E.: Covariance control for stochastic low-thrust trajectory optimization. In: AIAA SCITECH 2022 Forum (2022). https://doi.org/10.2514/6.2022-2474
5. Boone, S., Bonasera, S., McMahon, J.W., Bosanac, N., Ahmed, N.R.: Incorporating observation uncertainty into reinforcement learning-based spacecraft guidance schemes. In: AIAA SCITECH 2022 Forum. https://doi.org/10.2514/6.2022-1765
6. Bryson, A.E.: Applied Optimal Control: Optimization, Estimation and Control. Hemisphere Publishing Co., Washington, D.C. (1975)
7. Federici, L., Benedikter, B., Zavoli, A.: EOS: a parallel, self-adaptive, multi-population evolutionary algorithm for constrained global optimization. In: 2020 IEEE Congress on Evolutionary Computation (CEC), pp. 1–10 (2020). https://doi.org/10.1109/CEC48606.2020.9185800

8. Federici, L., Benedikter, B., Zavoli, A.: Deep learning techniques for autonomous spacecraft guidance during proximity operations. J. Spacecr. Rockets **58**(6), 1774–1785 (2021). https://doi.org/10.2514/1.A35076
9. Gaudet, B., Linares, R., Furfaro, R.: Deep reinforcement learning for six degree-of-freedom planetary landing. Adv. Space Res. **65**(7), 1723–1741 (2020). https://doi.org/10.1016/j.asr.2019.12.030
10. Haarnoja, T., Zhou, A., Abbeel, P., Levine, S.: Soft actor-critic: off-policy maximum entropy deep reinforcement learning with a stochastic actor. In: International Conference on Machine Learning, pp. 1861–1870. PMLR (2018)
11. Hargraves, C., Paris, S.: Direct trajectory optimization using nonlinear programming and collocation. J. Guidance Control Dyn. **10**(4), 338–342 (1987). https://doi.org/10.2514/3.20223
12. LaFarge, N.B., Miller, D., Howell, K.C., Linares, R.: Autonomous closed-loop guidance using reinforcement learning in a low-thrust, multi-body dynamical environment. Acta Astronaut. **186**, 1–23 (2021). https://doi.org/10.1016/j.actaastro.2021.05.014
13. Lantoine, G., Russell, R.P.: A hybrid differential dynamic programming algorithm for constrained optimal control problems. Part 2: application. J. Optim. Theory Appl. **154**(2), 418–442 (2012). https://doi.org/10.1007/s10957-012-0038-1
14. Miller, D., Englander, J.A., Linares, R.: Interplanetary low-thrust design using proximal policy optimization. Adv. Astronaut. Sci. **171**, 1575–1592 (2020)
15. Moritz, P., Nishihara, R., Wang, S., Tumanov, A., Liaw, R., Liang, E., Elibol, M., Yang, Z., Paul, W., Jordan, M.I., et al.: Ray: a distributed framework for emerging ai applications. In: 13th USENIX Symposium on Operating Systems Design and Implementation (OSDI 18), pp. 561–577 (2018)
16. Ozaki, N., Campagnola, S., Funase, R.: Tube stochastic optimal control for nonlinear constrained trajectory optimization problems. J. Guidance Control Dyn. **43**(4), 645–655 (2020). https://doi.org/10.2514/1.G004363
17. Rubinsztejn, A., Bryan, K., Sood, R., Laipert, F.: Using reinforcement learning to design missed thrust resilient trajectories. In: AAS/AIAA Astrodynamics Specialist Conference. No. AAS 20-453, Virtual Lake Tahoe (Aug. 2020)
18. Schulman, J., Wolski, F., Dhariwal, P., Radford, A., Klimov, O.: Proximal policy optimization algorithms. arXiv preprint arXiv:1707.06347 (2017)
19. Silver, D., Lever, G., Heess, N., Degris, T., Wierstra, D., Riedmiller, M.: Deterministic policy gradient algorithms. In: Proceedings of the 31st International Conference on Machine Learning, vol. 32, pp. 387–395 (2014)
20. Sims, J.A., Flanagan, S.N.: Preliminary design of low-thrust interplanetary missions. Adv. Astronaut. Sci. **103**(1), 583–592 (2000)
21. Tsitsiklis, J., Van Roy, B.: An analysis of temporal-difference learning with function approximation. IEEE Trans. Autom. Control **42**(5), 674–690 (1997). https://doi.org/10.1109/9.580874

Fault Detection Exploiting Artificial Intelligence in Satellite Systems

Nicola Ferrante, Gianluca Giuffrida, Pietro Nannipieri, Alessio Bechini, and Luca Fanucci

Abstract Mission control and fault management are fundamental in safety-critical scenarios such as space applications. To this extent, fault detection techniques are crucial to meet the desired safety and integrity level. This work proposes a fault detection system exploiting an autoregressive model, which is based on a Deep Neural Network (DNN). We trained the aforementioned model on a dataset composed of telemetries acquired from Mars Advanced Radar for Subsurface and Ionosphere Sounding (MARSIS). The training process has been designed as a sequence-to-sequence task, varying the length of input and output time series. Several DNN architectures were proposed, using both Long Short-Term Memory (LSTM) and Gated Recurrent Unit (GRU) as basic building blocks. Lastly, we performed fault injection modeling faults of different nature. The results obtained show that the proposed solution detects up to 90% of injected faults. We found that GRU-based models outperform LSTM-based ones in this task. Furthermore, we demonstrated that we can predict signal evolution without any knowledge of the underlying physics of the system, substituting a DNN to the traditional differential equations, reducing expertise and time-to-market concerning existing solutions.

Keywords Satellites · Artificial Intelligence · Deep learning · Fault detection

1 Introduction

Spacecrafts and satellites are examples of systems in which it is fundamental to achieve stringent robustness and reliability requirements.

Recently, the complexity of these systems has been increasing and, consequently, also the number of components needed, and the line of codes to be written for the

N. Ferrante (✉) · P. Nannipieri · A. Bechini · L. Fanucci
Department of Information Engineering, University of Pisa, 56122 Pisa, Italy
e-mail: nicola.ferrante@phd.unipi.it

G. Giuffrida
IngeniArs S.r.l., 56121 Pisa, Italy

C. Ieracitano et al. (eds.), *The Use of Artificial Intelligence for Space Applications*, Studies in Computational Intelligence 1088,
https://doi.org/10.1007/978-3-031-25755-1_10

controlling software. Firstly, this results in a growing area, power consumption, memory occupation, and computational resources needed. These factors have to be properly balanced since radiation-hardened devices are by themselves more resource-hungry than commercial off-the-shelf (COTS) devices [1–6].

Moreover, adding complexity means having an increasing number of candidate failure points and, consequently, a higher probability to experience a fault.

Furthermore, when dealing with systems relying on sensors measurements, error conditions and fault events can be estimated, but it is not possible to have a complete overview, since the input space can have a cardinality which makes unfeasible a full exploration of all the possible states of the system [7]. In addition, the occurrence of a fault on a sensor or actuator may lead to effects that can be left undetected by traditional techniques, this problem is also known in automotive as safety of the intended functionality (SOTIF).

To overcome the limitations of traditional approaches in improving the effectiveness of detection and the autonomy level of the system, Machine Learning (ML) and Artificial Intelligence (AI) techniques have been proved to be valid approaches [4–9]. In particular in tasks like prediction and anomaly detection [5, 6, 10, 11].

Fault detection techniques can be divided into two categories: signal-based and model-based [12]. The former exploits measurements from sensors, trying to identify deviations from a nominal pattern, the latter instead exploits a mathematical model of the system to predict its evolution.

Nowadays, the most widespread techniques for fault detection in the space domain are model-based techniques [13–15], exploiting a mathematical model which is frequently based on dynamic equations. These equations may be computationally intensive, and their writing requires expertise and deep knowledge of the system. One of the advantages of using Artificial Intelligence (AI) algorithms is the possibility to reduce the expertise needed, exploiting the ability of these algorithms to learn from data [6, 10].

The objective of this work is to provide a solution for fault detection by exploiting AI, in particular Deep Neural Networks (DNN), to predict the evolution of the system, using signals acquired from sensors. The prediction can be then compared to a real measurement, to understand if a faulty event has occurred.

The proposed solution should be able to work with measurements of different nature; for instance, voltage, current, and temperature. This is because there may be faults that are easily detectable by exploiting only a given sensor. An example is overheating, which can be detected using temperature sensors, but it is hardly observable from current and voltage measurements.

In addition, it should be possible to implement the solution on a hardware platform that can be placed on board a satellite system. This solution has several advantages with respect to the currently used ones:

- No physical knowledge needed: It is not necessary to have a deep understanding of the dynamic of the system since the model can learn the input-output relation. Moreover, it can be refined and updated by exploiting also measurements acquired during operation.

- On-board: Since it is on board it can operate without the need to communicate externally to the satellite or wait for remote commands.
- General: Even if the solution is intended to be deployed on a satellite system, can be extended to other application domains, in fact, it only needs sensors, which are employed in a wide spectrum of applications.

2 Related Works

The use of AI for Fault Detection in satellite systems has gathered increasing importance during the last decades and different solutions exploiting different AI algorithms have been proposed. In Guiotto et al. [8], it is presented a solution for Fault Detection Isolation and Recovery, developed by European Space Agency (ESA) and Politecnico di Milano. The detection module exploits Fuzzy Inductive Reasoning to predict the evolution of the signal, in this case, battery power and solar array housekeeping signals. In Codetta-Raiteri and Portinale [16] an onboard reasoning mechanism implemented using Dynamic Bayesian Networks has been proposed as a high-level failure detection mechanism, gathering the output of lower-level safety mechanisms. In Valdes et al. [17] an onboard fault detection mechanism based on Dynamic Neural Networks has been developed to detect thrusters' faults on satellites during formation flights, by monitoring the attitude control system. Ibrahim et al. [18] presented a Fault Detection Isolation and Recovery process based on Machine Learning (ML) applied to telemetry acquired from Egyptsat-1, using Support Vector Machines (SVM) Regression for performance analysis and K-means clustering for fault diagnosis. In OMeara et al. [10], it is described how a DNN-based system has been deployed at the German Space Observation Center (GSOC) to support ground control in fault detection and prediction using telemetry data.

Recent advancements in deep learning low power hardware accelerators enabled also the use of DNN in energy-constrained scenarios, such as satellites [4–6, 9, 11]. Concerning the use of DNN in on board Fault Detection, Ganesan et al. [19] exploited a Convolutional Neural Network for on board fault detection in a satellite's power subsystem.

Our solution allows a higher autonomy of the whole system from ground control while reducing field expertise and time-to-market. The first property is granted from the onboard placement of the fault detection system, the latter instead can be attributed to the fact that this mechanism has been designed to predict nominal behavior. This brings also the advantage that there is no need to have a dataset with fault-related data. Moreover, thanks to DL it has been possible to achieve our results without performing manual or offline feature extraction and working directly on raw data, only normalization it is performed.

3 Methods and Materials

The solution proposed has been developed by exploiting data acquired from MARSIS radar, which will be briefly described in the next section. Then we proceed to describe the preprocessing and analysis step applied to the dataset at our disposal and the system architecture of the proposed solution.

Successively, we will present the DNN architectures trained for our experiment and the fault injection techniques applied.

3.1 *Marsis*

Mars Advanced Radar for Subsurface and Ionosphere Sounding (MARSIS) is one of the instruments installed inside the payload of the ESA's mission Mars Express, launched on 2nd June 2003. Instrument's primary science objective was to map the distribution of solid and liquid water in the upper portions of Mars' crust. Three secondary science objectives are subsurface geologic probing, surface characterization, and ionospheric sounding [20, 21]. A high-level block scheme of MARSIS can be seen in Fig. 1. More details on MARSIS architecture and internals can be found in Orosei et al. [22].

The instrument consists of two assemblies and two antennas. From a functional standpoint it can be split into three subsystems:

- *Antenna Subsystem (ANT)*: Including the dipole and monopole antennas. The former is used for transmission and reception of sounder's pulses and the second is used for echoes reception.
- *Digital Electronic Subsystem (DES)*: including signal and timing generator, processing, and control units.
- *Radio-Frequency Subsystem (RFS)*: Including transmission (TX) and reception (RX) channels for primary dipole antenna and reception channel for monopole antenna.

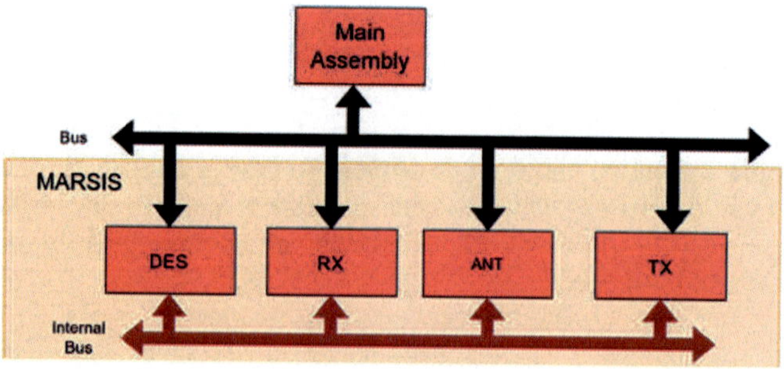

Fig. 1 MARSIS architecture block diagram

3.2 Dataset Analysis and Preprocessing

The dataset used for this work contains telemetry data of MARSIS. Telemetries are collected at each active orbit, that is each orbit when the instrument has been turned on.

The original dataset contains data about temperature, current, and voltage. Measurements are acquired from different components of MARSIS. In particular, Table 1 describes which measurements are available for each component of MARSIS architecture. Each quantity has a nominal and a redundant measurement, except for the MBUS measurement.

The original dataset was composed of separate comma-separated values (CSV) files for current, temperature, and voltage, in separate files for each orbit.

Furthermore, for each CSV file, there was a corresponding label file (LBL) containing all the related metadata, such as column names and a brief description for each column. The format used for label files was Planetary Data System (PDS) 3.0.

Moreover, each file contained a particular kind of timestamp, used in space applications, the Spacecraft Elapsed Time (SCET). This is the value of the S/C clock when the measurement has been taken and is computed as the number of seconds elapsed since the mission started [22].

Thanks to the SCET we established a total ordering of the samples across different files. This measure gave us the possibility also to infer additional information like the duration of an orbit (DURATION), and the interval between consecutive measurements (INTERVAL). To better handle the data and have a comprehensive overview of the dataset, it has been necessary to merge the information of the label and data files. To do this, it has been necessary to eliminate the empty data files and to discard all the ones where a mismatch occurred between the label and the data file, i.e., different number of columns or different ordering of them. Successively, data were divided into three different files, one for each different kind of measurement.

Exploiting the DURATION and INTERVAL it has been possible to filter out incomplete measurement sessions. We decided to consider a session complete looking at the distribution of the DURATION, and keeping all the sessions with

Table 1 Measurements available in the MARSIS housekeeping dataset provided

Quantity	Unit	Module	Component
Temperature	°C	RFS	TX module
Temperature	°C	RFS	RX module
Temperature	°C	DES	–
Temperature	°C	ANT	–
Voltage	V	DES	5.2 V line
Voltage	V	DES	3.3 V line
Voltage	V	RFS	TX module
Voltage	V	RFS	RX module
Voltage	V	-	MBUS

a duration in the interval $[m_D - d, m_D + d]$ with $d = 1.5 \times IQR_d$, where m_d is the mean value and IQR_d is the inter-quartile range. In this way, we were able to obtain a dataset containing only sessions containing all the state transitions required during an orbit, more details on the Finite State Machine describing instruments' operation can be found in Orosei et al. [22]. Furthermore, since during the mission it was not signaled any anomaly or failure from the dataset provider, we assumed to have a dataset representing the nominal behavior of MARSIS. This assumption is fundamental since we had no information on fault modes of MARSIS, and we would not be able to discriminate faults if present. To use this data for the training of a neural network we divided data into three disjoint sets: training, validation, and test set. The partitioning was done while maintaining the original temporal order. The first 70% of the dataset was used as the training set, the next 10% as the validation set, and the remaining 20% as the test set. Data were normalized in the range [0,1] using min-max normalization. Each time series was split into many consecutive windows. The length of a window is determined by the following factors:

- The number of measurements to use as input
- The number of measurements desired as output
- The distance between the last input measurement and the first output measurements (the distance is expressed in terms of samples and not using the SCET) is called lag.

As can be seen in Fig. 2, consecutive windows were acquired sliding forward of a number of measurements equal to the number of input steps. Each window then was divided into two parts, the first one was the input (or sample) used to feed the network during training, and the second one was the expected output (or target) to be compared to the actual output of the network to compute the loss function and update the weights.

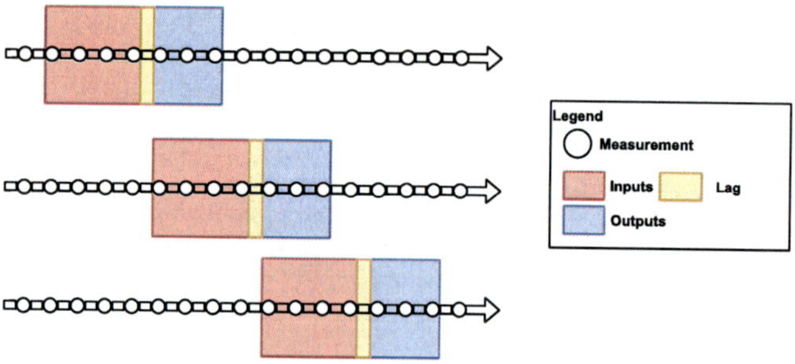

Fig. 2 Windowing process diagram

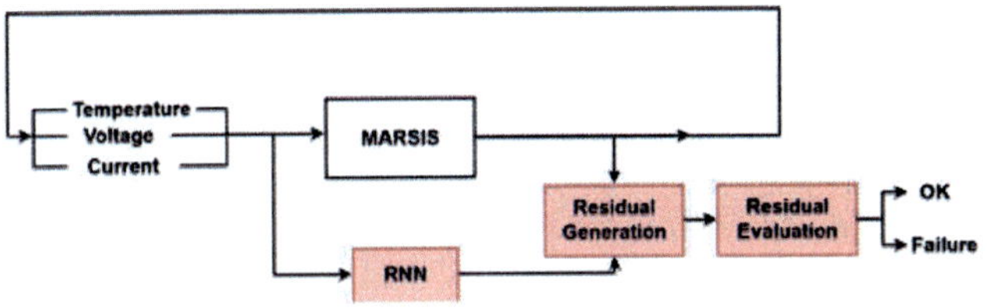

Fig. 3 System architecture block diagram

3.3 System Architecture

The proposed solution exploits a system architecture similar to Parity Equation models [12], substituting the differential equations with an RNN. The system should be able to predict the evolution of a housekeeping signal, using as input the signal themselves. This task is called autoregression. Hereafter, referring to Fig. 3, the highlighted blocks have the following functions:

- *RNN*: The RNN will be used to predict the evolution of the system, the input of the network will be a number n of measurements for each different signal (one or more signals can be used), and the output will be a number m of predictions (again it can be for one or more signals), note that could be $n \neq m$.
- *Residual Generation*: After the predictions are made the system will wait for m measurements from sensors and then predictions and measurements will be used to compute residuals. Supposing that y is the real measurements and $\hat{y}$ is the prediction coming from the RNN, residuals are computed in the following way:

$$\mathrm{r_i} = \left|y_i - \hat{y}_i\right|, \forall \mathrm{i} = 1 \ldots m \tag{1}$$

  That is the element-wise absolute deviation between y and $\hat{y}$.

- *Residual Evaluation*: Evaluate the residuals obtained to detect anomalous behaviors of the system. The check applied for the detection of a fault is on the absolute value of the mean of the residuals $\hat{y}$. If this value is greater than a given tolerance ε it is classified as an anomaly, otherwise as correct behavior.

3.4 DNN Architectures and Training

The core of our fault detection system is the RNN architecture, which will be used as a model for our system. Plain RNNs have been proven to suffer many training and performance issues, mainly due to the vanishing and exploding gradient problem [23]; for this reason, they have not been taken into consideration as the basic building block for the network. Furthermore, since there is no evidence of benefits coming from using LSTM [24] or GRU [25], we decided to evaluate both as possible solutions. The framework used to develop the neural network architecture is Tensorflow 2.2. The training has been done using an NVIDIA Tesla T4 GPU.

We defined three meta-architectures, with two variants each. One of the variants used LSTM, while the other used GRU. The meta-architectures defined are the following ones:

- *2X*: This is the simplest architecture proposed. It is composed of only two layers of LSTM (GRU) cells, the first layer can have different sizes, whereas the second's size is always equal to the number of output time steps required.
- *stacked2X*: This architecture is composed of two layers of LSTM (GRU) cells of the same size and a fully connected layer with a dimension equal to the number of output time steps.
- *funnel4X*: This is the deepest meta-architecture proposed. It is composed of four LSTM (GRU) layers, with decreasing width. Width is halved at each successive layer as long as the resulting halved width is larger than the output layer's size. The output layer is a fully connected one having as width the number of required time steps.

X is a placeholder that is equal to the name of the basic building block used in the realization of the meta-architecture (LSTM or GRU).

After each LSTM layer, it has been placed an activation layer, with a LeakyReLU activation function. This function has been chosen since there is evidence that it improves deep learning model performance, in particular, it makes the training faster, and mitigates the vanishing gradient problem [26].

The equation of this activation function is the following:

$$LeakyRelu(x) = \begin{cases} x, & x > 0 \\ \frac{x}{100}, & x \leq 0 \end{cases} \tag{2}$$

Moreover, when a Dense layer is present, it has a ReLU activation function. We used as activation function for the output gate of each LSTM (GRU) layer a hyperbolic tangent (*tanh*), as in the original version proposed by the authors [24, 25]. Another motivation for this choice for the output activation function of LSTM layers is that it allows using the optimized kernel from NVIDIA CuDNN libraries instead of generic ones. This gave us the possibility to speed up both the training and the inference phase of the network. In the training phase, we made several experiments to find which one gives the lowest error. Experiments were done by varying the number of input steps, output steps, lag, and architecture Table 2.

The choice for the hyperparameters has been done by trial and error, using a 2LSTM architecture with the first layer composed of 64 LSTM cells, with 16 input timesteps and 4 output timesteps, and training it for 20 epochs. The error metrics used are mean absolute error (MAE), mean squared error (MSE) and root mean squared error (RMSE). In Table 3 we reported the configuration of hyperparameters that minimizes the error metrics.

Table 2 Network parameters for the DNN architectures proposed

Parameter	Value
Width	8,16,32,64
Input steps	4,8,16,32
Output steps	1, input/2, input
Lag steps	1, input/2, input

Table 3 Hyperparameter values used during training of the DNN architectures proposed

Hyperparameter	Value
Loss function	MAE
Epochs	100
Batch size	64
Weight initializer	Glorot normal
Optimizer	Adadelta
Decay rate	0.8
Stability constant	10^{-4}
Learning rate	0.01
Dropout	None
Early stopping	Yes
Tolerance	10^{-4}
Patience	5

The number of epochs was set to 100. This choice was made by training the same network used for hyperparameters selection for 1000 epochs, observing that MAE at that point had reached a value around 0.1 which was sufficient for the objective of this work. We decided to focus on temperature measurements of Antenna PCB (ANT) and the Digital Electronic Subsystem (DES) of MARSIS. The rationale behind this choice is that the sampling time of measurements at our disposal is relatively long (from 16 to 64 s), and the temperature is the only measure having a variation speed comparable to the sampling time.

3.5 *Fault Simulation*

Since we did not dispose of labeled anomalies in our dataset, it has been necessary to introduce artificially generated ones on the test set. We decided to perform multiple fault simulation tests, simulating different types of faults.

The approach used was to apply a periodic distortion on measured signals. Eight different tests have been performed for each model selected. In each test we applied a different fault type, the same fault type has been applied in a time window, periodically, varying the category, nature, and sign of the distortion introduced by the

Table 4 Fault typologies used for fault injection tests performed on temperature signals

Category	Nature	Sign
Additive	Step	+
Additive	Incipient	+
Multiplicative	Step	+
Multiplicative	Incipient	+
Additive	Step	–
Additive	Incipient	–
Multiplicative	Step	–
Multiplicative	Incipient	–

fault on the reference signal. Table 4 shows the different types of faults simulated. We considered both step like and incipient faults, having a multiplicative or additive nature. We also reproduced increasing and decreasing step and positive or negative slope changing the sign of variations.

4 Results

4.1 Signal Prediction and Fault Detection

After the training phase, trained models have been evaluated on the test set. Looking at the results of these evaluations we have understood that there is no preferable architecture among the proposed one, since for each configuration of width, lag, input steps, and output steps the results are different.

Furthermore, focusing on lag, what emerges from Fig. 4 is that increasing it leads, on average, to an increasing error. This may be accounted to the fact that the information content of input data may be not sufficient to predict values far in time from the input data.

We observed that the error increases as the number of output steps increase. This phenomenon may be related to the problem experienced with the lag. What we inferred from these results is that the proposed architectures perform better when the output is immediately consecutive to the input.

Among the trained networks we choose as predictors in our system, evaluating for each configuration of input, output, and lag the architectures that showed the lowest error using MAE as the metric.

Sensitivity and Specificity. It has been decided to discard solutions having an MAE higher than 0.1, even if it is the minimum error for that configuration. This is because, since the error is measured on data normalized in [0,1], an MAE of 0.1 means that there is a deviation of $\pm 10\%$ from the target value. It has been evaluated, during hyperparameter selection, that, under this threshold, the architectures'output shows

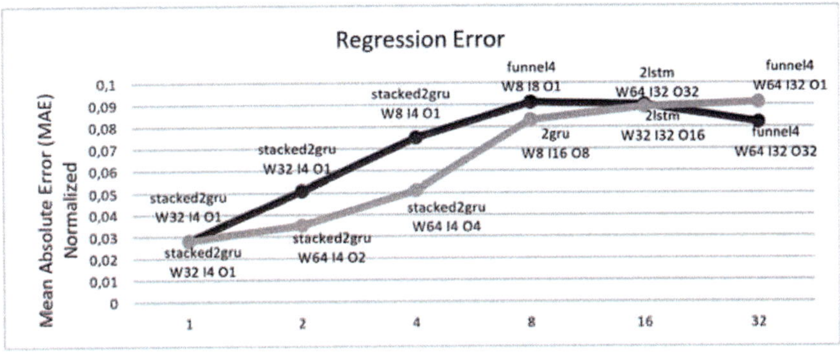

Fig. 4 Regression error varying Output Width and Lag. Each point in the diagram represents the network architecture achieving the lowest error (MAE). The name of the architecture is indicated, with the values of its parameters: width (W), input (I), and output (O), the results refer to networks trained on ANT data

a trend similar to the target, whereas above this threshold the trend is too noisy, thus making its use for anomaly detection unfeasible. Other than the error the ratio between input and output size is another important evaluation metric. Networks having an input-output ratio equal to 1 are preferable since they can predict the same amount of timesteps given in input. Table 5 resumes the features of the models we choose for the experiments. The decision was made following the evaluation criteria discussed above.

Before testing the model there is the need to fix the tolerance ε to be used in the residual estimation phase described previously. Since the residual is the MAE between the real value and the prediction, it has been decided to use as tolerance the double of the MAE of the chosen model on the training set.

As shown in Table 6, the percentage of true positive achieved is over 90% for the majority of the typologies of fault injected. The worst TP rate is achieved on incipient additive faults for both models, even if the model trained on ANT data performs worse on perturbations injected with a positive sign, whereas the other is less accurate on the negative ones. We noticed that the model with the lowest MAE has lower overall sensitivity compared to the other model. However, the specificity is worse in the DES model, this means that this model will result in a higher number of false alarms.

Table 5 Chosen architectures for fault injection tests with parameters' values (width, input, and output size) and MAE

Network	Signal	Input size	Output size	Width	MAE on training set
stacked2gru	ANT	4	4	64	0.0514
2gru	DES	4	4	32	0.0709

Table 6 Results of experiments performed with selected DNN architecture on ANT e DES data

Experiment	DES		ANT	
	TP rate	FP rate	TP rate	FP rate
Step additive negative	0.82	0.67	0.92	0.49
Step additive positive	0.98	0.54	0.9	0.52
Step multiplicative negative	1.0	0.63	0.98	0.47
Step multiplicative positive	0.98	0.63	0.92	0.53
Incipient additive negative	0.64	0.73	0.86	0.52
Incipient additive positive	0.96	0.55	0.76	0.58
Incipient multiplicative negative	1.0	0.63	0.94	0.49
Incipient multiplicative positive	0.98	0.62	0.8	0.56

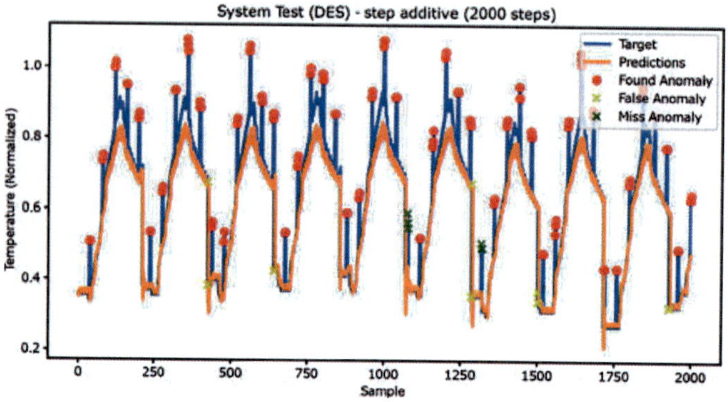

Fig. 5 Step additive faults injection test on DES temperature signal

It is worth to be noted that the specificity of both models is poor (FP rate around 50%) For the model. Looking at Figs. 5 and 6 it can be seen that the False Anomalies are mainly concentrated after the injection of a fault, more precisely, after the fault is removed. The model uses the anomalous data to make its next prediction, which will diverge from the nominal behavior of the system. The lower specificity of the model trained on the DES data is due to the higher error of the model, it can be seen that around the peaks of the signal the gap between the predictions and the real measurements is higher, and this results in false alarms.

4.2 *Model Complexity*

Inference Time. The inference time has been evaluated for both models using an AMD A10-9600P CPU. Results show that for the 2gru model with a first layer width of 32 the mean inference time is 12,83 ms ±0,09 ms, (confidence interval at

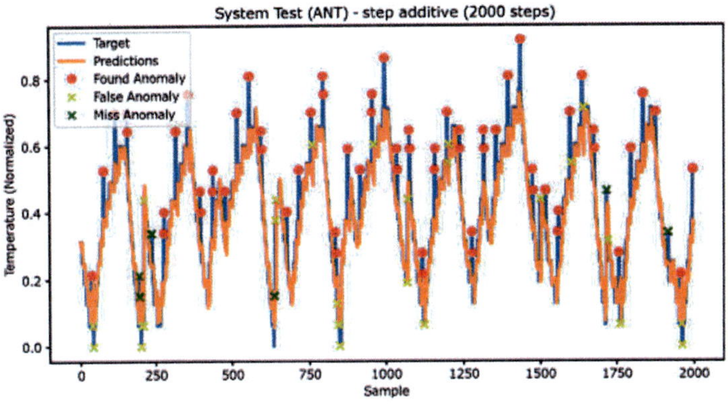

Fig. 6 Step additive faults injection test on the ANT temperature signal

95%). For the other model, stacked2gru, having GRU layers width of 64, the mean inference time is 21,61 ms ±0,11 ms (confidence interval at 95%). Both models have an inference time lower than the minimum sampling time of sensors [22].

Network Size. The number of weights for both models is quite small: 38084 for the stacked2gru and 3816 for the 2gru. Supposing that weights are represented as 32bit floating-point numbers, they require respectively 152kiB and 15kiB, this amount of memory is sufficiently small to fit the memory constraints of MARSIS [22], which are more restrictive than the majority of low power hardware accelerators ones [4–6, 27]. These results have been achieved without using any technique to reduce the network size, i.e., pruning and quantization. Exploiting these techniques, it would be possible to further reduce the amount of memory needed for the model, increasing also its energy efficiency while lowering inference time. However, these techniques may also penalize the accuracy of the model.

4.3 Prediction Horizon

Other than the error, the ratio between input and output size is another important evaluation metric. Networks with a ratio equal to 1 are preferable since they can predict the same amount of time steps that have been given in input. Among the architectures trained there are some with this characteristic that has achieved a sufficiently low error, in particular, the lowest error for both time series in networks with an input-output ratio of 1 is achieved by those with 4 inputs and 4 outputs, which are a stacked2gru with GRU layers width 64 for the ANT time series and a 2gru with first GRU layer width 32 for the DES time series. These two networks have been selected as the RNN models to be used in our fault detection system since they give us a low error and the ideal input-output ratio.

5 Conclusions

The objective of this work was to build an on-board fault detection system for satellite systems exploiting AI techniques. AI enables to decrease of expertise and computational complexity needed in traditional approaches. The solution proposed exploits a model-based approach for fault detection, focusing on temperature measurements coming from MARSIS telemetry data. The model-based approach used differs from the traditional ones since it uses an RNN architecture instead of differential equations. Two of the trained RNN architectures have been selected, one trained on Antenna controller (ANT) temperature measurements and the other on Digital Electronic Subsystem (DES) temperature measurements, and they were used to predict the evolution of signals on which they were trained, thus acting as a mathematical model.

Results obtained from fault injection tests performed on both architectures highlight that the system can achieve satisfactory results in terms of detected faults. Both models can detect 93% of step-like faults, whereas the number of incipient faults detected is lower (84% for the ANT model and 89,5% for the DES model).

Moreover, the system has shown to achieve these results respecting the constraint imposed on the inference time, which should be lower than 4 s and is 21.61 ms for the ANT model, which is the slower one.

Furthermore, given that the number of weights is relatively small (38084 for the larger model, which means that using 32-bit floating-point encoding needs 152kiB of memory), there is the possibility to implement the system exploiting FPGA-based or ASIC-based low-power hardware accelerators, which are also compliant with the power budget of MARSIS (64.5 W). The proposed solution is also generalizable to different signals and application scenarios, given that the same approach has been used for two different time series, achieving similar results. We noticed that the high FP rate of the model depends mainly on the implementation of the fault detection algorithm, and the accuracy of the model.

In this work, a simple threshold-based algorithm has been used for detection, but more advanced techniques can be evaluated to reduce the FP rate, possibly increasing also the TP rate. The reduced domain expertise required using deep learning models instead of differential equations can grant remarkable advantages for the development of FDIR systems, such as a reduced time-to-market, the possibility to improve the model as soon as new data or a better architecture is available, and the possibility to apply the same approach to different use cases, without the need to change the development process.

References

1. Ginosar, R.: Survey of Processors for Space. Data Systems in Aerospace (DASIA), Eurospace (2012)
2. Lentaris, G., Maragos, K., Stratakos, I., et al.: High-performance embedded computing in space: evaluation of platforms for vision-based navigation. J. Aerosp. Inf. Syst. **15**, 178–192 (2018). https://doi.org/10.2514/1.I010555
3. Iturbe, X., Venu, B., Ozer, E., et al.: The arm triple core lock-step (TCLS) processor. ACM Trans. Comput. Syst. **36**, 1–30 (2019). https://doi.org/10.1145/3323917
4. Reuther, A., Michaleas, P., Jones, M., et al.: Survey of Machine Learning Accelerators. https://ieeexplore.ieee.org/Xplore/home.jsp
5. Giuffrida, G., Diana, L., de Gioia, F., et al.: CloudScout: a deep neural network for on-board cloud detection on hyperspectral images. Remote Sens. **12**, 2205 (2020). https://doi.org/10.3390/rs12142205
6. Rapuano, E., Meoni, G., Pacini, T., et al.: An FPGA-based hardware accelerator for CNNs inference on board satellites: benchmarking with Myriad 2-based solution for the CloudScout case study. https://www.mdpi.com/ (2021). https://doi.org/10.3390/rs13081518
7. Ecoffet, A., Huizinga, J., Lehman, J., et al.: Go-Explore: A New Approach for Hard-Exploration Problems (2019). https://arxiv.org/
8. Guiotto, A., Martelli, A., Paccagnini, C., Lavagna, M.: SMART-FDIR: Use of Artificial Intelligence in the Implementation of a Satellite FDIR
9. Giuffrida, G., Fanucci, L., Meoni, G., et al.: The Φ-Sat-1 mission: the first on-board deep neural network demonstrator for satellite earth observation. IEEE Trans. Geosci. Remote. Sens. **60** (2022). https://doi.org/10.1109/TGRS.2021.3125567
10. OMeara, C., Schlag, L., Wickler, M.: Applications of deep learning neural networks to satellite telemetry monitoring. In: 2018 SpaceOps Conference. American Institute of Aeronautics and Astronautics, Reston, Virginia (2018)
11. di Mascio, S., Menicucci, A., Ottavi, M., et al.: On-Board satellite telemetry forecasting with RNN on RISC-V based multicore processor. In: IEEE International Symposium on Defect and Fault Tolerance in VLSI and Nanotechnology Systems (DFT) (2020). https://doi.org/10.1109/DFT50435.2020.9250796
12. Miljković, D.: Fault detection methods: a literature survey. In: 2011 Proceedings of the 34th International Convention MIPRO (2011)
13. Zolghadri, A.: Advanced model-based FDIR techniques for aerospace systems: today challenges and opportunities. Prog. Aerosp. Sci. **53**, 18–29 (2012). https://doi.org/10.1016/j.paerosci.2012.02.004
14. Zolghadri, A.: The challenge of advanced model-based FDIR for real-world flight-critical applications. Eng. Appl. Artif. Intell. (2018)
15. Schulte, P., Spencer, D.A.: On-Board model-based fault diagnosis for autonomous proximity operations. In: 69th International Astronautical Congress (IAC) (2018)
16. Codetta-Raiteri, D., Portinale, L.: Dynamic Bayesian networks for fault detection, identification, and recovery in autonomous spacecraft. IEEE Trans. Syst. Man, Cybern. Syst. **45**, 13–24 (2014). https://doi.org/10.1109/TSMC.2014.2323212
17. Valdes, A., Khorasani, K., Ma, L.: Dynamic neural network-based fault detection and isolation for thrusters in formation flying of satellites. In: Lecture Notes in Computer Science (Including Subseries Lecture Notes in Artificial Intelligence and Lecture Notes in Bioinformatics), vol. 5553. LNCS, pp. 780–793 (2009). https://doi.org/10.1007/978-3-642-01513-7_85
18. Ibrahim, S.K., Ahmed, A., Zeidan, M.A.E., Ziedan, I.E.: Machine learning techniques for satellite fault diagnosis. Ain Shams Eng. **11**, 45–56 (2020). https://doi.org/10.1016/j.asej.2019.08.006
19. Ganesan, M., Lavanya, R., Nirmala Devi, M.: Fault detection in satellite power system using convolutional neural network. Telecommun. Syst. **76**, 505–511 (2021). https://doi.org/10.1007/S11235-020-00722-5

20. Picardi, G., Biccari, D., Cartacci, M., et al.: MARSIS, a radar for the study of the Martian subsurface in the Mars Express mission. Mem. Della Soc. Astron. Ital. Suppl. **11**, 15 (2007)
21. Orosei, R., Jordan, R.L., Morgan, D.D., et al.: Mars advanced radar for subsurface and ionospheric sounding (MARSIS) after nine years of operation: a summary. Planet. Space Sci. **112**, 98–114 (2015). https://doi.org/10.1016/j.pss.2014.07.010
22. Orosei, R., Huff, R.L., Ivanov, A.B., et al.: Mars Express-MARSIS to Planetary Science Archive Interface Control Document (2007)
23. Hochreiter, S.: The vanishing gradient problem during learning recurrent neural nets and problem solutions. Int. J. Uncertain. Fuzziness Knowl. Based Syst. **6**, 107–116 (1998). https://doi.org/10.1142/S0218488598000094
24. Hochreiter, S., Schmidhuber, J.: Long short-term memory. Neural Comput. **9**, 1735–1780 (1997). https://doi.org/10.1162/neco.1997.9.8.1735
25. Chung, J., Gulcehre, C., Cho, K., Bengio, Y.: Empirical Evaluation of Gated Recurrent Neural Networks on Sequence Modeling (2014). https://arxiv.org/
26. Maas, A.L., Hannun, A.Y., Ng, A.Y.: Rectifier nonlinearities improve neural network acoustic models. In: Proceedings of the 30th International Conference on Machine Learning (2013)
27. Furano, G., Meoni, G., Dunne, A., et al.: Towards the use of Artificial Intelligence on the edge in space systems: challenges and opportunities. IEEE Aerosp. Electron. Syst. Mag. **35**, 44–56 (2020). https://doi.org/10.1109/MAES.2020.3008468

ISS Monocular Depth Estimation Via Vision Transformer

Luca Ghilardi, Andrea Scorsoglio, and Roberto Furfaro

Abstract Monocular depth estimation can be used as an energy-efficient and lightweight backup system to the onboard light detection and ranging (LiDAR) instrument. Unfortunately, the monocular depth estimation is an ill-posed problem. Therefore typical methods resort to statistical distributions of image features. In this work, a deep neural network that exploits Vision Transformer in the encoder is trained in a supervised fashion to solve the regression problem. Specifically, the problem considered is monocular depth estimation with images taken at different positions and distances from the International Space Station (ISS) to measure the network performance in a rendezvous maneuver.

Keywords Computer vision · Rendezvous · Transformers · Machine learning · Deep learning

1 Introduction

Understanding the 3D scene geometry is crucial in many applications, such as autonomous navigation [11, 16], virtual reality [12], scene understanding and segmentation [9]. A 3D map of the surrounding environment is normally obtained with LiDAR instruments or with stereo cameras [1]. LiDAR are active instruments that emits laser pulses to measure range, and their use for space applications has increased significantly with the development of the technology [2]. According to NASA[19], modern spacecrafts used for docking and rendezvous maneuver the Triangulation & Light Detection and Ranging Automated Rendezvous & Docking (TriDAR) for

L. Ghilardi (✉) · A. Scorsoglio · R. Furfaro
University of Arizona, Tucson, AZ 85721, USA
e-mail: lucaghilardi@arizona.edu

A. Scorsoglio
e-mail: andreascorsoglio@arizona.edu

R. Furfaro
e-mail: robertof@arizona.edu

C. Ieracitano et al. (eds.), *The Use of Artificial Intelligence for Space Applications*, Studies in Computational Intelligence 1088,
https://doi.org/10.1007/978-3-031-25755-1_11

the autonomous resupply of the ISS, which has at its core the LiDAR sensors for 3D imaging. However, these systems also have drawbacks [2], like moving parts or calibration issues depending on the system installed. Optical cameras for space applications, on the other hand, have been proved to be an energy-efficient and lightweight technology. This makes generating high-quality depth of field estimation from a single camera attractive because it can complement LiDAR sensors inexpensively.

In this paper, we examine the problem of Monocular Depth Estimation (MDE), which is the task of estimating depth pixel-wise from a single image. MDE is an ill-posed problem [21], which makes it challenging. However, humans can perform MDE by exploiting features in the image such as perspective, relative scaling, occlusion, lighting, and shading [10]. Modern computer vision and deep learning methods can efficiently discern those features, allowing us to approach this problem from a data-driven perspective rather than a geometrical one. In particular, deep convolutional neural networks with an encoder-decoder architecture have significantly improved their performances on this problem [5, 7, 14]. The encoder gradually reduces the spatial resolution to learn more information under a larger receptive field, while the decoder superimposes this information on the corresponding resolution. However, making a pixel-wise prediction using the global context is still challenging. Transformers [20] are promising to mitigate this issue. Transformers became popular due to their performances with the task of modeling sequences in natural language processing, and their recent applications in the field of computer vision, Vision Transformer (ViT) [4, 22], surpassed the performance of convolutional neural networks on large scale datasets.

In this work, we adopt a Vision Transformer architecture for dense MDE [15] of the ISS. The dataset to train and test the network is created using the 3D model software, Blender.[1] Within Blender, we used the 3D model of the ISS, made available by NASA[2] to render the images using the Cycles open-source Blender package. Cycles is a physically-based path tracer for production rendering that computes light traces from all the light sources in the scene and their interactions with the objects. Blender can easily interface with python, and the authors tested its flexibility and performance for optical hazard detection and navigation in case of moon landing [8, 17, 18] or binary asteroids [6].

2 Method

As previously mentioned, we adopt a Vision transformer architecture to compute dense regression of the depth of field of the images. Dense prediction is the task of assigning a label, or in our case, a value to every pixel of an input image. Every depth value is the object's distance in a specific pixel from the camera.

[1] Blender is an open-source software for modeling and physically-based rendering.

[2] https://solarsystem.nasa.gov/resources/2378/international-space-station-3d-model/.

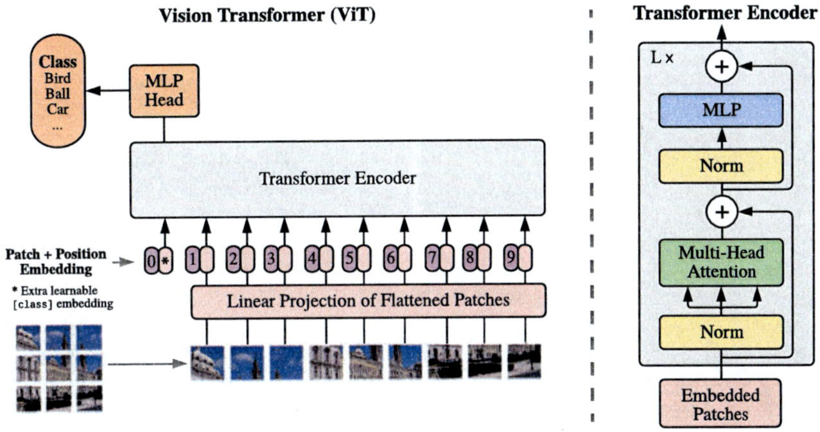

Fig. 1 Transformer layer [4]

2.1 *Architecture*

In this section, the vision transformer architecture is introduced. In a typical convolutional neural network, the increase of receptive field in the deepest layers of the encoder comes at the cost of the resolution of the image, therefore a loss in details. On the other hand, vision transformers promise to maintain a global receptive field with high feature resolution throughout the encoder. This is possible thanks to the transformers layers in the encoder.

The transformers layers were first introduced by Dosovitskiy et al. [4] as a novel method for image classification, Fig 1. For this project, the author adopted the architecture created by Belkar et al.,[3] which promises to be more customizable and easy to train.

Before being input into the transformer encoder, the image is subdivided into patches called tokens. Those tokens are linearly embedded into a feature space to identify the position and class of each token. Once embedded, the patches are passed into the transformer encoder, which returns a value for each class token. The encoder adopted is the ViT-Base [4] that has been pre-trained on ImageNet [3]. Most importantly, for MDE applications, the transformer encoder does not use down-sampling operations, which preserve all the image details. Into the encoder, multiple transformers layers are applied on cascade. Once the encoder has processed the tokens, they must be reassembled to complete the dense prediction.

As can be seen in Fig. 2, there are representation extractions in the encoder, these representations are combined in the Reassemble block that will aim to that will aim to transform these representations in a higher space that can be used in the Fusion block. Each Reassemble block is composed of 3 sub-blocks:

[3] https://github.com/antocad/FocusOnDepth.

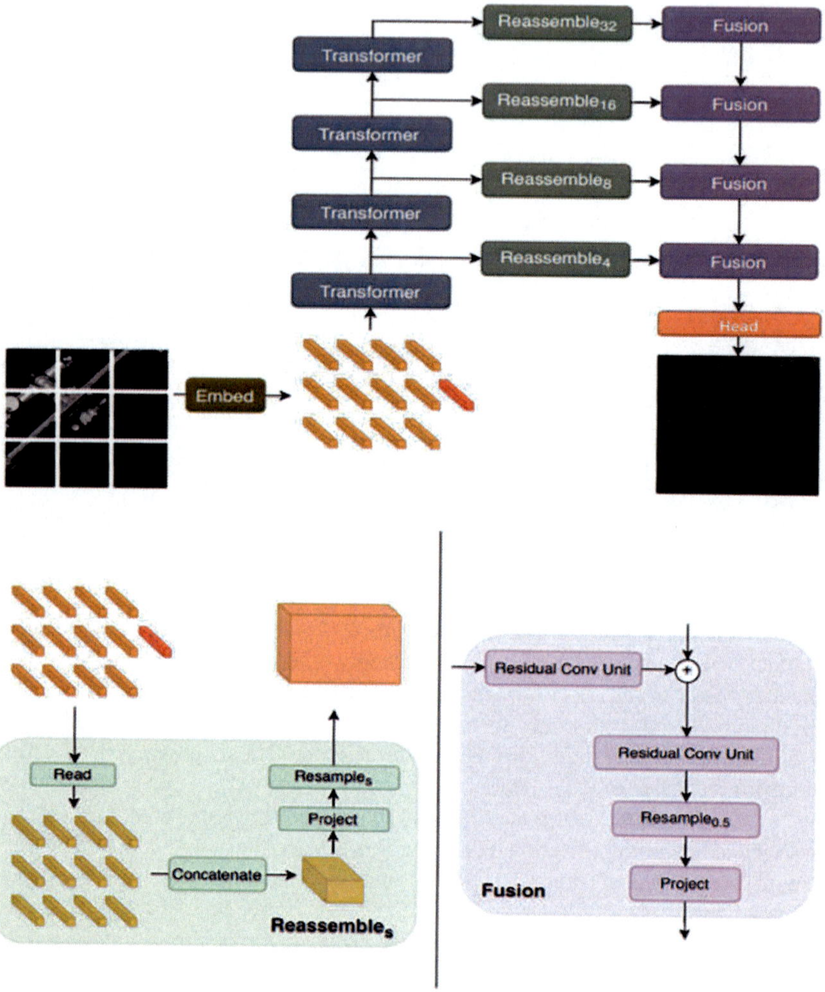

Fig. 2 Vision Transformer architecture [4] with Bi-Head for depth estimation and segmentation

- Read Block: it reads the input to be mapped into a representation size
- Concatenate Block: in this block, the representations are combined. The step consists in concatenating each representation following the order of the patches. This yields an image-like representation of this feature map.
- Resample Block: This block consists of applying a 1×1 convolution to project the input image-like representation into a space $\hat{D}$ of dimension 256. This convolutional block is followed by another convolution where the inner parameter changes to the layers of the ViT encoder.

The Fusion block takes the representations from the Reassemble and the previous Fusion block to sum them. Once summed, we apply two successive convolutional units and upsample the predicted representations.

We modified the architecture to process *tif* images, which supports 32-bit information per pixel. This is necessary to avoid approximation errors. In the original architecture, the depth images are imported as *uint8* data, which is limited to 256 values. The depth head of the model has been modified as well. The original model has a sigmoid layer, which squishes the output values between 0 and 1. This approach works for subjects' discrimination in the image, but our goal is to do regression on the distance values. Therefore, the final layer has been substituted with a Relu activation layer, which is linear for positive values.

2.2 *Simulation Environment*

This section will explain how the dataset and ground truth are created. As mentioned in the introduction, we adopted the open-source 3D model software Blender to render the images and create the corresponding depth maps. The scene was created by importing the model of the ISS provided by NASA and creating a high-resolution spherical image (known as HDRI) of the Earth with the Sun. On top of that, we develop a tool to replicate the measurement of a LiDAR sensor within Blender. The images and LiDAR measurements are taken at different positions and distances from the ISS. Specifically, from a distance of 60 m to a maximum of 300 m, with distance steps of 24 m, and with arches of 20° to complete a sphere around the station. This gives us 1620 images, which are not many for deep learning training. However, keeping the number of samples low became necessary since the LiDAR tool is computationally expensive, as it takes around 2 minutes to compute each image. Figure 3 is shown as an example of the rendered image and the respective LiDAR measurement. To increase the diversity of the dataset, a data augmentation transformation has been applied to random images. In the next section, the transformation applied is described in more detail.

It is worth noticing that both the Sun and the Earth with its albedo are present in the scene in this dataset. However, the illumination of the model is static, and lens effects like lens flare and chromatic aberration have not been implemented.

3 Parameters

The images were rendered with a resolution of 384×384 pixels in RGB with a camera focal length of 50 mm. The resolution as been choosen to match the resolution of the pre-trained encoder. This prevents the need of an interpolation up-scaling of the image. The dataset has been augmented using random rotations between -10 and

(a) Camera View (b) LiDAR View

Fig. 3 Dataset sample

+10 degrees, random crops of the image with a minimum resolution of 256 × 256 and flipping the images.

60% of the dataset is randomly selected for the training, 20% for the validation, and the last 20% is used for the testing. The loss function adopted is the mean squared error loss function L_{mse} for the depth estimator, which is the common choice for regression problems.

The training has been performed with the following parameters: Optimizer: ADAM with learning rate: 1e-5. The batch size adopted is only 1 due to memory limitations of the GPU. The model has been trained for 100 Epochs.

4 Results

In this section, we show the model's performance evaluated using 20% of the dataset for a total of randomly selected 324 images. For the depth estimation the metric considered are the mean squared error $MSE = \frac{1}{n}\sum_{i=1}^{n}(y_i - \hat{y}_i)^2$ and the absolute error. Lastly, a comparison between our model and an architecture based on an encoder with down-sampling operations is shown. In particular, we adopted a UNet [13] for this comparison due to its flexibility and performance with small datasets.

Figure 4 shows the predictions of the vision transformer on some of the most challenging examples from the test set. In the first row, there is a close approach to the station, while in the second row, there is a scene with the sun in the frame of the camera, and in the last row, a scene with the earth as a background. The model performs well in this scenarios, distinguishing the space station from the rest of the scene.

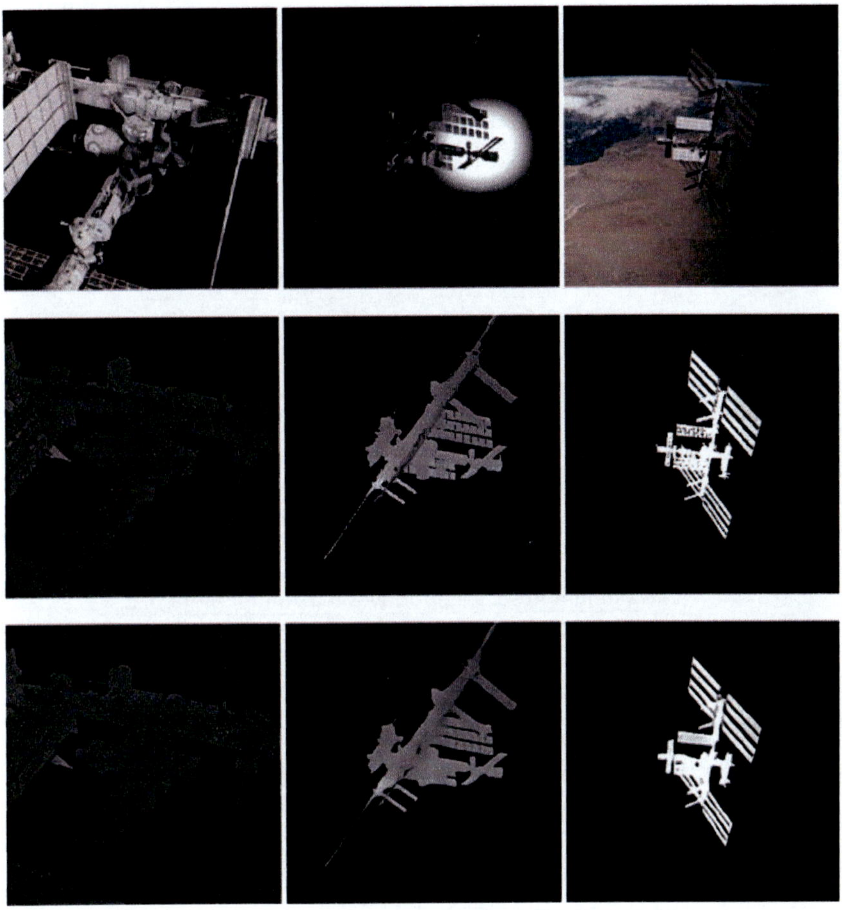

Fig. 4 In the sample test dataset, in the first row, the input images are shown; in the second, the target predictions; and in the last, the model predictions

The following graphs show the model's performance on all test sets. It is important to notice that the test images have been sorted depending on the range from the ISS from the nearest to the further. Therefore, for Fig. 5 on the x-axis the first images are around 60 m and the last ones are around 300 m.

In Figs. 5 and 6 the error increases with the distance from the station, which is expected since the features of the station are less visible. The maximum absolute error, Fig. 6, has been computed considering the entire image or only the target mask of the ISS. This has been done to highlight the distance evaluation of the model over the segmentation performance. In other words, if a pixel that is part of the background in the ground truth is misclassified as part of the station, that error is related to the segmentation performance of the model. Figure 6 shows that the maximum absolute

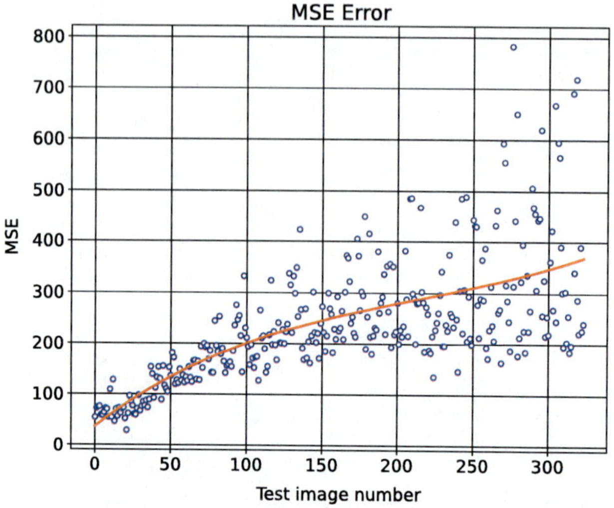

Fig. 5 MSE error on the test dataset

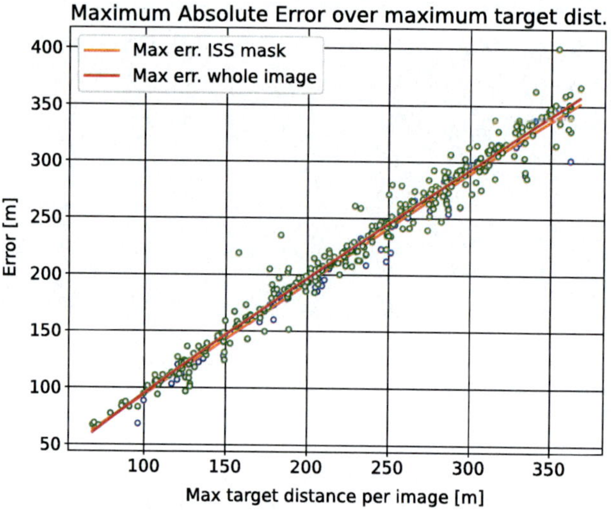

Fig. 6 Maximum absolute error on the test dataset. The red line represent the interpolation of the maximum error computed on the whole image, while the orange one is the interpolation of the maximum error computed on the target mask of the ISS

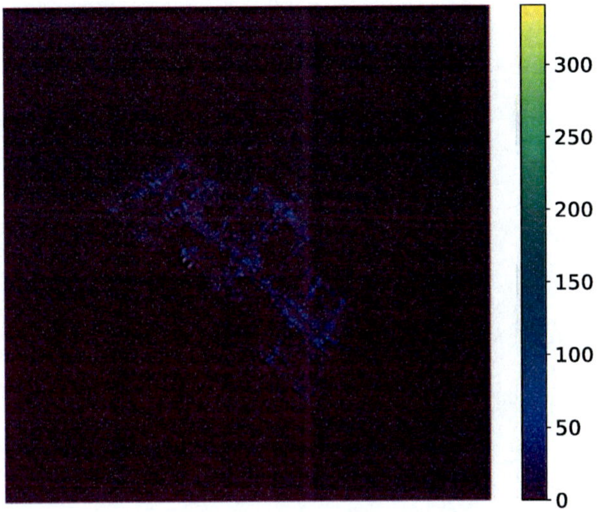

Fig. 7 Maximum absolute error heat map on a sample test image

error does not change significantly in the case of the entire image or only the target mask. However, if we observe Fig. 8 we can see that the maximum predicted value in each image is close to the maximum target value, which is also similar to the maximum absolute error. Additionally, Fig. 7 shows that the more significant errors are concentrated on the borders of the image. This lets us infer that some of the pixel belonging to the ISS has been classified as part of the background, therefore with a distance of 0 m.

In Fig. 9, we can see the mean error of the model computed on the entire image and the ISS target mask. The methods have been divided to avoid diluting the mean error with many pixels belonging to the background, especially higher distances. The median error has been computed as well on the ISS target mask to filter out the outlier values. The error increases with the distance and, on average, is below 10% of the distance from the ISS.

The model has been also compared with a UNet's architecture shown in Table 1. The UNet has been trained on the same dataset of the ViT with the same data augmentation. The parameters used for the training are learning rate: 1e-5, optimizer: ADAM, batch-size: 20, and epochs: 100. However, we saved the model weights only when the validation loss decreased to the previous epoch. The validation minimum was at the 26th epoch, which means that the model was over-fitting after that epoch.

In Fig. 10 are shown some examples of the test set. The UNet struggle to discriminate against the ISS from the background, while the ViT model does a better job. Additionally, the pixel distance on the ViT model is more uniform along the entire ISS with respect to the UNet. This is visible from Fig. 10, where the predictions of the UNet look like having a salt & pepper noise in them.

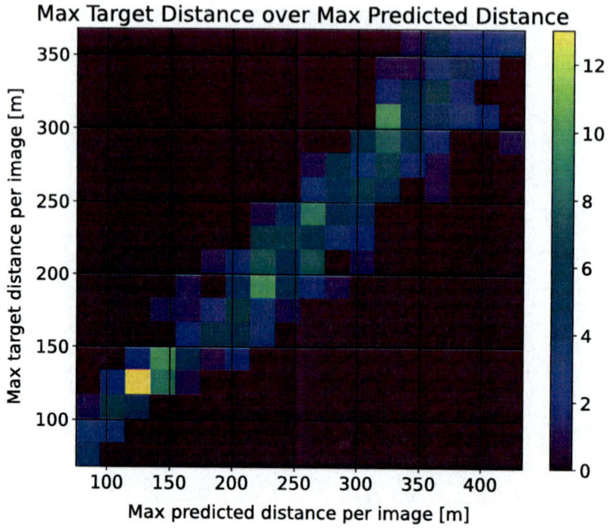

Fig. 8 Maximum pixel value on the prediction per image over maximum pixel value on the target per image. The colorbar represent the number of images that share those bins

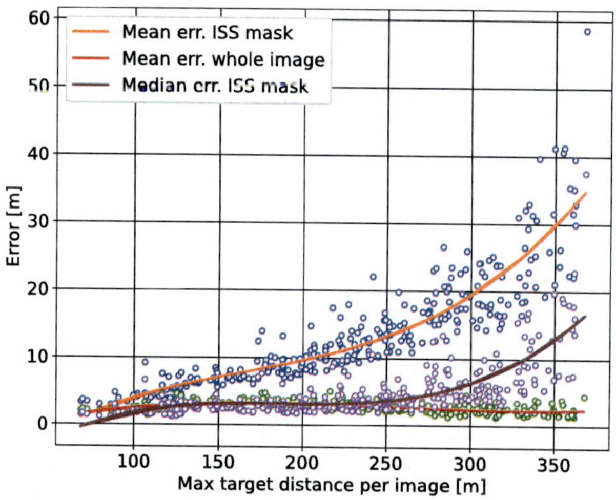

Fig. 9 The red line represent the interpolation of the mean error computed on the whole image, while the orange one is the interpolation of the mean error computed on the target mask of the ISS, and the purple line is the interpolation of the median error computed on the target mask of the ISS

Table 1 UNet architecture

Blocks	Channels
Double Conv.	3 → 64
Down1	64 →128
Down2	128 → 256
Down3	256 → 512
Down4	512 → 1024
Up1	1024 → 512
Up2	512 → 256
Up3	256 → 128
Up4	128 → 64
Regression Head	64 → 1
Blocks	**Layers**
Double Conv.	Conv2D
	BatchNorm2D
	ReLU
	Conv2D
	BatchNorm2D
	ReLU
Down	MaxPool2D
	Double Conv.
Up	Bilinear Up-sampling
	Double Conv.
Regression Head	Conv2D
	ReLU
	Conv2D
	ReLU

The performance gap between the two models is even more visible in Fig. 11, where the mean and max error is computed for both models on the ground truth mask of the ISS. The maximum error of both models is similar, and it is linear with the distance, which means that both models learned to recognize the distance of the ISS relatively well. However, the mean error on the ISS mask of the ViT model is lower. This is because the ViT has more consistency and precision in discriminating the ISS and evaluating the pixels assigned to it. It is worth noting that the UNet is a smaller model regarding ViT. Our UNet has around 17 million learnable parameters, while even the base model of ViT has more than 105 million. On top of that, the ViT encoder was pre-trained on a large dataset, while the UNet was trained from scratch. All of this can have an impact on the overall performance of the models.

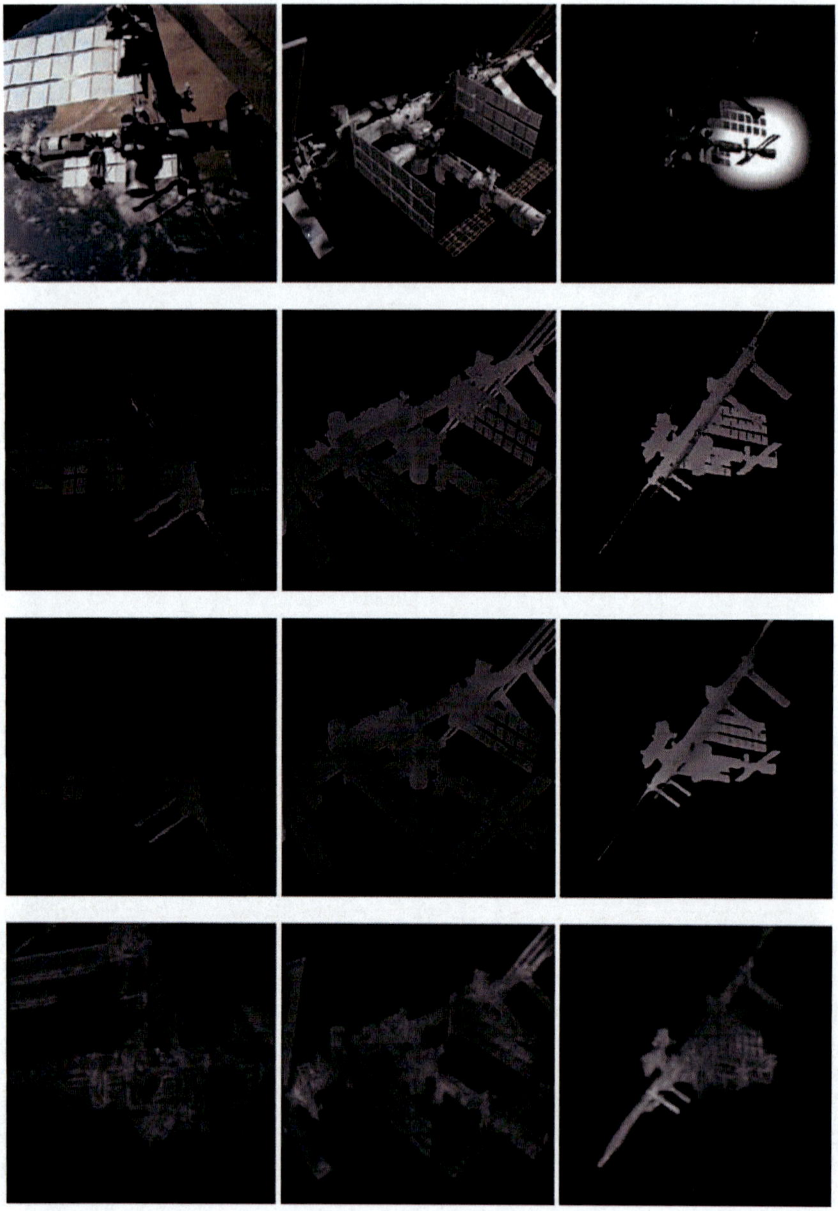

Fig. 10 In the sample test dataset, the input images are shown in the first row; in the second, the target predictions; in the third row, the Vision Transformer model's predictions; and in the last row, the UNet's predictions

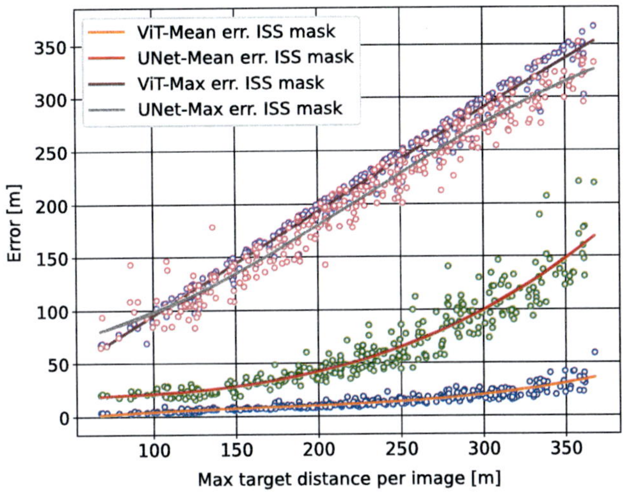

Fig. 11 The red line represents the interpolation of the ViT mean error computed on the target ISS's mask, while the orange one is the interpolation of the UNet mean error computed on the target ISS's mask. The purple line is the interpolation of the ViT max error computed on the target ISS's mask, and the grey line is the interpolation of the UNet max error computed on the target ISS's mask

5 Conclusions

In this work, the MDE problem in the case of a rendezvous with the ISS has been approached with a regression model that exploits vision transformers. Vision transformer are a new architecture for image encoders that promises to solve the problem of image degradation at the increase of the receptive field. The model makes dense predictions taking in input monocular RGB images of the ISS at various distances and returns the distance of the station's components for each pixel as an output.

We noticed an increase in the error with the distance in our test. The more challenging areas of the image for the model are the edges of the station's components. The mean error stays below 10% of the camera's distance from the station. The model shows consistent results also with the Sun or the Earth in the camera frame. This means that the model learned how to discern the ISS components.

Our model has been compared to a classic down-sampling CNN encoder, specifically the UNet architecture. ViT shows better results in the uniformity of the distance distribution on the target, and a much superior capacity to discriminate the subject from the background.

The MDE problem is challenging, especially when specific distances are required. The model does not reach the precision of active LiDAR sensors, nor can it be used at large distances from the target without significantly increasing the resolution. Even if the computational speed has not been taken into account in this work, an increase in resolution would drastically increase the computational cost of the model. Nevertheless, the model is powerful, and it is easy to implement as a backup or aid system to the more complex LiDAR sensors.

References

1. Cao, Z.L., Yan, Z.H., Wang, H.: Summary of binocular stereo vision matching technology. J. Chongqing Univ. Technol. (Nat. Sci.) **29**(2), 70–75 (2015)
2. Christian, J.A., Cryan, S.: A survey of lidar technology and its use in spacecraft relative navigation. In: AIAA Guidance, Navigation, and Control (GNC) Conference, p. 4641 (2013)
3. Deng, J., Dong, W., Socher, R., Li, L.J., Li, K., Fei-Fei, L.: Imagenet: A large-scale hierarchical image database. In: 2009 IEEE Conference on Computer Vision and Pattern Recognition, pp. 248–255. IEEE (2009)
4. Dosovitskiy, A., Beyer, L., Kolesnikov, A., Weissenborn, D., Zhai, X., Unterthiner, T., Dehghani, M., Minderer, M., Heigold, G., Gelly, S., et al.: An image is worth 16x16 words: Transformers for image recognition at scale. ArXiv preprint arXiv:2010.11929 (2020)
5. Eigen, D., Puhrsch, C., Fergus, R.: Depth map prediction from a single image using a multi-scale deep network. Adv. Neural Inf. Proc. Syst. **27** (2014)
6. Federici, L., Scorsoglio, A., Ghilardi, L., D'Ambrosio, A., Benedikter, B., Zavoli, A., Furfaro, R.: Image-based meta-reinforcement learning for autonomous terminal guidance of an impactor in a binary asteroid system. In: AIAA SCITECH 2022 Forum, p. 2270 (2022)
7. Fu, H., Gong, M., Wang, C., Batmanghelich, K., Tao, D.: Deep ordinal regression network for monocular depth estimation. In: Proceedings of the IEEE Conference on Computer Vision and Pattern Recognition, pp. 2002–2011 (2018)
8. Ghilardi, L., D'ambrosio, A., Scorsoglio, A., Furfaro, R., Linares, R., Curti, F.: Image-based optimal powered descent guidance via deep recurrent imitation learning. In: AIAA/AAS Astrodynamics Specialist Conference (2020)
9. Hazirbas, C., Ma, L., Domokos, C., Cremers, D.: Fusenet: Incorporating depth into semantic segmentation via fusion-based cnn architecture. In: Asian Conference on Computer Vision, pp. 213–228. Springer (2016)
10. Howard, I.P.: Perceiving in Depth, vol. 1: Basic mechanisms. Oxford University Press (2012)
11. Hussain, R., Zeadally, S.: Autonomous cars: Research results, issues, and future challenges. IEEE Commun. Surv. Tutorials **21**(2), 1275–1313 (2018)
12. Lee, W., Park, N., Woo, W.: Depth-assisted real-time 3d object detection for augmented reality, vol. 11. In: ICAT, pp. 126–132 (2011)
13. Li, X., Chen, H., Qi, X., Dou, Q., Fu, C.W., Heng, P.A.: H-denseunet: hybrid densely connected unet for liver and tumor segmentation from ct volumes. IEEE Trans. Med. Imaging **37**(12), 2663–2674 (2018)
14. Liu, F., Shen, C., Lin, G., Reid, I.: Learning depth from single monocular images using deep convolutional neural fields. IEEE Trans. Pattern Anal. Mach. Intell. **38**(10), 2024–2039 (2015)
15. Ranftl, R., Bochkovskiy, A., Koltun, V.: Vision transformers for dense prediction. In: Proceedings of the IEEE/CVF International Conference on Computer Vision, pp. 12179–12188 (2021)
16. Rasouli, A., Tsotsos, J.K.: Autonomous vehicles that interact with pedestrians: A survey of theory and practice. IEEE Trans. Intell. Transp. Syst. **21**(3), 900–918 (2019)
17. Scorsoglio, A., D'Ambrosio, A., Ghilardi, L., Furfaro, R., Gaudet, B., Linares, R., Curti, F.: Safe lunar landing via images: A reinforcement meta-learning application to autonomous hazard avoidance and landing. In: Proceedings of the 2020 AAS/AIAA Astrodynamics Specialist Conference, Virtual, pp. 9–12 (2020)

18. Scorsoglio, A., D'Ambrosio, A., Ghilardi, L., Gaudet, B., Curti, F., Furfaro, R.: Image-based deep reinforcement meta-learning for autonomous lunar landing. J. Spacecraft Rocket. **59**(1), 153–165 (2022)
19. Thumm, T.L., Robinson, J.A., Buckley, N., Johnson-Green, P., Kamigaichi, S., Karabadzhak, G., Nakamura, T., Sabbagh, J., Sorokin, I., Zell, M.: International space station benefits for humanity. In: 63rd International Astronautical Congress (IAC2012). No. JSC-CN-25979 (2012)
20. Vaswani, A., Shazeer, N., Parmar, N., Uszkoreit, J., Jones, L., Gomez, A.N., Kaiser, Ł., Polosukhin, I.: Attention is all you need. Adv. Neural Inf. Proc. Syst. **30** (2017)
21. Zhao, C., Sun, Q., Zhang, C., Tang, Y., Qian, F.: Monocular depth estimation based on deep learning: An overview. Sci. China Technol. Sci. **63**(9), 1612–1627 (2020)
22. Zheng, S., Lu, J., Zhao, H., Zhu, X., Luo, Z., Wang, Y., Fu, Y., Feng, J., Xiang, T., Torr, P.H., et al.: Rethinking semantic segmentation from a sequence-to-sequence perspective with transformers. In: Proceedings of the IEEE/CVF Conference on Computer Vision and Pattern Recognition, pp. 6881–6890 (2021)

RobDT: AI-enhanced Digital Twin for Space Exploration Robotic Assets

Marco Bozzano, Riccardo Bussola, Marco Cristoforetti, Srajan Goyal, Martin Jonáš, Konstantinos Kapellos, Andrea Micheli, Davide Soldà, Stefano Tonetta, Christos Tranoris, and Alessandro Valentini

Abstract Model-Based System and Software Engineering (MBSE) technology such as simulation has been adopted for decades by the space industry. During the lifecycle of a space mission a number of models are developed to support simulation and other analysis capabilities addressing needs specific for the different project phases. Typical concerns are: feasibility assessment, design optimization and validation, system performance and safety assessments, detail design verification and on-board

M. Bozzano · R. Bussola · M. Cristoforetti · S. Goyal · M. Jonáš · A. Micheli · D. Soldà · S. Tonetta (✉) · A. Valentini
Fondazione Bruno Kessler, Trento, Italy
e-mail: tonettas@fbk.eu

M. Bozzano
e-mail: bozzano@fbk.eu

R. Bussola
e-mail: rbussola@fbk.eu

M. Cristoforetti
e-mail: mcristofo@fbk.eu

S. Goyal
e-mail: sgoyal@fbk.eu

M. Jonáš
e-mail: mjonas@fbk.eu

A. Micheli
e-mail: amicheli@fbk.eu

D. Soldà
e-mail: dsolda@fbk.eu

A. Valentini
e-mail: alvalentini@fbk.eu

K. Kapellos · C. Tranoris
Trasys International, Bruxelles, Belgium
e-mail: konstantinos.kapellos@trasysinternational.com

C. Tranoris
e-mail: christos.tranoris@trasysinternational.com

C. Ieracitano et al. (eds.), *The Use of Artificial Intelligence for Space Applications*, Studies in Computational Intelligence 1088,
https://doi.org/10.1007/978-3-031-25755-1_12

software validation. In this context, symbolic and data-driven AI techniques can provide advanced capabilities to support the online operations of space missions. One of the main challenges to enable AI in the virtual flight segment is the problem of combining heterogeneous models in a common framework. The ROBDT project aims at developing a Robotic Digital Twin framework that combines data-driven models, physics-based and symbolic models and uses online data and data analytics to adapt the models at runtime. The digital twin will support the robotic asset operations by providing timing and reliable prediction and by supporting what-if analysis to assess multiple scenarios. In this paper, we present the architecture of the ROBDT framework and the preliminary achievements.

Keywords Digital twins · Space domain · Planning and scheduling · Diagnosis · Monitoring

1 Introduction

In this paper, we present the objectives and the early achievements of the ROBDT system, which is under development by the "Robotic Digital Twin" Activity funded by ESA and led by TRASYS in collaboration with FBK and GMV. In robotics, Digital Models of the target systems are traditionally used in all phases of a mission under the name 'Virtual Flight Segment' ranging from design to the development and the operations. These digital systems struggle to fully support their objectives in particular when the involved models are not able to capture the complete physical reality. This is particularly true when the operations environment evolves during the mission (e.g., when discovering a new planetary area).

The ROBDT activity proposes a new framework (see Fig. 1) where engineering methods and AI techniques are integrated into a coherent Robotic Digital Twin Framework in order to allow:

- *On-line update of the system models*: The appropriate combination of data-driven and physics-based simulation models enables the application of online data analytics for adapting at runtime the models of the virtual asset guaranteeing a high-fidelity representation of the physical asset and its environment.
- *Planning and what-if analyses*: A digital twin enables planning of actions and what-if analyses based on more reliable models. These analyses allow to synthesize unexpected scenarios and study the response of the system as well as the corresponding mitigation strategies. This kind of analysis without jeopardizing the real asset is only possible via a digital twin.
- *Plan monitoring and fault diagnosis*: Telemetry data are monitored to detect and identify anomalies. Diagnosis is performed to enable a retrospective analysis to extract the root causes of the observed failures. This is essential in order to support timely recovery from problematic situations and/or safely operate the real asset in a degraded mode of operation.

In this paper, we present the proposed software architecture to build the ROBDT system and the corresponding functionalities, with particular attention to the interplay and synergies between engineering methods, symbolic and data-driven AI. E.g., in a planetary exploration mission, the terrain model used for planning is the same as the one used for simulation, and it is adapted with machine learning algorithms based on the telemetry data. Finally, we describe the preliminary achievement in the demonstrator based on the ExoMars planetary exploration mission.

The rest of the paper is organized as follows: Sect. 2 discusses the related work; Sect. 3 details the framework's components; Sect. 4 describes the rover demonstrator; and Sect. 5 draws the conclusions.

2 Related Work

Analysed at its roots, NASA pioneered the concept of twins in the 1960s with Apollo project where two identical spacecraft were built, with the one on Earth called twin which reflected (or mirrored) the status of the spacecraft on a mission. Since then, *digital twin* (DT) concept has been applied extensively in the fields of manufacturing and robotics [2, 9, 10, 33, 35, 36, 43]. Many research works discussed the connotation, definition of digital twin concept, independently of the industry field [11, 16, 19, 22, 31, 34]. The term is not always used consistently and is explained in various ways from different perspectives. Considering the scope of this study, the following definition from [34] (adapted from [31]) is considered as the baseline:

"A Digital Twin is defined as a dynamic and self-evolving digital/virtual model or simulation of a physical asset (part, machine, process, human, etc.) representing its exact state at any given point of time enabled through real-time bi-directional data assimilation as well as keeping the historical data, for real-time prediction, monitoring, control and optimization of the asset to improve decision making throughout the life cycle of the asset."

Combination of data with models is at the heart of any DT. However, the specific types of models and integration of data strongly depend on the required services built on top of the DT.

With regards to online model adaptation, different methods have been proposed in literature to handle various uncertainties and partial observability (cfr., e.g., [8, 15, 23]). Digital twin frameworks based on supervised/unsupervised ML methods, such as dynamic Bayesian networks [13], particle filters [41], stacked denoising autoencoders (SDA) [24], etc., have been shown to enable continuous adaptation of physics based models for end-to-end uncertainty quantification, optimal decision-making, anomaly detection and the prediction of future conditions.

Hybrid modeling approaches have been widely used in scientific applications to embed the knowledge from simplified theories of physics based models directly into an intermediate layer of the neural network (see for example [4]). Within this paradigm, physics-informed learning (PIL) [14, 30, 39] is based on regularization design for discriminative properties, while physics-augmented learning (PAL) [20] is based on model design for generative properties.

Reinforcement learning (RL) algorithms typically replace the traditional model-based planning and control processes. Recent scientific applications [32, 37, 42] have utilized the potential of RL for the inference of physics-based model parameters. They have been shown to: provide accurate real-time dynamical calibration, adapt to new scenarios, scale to large datasets and high-dimensional spaces, and to be robust to observation and model uncertainty. Within the context of DT, RL algorithms have been applied in the field of manufacturing [43], robotics [27, 32] and autonomous driving [28, 42], to provide services such as real-time model adaptation, anomaly detection, or what-if analysis.

In the context of planning, [26] proposes a production system control concept where a digital twin and an automated AI planner are tightly integrated together as one smart production planning and execution system. In the context of run-time monitoring, [25] proposes a formal specification framework to facilitate and automate specification extraction in natural language from documentation, as well as formalizing them for the digital twin. [21] proposes an approach for active monitoring of a neural network (NN) deployed in the real-world, which detects previously unseen inputs and creates interpretable queries for incremental NN adaptation to ensure safe behaviour.

Overall, with respect to the above mentioned works, ROBDT presents some novel contributions. On one side, it is focused on space robotic systems, with the relevant peculiarities as for example the communication delays. On the other side, it provides a unique combination of symbolic automated reasoning, simulation, and machine learning techniques.

3 The ROBDT Framework

3.1 System Architecture

Figure 1 shows the ROBDT architecture as a first layer decomposition of the software into the following components:

Simulation As A Service (SimAAS): consists of multiple simulators which simulate the functionality of the robotic assets as well as the mechanisms for their upload to the system, their configuration and management (start, stop, delete), and finally the monitoring of their status. It involves the use of Kubernetes [18] and associated services.

Model As A Service (MAAS): provides ML Ops capabilities to the ROBDT by facilitating tasks such as data preparation, model training and model serving, while also enabling easy, repeatable, portable deployments on diverse infrastructure. It is based on the Kubeflow [17] open-source project. In the demonstrator, two

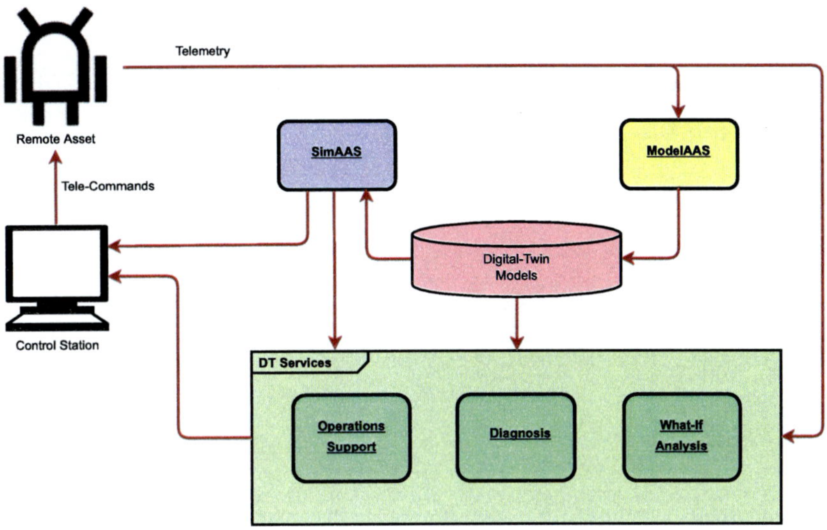

Fig. 1 The ROBDT architecture

such models and the corresponding pipelines are proposed: the wheel-terrain interaction model and the Data Handling Subsystem (DHS) update model.

Digital Twin Models: the models are handled by a Digital Twin Manager (DTM), which manages DT definitions and facilitates operations associated with launching, monitoring and stopping the corresponding simulations by hiding the complexities of Kubernetes' APIs that are used in SimAAS to perform the same operations.

What-if Analysis (WIA): allows simulating the system from its current state or from a hypothetical state according to a given scenario with the additional possibility to check whether a certain goal condition is satisfied, or it is violated. In the context of this activity the WIA component focuses on the simulation of activity plans under various conditions with the additional capability of automatic activity plan generation,

Diagnosis (FDIRPM): allows detecting faults in the current execution or on historic data, identifying causes of the faults based on their models, and providing the corresponding feedback to the operators. In the context of this activity, we propose to focus on the detection of faults during the execution of an activity plan and to propose a recovery action by generating an alternative one.

Operations Support (SCIDET): supports engineering or science operations planning and assessment. In this activity, a "scientific agent" is integrated to detect predefined patterns of interest or novelty on on-line or historical images acquired by the robotic asset.

These components use various heterogeneous models and there is a strong interplay between data-driven models, physics-based, and symbolic models, as summa-

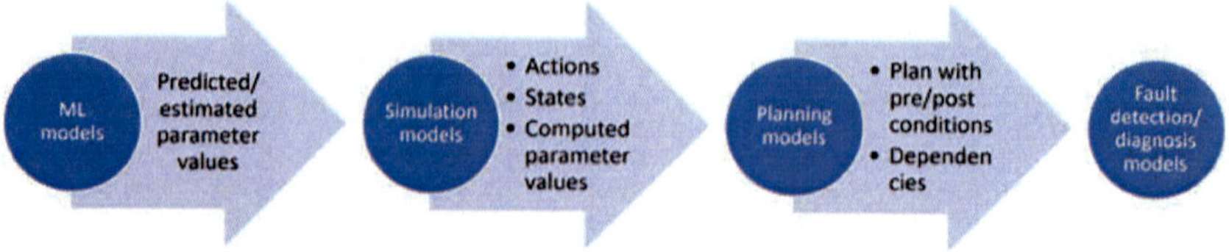

Fig. 2 Information exchanged between ROBDT models

rized in Fig. 2. In the following sections, we describe in more detail the components using these models.

3.2 Simulator

The simulation capabilities are provided by the instantiation of the SIMROB multi-asset space robotics simulator [12]. A high-level breakdown in models of SIMROB is depicted in Fig. 3. It includes:

- The **Assets models** that allow to compute the state evolution of the subsystems of the assets that are under control. For each asset, it mainly includes models of its mechanical, electrical and thermal dynamics, of its data handling, communications and payload subsystems as well as the model of its control software. In particular:
  - The *Mechanical Dynamic Model* provides the state evolution of a multi-body actuated and sensorised mechanism possibly subject to external forces. Main elements of the models are the bodies (rigid or flexible) with their inertial param-

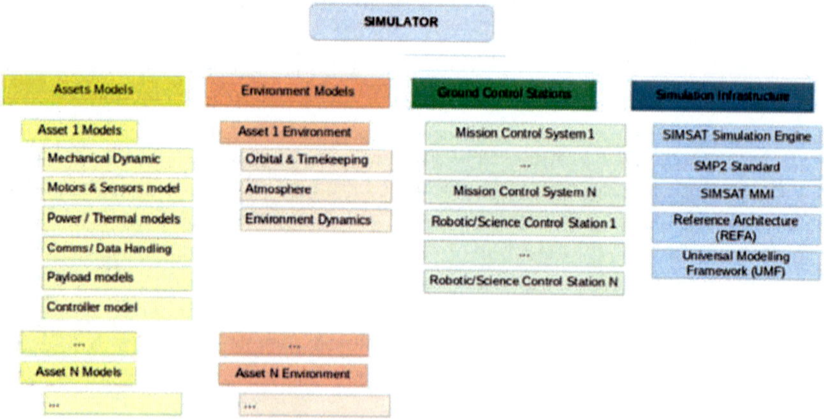

Fig. 3 The SIMROB simulator breakdown

eters, the joints (actuated or passive) including fixed, prismatic, rotoid and universal joints, and finally their topology.

  - The *Power Generation & Distribution Model* predicts the instantaneous energy flow in the power conditioning and distribution network in terms of current, voltages and losses at each relevant node, the state of charge and the behaviour of the battery, replicates the PCDE control logic and behaviour and finally simulates the power interfaces conditions (voltage, current) for all connected units.
  - The *Controller Model* is a model of particular importance as it reproduces the role of the onboard flight software ranging from asset control to instruments control and mission management. It is structured in a three-layered architecture. At the lowest level, close to the actuators and sensors, the Functional layer implements the Actions representing the elementary Activities of the asset. The Executive layer implements the Tasks defined as a logical and temporal composition of Actions. Finally, the Mission layer handles mission planning and scheduling aspects executing Activity Plans either created on-ground or on-board.

- The **Environment models** are in charge of the simulation of the environments that interact with the assets. These models encompass the reproduction of planetary and orbiter ephemerides, the provision of planetary atmospheric data as well as the morphology and the topology of the environment in which the assets evolve. In particular:
  - The *Atmosphere Model* provides at any time and location of the asset the relevant atmospheric characteristics. For example, for Mars surface operations, elements such as the air temperature, the surface temperature, the air specific heat capacity, the wind velocity, the dust optical depth, the atmospheric chemistry, as well as the radiation flux at ground level are important inputs for the simulation.
  - The *Environment Dynamic Model* represents the topological, the morphological and the mechanical dynamics aspects of the environment in which the assets evolve. It concerns exclusively the assets which operate in close loop with their environment: a rover moving on a terrain, a robotic arm grasping an object, a robotic system using exteroceptive sensors (e.g. imagers) to perceive its environment. The outputs of this model are forces/torques applied at a given point, distances measured from a given point, images generated from a given point-of-view, etc. These measures are used by the mechanical dynamic model of the asset to evaluate its state after impact/contact, and by the sensor models (imagers, distance, force/torque sensors, etc) to compute their outputs.

- The **Simulation framework** is based on SIMULUS. It covers both the execution of a spacecraft simulation and the simulator architectural design. The SIMSAT Simulation Engine is the 'engine' of the simulator, the SIMSAT Man-Machine Interface provides the interface between the user and the simulator, the Generic Models (GENM) comprise a suite of reusable generic simulation models, the SMP2 standard enables reuse and portability of the simulation models, the

Reference Architecture (REFA) establishes a suitable breakdown of simulators into models and finally the Universal Modelling Framework (UMF) supports an efficient and smooth approach of software development for SMP2 simulations.

3.3 Model Updaters

In the ROBDT architecture, the Model As A Service (MAAS) is a component whose goal is to provide ML Ops capabilities to the system. The MAAS facilitates data preparation, model training, and model serving and enables easy, repeatable, portable deployments on diverse infrastructure. Serving the models from MAAS allows specific digital twin components to access the model predictions for different use such as simulation and diagnosis.

Infrastructure The MAAS is based on a specific component of the Kubeflow open-source project: Kubeflow Pipeline. Kubeflow is a service capable of making deployments of machine learning (ML) workflows on Kubernetes simple, portable and scalable. The Kubeflow Pipeline is a platform in Kubeflow that has a UI and an API for triggering and tracking experiments, jobs, and runs. It also contains the engine used to manage and execute the ML workflows created by the Model Training Pipeline steps. The Kubeflow Pipelines is a platform for building and deploying portable, scalable machine learning (ML) workflows based on Docker containers. The role of the ROBDT's private Docker Registry is to manage the Docker images into which the various steps of model training pipelines are packaged. Based on Seldon's Core, the Model Server exposes trained models as web APIs, providing its clients the ability to make predictions and get information on the models. The MAAS offers interaction with the Monitoring and Control Station (MCS) component to obtain the historical and real-time telemetries. Those data are requested from the model updaters and used to build the training dataset.

Model Updater Component Based on the functionality provided by Kubeflow it is possible to trigger the training of a model manually or automatically at regular intervals. Both can be performed and configured via Kubeflow's Pipeline UI or the corresponding API. Once the pipeline is triggered, Kubeflow Pipeline's orchestration engine starts executing the described workflow. The pipeline is made up of Docker images related to each other as a graph through input and output files dependencies. The pipeline steps are the following:

- *Telemetry acquisition*: the first step of the pipeline is responsible for the telemetry acquisition. The operation is done by contacting the MCS component by REST API, which provides the requested data in JSON format.
- *Data pre-processing*: the historical telemetries of the mission are pre-processed for the training and testing phase of the model.

- *Training*: the training dataset is obtained from the previous step and loaded using a custom DataLoader. The model is instantiated and trained; the model parameters are saved on a dedicated, persistent volume. As a result of this step, the best performing model is available for the next steps.
- *Test*: the model runs on the testing dataset to validate the performance. The score is then passed to the next step to provide information about the behavior of the updated model.
- *Deployment*: to enable inference service to the other ROBT components, this final step provides the configuration for the Seldon Core platform about the model parameters and the Python's handler script.

3.4 Planner and What-if Analysis

One of the main problems when managing a remote asset is to plan the activities to be performed ahead-of-time, because a significant time delay can hinder direct tele-commanding. For example, it is not possible to tele-command an asset on the surface of Mars due to the communication delay that is in the order of tens of minutes. For this reason, the remote assets are equipped with autonomous executors of mission plans that need to be properly formulated on-ground ahead of time and uploaded before execution. Designing activity plans that are robust and achieve a desired objective is no trivial task when the complexity of the target system is more than trivial. For this reason, automated planning technologies have been historically employed in space applications. The usefulness of such technique is not limited to the generation of plans for the immediate future, but also to investigate hypothetical situations and to perform so-called "what-if analyses" (WIA): before committing to a specific plan or addressing a problematic situation, a planning and simulation system allows the study of different plans and objectives in different real or hypothetical situations. Finally, WIA allows for retrospective analyses: in light of new model information, one can re-assess past decision in order to improve future decision-making.

Within the ROBDT framework, WIA is seen as a service that takes advantage of the superior precision of digital twin models: thanks to the strong alignment between the models and the physical assets it is possible to provide more realistic estimations of the costs of a certain plan and ultimately to provide a better support for decision-making. Moreover, an interesting and pivotal feature offered by digital twins is the evolution of models, making it possible to adapt automated planning to the degradation of capabilities or to the evolving conditions of the environment.

On the practical side, the overall idea behind our digital-twin-enhanced WIA is to maintain a model-based approach for planning and high-level simulation, but to allow for parameters that are to be estimated/learned from the telemetry data within the digital twin. Concretely, we will study the behavior of automated planning when some parameters (in particular the duration of some activities) are estimated by means of ML models. Ideally, having a more precise timing model of the system will allow a less conservative planning which in turn will allow to fully exploit the remote asset capabilities.

3.5 Fault Detection and Diagnosis

Diagnosing faults is essential in order to detect and identify anomalies that could endanger the real asset. To this aim, the system is equipped with a Fault Detection, Isolation, and Recovery (FDIR) component that monitors the telemetry data and the execution of the activities, in their initial, in progress and terminating phase. A relevant aspect strictly related to the monitoring phase is the ability to perform its task even in absence of complete information on the state; for example, both the state of some components and the currently executed action may be unknown. In such a case, a set of belief states that are compatible with the observations, strictly contained in the set of all possible states, is considered and the diagnosis phase can be employed with this partial information. With the aim of addressing the partial observability problem we have used NuRV [5–7]; this tool is able to generate a monitor for a specific LTL property to be monitored on a given system. In this case the system consists of the plan, which can be seen as a loop-free algorithm, synthesized by the planner component, while the conditions to be monitored are pre, in progress and post conditions. In case an anomaly is detected, a retrospective analysis based on the DT models and the historical information on past states, is carried out to localize the possible faults and identify the root causes of the failure. Diagnosis is based on a fault model which describes the effects of faults, their dependencies and the fault propagation rules. The adaptation of the DT models at run-time can improve the situational awareness of the real asset and provide a more precise analysis with respect to the FDIR capabilities of the physical system alone. When an anomaly is detected, the FDIR component can trigger a reconfiguration of the system, e.g. to continue operation in a degraded mode. FDIR can also be used to aid predictive maintenance by supporting the detection of the performance degradation of some component. Finally, the FDIR component provides a service to the what-if analysis functionality, namely it supports the planning activities by monitoring the plan execution in order to detect and identify unexpected outcomes of the actions.

4 The ROBDT Case Study

For the demonstration of the ROBDT framework, we chose to work on a planetary robotic asset provided by TRASYS. Because of the inherent uncertainty of the robot-environment interaction, data-based models are well suited. Moreover, it is an ideal case study for path planning and monitoring. Let us consider a typical scenario prepared for a 'sol' execution from the ExoMars planetary exploration mission:

the 'Drilling site approach and surface sample acquisition'. The following activities shall be performed autonomously under the constraints of the available power, memory capacity for data storage, and duration (single sol). Initially, the rover waits the transition Night to Day to wake-up (driven by the rover PCDE, when the solar panels generate power greater than a threshold (20W)) and configures accordingly the rover for the day activities. In particular, the subsystems involved for travelling are warmed-up and moved to a 'standby' state. These steps involve several uncertainties, mainly the exact local Mars time at which the rover wakes-up as well as the warm-up durations, which all depend on the external conditions (e.g., atmospheric temperatures, relative orientation of the rover solar panels with respect to the Sun, etc.).

At completion of the rover configuration, the rover starts traveling to reach the outcrop whose position has been identified from ground. Although the duration of the travel depends on the topology and characteristics of the encountered terrain, it can be *estimated* at planning time. At arrival at the outcrop, the rover is unconfigured from travelling operations and configured for drilling: travel related units are switched off while the drill box and the drill are warmed-up and moved to the 'standby' state. Again, the expected durations (and therefore power consumptions) have to be estimated as they depend on the time in the 'sol' that the rover reached the outcrop. At the next step the drill box is deployed, the drill is initialised, and reaches the soil to collect the surface sample (10 cm depth). Afterwards, the drill retracts. The duration of the sampling procedure, and therefore the power consumption, depends on the hardness of the soil. Finally, images of the environment shall be acquired and downloaded to guarantee that the ground planning team has enough information for planning for the next sol. After establishing the communications with the orbiter and transferring the acquired data, the rover waits the transition Day to Night (driven by the rover PCDE, when the solar panels generate power less than the threshold of 20 W), configures the Rover for night and 'sleeps' waiting for the next plan to be uploaded for execution.

In this scenario, the use case foresees:

- For the ROBDT system, there are two specific machine learning models that are adapted online: the wheel terrain interaction (WTI) model and the DHS model that predicts the Actuator Drive Electronics (ADE) warm-up time. Those elements run on top of the MAAS component taking advantage of the Kubeflow Pipeline architecture. In both cases, the approach used to fix the model architecture is the same. As an example, we present here the case of the WTI model. The purpose of the WTI model is to estimate the drawbar pull of the robotic asset and the power consumption of the motors. These calculations are made from the terrain and rover characteristics given as input from pre-processed telemetries. In particular, the inputs of the model are the terrain type (class), slip ratio (%), stiffness (%), and normal axis load (Newton). The outputs are the average drawbar pull (Newton) and the power consumption (Watt). Given the small number of input features of the model, the advantage of using Deep Learning solutions is not clear, as they usually require many parameters and are therefore heavy to train. For this reason, we are

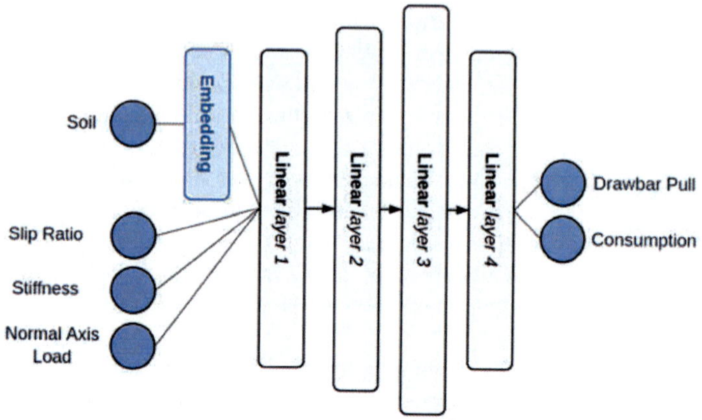

Fig. 4 The architecture of the Fully connected neural network for the WTI model

testing two different solutions to achieve the described functionality: the first is based on a Deep Learning Fully Connected Neural Network, and the second uses Boosted Decision Tree. While both can fit in the pipeline based on the MAAS, predictive power and computational resources required for the online adaptation can be different and must be carefully evaluated. The framework adopted for the development of the fully connected neural network is Pytorch. The neural network solution is composed of an initial embedding layer that maps the terrain class to a higher dimensional space. Then, this embedding is merged with the other features forming the input of the first layer of the network. A full representation of the network architecture is presented in Fig. 4. For the boosted decision tree we used the Catboost Python library [29].

- For the planning part, we are setting up and experimenting with a fully-integrated solution, which is capable of planning with semi-opaque models. In particular, we have access to a model of the activities and tasks of the rover and we need to automatically synthesize activity plans. However, some of the actions have simulated effects, meaning that the consequences of applying such actions are not modeled but can only be simulated, and the duration of some actions are also not formally modeled. Most notably, among these "evaluatable" quantities, we have the ADE warm-up timing that is estimated by a learned and evolving ML model as discussed above. We use the AIPlan4EU Unified Planning library [1] to model simulated effects and the TAMER planner[1] [38] for the actual plan generation. Preliminary results show that the approach is capable of generating valid plans quickly, and we are currently working on experimenting with the quantities estimated by means of ML.
- As for the fault detection part, we are monitoring the execution of the plan. The telemetry provides complete information about the state components, but no infor-

[1] https://tamer.fbk.eu.

mation about the state of the task execution. Looking at the case study, the plan provided by the planner requires first to run a task that waits for the rover to warm up and switch on a set of subsystems for traveling. Upon completion of this action, a state component is modified. This state component acts as a precondition for a task that is used for updating the rover heading estimate with a value provided by ground. If the monitor notices that the heating level has just been changed, but the precondition state component of the heating level update task has not been satisfied before, it can decide that the heating update task has been executed violating its preconditions in one of the possible belief states will report this violation to the operator and to the diagnostic component.

- Given the anomalies identified by the monitoring component, the goal of the diagnosis component is to provide a list of most probable explanations for these anomalies. The explanations are identified using a fault propagation graph (FPG), which describes how failures of one subsystem or component of the rover can cause failures of other components. In particular, for ROBDT, we construct the FPG as follows. First, for each task, we use the DT specification to identify the actions of other subsystems that can cause a failure of the given task. For example, the warm-up task that prepares the rover for travel depends on warming up the navigation cameras, localization cameras, actuator drive etc. Second, we use the description of hardware implementation and FMECA tables to describe how failures in the hardware components can cause failures of the higher level subsystems. For example, the actuator drive depends on working hold-down release mechanism, which in turn depends on working motors, motor heaters, etc. We then use efficient techniques rooted in formal methods [3] that for each set of identified failures list all the possible root causes.

  As a result, if the monitoring component reports an anomaly in the warm-up task, we can list failure of motors as one of the root causes (among many others). More interestingly, if the monitoring component reports several anomalies, which all transitively depend on the motors, the diagnosis component can report the motor failure as the most probable root cause as it is more probable than multiple separate failures of independent subsystems.

5 Conclusions

In this paper, we presented the architecture of the Robotic Digital Twin, which is under development as part of the ROBDT study funded by ESA. We detailed the planning and monitoring functionalities and the related combination of data-driven physics-based, and symbolic models. Finally, we described the preliminary achievement in the demonstrator based on the ExoMars planetary exploration mission. In the remaining activities of the project, which will conclude within 2022, we will complete the prototype implementation and will evaluate it on the described scenario, thus identifying strengths and weaknesses of the approach. In the future, we are going to use the same infrastructure to validate and verify autonomous systems with AI/ML components with a simulation-based system level approach. This is part of VIVAS, another ESA-funded study started in May 2022 [40].

References

1. AIPlan4EU.: The AIPlan4EU Unified Planning Library. https://github.com/aiplan4eu/unified_planning
2. Booyse, W., Wilke, D.N., Heyns, S.: Deep digital twins for detection, diagnostics and prognostics. Mech. Syst. Sig. Proc. **140**, 106612 (2020)
3. Bozzano, M., Cimatti, A., Pires, A.F., Griggio, A., Jonáš, M., Kimberly, G.: Efficient SMT-based analysis of failure propagation. In: Silva, A., Leino, K.R.M. (eds.) Computer Aided Verification–33rd International Conference, CAV 2021, July 20–23, Proceedings, Part II. Lecture Notes in Computer Science, vol. 12760, pp. 209–230. Springer (2021). https://doi.org/10.1007/978-3-030-81688-9_10
4. Chao, M.A., Kulkarni, C., Goebel, K., Fink, O.: Fusing physics-based and deep learning models for prognostics. Reliab. Eng. Syst. Saf. **217**, 107961 (2022)
5. Cimatti, A., Tian, C., Tonetta, S.: Assumption-based runtime verification with partial observability and resets. In: International Conference on Runtime Verification, pp. 165–184. Springer (2019)
6. Cimatti, A., Tian, C., Tonetta, S.: Nurv: a nuxmv extension for runtime verification. In: International Conference on Runtime Verification, pp. 382–392. Springer (2019)
7. Cimatti, A., Tian, C., Tonetta, S.: Assumption-based runtime verification of infinite-state systems. In: International Conference on Runtime Verification, pp. 207–227. Springer (2021)
8. Damianou, A., Lawrence, N.D.: Deep gaussian processes. In: Artificial Intelligence and Statistics, pp. 207–215. PMLR (2013)
9. Glaessgen, E., Stargel, D.: The digital twin paradigm for future nasa and us air force vehicles. In: 53rd AIAA/ASME/ASCE/AHS/ASC Structures, Structural Dynamics and Materials Conference 20th AIAA/ASME/AHS Adaptive Structures Conference 14th AIAA, p. 1818 (2012)
10. Hochhalter, J., Leser, W.P., Newman, J.A., Gupta, V.K., Yamakov, V., Cornell, S.R., Willard, S.A., Heber, G.: Coupling damage-sensing particles to the digitial twin concept. Tech, Rep (2014)
11. Jones, D., Snider, C., Nassehi, A., Yon, J., Hicks, B.: Characterising the digital twin: A systematic literature review. CIRP J. Manuf. Sci. Technol. **29**, 36–52 (2020)
12. Kapellos, K.: Simulus based simulations of human-robotic operations. ESAW (2021)
13. Kapteyn, M.G., Pretorius, J.V., Willcox, K.E.: A probabilistic graphical model foundation for enabling predictive digital twins at scale. Nat. Comput. Sci. **1**(5), 337–347 (2021)
14. Karniadakis, G.E., Kevrekidis, I.G., Lu, L., Perdikaris, P., Wang, S., Yang, L.: Physics-informed machine learning. Nat. Rev. Phys. **3**(6), 422–440 (2021)
15. Kennedy, M.C., O'Hagan, A.: Bayesian calibration of computer models. J. Royal Stat. Soc.: Ser. B (Stat. Methodol.) **63**(3), 425–464 (2001)
16. Kritzinger, W., Karner, M., Traar, G., Henjes, J., Sihn, W.: Digital twin in manufacturing: A categorical literature review and classification. IFAC-PapersOnLine **51**(11), 1016–1022 (2018)
17. Kubeflow. https://www.kubeflow.org
18. Kubernetes. https://www.kubernetes.io
19. Liu, M., Fang, S., Dong, H., Xu, C.: Review of digital twin about concepts, technologies, and industrial applications. J. Manuf. Syst. **58**, 346–361 (2021)
20. Liu, Z., Chen, Y., Du, Y., Tegmark, M.: Physics-augmented Learning: A New Paradigm Beyond Physics-informed Learning. arXiv:2109.13901 (2021)

21. Lukina, A., Schilling, C., Henzinger, T.A.: Into the unknown: Active monitoring of neural networks. In: Feng, L., Fisman, D. (eds.) Runtime Verification, pp. 42–61. Springer International Publishing, Cham (2021)
22. Madni, A.M., Madni, C.C., Lucero, S.D.: Leveraging digital twin technology in model-based systems engineering. Systems **7**(1), 7 (2019)
23. Marmin, S., Filippone, M.: Variational Calibration of Computer Models. ArXiv preprint arXiv:1810.12177 (2018)
24. Meraghni, S., Terrissa, L.S., Yue, M., Ma, J., Jemei, S., Zerhouni, N.: A data-driven digital-twin prognostics method for proton exchange membrane fuel cell remaining useful life prediction. Int. J. Hydrogen Energy **46**(2), 2555–2564 (2021)
25. Naumchev, A., Sadovykh, A., Ivanov, V.: VERCORS: Hardware and software complex for intelligent round-trip formalized verification of dependable cyber-physical systems in a digital twin environment (position paper). In: Lecture Notes in Computer Science (Including subseries Lecture Notes in Artificial Intelligence and Lecture Notes in Bioinformatics), 11771 LNCS, 351–363 (Oct 2019). https://doi.org/10.1007/978-3-030-29852-4_30, https://link.springer.com/chapter/10.1007/978-3-030-29852-4_30
26. Novák, P., Vyskočil, J., Wally, B.: The digital twin as a core component for industry 4.0 smart production planning. IFAC-PapersOnLine **53**(2), 10803–10809 (2020). https://doi.org/10.1016/j.ifacol.2020.12.2865, https://www.sciencedirect.com/science/article/pii/S2405896320336314
27. Oh, M.h., Iyengar, G.: Sequential anomaly detection using inverse reinforcement learning, p. 1480–1490. KDD '19, Association for Computing Machinery, New York, NY, USA (2019). https://doi.org/10.1145/3292500.3330932
28. Pires, F., Ahmad, B., Moreira, A.P., Leitão, P.: Recommendation system using reinforcement learning for what-if simulation in digital twin. In: 2021 IEEE 19th International Conference on Industrial Informatics (INDIN), pp. 1–6 (2021). https://doi.org/10.1109/INDIN45523.2021.9557372
29. Prokhorenkova, L., Gusev, G., Vorobev, A., Dorogush, A.V., Gulin, A.: Catboost: unbiased boosting with categorical features. Adv. Neural Inf. Proc. Syst. **31** (2018)
30. Raissi, M., Perdikaris, P., Karniadakis, G.E.: Physics-informed neural networks: A deep learning framework for solving forward and inverse problems involving nonlinear partial differential equations. J. Comput. Phys. **378**, 686–707 (2019)
31. Rasheed, A., San, O., Kvamsdal, T.: Digital twin: Values, challenges and enablers from a modeling perspective. IEEE Access **8**, 21980–22012 (2020)
32. Regatti, J.R., Deshmukh, A.A., Cheng, F., Jung, Y.H., Gupta, A., Dogan, U.: Offline rl with resource constrained online deployment (2021). https://doi.org/10.48550/ARXIV.2110.03165, https://arxiv.org/abs/2110.03165
33. Resman, M., Protner, J., Simic, M., Herakovic, N.: A five-step approach to planning data-driven digital twins for discrete manufacturing systems. Appl. Sci. **11**(8), 3639 (2021)
34. Singh, M., Fuenmayor, E., Hinchy, E.P., Qiao, Y., Murray, N., Devine, D.: Digital twin: Origin to future. Appl. Syst. Innovation **4**(2), 36 (2021)
35. Söderberg, R., Wärmefjord, K., Carlson, J.S., Lindkvist, L.: Toward a digital twin for real-time geometry assurance in individualized production. CIRP Ann. **66**(1), 137–140 (2017)
36. Sun, X., Bao, J., Li, J., Zhang, Y., Liu, S., Zhou, B.: A digital twin-driven approach for the assembly-commissioning of high precision products. Robot. Comput. Integr. Manuf. **61**, 101839 (2020)
37. Tian, Y., Chao, M.A., Kulkarni, C., Goebel, K., Fink, O.: Real-time Model Calibration with Deep Reinforcement Learning. ArXiv preprint arXiv:2006.04001 (2020)
38. Valentini, A., Micheli, A., Cimatti, A.: Temporal planning with intermediate conditions and effects. In: The Thirty Fourth AAAI Conference on Artificial Intelligence, AAAI 2020, The Thirty-Second Innovative Applications of Artificial Intelligence Conference, IAAI 2020, The Tenth AAAI Symposium on Educational Advances in Artificial Intelligence, EAAI 2020, New York, NY, USA, February 7–12, 2020. pp. 9975–9982. AAAI Press (2020). https://ojs.aaai.org/index.php/AAAI/article/view/6553

39. Viana, F.A., Nascimento, R.G., Dourado, A., Yucesan, Y.A.: Estimating model inadequacy in ordinary differential equations with physics-informed neural networks. Comput. Struct. **245**, 106458 (2021)
40. VIVAS. https://es.fbk.eu/index.php/projects/vivas/
41. Ward, R., Choudhary, R., Gregory, A., Jans-Singh, M., Girolami, M.: Continuous calibration of a digital twin: Comparison of particle filter and bayesian calibration approaches. Data-Cen. Eng. **2** (2021)
42. Wu, J., Huang, Z., Hang, P., Huang, C., De Boer, N., Lv, C.: Digital twin-enabled reinforcement learning for end-to-end autonomous driving. In: 2021 IEEE 1st International Conference on Digital Twins and Parallel Intelligence (DTPI), pp. 62–65 (2021). https://doi.org/10.1109/DTPI52967.2021.9540179
43. Xia, K., Sacco, C., Kirkpatrick, M., Saidy, C., Nguyen, L., Kircaliali, A., Harik, R.: A digital twin to train deep reinforcement learning agent for smart manufacturing plants: Environment, interfaces and intelligence. J. Manuf. Syst. **58**, 210–230 (2021)

An Overview of X-TFC Applications for Aerospace Optimal Control Problems

Enrico Schiassi, Andrea D'Ambrosio, and Roberto Furfaro

Abstract This paper is an overview of Optimal Control Problems (OCPs) for aerospace applications tackled via the indirect method and a particular Physics-Informed Neural Networks (PINNs) framework, developed by the authors, named Extreme Theory of Functional Connections (X-TFC). X-TFC approximates the unknown OCP solutions via the Constrained Expressions, which are functionals made up of the sum of a free-function and a functional that analytically satisfies the boundary conditions. Thanks to this property, the framework is fast and accurate in learning the solution to the Two-Point Boundary Value Problem (TPBVP) arising after applying the Pontryagin Minimum Principle. The applications presented in this paper regard intercept problems, interplanetary planar orbit transfers, transfer trajectories within the Circular Restricted Three-Body Problem, and safe trajectories around asteroids with collision avoidance. The main results are presented and discussed, proving the efficiency of the proposed framework in solving OCPs and its low computational times, which can potentially enable a higher level of autonomy in decision-making for practical applications.

Keywords Aerospace optimal control problems · Extreme learning machines · Machine learning · Physics-informed neural networks

E. Schiassi (✉) · R. Furfaro
Department of Systems and Industrial Engineering, The University of Arizona, Tucson, AZ, USA
e-mail: eschiassi@arizona.edu

A. D'Ambrosio
Department of Aeronautics and Astronautics, Massachusetts Institute of Technology, Cambridge, MA, USA

R. Furfaro
Department of Aerospace and Mechanical Engineering, The University of Arizona, Tucson, AZ, USA

C. Ieracitano et al. (eds.), *The Use of Artificial Intelligence for Space Applications*, Studies in Computational Intelligence 1088,
https://doi.org/10.1007/978-3-031-25755-1_13

1 Introduction

The continued and exponential increase in computational capabilities can enable a new level of autonomy and decision-making, especially for aerospace systems. In particular, the ability to generate real-time trajectories may be considered a technology of utmost importance for managing autonomous systems within complex environments. Similarly, new analytical and data-driven methods for real-time synthesis of closed-loop optimal controllers may integrate autonomy with robustness and stability. Optimal trajectory generation and closed-loop control rely on optimal control theory, a well-established discipline. Indeed, given a prescribed model of the dynamical system under consideration, it aims at finding optimal policies and trajectories by minimizing a cost functional that accounts for all constraints and design objectives.

Overview on Optimal Control Problems. Usually, two methods are available to solve Optimal Control Problems (OCPs), i.e., direct and indirect methods.

Direct methods are based on discretizing the continuous states and controls to transform the continuous problem into a Non-Linear Programming (NLP) problem [11, 16, 46]. The latter can be posed as a finite constrained optimization problem, which can be solved via any numerical algorithms able to find a local minimum, such as the thrust region method [4]. Although direct methods have been applied to solve a large variety of aerospace optimal control problems [20, 23, 26, 37], the general NLP problem is considered to be NP-hard, i.e., non-deterministic polynomial-time hard. NP-hard problems imply that the required computational time needed to find the optimal solution does not have a predetermined bound. Consequently, the lack of assured convergence raises questions on the reliability of the proposed approach. More recently, researchers have been experimenting with transforming optimal control problems from non-convex optimization problems into convex optimization problems [1, 3]. Here, the goal is to take advantage of the assured convex convergence properties. Indeed, convex optimization problems are computationally tractable as their related numerical algorithms guarantee convergence to a globally optimal solution in a polynomial time. The general convex methodology requires that the original problem is formulated as convex optimization using convexification techniques. Such methodologies have been proposed and applied to solve optimal guidance and control via direct method in a large variety of aerospace problems including, planetary landing [1, 3], entry atmospheric guidance [56, 57], rocket ascent guidance [59], low thrust [58], and rendezvous [19].

Conversely, OCP can be solved by *indirect method*. The latter derives the first-order necessary conditions by direct application of *Pontryagin Maximum Principle or PMP*. The necessary conditions result in a Two-Point Boundary Value Problem (TPBVP) in the state-costate pair that is generally tackled via numerical techniques such as single and multiple shooting methods [28, 53], orthogonal collocation [43], or pseudo-spectral methods [15]. Generally, the indirect method provides a solution that is guaranteed to be optimal by definition. This is why the indirect method should be preferred to the direct method. However, the solution of the TPBVPs can be very cumbersome because of the strong sensitivity of the solution to the initial guesses when facing nonlinear problems. Moreover, it is also challenging to provide good initial guesses for the costates since they do not represent any physical quantity for most applications. On the contrary, although the direct method may provide solutions that are not necessarily optimal, the solution is more straightforward. Thus it is much more spread within the scientific community. In principle, the two approaches can also be combined. For instance, the solution obtained with the direct method can be exploited as an initial guess for the indirect method, which can better refine the solution and, at the same time, guarantee optimality.

A complementary approach to solving OCP comes from the application of the *Bellman Principle of Optimality or BPO*. The latter provides the necessary and sufficient conditions for optimality which yields in the so-called *Hamilton-Jacobi-Bellman Equation or HJB* equation. The HJB is a nonlinear, high dimensional Partial Differential Equation (PDE), crucial in the optimal control theory. The unique solution of the HJB equation is the *value function* in all the time-state space. The value function is the optimal *cost-to go* for a dynamical system with an associated cost function. Once the value function is known, the optimal control is also known as a function of the gradient of the value function with respect to the state variables. For many practical applications is impossible to find an analytical solution to the HJB equation. Thus, one has to rely on numerical methods. The main issue when facing the HJB equation is that its space complexity increases exponentially to the number of dimensions of the system (e.g., the number of independent variables), i.e., the problem is affected by the *curse of dimensionality* [44].

Physics-Informed Machine Learning and Neural Networks. Over the past decade, significant advances in Machine Learning (ML) have dramatically transformed many scientific fields, including aerospace engineering. Indeed, enabled by the reduction in the cost of sensors, storage, and computational resources, a new class of data-driven discovery methods have contributed to innovative results in characterizing complex nonlinear relationships in high-dimensional experimental data. However, discovering underlying physical laws and the associated governing equations from available Spatio-temporal data is a much less understood problem and is currently an active field of research. Governing equations are generally derived from first principles and are rooted in physical laws, such as conservation laws (e.g., conservation of mass, momentum, energy). The application of such principles results in mathematical models, usually ordinary or partial Differential Equations (DEs), which are found in any field of science and engineering. However, complete and accurate models of complex and interconnected systems are challenging. Physics-Informed Machine Learning (PIML) may play an essential and critical role in this context. Rather than optimizing a solely data-driven loss function, PIML constraints the loss function and then the optimization procedure to obey a prescribed physics law [27]. Recently, a new framework named *Physics-Informed Neural Networks* (PINNs) has been introduced by Raissi et al. [45] which refers to Neural Networks (NNs) that are trained by data and constrained to satisfy specific physical laws. Traditional ML methods are *data-driven*, i.e., they are solely trained on data to model the unknown nonlinear relationship between inputs and outputs. Although successful in discovering patterns and extracting system behavior, they do not incorporate explicit information about the collected data's physical processes. Thus, one wants to define the proper model based on how much data is available and how much physics is known. PINN fits in the PIML space [27], i.e., using the universal approximating power of NN to constrain data by the physical laws (forward problem) or discover governing equations/physical laws and/or unknown parameters from the data (inverse problem). Furthermore, PINN employs recent advancements in Automatic Differentiation (AD) as implemented in current ML libraries (e.g., Tensorflow, Pytorch) [34] to directly compute derivatives of NN models that are explicitly employed to satisfy dynamical models during the training process. The PINN task is, for a given training set $\mathcal{S} = \{\boldsymbol{x}_i, u_i\}_{i=1}^N$ of size N, to capture the (non-)linear function underlying the data $u(\boldsymbol{x})$ by using a NN $u^{\boldsymbol{\theta}} = u(\boldsymbol{x}; \boldsymbol{\theta})$ that approximates the data while constraining the training process to satisfy a governing DE,

$$G\left(\boldsymbol{x}; \frac{\partial u}{\partial x_1}, ..., \frac{\partial u}{\partial x_d}; \frac{\partial^2 u}{\partial x_1 x_d}, ..., \frac{\partial^2 u}{\partial x_1 x_d}; ...; \boldsymbol{\Gamma}\right) = 0 \quad \forall \boldsymbol{x} \in \Omega, \; \mathcal{B}(u, \boldsymbol{x}) = 0, \forall \boldsymbol{x} \in \partial\Omega \tag{1}$$

Here, $\boldsymbol{\Gamma}$ is a set of parameters that may be estimated during the training process, where the following loss, which is data-physics-driven, is minimized:

$$\mathcal{L}(\mathcal{S}; \boldsymbol{\theta}) = \lambda_{data}\mathcal{L}_{data}(\mathcal{S}_\Omega; \boldsymbol{\theta}) + \lambda_\Omega \mathcal{L}_\Omega(\mathcal{S}_\Omega; \boldsymbol{\theta}) + \lambda_\mathcal{B}\mathcal{L}_\mathcal{B}(\mathcal{S}_\Omega; \boldsymbol{\theta}) \tag{2}$$

The loss associated to the available data is $\mathcal{L}_{data}(\mathcal{S}_\Omega; \boldsymbol{\theta})$, while $\mathcal{L}_\Omega(\mathcal{S}_\Omega; \boldsymbol{\theta})$ and $\mathcal{L}_\mathcal{B}(\mathcal{S}_\Omega; \boldsymbol{\theta})$ are the losses inside the domain Ω and at the boundary $\partial\Omega$ imposed by the governing equation, respectively. The loss involves partial derivatives of the NN approximation which can be computed with an AD package. One is generally interested in two problems. Data-physics-driven DE solution (i.e., forward problem): given the parameter $\boldsymbol{\Gamma}$, one would like to infer the true function $u(\boldsymbol{x})$ underlying the data and satisfying the governing Eq. (1). To be noticed that in this scenario, the data term in Eq. (2) may not be present. In this case, we talk about physics-driven DE solution). Data-physics-driven DE parameters discovery/estimation (i.e., inverse problem): find the parameter $\boldsymbol{\Gamma}$ that best describes the observed data. PINN frameworks can be used to learn the solutions of OCPs in a (data)-physics-driven fashion via the application of both the PMP and the BPO. Importantly, PINNs allow overcoming the curse of dimensionality for the solution of HJB equations. This is possible because of the NN mathematical structure and because the training points do not necessarily need to be collocated on a grid. Indeed any collocation scheme could be adopted to collocate the training points within the training domain [38]. Traditionally, PINNs are trained via gradient-based methods [5, 29], which are computationally expensive, especially when dealing with large-scale problems. In addition, the loss function must include other terms to satisfy the DE constraints (e.g., the initial and/or boundary conditions). Satisfying the DE constraints is a complex task that can affect the result of the training both in terms of accuracy and computational time [52]. Indeed, this is the major drawback of the classic (or traditional) PINN frameworks [45], where DE constraints are not analytically satisfied and must be included in the loss function with the DE residual within the domain. Hence, in the PINN training, we have competing objectives: (1) to minimize the DE residual within the domain (which is an unsupervised learning task), and (2) to minimize the discrepancy of the PINN approximation of the DE solution on the boundaries with the given DE constraints (which is a supervised learning task). As studied in [54], having competing objectives in the loss function cause stiffness in the gradient flow dynamics, which can lead to unbalanced gradients during the PINN training via gradient-based techniques. This represents the primary PINN failure mode in learning the DE solutions. It is indeed known that gradient-based methods may get stuck in limit cycles or diverge in the presence of multiple competing objectives [2, 36].

Physics-Informed Neural Networks and Functional Interpolation. Leake and Mortari in [33] and Schiassi et al. [51], overcomes the issue of competing objectives by combining NNs with a mathematical framework for functional interpolation, called *Theory of Functional Connections (TFC)* [40]. TFC is a general mathematical method for functional interpolation where functions are modeled via functionals called Constrained Expressions (CEs). A CE is a sum of a free-function and a functional that analytically satisfies any given linear constraints regardless to the choice of the free-function [24, 30, 31, 42]. That is,

$$f(\boldsymbol{x}) \simeq f_{CE}(\boldsymbol{x}, g(\boldsymbol{x})) = A(\boldsymbol{x}) + B(\boldsymbol{x}, g(\boldsymbol{x}))$$

where $f(\boldsymbol{x})$ is the true underlying function to be interpolated, $f_{CE}(\boldsymbol{x}, g(\boldsymbol{x}))$ is the CE, $A(\boldsymbol{x})$ is the functional that analytically satisfies any given linear constraint (more details are given in Refs. [24, 31]), and $B(\boldsymbol{x}, g(\boldsymbol{x}))$ projects the free-function $g(\boldsymbol{x})$, which is a real function that must exist on the constraints, into the space of functions that vanish at the constraints [33]. TFC can potentially be used for any mathematical problems subject to linear constraints, such as, for example, quadratic and nonlinear programming subjects to equality constraints [35], and to homotopy continuation algorithm for control and dynamics problems [55]. As of now, TFC has widely been used for approximating the solution of DEs [32, 39, 41]. In addition, TFC has already been employed to tackle several classes of aerospace optimal control problems via the PMP approach, such as energy optimal landing on large and small planetary bodies [10, 18], fuel optimal landing on large planetary bodies [25], and energy optimal relative motion problems subject to Clohessy-Wiltshire dynamics [14]. The original TFC method, also known as Vanilla-TFC (V-TFC) [24, 31], for solving DE, employs a linear combination of orthogonal polynomials [39, 41] as free-function. However, this free-function choice becomes cumbersome when solving PDEs with more than two independent variables as V-TFC is affected by the curse of dimensionality [32]. To overcome this limitation, PINN-TFC-based methods use NNs as free-function. There are two PINN-TFC-based methods: Deep-TFC and Extreme-TFC (X-TFC). Deep-TFC [33] uses deep NN as free-function, while X-TFC employs shallow NN trained via Extreme Learning Machine algorithm [51]. Both methods can be used to learn the solution to problems involving DEs and, therefore, the solution to optimal control problems, both via the indirect method and via the BPO. Nevertheless, for optimal control problems, especially for aerospace systems, where autonomy is crucial, X-TFC is the most suitable one. Indeed, thanks to the ELM algorithm characteristics [21, 22], the training is fast and accurate enough to potentially enable real-time applications, both for open-loop and closed-loop controllers. X-TFC has been already applied to solve problems in different disciplines, such as rarefied gas dynamics [13], radiative transfer [12], nuclear reactor dynamics [50], and epidemiological compartmental models [49]. This article will show how X-TFC is applied to aerospace OCPs. More specifically, we will provide an overview of the current status of the X-TFC applications for aerospace OCPs.

2 X-TFC for Aerospace Optimal Control Problems

As previously stated, X-TFC is a PINN TFC-based framework where the free-function $g(\boldsymbol{x})$ is a shallow NN trained via ELM algorithm. That is,

$$g(\boldsymbol{x};\boldsymbol{\beta})) = g^{\boldsymbol{\beta}}(\boldsymbol{x}) = \sum_{j=1}^{L} \beta_j \sigma\left(\boldsymbol{w}_j^{\mathrm{T}} \boldsymbol{x} + b_j\right) \tag{3}$$

where, the input weights, $\boldsymbol{w}_j$, and bias, b_j, with $j = 1, \dots, L$, are randomly sampled and not adjusted in the training. The outputs weights $\boldsymbol{\beta} \in \mathbb{R}^L$ are the only left learnable parameters, which are trained via linear least-squares for linear DEs and iterative least-squares for nonlinear ones, as explained in details in [51].

Regarding the application to aerospace optimal control problems, X-TFC is suitable to tackle these problems either via the PMP or the BPO application. Indeed, thanks to the choice of the free-function, X-TFC is accurate and fast, which is an essential requirement to enable autonomy when tackling the OCPs via the PMP application, usually allowing to obtain open-loop solutions. Moreover, it is not affected by the curse of dimensionality, which is a crucial requirement when tackling the OCPs via the BPO application, allowing to obtain closed-loop solutions. One in principle could also combine the BPO and the PMP using coarse HJB solutions to initialize the TPVBP solutions [6]. One could eventually guarantee that the optimal control actions would be retrieved from the TPVBP. Then, for real-time applications, one would need to compute open-loop solutions fast enough to generate a closed-loop solution with a series of open-loop ones. Conversely, open-loop optimal control actions can be generated within a specific time-state space and then used during the NN training to learn the closed-loop solution in a data-physics-driven fashion within the same time-state space. More precisely, if the PMP approach is used, the goal is to model a NN representation of the state-costate pair, which are the solutions of the arising TPBVP. We will refer to this X-TFC model as *Portraying Neural Networks or PoNN.*

Likewise, if the BPO approach is used, the goal is to model a NN representation of the value function, which is the solution to the arising HJB equation. We will refer to this PINN model as *Bellman Neural Networks or BeNN*.

The current state-of-the-art sees PoNNs applied to several interesting OCPs in aerospace, such as intercept problem, both in its unconstrained and control-constrained versions [9], planar orbit transfer [47], circumnavigation trajectories around asteroids with obstacles avoidance [8], transfer trajectories within the Circular Restricted Three-Body Problem (CR3BP) [7]. All these problems can be cast as OCPs with the following general formulation,

$$\min_{\boldsymbol{u}\in\mathcal{U}} J(t,\boldsymbol{x},\boldsymbol{u}) = \phi(\boldsymbol{x}(t_f)) + \int_{t_0}^{t_f} \mathcal{L}(t,\boldsymbol{x},\boldsymbol{u}) \quad \text{subject to:} \begin{cases} \dot{\boldsymbol{x}} = \boldsymbol{f}(t,\boldsymbol{x},\boldsymbol{u}) \\ \boldsymbol{x}(t_0) = \boldsymbol{x_0} \\ \boldsymbol{x}(t_f) \in \mathcal{C} \\ t \in [t_0, t_f] \end{cases} \tag{4}$$

where $\boldsymbol{x} \in \Omega \subseteq \mathbb{R}^n$, $\boldsymbol{u} \in \mathcal{U} \subseteq \mathbb{R}^m$, and $\mathcal{C} \subset \Omega$ are the states, control (that eventually can be subject also to inequality constraints), and terminal conditions, respectively. The cost function considers both the Meyer cost (as a function on the final state and time) and the Lagrangian cost (as a function of the states and control related to each time instant). In general, the final time t_f may be fixed or it might be a function of the initial states $\boldsymbol{x}_0$ (e.g. for the time-free problems). Via the PMP application, the OCP reduces to the following TPBVP,

$$\begin{cases} H_{\boldsymbol{u}}(t,\boldsymbol{x},\boldsymbol{\lambda}) = \boldsymbol{0} \\ \dot{\boldsymbol{x}} = H_{\boldsymbol{\lambda}}(t,\boldsymbol{x},\boldsymbol{\lambda}) \\ \dot{\boldsymbol{\lambda}} = -H_{\boldsymbol{x}}(t,\boldsymbol{x},\boldsymbol{\lambda}) \end{cases} \quad \text{subject to:} \begin{cases} \boldsymbol{x}(t_0) = \boldsymbol{x_0},\ \boldsymbol{\lambda}(t_0) = -\frac{\partial J}{\partial \boldsymbol{x_0}},\ H(t_0) = \frac{\partial J}{\partial t_0} \\ \boldsymbol{x}(t_f) \in \mathcal{C},\ \boldsymbol{\lambda}(t_f) = \frac{\partial J}{\partial \boldsymbol{x}_f},\ H(t_f) = -\frac{\partial J}{\partial t_f} \end{cases} \tag{5}$$

with $t \in [t_0, t_f]$, where H is the Hamiltonian function of the problem, $H_{\boldsymbol{u}} = \frac{\partial H}{\partial \boldsymbol{u}}$, $H_{\boldsymbol{x}} = \frac{\partial H}{\partial \boldsymbol{x}}$, $H_{\boldsymbol{\lambda}} = \frac{\partial H}{\partial \boldsymbol{\lambda}}$, with $\boldsymbol{\lambda}$ being the costates. In Eq. (5), we considered all the possible transversality conditions. However, the transversality conditions to be applied are generally problem dependent. More details can be found in [48]. This section reports the main results for the OCPs mentioned above. For each problem, we write just the cost function and the dynamics to give the idea of the complexity of the problem, together with the main plots describing the results.

(1) The **constrained optimal intercept problem** is cast as follows [9],

$$\min \quad J = \Gamma t_f + \frac{1}{2}\int_{t_0}^{t_f} (\boldsymbol{a}_M^T \boldsymbol{a}_M)\, dt + \epsilon \int_{t_0}^{t_f} \mathbf{w}^T \mathbf{w}\, dt \quad \text{subject to:} \begin{cases} \dot{\boldsymbol{r}} = \boldsymbol{v} \\ \dot{\boldsymbol{v}} = \boldsymbol{a}_T - \boldsymbol{a}_M \\ t_0 \le t \le t_f \\ \boldsymbol{r}(t_0) = \boldsymbol{r}_0 \\ \boldsymbol{v}(t_0) = \boldsymbol{v}_0 \\ \boldsymbol{r}(t_f) = \boldsymbol{0} \\ u_{min,i} \le a_{M,i} \le u_{max,i} \end{cases} \tag{6}$$

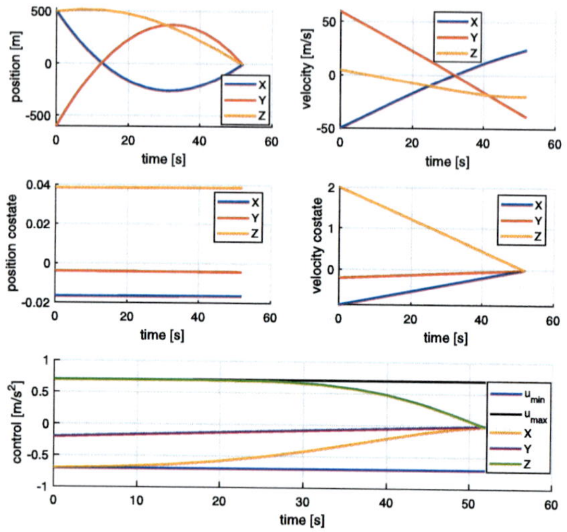

Fig. 1 Constrained intercept problem with penalty $\Gamma = 1$

where a time-energy optimal trajectory with control constraints is considered. The arising TPBVP is given in [9]. The results are illustrated in Fig. 1, where the time history of the states, costates and control is provided. As can be seen, the intercept succeeds, since the components of the relative position vector at final time are 0, while the control is constrained within a user-defined range. Moreover, the position costates are constant, as expected from the theory, proving the accuracy of the obtained results.

(2) In Ref. [47], three optimal **planar orbit transfer** problems were considered: maximum radius low-thrust transfer, minimum time low-thrust transfer, and minimum time solar sail transfer. In this paper, the maximum radius low-thrust transfer problem is proposed and it is cast as follows,

$$\min \mathcal{J} = -x_{1,f} \qquad \text{subject to:} \quad \begin{cases} \dot{x_1} = x_2 \\ \dot{x_2} = \frac{x_3^2}{x_1} - \frac{\gamma}{x_1^2} + \phi(t)\sin(u) \\ \dot{x_3} = -\frac{x_2 x_3}{x_1} + \phi(t)\cos(u) \\ t_0 \le t \le t_f \\ x_1(t_0) = x_{1,0} \\ \dot{x}_{1,0} = x_2(t_0) = x_{2,0} \\ x_3(t_0) = x_{3,0} \\ x_1(t_f) = x_{1,f} \\ \dot{x}_{1,f} = x_2(t_f) = x_{2,f} \\ x_{3,f} = (x_{1,f})^{-1/2} \end{cases} \tag{7}$$

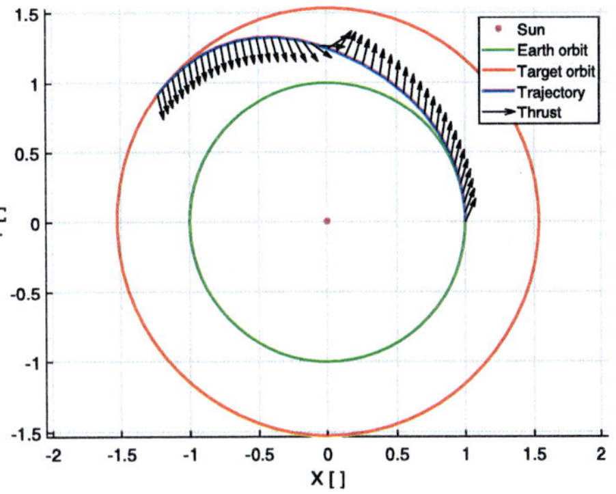

Fig. 2 Trajectory of the maximum radius orbit transfer problem (the thrust is represented in the heliocentric frame). The quantities in the plot are dimensionless

The arising TPBVP, along with the other analyzed problems, can be found in [47]. An Earth-Mars low thrust planar orbit transfer is taken into account. The maximum final radius found by X-TFC is 1.5318 with a precision of the dynamics on the order of 10^{-3}. In particular, the trajectory is shown in Fig. 2, where the arrows indicate the thrust direction at each discretization point. Additionally, the computational time needed to solve the problem with our algorithm is about 15 milliseconds, which is potentially suitable for real-time applications for this kind of orbit transfer problem. Moreover, the software GPOPS-II has been run for this problem as a further comparison. The computed maximum radius was 1.5189, which is lower than the values found with the proposed method.

(3) The **energy optimal trajectory problem of a spacecraft orbiting around an asteroid** can be cast as,

$$\min J = \frac{1}{2}\int_{t_0}^{t_f} \boldsymbol{a}_c^{\mathrm{T}}\boldsymbol{a}_c dt \quad \text{subject to:} \begin{cases} \dot{\boldsymbol{r}} = \boldsymbol{v} \\ \dot{\boldsymbol{v}} = \boldsymbol{g}(\boldsymbol{r}_0) + \boldsymbol{G}(\boldsymbol{r}_0)\cdot(\boldsymbol{r}-\boldsymbol{r}_0) + \boldsymbol{M}\boldsymbol{v} + \boldsymbol{N}\boldsymbol{r} + \boldsymbol{a}_c \end{cases} \tag{8}$$

where a linearized gravitational field and an asteroid body-fixed reference frame are considered to write the dynamics equations. In order to design a safe transfer trajectory around the asteroid avoiding any impacts with its surface, a Rapidly-Explored Random Tree (RRT*) algorithm is adopted, where all the nodes are connected with optimal trajectories represented by Eq. (8) and solved via X-TFC. Therefore, a multi-arch trajectory is obtained. This approach to compute safe trajectory is actually feasible due to the low computational times of X-TFC in solving TPBVPs and its low sensitivity to the initial guess of the solution. The results of a safe circumnavigation trajectory around asteroid Bennu is shown in Fig. 3, where the obstacle to avoid is a sphere surrounding Bennu's surface. As can be seen after 800 iterations of the RRT*, the spacecraft is able to reach the final position (on the opposite pole with respect to the initial position) without impacting with the surface of Bennu. The total cost and time of flight of the trajectory are 0.00306631 and 4000 s, respectively. The fuel consumption of the entire maneuver is 1.91453 kg. For what concerns the precision

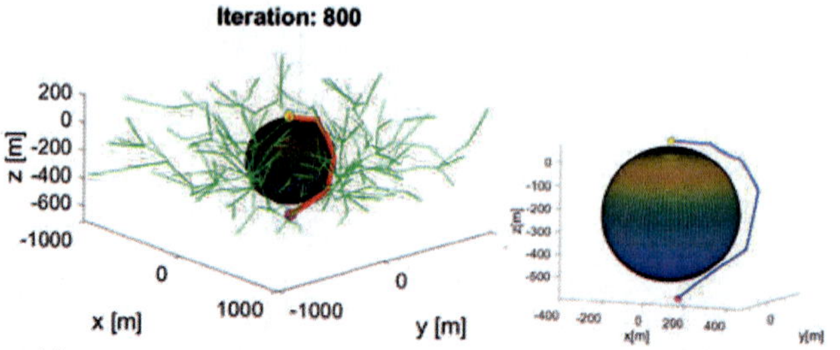

(a) Tree representation at the last iteration (b) Final trajectory of RRT*

Fig. 3 Safe trajectory, around Bennu, obtained at the end of the RRT* algorithm iterations

of the dynamics and the optimality of the reults, the norm of the loss vector and the Hamiltonian are both on the order of 10^{-6}.

(4) A **time-energy optimal transfer trajectory within the CR3BP framework** with control constraints can be cast as follows [7],

$$\min J = \Gamma t_f + \frac{1}{2}\int_{t_0}^{t_f} \mathbf{u}^T \mathbf{u}\, dt + \epsilon \int_{t_0}^{t_f} \mathbf{w}^T \mathbf{w}\, dt \quad \text{subject to:} \quad \begin{cases} \dot{\mathbf{r}} = \mathbf{v} \\ \dot{\mathbf{v}} = \nabla U(\mathbf{r}) + \mathbf{M}\mathbf{v} + \mathbf{u} \\ t_0 \le t \le t_f \\ \mathbf{r}(t_0) = \mathbf{r}_0 \\ \mathbf{v}(t_0) = \mathbf{v}_0 \\ \mathbf{r}(t_f) = \mathbf{r}_f \\ \mathbf{v}(t_f) = \mathbf{v}_f \\ u_{\min} \le u_i \le u_{\max} \end{cases} \tag{9}$$

The entire derivation of the associated TPBVP is given in [7]. The results reported in Fig. 4 show a transfer trajectory to pass from a Halo orbit around L1 to a Halo orbit around L2 within the Eart-Moon CR3BP. Figure 5a shows the plots of the time history of all the components of position and velocity vectors in dimensionless units. Moreover, Fig. 5b illustrates the components of the control acceleration vector along each axis in dimensionless units. As can be seen, the imposed control constraints are always fulfilled. The computed time of flight is 9.98 days, and the corresponding ΔV is about 0.36 m/s. The average value of the loss vector norm, representing the precision of the dynamics, is about $6.6 \cdot 10^{-4}$. The Hamiltonian is quite constant during the whole trajectory, apart from some points at the end. The value of the cost function, represented in Eq. (9) without considering the term due to Γt_f, is 0.034216208833413. The part due to the additional regularization term considering $\epsilon = 10^{-7}$ is about 0.002538402412614. Therefore it represents the 7% of the entire cost function.

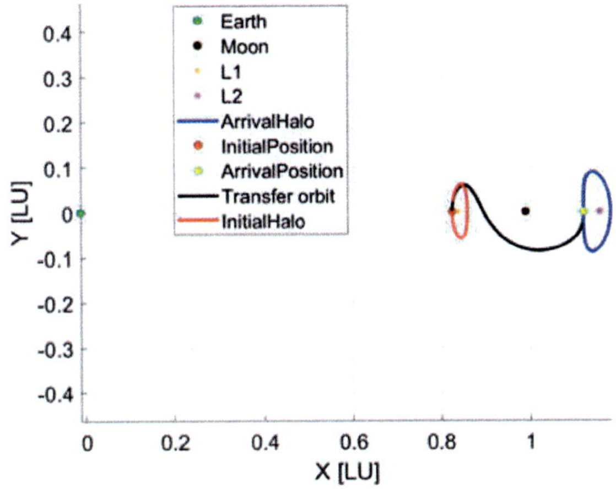

Fig. 4 L1–L2 Halo transfer trajectory with $\epsilon = 10^{-6}$

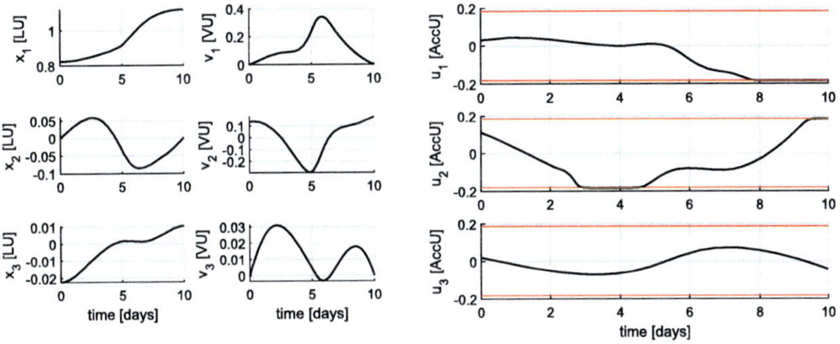

(a) Position and velocity vectors components.
(b) Control acceleration vector components.

Fig. 5 States and control (in dimensionless units) for L1–L2 Halo transfer trajectory with $\epsilon = 10^{-6}$

3 Conclusions and Outlooks

This work presented an overview of the X-TFC applications for aerospace OCPs. The current state-of-the-art sees X-TFC applied several interesting aerospace OCPs faced via the indirect method. As the indirect method relies on the PMP application, these X-TFC-based models are called *Pontryaging Neural Networks or PoNN*. The results show the accuracy and the computational efficiency of PoNNs in learning the optimal control actions in an open-loop fashion, which, for at least some of the problems considered, could result in the employment of PoNNs for real-time applications. Although all the applications are different, for all of them, the resulting optimal control is continuous.

Works are in progress to extend PoNNs to OCPs with discontinuous control (e.g., bang-bang type control), such as fuel optimal problems. Moreover, efforts are in progress to apply BeNNs for a series of aerospace OCPs with integral quadratic cost [17] such as attitude control problems (both four and six degrees of freedom) and energy optimal landing. Future works will focus on the BeNN application to problems with discontinuous control and on integrating PoNN and BeNN to have both data-physics-driven PoNN training and data-physics-driven BeNN training.

References

1. Acikmese, B., Ploen, S.R.: Convex programming approach to powered descent guidance for mars landing. J. Guidance Control Dyn. **30**(5), 1353–1366 (2007)
2. Balduzzi, D., Racaniere, S., Martens, J., Foerster, J., Tuyls, K., Graepel, T.: The mechanics of n-player differentiable games. In: International Conference on Machine Learning, pp. 354–363. PMLR (2018)
3. Blackmore, L., Acikmese, B., Scharf, D.P.: Minimum-landing-error powered-descent guidance for mars landing using convex optimization. J. Guidance Control Dyn. **33**(4), 1161–1171 (2010)
4. Byrd, R.H., Gilbert, J.C., Nocedal, J.: A trust region method based on interior point techniques for nonlinear programming. Math. Program. **89**(1), 149–185 (2000)
5. Byrd, R.H., Lu, P., Nocedal, J., Zhu, C.: A limited memory algorithm for bound constrained optimization. SIAM J. Sci. Comput. **16**(5), 1190–1208 (1995)
6. Cristiani, E., Martinon, P.: Initialization of the shooting method via the hamilton-jacobi-bellman approach. J. Optim. Theory Appl. **146**(2), 321–346 (2010)
7. D' Ambrosio, A., Schiassi, E., Curti, F., Furfaro, R.: Physics-informed neural networks applied to a series of constrained space guidance problems. In: 31st AAS/AIAA Space Flight Mechanics Meeting (2021)
8. D' Ambrosio, A., Schiassi, E., Curti, F., Furfaro, R.: Physics-informed neural networks for optimal proximity maneuvers with collision avoidance around asteroids. In: 2021 AAS/AIAA Astrodynamics Specialist Conference (2021)
9. D'Ambrosio, A., Schiassi, E., Curti, F., Furfaro, R.: Pontryagin neural networks with functional interpolation for optimal intercept problems. Mathematics **9**(9), 996 (2021)
10. D'Ambrosio, A., Schiassi, E., Johnston, H., Curti, F., Mortari, D., Furfaro, R.: Time-energy optimal landing on planetary bodies via theory of functional connections. Adv, Space Res (2022)
11. Darby, C.L., Hager, W.W., Rao, A.V.: An hp-adaptive pseudospectral method for solving optimal control problems. Optim. Control Appl. Methods **32**(4), 476–502 (2011)
12. De Florio, M., Schiassi, E., Furfaro, R., Ganapol, B.D., Mostacci, D.: Solutions of chandrasekhar's basic problem in radiative transfer via theory of functional connections. J. Quant. Spectrosc. Radiat. Transfer **259**, 107384 (2021)
13. De Florio, M., Schiassi, E., Ganapol, B.D., Furfaro, R.: Physics-informed neural networks for rarefied-gas dynamics: Thermal creep flow in the bhatnagar-gross-krook approximation. Phys. Fluids **33**(4), 047110 (2021)
14. Drozd, K., Furfaro, R., Schiassi, E., Johnston, H., Mortari, D.: Energy-optimal trajectory problems in relative motion solved via theory of functional connections. Acta Astronaut. (2021)
15. Fahroo, F., Ross, I.: Trajectory optimization by indirect spectral collocation methods. In: Astrodynamics Specialist Conference, p. 4028 (2000)
16. Fahroo, F., Ross, I.M.: Direct trajectory optimization by a chebyshev pseudospectral method. J. Guidance Control Dyn. **25**(1), 160–166 (2002)
17. Furfaro, R., D'Ambrosio, A., Schiassi, E., Scorsoglio, A.: Physics-informed neural networks for closed-loop guidance and control in aerospace systems. In: AIAA SCITECH 2022 Forum, p. 0361 (2022)

18. Furfaro, R., Mortari, D.: Least-squares solution of a class of optimal space guidance problems via theory of connections. Acta Astronaut. (2019). https://doi.org/10.1016/j.actaastro.2019.05.050, http://www.sciencedirect.com/science/article/pii/S0094576519302292
19. Gonzalez, R., Rofman, E.: On deterministic control problems: An approximation procedure for the optimal cost ii. The nonstationary case. SIAM J. Control Optim. **23**(2), 267–285 (1985)
20. Graham, K.F., Rao, A.V.: Minimum-time trajectory optimization of multiple revolution low-thrust earth-orbit transfers. J. Spacecraft Rocket. **52**(3), 711–727 (2015)
21. Huang, G.B., Chen, L., Siew, C.K.: Universal approximation using incremental constructive feedforward networks with random hidden nodes. IEEE Trans. Neural Netw. **17**(4), 879–892 (2006)
22. Huang, G.B., Zhu, Q.Y., Siew, C.K.: Extreme learning machine: Theory and applications. Neurocomputing **70**(2006), 489–501 (2006). https://doi.org/10.1016/j.neucom.2005.12.126
23. Jiang, X., Li, S., Furfaro, R.: Integrated guidance for mars entry and powered descent using reinforcement learning and pseudospectral method. Acta Astronaut. **163**, 114–129 (2019)
24. Johnston, H.: The theory of functional connections: A journey from theory to application. ArXiv preprint arXiv:2105.08034 (2021)
25. Johnston, H., Schiassi, E., Furfaro, R., Mortari, D.: Fuel-efficient powered descent guidance on large planetary bodies via theory of functional connections. J. Astronaut. Sci. (under review)
26. Josselyn, S., Ross, I.M.: Rapid verification method for the trajectory optimization of reentry vehicles. J. Guidance Control Dyn. **26**(3), 505–508 (2003)
27. Karniadakis, G.E., Kevrekidis, I.G., Lu, L., Perdikaris, P., Wang, S., Yang, L.: Physics-informed machine learning. Nat. Rev. Phys. **3**(6), 422–440 (2021)
28. Keller, H.B.: Numerical solution of two point boundary value problems, vol. 24. SIaM (1976)
29. Kingma, D.P., Ba, J.: Adam: A method for stochastic optimization. ArXiv preprint arXiv:1412.6980 (2014)
30. Leake, C., Mortari, D.: An explanation and implementation of multivariate theory of connections via examples. In: 2019 AAS/AIAA Astrodynamics Specialist Conference, Portland, MN, August 11–15, 2019. AAS/AIAA (2019)
31. Leake, C.: The multivariate theory of functional connections: An n-dimensional constraint embedding technique applied to partial differential equations. ArXiv preprint arXiv:2105.07070 (2021)
32. Leake, C., Johnston, H., Mortari, D.: The multivariate theory of functional connections: Theory, proofs, and application in partial differential equations. Mathematics **8**(8), 1303 (2020)
33. Leake, C., Mortari, D.: Deep theory of functional connections: A new method for estimating the solutions of partial differential equations. Mach. Learn. Knowl. Extr. **2**(1), 37–55 (2020)
34. Lu, L., Meng, X., Mao, Z., Karniadakis, G.E.: Deepxde: A deep learning library for solving differential equations. SIAM Rev. **63**(1), 208–228 (2021)
35. Mai, T., Mortari, D.: Theory of functional connections applied to quadratic and nonlinear programming under equality constraints. J. Comput. Appl. Math. **406**, 113912 (2022)
36. Mertikopoulos, P., Papadimitriou, C., Piliouras, G.: Cycles in adversarial regularized learning. In: Proceedings of the Twenty-Ninth Annual ACM-SIAM Symposium on Discrete Algorithms, pp. 2703–2717. SIAM (2018)
37. Miller, A.T., Rao, A.V.: Rapid ascent-entry vehicle mission optimization using hp-adaptive gaussian quadrature collocation. In: AIAA Atmospheric Flight Mechanics Conference, p. 0249 (2017)
38. Mishra, S., Molinaro, R.: Estimates on the generalization error of physics-informed neural networks for approximating a class of inverse problems for pdes. IMA J. Numer, Anal (2021)
39. Mortari, D.: Least-squares solution of linear differential equations. Mathematics **5**(48), 1–18 (2017). https://doi.org/10.3390/math5040048, http://www.mdpi.com/2227-7390/5/4/48
40. Mortari, D.: The theory of connections: Connecting points. MDPI Math. **5**(57) (2017)
41. Mortari, D., Johnston, H., Smith, L.: High accuracy least-squares solutions of nonlinear differential equations. J. Comput. Appl. Math. **352**, 293–307 (2019). https://doi.org/10.1016/j.cam.2018.12.007, http://www.sciencedirect.com/science/article/pii/S0377042718307325

42. Mortari, D., Leake, C.: The multivariate theory of connections. Mathematics **7**(3) (2019). https://doi.org/10.3390/math7030296, https://www.mdpi.com/2227-7390/7/3/296
43. Oh, S., Luus, R.: Use of orthogonal collocation method in optimal control problems. Int. J. Control **26**(5), 657–673 (1977)
44. R.E., B.: Dynamic Programming. Princeton University Press, Princeton, NJ (1957)
45. Raissi, M., Perdikaris, P., Karniadakis, G.E.: Physics-informed neural networks: A deep learning framework for solving forward and inverse problems involving nonlinear partial differential equations. J. Comput. Phys. **378**, 686–707 (2019)
46. Ross, I.M., Fahroo, F.: Pseudospectral knotting methods for solving nonsmooth optimal control problems. J. Guidance Control Dyn. **27**(3), 397–405 (2004)
47. Schiassi, E., D'Ambrosio, A., Drozd, K., Curti, F., Furfaro, R.: Physics-informed neural networks for optimal planar orbit transfers. J. Spacecraft Rocket. 1–16 (2022)
48. Schiassi, E., D'Ambrosio, A., Scorsoglio, A., Furfaro, R., Curti, F.: Class of Optimal Space Guidance Problems Solved via Indirect Methods and Physics-informed Neural Networks
49. Schiassi, E., De Florio, M., D'ambrosio, A., Mortari, D., Furfaro, R.: Physics-informed neural networks and functional interpolation for data-driven parameters discovery of epidemiological compartmental models. Mathematics **9**(17), 2069 (2021)
50. Schiassi, E., De Florio, M., Ganapol, B.D., Picca, P., Furfaro, R.: Physics-informed neural networks for the point kinetics equations for nuclear reactor dynamics. Ann. Nucl. Ener. **167**, 108833 (2022)
51. Schiassi, E., Furfaro, R., Leake, C., De Florio, M., Johnston, H., Mortari, D.: Extreme theory of functional connections: A fast physics-informed neural network method for solving ordinary and partial differential equations. Neurocomputing (2021)
52. Shin, Y., Darbon, J., Karniadakis, G.E.: On the convergence and generalization of physics-informed neural networks. ArXiv preprint arXiv:2004.01806v1 (2020)
53. Stoer, J., Bulirsch, R.: Introduction to Numerical Analysis, vol. 12. Springer Science & Business Media (2013)
54. Wang, D.Y.: Study Guidance and Control for Lunar Soft Landing (Ph.D. Dissertation). School of Astronautics, Harbin Institute of Technology, Harbin, China (2000)
55. Wang, Y., Topputo, F.: A tfc-based homotopy continuation algorithm with application to dynamics and control problems. J. Comput. Appl. Math. **401**, 113777 (2022)
56. Wang, Z., Grant, M.J.: Constrained trajectory optimization for planetary entry via sequential convex programming. In: AIAA Atmospheric Flight Mechanics Conference, p. 3241 (2016)
57. Wang, Z., Grant, M.J.: Autonomous entry guidance for hypersonic vehicles by convex optimization. J. Spacecraft Rocket. **55**(4), 993–1006 (2018)
58. Wang, Z., Grant, M.J.: Minimum-fuel low-thrust transfers for spacecraft: A convex approach. IEEE Trans. Aerosp. Electron. Syst. **54**(5), 2274–2290 (2018)
59. Zhang, K., Yang, S., Xiong, F.: Rapid ascent trajectory optimization for guided rockets via sequential convex programming. In: Proceedings of the Institution of Mechanical Engineers, Part G: Journal of Aerospace Engineering, p. 0954410019830268 (2019)

Innovative ML-based Methods for Automated On-board Spacecraft Anomaly Detection

Carlo Ciancarelli, Eleonora Mariotti, Francesco Corallo, Salvatore Cognetta, Livia Manovi, Alex Marchioni, Mauro Mangia, Riccardo Rovatti, and Gianluca Furano

Abstract Spacecraft health monitoring is an important task to assure the mission operational life. For this purpose, a variety of telemetry data are analyzed to detect anomalies that can lead to failures and cause irreversible damage to on-board devices. In this paper we propose and analyze different ML-based methods that contribute to the generation of an intelligent anomaly detector capable of identifying anomalies in spacecraft telemetries, with a particular attention to the memory footprint of each method. Finally, we investigated how to model the abnormal behaviors during the validation/test phase, exploiting different families of possible configurations for anomaly injection. The achieved results, after several tuning setups, suggest that all of the adopted methods are suitable for implementation when observing relatively

C. Ciancarelli (✉) · E. Mariotti · F. Corallo · S. Cognetta
Thales Alenia Space Italia S.p.A., Rome, Italy
e-mail: carlo.ciancarelli@thalesaleniaspace.com

E. Mariotti
e-mail: eleonora.mariotti@thalesaleniaspace.com

F. Corallo
e-mail: francesco.corallo@thalesaleniaspace.com

S. Cognetta
e-mail: salvatore.cognetta-somministrato@thalesaleniaspace.com

L. Manovi · A. Marchioni · M. Mangia · R. Rovatti
Alma Mater Studiorum–Universitá di Bologna, Bologna, Italy
e-mail: livia.manovi@unibo.it

A. Marchioni
e-mail: alex.marchioni@unibo.it

M. Mangia
e-mail: mauro.mangia@unibo.it

R. Rovatti
e-mail: riccardo.rovatti@unibo.it

G. Furano
European Space Agency–ESTEC, Noordwijk, Netherlands
e-mail: Gianluca.Furano@esa.int

C. Ieracitano et al. (eds.), *The Use of Artificial Intelligence for Space Applications*, Studies in Computational Intelligence 1088,
https://doi.org/10.1007/978-3-031-25755-1_14

short intervals of time. Instead, when longer time windows are used, the score-based detectors outperform the proximity/density-based ones, despite being more resource hungry.

Keywords Spacecraft · Health monitoring · Anomaly detection · On-board machine learning

1 Introduction

Spacecraft Health Monitoring (HM) systems, also referred to as FDIR (Fault Detection, Isolation and Recovery), are a key component of any space mission design and play a relevant role in the definition of the spacecraft availability, reliability and safety objectives.

Modern space systems, due to their size, autonomy requirements, and complexity of the required operations have a very heterogeneous coupling between onboard cyber (processing, communication) and physical (sensing, actuation) elements to cope also with the impossibility of repairs and the harsh extraterrestrial environment. The generated telemetries detail different health status of the spacecraft, like temperature, voltage, absorbed current, etc. which need to be monitored given the overall system complexity and cost: a failure not correctly detected could results in a potential loss of the spacecraft itself. Fault and anomaly detection systems are needed to alert space operations engineers of anomalous behaviour and prevent significant failures.

Current health monitoring systems require an expert knowledge to be developed and maintained, in order to comply the increasing scale and complexity of spacecraft coupled with the difficulty of providing cost-effective ground operations. The state of the art anomaly detection methods for spacecraft telemetries consist of verifying whether telemetry values are within pre-defined limits or stray outside of them, also referred to "Out-Of-Limits" (OOL) technique. These approaches have different limitations, alongside the costly expert knowledge to define and update the nominal ranges/threshold values and tables for each telemetry sensor, there is the need of ongoing analysis of telemetry data which can be translated in workload for space operations engineers.

The lesson learnt from last decades in Space Segment Operations at Thales Alenia Space Italia (TAS-I) highlights that the increasing operational demands to analyse spacecraft telemetries can no longer be met with traditional approaches, but innovative and advanced monitoring systems are required to detect anomalous behaviours, reduce the space operations engineers workload and satellite's operational risk. The generated data streams, in fact, can be used in advantage for a data driven approach to automate the above-mentioned tedious and complex tasks, like failure detection and prognostic. Artificial Intelligence and Machine Learning techniques can be exploited to model and predict the system behaviour, contextually detect if there is an anomaly in the system itself.

1.1 Related Works

Several attempts using AI have been implemented for aerospace FDIR since the 90s [19], however, on board FDIR solutions are today dominated by simple OOL checks. This result highlighted the lack of maturity and reliability of the techniques, implementation complexity and computational power required not matching the available HW.

More maturity on anomaly detection techniques can be observed for on-ground applications where NASA, ESA or JAXA [16] have successfully deployed deep learning to assist their operators. Activities such as this study shall aim at adapting and closing the gap for On-board adaptation of these techniques.

The vast majority of reviewed work tackles the anomaly detection but not the isolation problem which is deemed a fundamental enabler for the action-ability of the on board intelligence. When tackling local FDIR use cases (limited to the processing of 5–20 variables) embedding the AI algorithm along the rest of the software running on modern On-Board Computer (OBC) CPUs can suffice. However, when aiming at a large scale TM/TC processing the ML models may no longer fit in OBC memory and execution time budgets and dedicated processing units may be necessary.

The availability of high volumes of high-fidelity synthetic/simulated data prior to flight is restricted to small subset of equipment for which there are functional simulators, e.g. payload or Attitude Orbit Control System (AOCS) equipment. However, for a wider applicability of an AI-based FDIR the system needs to keep "training" with safe criteria once in orbit.

The resulting datasets lack of labeled anomaly and, fortunately, are highly unbalanced in favour of nominal samples. For these reasons anomaly detection methods are mainly unsupervised or semi-supervised techniques; we analyzed different approaches in literature like clustering, Support Vector Machines (SVM), decision trees and Artificial Neural Networks (ANN or NN).

One of the most popular clustering technique is K-means [12, 13] which was applied for industrial robot health monitoring and anomaly detection in [8]. The detection phase consisted in the comparison between the test data and the cluster centres computed by the K-means algorithm after the training procedure: if the distance from the new data points to the cluster centres exceeded a predefined threshold the data point was flagged as an anomaly.

Another approach for fault and anomaly detection was developed using Least Squares-SVM in [18], where the authors proposed an LS-SVM regression model for spacecraft telemetry to identify contextual and collective anomalies ([6]) of in-orbit satellite telemetries. The resulted algorithm consists in an improved version of the classical SVM method, obtaining very good classification and generalization ability for anomaly detection of spacecraft in-orbit telemetries.

Random Forest (RF), a Decision Tree approach, was exploited in [3] for condition monitoring and early detection of failures for industrial plant equipment. After the data pre-processing and feature extraction phases, which include the transformation in the frequency domain via the Fast Fourier Transform (FFT), the authors used the extracted features to train the RF model and construct N decision trees trained on N bootstrapped samples of the training data. The resulting model was able to predict the development of the industrial plant health index with better performance of a persistence technique used as benchmark.

The framework [1] exploited, distinctly, two different approaches VAE and GANomaly [2, 11] trained on nominal data generated by LUNASIM [10]. Hence, anomalies are detected by measuring the reconstruction error for newly input data. If nominal and abnormal samples are not clearly different, the VAE obtains noisy output conversely GANomaly could discriminate abnormal samples more effectively because it utilizes AE to reconstruct input data and GAN to match a distribution of generated samples to the nominal distribution.

The paper is organized as follows. In Sect. 2 a background on the satellite FDIR mechanisms is provided, as an introduction to the proposed operational use case. Section 3 is devoted to the mathematical modeling of the adopted anomaly detection ML techniques and the identification of the critical hyperparameters to be tuned, alongside the definition of the score generation mechanisms. Finally, the memory footprint of each method is estimated. In Sect. 4 the experimental settings are described and detection performances in the reference scenario are discussed and compared. Finally, some conclusions are drawn.

2 Identified Scenario and System Requirement

Until now, the OBCs used for space applications lag several generations behind corresponding terrestrial embedded systems, being designed more for robustness than for performance [9]. Most of the tasks executed by processors in space data systems are thus non compute-intensive workloads, i.e. they perform a little amount of operations per byte read and written from memory. The issue for the space industry is that it is not possible to reuse in a straightforward way the hardware platforms for terrestrial application, given the specific constraints of satellite data systems, especially in terms of availability and reliability effects of ionizing radiation. For this reason, the short term solution for AI on the edge in space depends on both computing intensities needed and the use of ML model on mission critical applications. Thanks to the most recent improvements in space processor performance and availability of dedicated vector acceleration [7], use of new OBCs with enhanced performances on ML-generated applications can be envisaged targeting systems with medium to high dependability requirements, without the need that ML inference is supervised by a fault tolerant engine. When the ML algorithm is designed, several hardware requirements shall be taken into account in order to avoid excessive computing effort, such as computational capabilities, maximum volatile and non-volatile memories constraints, the maximum stack of any algorithm function, and so on. It should be also necessary to investigate about the possibility to optimize the software coding in order to reduce the computational needs, paying attention to the use of external AI and ML libraries that should be no hardware dependent or adaptive to space software compilers.

Before going into details about the proposed solutions it's necessary to briefly introduce the FDIR mechanisms. Nowadays, satellite FDIR is organized according to a hierarchical architecture with the aim of isolating and recovering faults at unit, subsystem or equipment level. This architecture is structured in such a way that it tries to confine failures at the lower FDIR levels to minimize outages and provide high system availability. It is a deterministic approach based on several predefined tables containing selected monitoring items and relative recoveries. These tables are designed according to experience and then implemented in the Avionic SW (ASW), they come from different satellite subsystems and each of these stores a set of parameters. The goal of this approach is to verify that a proper set of parameters, computed by the ASW, does not exceed the operational thresholds. The detection of the violation of a monitoring criterion foresees a recovery action. The recovery is usually not immediate since it is required to detect the same violation several consecutive times ("filter" configuration data) before raising an alarm on it (failure confirmation), in order to eliminate spurious or transient events. Due to the predefined nature of this table-driven FDIR strategy, since both operational limits and confirmation time are decided a priori by design, it could affect the FDIR performance. Furthermore, the approach is not flexible and does not guarantee any preventive maintenance. The satellite FDIR is usually handled by software functionalities, but the highest level of the FDIR hierarchy is in charge to the Ground Control Centre (GCC). The GCC is able to send telecommands in order to enable/disable the on-board autonomous FDIR operations, or for setting the FDIR configuration parameters and logics. However, it could happen the satellite will be affected by unexpected recoveries due to lack of prediction capabilities, and GCC shall investigate about the anomaly causes and it could take long times to restore the satellite functionalities.

The proposed future use of ML algorithms can significantly enhance the performance of the on-board FDIR, especially in identifying and isolating failures at the lowest level possible (equipment level) thus fostering equipment/software reuse, mission availability and autonomy. Indeed, once the failure has been detected and localised to a given equipment (i.e. isolated), the relevant recovery actions can be put in place in order to not discontinue operations. Note that any strategy relying only on an enhanced Failure Detection (FD) function would not have any impact on the availability and/or autonomy if the detection cannot be traced down to a particular equipment. Besides, a perfectly efficient and reliable FDIR strategy makes it possible to put in place simple Recovery strategies leading to Fail Operational behaviour, with no further need for complex algorithms for it.

In order to verify and test the capabilities of the algorithms in the task of identifying anomalies on-board, thus reducing the on-board reaction time of the FDIR function, a real operational use case is proposed hereafter. For decades spacecrafts have been making use of Reaction Wheels (RWs) as actuators of the on-board attitude control function: these are typically selected in configurations of three or more units for redundancy, and they are generally used to apply a continuous fine attitude control over the orbit during the nominal operations, or to react to the environmental

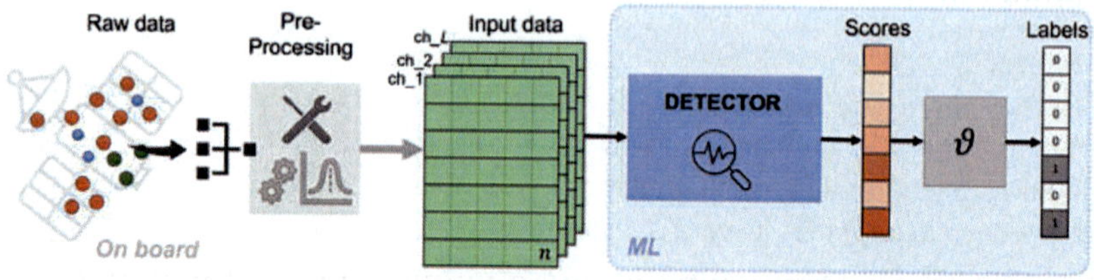

Fig. 1 Signal acquisition and processing chain, score computation scheme and threshold-based binary labels generation

disturbance torque. A RW is generally composed by an external part fixed to the spacecraft, and an internal rotating flywheel which is driven by an electric motor. The dataset used in this work was retrieved from real telemetry data collected from in-flight RW unit.

Given all the above premises, the main contributions of this work consist in: characterization of the score-based detection mechanism provided by the selected ML techniques; modeling of the possible faulty conditions through Gaussian sources; description the adopted training and test procedures; estimation of the memory footprint of each model in terms of the amount of digital quantities to be locally stored; discussion of the achieved results and identification of the methods performing properly given the system requirements.

3 ML Models for Anomaly Detection

In order to make the detector effective, a classical signal processing chain is considered, as depicted in Fig. 1. On-board sensors readings, pre-processed in order to transform heterogeneous sources of information into homogeneous time-series,[1] are examined by employing a time window-based partition, so that it is more likely to accomplish the task of uncovering sub-sequences of unusual behavior. Depending on how many sensors are included in the monitoring system, each score value is computed starting from more than one windows of samples. Final labels are computed by matching the detectors outputs, i.e., the scores, with predefined thresholds.

In the optic of not being able to simulate every faulty condition that might occur in a complex system, the monitoring issue, and so the definition of these kinds of detectors, can be addressed by training phases realising on the recognition of those behaviors which don't exceed the boundary of a nominal one. We here focus on four possible detectors which differ in the concept of similarity they rely on, i.e., on how they define the boundary representing the nominal behaviour:

[1] Actions performed in the pre-processing stage could include: missing data imputation, filtering/cleaning, re-sampling, re-scaling and low-level feature extraction.

(*i*) a proximity-based method, the approach called Local Outlier Factor, LOF, originally proposed in [5], which combines the k-nearest neighbors learning method with some data-dependent similarity measure to address the issue of local density variation. The local distance behavior is well accounted for by using distance ratios in the definition of the score;

(*ii*) a linear method as Principal Component Analysis, PCA, which allows to extract a low dimensional representation of the data by projecting the windows of samples onto the subspace in which the information content of the signal concentrates;

(*iii*) a first non-linear method, One-Class Support Vector Machines, OCSVM. It takes advantage of non-linear mappings to compute similarity measures in what is called a feature space;

(*iv*) a second non-linear method representing the generalization of PCA, a deep Neural Network architectures named Autoencoder, AE. Here the goal is to find the manifold which better represents the input signals in a low-dimensional space [14].

Here follows a mathematical modeling aimed at explaining, for each proposed method, the characteristic parameters' meaning alongside with the score generation mechanism.

3.1 Local Outlier Factor

Let

$$t_1, ..., t_{N_T} \in \mathcal{T} \tag{1}$$

be a reference data set, where $N_T \in \mathbb{N}$ is the number of training examples and $\mathcal{T} \subset \mathbb{R}^n$, and

$$x_1, ..., x_N \in \mathcal{X} \tag{2}$$

be a data set containing both normal and anomalous instances, where $N \in \mathbb{N}$ is the number of observations and $\mathcal{X} \subset \mathbb{R}^n$.

Given an integer parameter $k > 0$ and a suitable distance function $d(\cdot, \cdot)$, the k-distance of an observation x_i, denoted as $d_k(x_i)$, is the distance between x_i and its nearest neighbor t_i in the training set $\mathcal{T}$, which is compared to distance $d_k(t_i)$ between t_i and its neighbors in the training set. Values of $d_k(x_i)$ lower or comparable to $d_k(t_i)$ imply a higher likelihood of x_i belonging to a normal instance. Let us also denote as $N_k(x_i)$ the set of all points within the k-distance of x_i. The cardinality of the set is $|N_k(x_i)| \geq k$, as it may contain duplicate points. Provided with this information, one is able to compute the so-called *reachability distance* of object x_i from object t_j as

$$r_k(x_i, t_j) = \max\{d(x_i, t_j), d_k(x_i)\}. \tag{3}$$

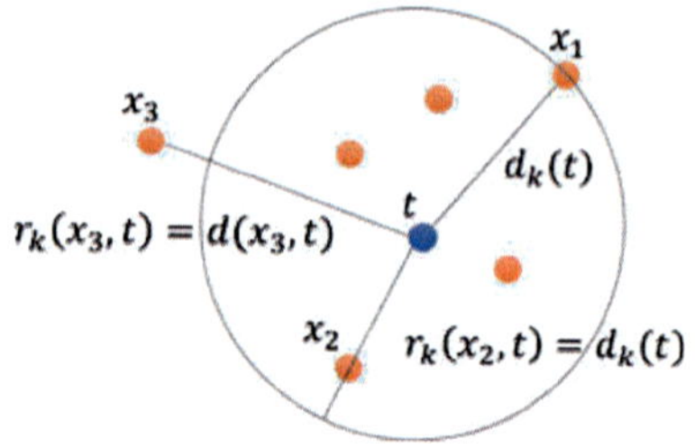

Fig. 2 k-distance and reachability distances, for $k = 5$

Such a quantity introduces a smoothing effect, controlled by k, reducing the statistical fluctuations of the actual distance for all the points x_i close to t_j, in that all points belonging to a neighborhood are considered to be equally distant (refer to Fig. 2).

Then, by taking the inverse of the average r_k of object x_i from its neighborhood, one derives the *local reachability density*, namely $l_k(x_i)$, which is the distance at which x_i can be reached by its neighbors: the quantity is small when x_i is isolated from the surrounding neighbors.

Finally, the local outlier factor score can be derived by comparing the local reachability densities of the neighbors and the object's own value as

$$\mathrm{LOF}_k(x_i) = \frac{1}{|N_k(x_i)|} \sum_{y_i \in N_k(x_i)} \frac{l_k(y_i)}{l_k(x_i)}. \tag{4}$$

In general, if $\mathrm{LOF}_k(x_i) < 1$, the input sample x_i presents a higher reachability density with respect to its neighbors, meaning it can be considered as normal as also for the case $\mathrm{LOF}_k(x_i) \sim 1$. Conversely, a value $\mathrm{LOF}_k(x_i) \gg 1$ indicates that the object in exam shows lower density than the surrounding ones: most surely it will be the case of an outlier.

3.2 *Principal Component Analysis*

The main goal of this method is to find the so called *principal components*, i.e, a set of $m < n$ vectors along which the energy of the input signals concentrates in average. This means that each incoming input signal $x_i \in \mathbb{R}^n$ is mapped into a lower dimensional subspace where the projection is $z_i \in \mathbb{R}^m$, with $m < n$. The input signal can then be recovered by means of a decoding stage as $\hat{x}_i \in \mathbb{R}^n$ from the low-dimensional representation (see Fig. 3).

Linear projection is done by means of a multiplication through a matrix $U \in \mathbb{R}^{n \times m}$ in a way that

$$z_i = U^\top x, \quad \hat{x}_i = U z_i. \tag{5}$$

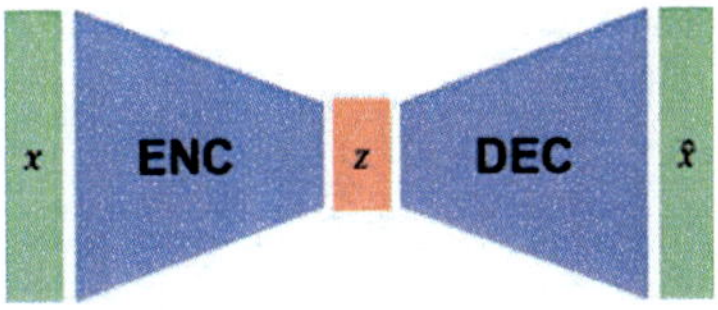

Fig. 3 Working principle of PCA and AE

where U is the matrix minimizing the expectation of the error $\|x_i - \hat{x}_i\|_2$ and it can be computed by arranging as columns of U the eigenvectors of the input signal correlation matrix associated to la largest eigenvalues. Here $\|\cdot\|$ indicates the l_2-norm of a vector

Starting from the knowledge of U it is possible to define two distinct outlier scores to catch unusual variability both within and outside the learned principal subspace.

1. *Squared Prediction Error*, SPE: it represents both the reconstruction error for x_i and the level of inconsistency with respect to the identified best linear fit, i.e., the amount of energy in the minor subspace.

$$\mathrm{SPE}(x_i) = \|x\_\hat{x}_i\|_2^2 = \|x - UU^T x\|_2^2 \tag{6}$$

2. *Hotelling's* T^2: it consists in a statistical test which provides the level of similarity between the energy content of the principal subspace and that of a specific input sample x_i and is defined as:

$$T^2(x_i) = \sum_{l=1}^{m} \frac{(U^T x_i)_l^2}{\lambda_l}, \tag{7}$$

where $(\cdot)_l$ represents the l-th element of a vector and λ_l is the l-th highest eigenvalue of the input signal correlation matrix.

3.3 One-Class Support Vector Machines

The detailed explanation of the mathematical model defining OCSVM is beyond the scope of this paper, therefore, the reader is invited to refer to sources as [4, 17] for more insights on the statistical theory of learning, to which the core of this method belongs, linear modeling and its extension to the non-linear case with the adoption of kernel functions.

The basic idea behind a SVM-based detection mechanism is summarized with the schematic in Fig. 4. It resembles the architecture of a simple neural network, in which the weights α_j are derived from a subset of the training examples.

The approach allows to find a hyperplane separating the *single-class* normal data examples from the origin with maximal margin ρ. The attempt of minimizing the

Fig. 4 General architecture of a SVM

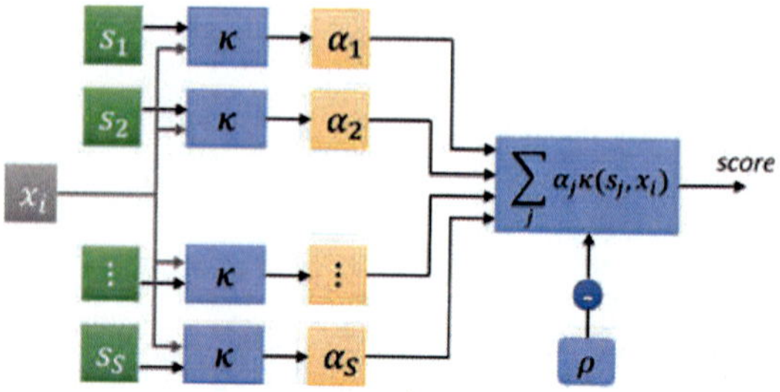

structural error of such a hyperplane results in the definition of a quadratic programming problem [15].

In order to make the approach more effective, all SVM-based methods adopt the so called kernel trick such that the scalar product between two vectors is generalized as follows

$$\kappa(x, y) = \exp -\gamma|x - y|^2, \tag{8}$$

where this represents the case of a Gaussian kernel function.

Within this framework, for a new observation x_i, the score is computed by

$$OCSVM(x_i) = \sum_{j=1}^{S} \alpha_j \kappa(s_j, x_i) - \sum_{i=1}^{S} \sum_{j=1, j\neq i}^{S} \alpha_j \kappa(s_j, s_i) \tag{9}$$

where $\rho = \sum_{i=1}^{S} \sum_{j=1, j\neq i}^{S} \alpha_j \kappa(s_j, s_i)$ is the maximal marginal and $s_1, \ldots, s_S$ is a set of examples in the training set properly selected in the training phase, named support vectors, and for which the cardinality of the set is controlled by a tunable parameter $\nu \in [0, 1]$ [15].

3.4 Autoencoder

As for PCA, an Autoencoder-based detector encodes some meaningful information extracted, by non-linear projection of the input observation x_i, into a manifold living in a lower dimensional space $z_i \in \mathbb{R}^m$, with $m < n$. This is also the input of a decoder stage which produces as output a vector $\hat{x}_i \in \mathbb{R}^n$ designed to be a replication of x_i (see Fig. 4).

Both encoding and decoding stages involve a neural network represented by the two non-linear functions $f(x_i)$ and $g(z_i)$ such that $\hat{x}_i = g(f(x_i))$. More in detail, in the encoder stage there are a convolutional layer (2 filters, strides equal to 4 and a kernel size equal to 40) and a fully connected layer producing z_i and preceded by a layer performing a flattening operation. The decoder is done by transposing the encoder architecture.

The networks' parameters are trained by minimizing a loss function given by the *Mean Squared Error, MSE* computed as the expectation of

$$\|x - \hat{x}\|_2^2. \tag{10}$$

MSE is also used as anomaly score.

3.5 *Model Tuning for On-board Implementation*

The training phase allows to derive some confidence intervals for the scores assigned to observations known to be within normal range. At inference time, the same scoring mechanism (detailed for each method in Sects. 3.1–3.4) is applied to previously unseen instances.

Based on the previous discussion, an estimation of the memory footprint of each method is given in terms of how many digital quantities must be stored. This is a quantity depending on n (the number of samples in each observation), the parameters characterizing each method and also, in same cases, on N_T, i.e., the amount of training examples.

The critical parameter which has to be tuned for LOF is k, the cardinality of the neighborhoods along with the number of examples in the training set $\mathcal{T}$. For each of the N_T observations, the values of the k distances of the nearest neighbors need to be stored. Therefore, given the same prediction capabilities, the smallest effective k would be the preferable choice. The number of digital quantities to be stored is then computed as

$$\mathrm{M}_{LOF} = k \times N_T, \quad \text{with} \quad N_T \gg n$$

In the case of OCSVM, the critical parameters are γ, which modulates the kernel width (e.g., the Gaussian kernel in (8)), and ν, which sets a minimum on the number of support vectors and a maximum on the tolerated amount of outliers. Nevertheless, only the former impacts on the required amount of digital words to be locally stored

$$\mathrm{M}_{OCSVM} = n \times \nu N_T + \nu N_T, \quad \text{with} \quad N_T \gg n \quad \text{and} \quad \nu \in (0, 1]$$

Regarding the detector based on PCA, the matrix U is the only entity to be stored on board such that

$$\mathrm{M}_{PCA} = n \times m, \quad \text{with} \quad m < n$$

Regarding the number of parameters characterizing the AE detector, this is a value that strongly depends on the adopted architecture.

$$\mathrm{M}_{AE} = n \times m + m + n/2 + 163, \quad \text{with} \quad m < n$$

As an example, with $n = 1000$ and $m = 250$ we have $M_{AE} = 250913$ parameters.

As a final remark, all methods are trained having a reference data set consisting of one year of telemetry measurements. This data base is scanned by sliding the position of a window of fixed length n over successive sequences of time samples in order to highlight temporal correlation, producing a high dimensional input matrix $X \in \mathbb{R}^{N_T \times n}$. Telemetry data is incline to showing periodical behaviors during correct operation, although the periodicity may occur at different rates. As a consequence, the the task of determining an optimal window length is not trivial: a dominant periodicity can make it challenging to detect faster variations, which may be instead significant from the behavioral interpretation point of view. Therefore, since outlier scores are computed based on the comparison between groups of samples located in subsequent intervals of time, the size n of the adopted windows affects the achievable performance in terms of abnormal behaviour detectability.

For what concerns the model we adopt to represent anomalous events, the presented analysis is specialized to the case of anomalies originated by Gaussian sources. Such kinds of anomalies could represent the effect of a large variety of possible system degradation conditions as well as it has been proved that white noise acts as a reference case in modeling the statistics of any possible class of anomalies [14].

All the models are chosen accordingly to hw constraints: e.g. a DNN cannot be selected because only few MB of volatile and non volatile memory are available, while the available stack is in the order of ten KB. Moreover, to increasing complexity of the model corresponds increased processing time during inference.

4 Numerical Evidences

In the training phase, for LOF and OCSVM, the model is fitted in three different configurations based on the size N_T of the training set, using references containing approximately $N_{T_1}n \approx 300\,\mathrm{K}$ samples, $N_{T_2}n \approx 1.3\,\mathrm{M}$ samples, $N_{T_3}n \approx 6.5\,\mathrm{M}$ samples, respectively. Whereas for PCA and AE, since their memory footprint is not directly affected by the amount of training examples, only the third reference set and a set containing $N_{T_4}/n \approx 15.5\,\mathrm{M}$ samples have been considered. We refer to a test set including $N_S n$ samples, that is in the order of $\approx 300K$ sensor readings, for the test phase. Both training and test set are passed to a pre-processing stage where filtering and data standardization are performed. For the identified reference scenario we focus on five different cases, described in Table 1.

Detectors' accuracy is assessed by showing histograms of the score values for each method and for all mentioned cases. We also adopt the area under the Receiver Operating Characteristic curve, ROC, briefly referred to as AUC will be considered to remark for which condition it is possible to detect each anomalous signal without any false positive. In these cases, the AUC value reaches the maximum value 1. The AUC metric allows to be independent from a specific threshold which might be used in order to convert numerical scores into binary labels.

Table 1 Reference cases in which anomalies are modeled by Gaussian sources

Case	Description
C1	The clean signal is compared to pure white Gaussian noise sharing the signal energy (WGN)
C2	Additive white Gaussian noise is considered, (AWGN), where noise and clean data have the same energy
C3	AWGN as for C2 but with a medium noise intensity, i.e., the noise contribution exhibits an energy level that is a half with respect to the clean signal
C4	AWGN with low noise intensity modelled by an energy ratio between signal and noise that is equal to 10
C5	The abnormal behaviour is represented by a whitened version of the input signal, presenting an increase in the noise-floor while keeping the energy of each instance constant

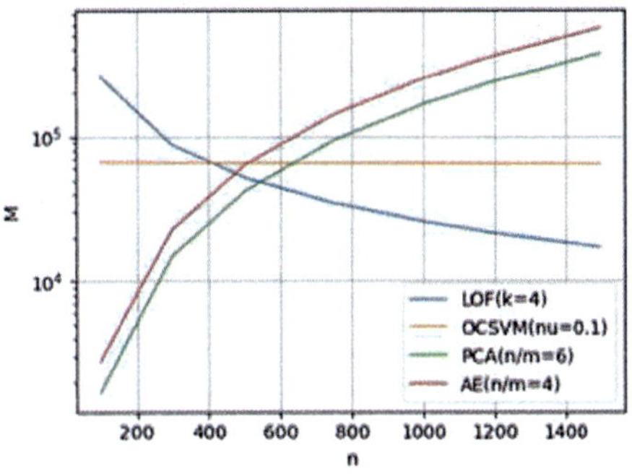

Fig. 5 Amount of digital words required by LOF, OCSVM PCA and AE where LOF and OCSVM adopt the N_{T_3} configuration

In order to include in the analysis the required memory footprint Fig. 5 provides an estimation of the amount of digital words to be stored. We limit the analysis to the cases $n \geq 100$ since all detectors reach AUC $= 1$ for cases C1 and C2.

Figure 6 shows a comparison between all detectors except AE in their respective least expensive setting when scores of clean signal are compared with the ones for C1 and C2 when $n \leq 500$. LOF is set up with $k = 4$ as number of nearest neighbors, OCSVM with $\gamma = 1/n$, $\nu = 0.01$ and $n/m = 6$ for the PCA. Moreover, the first two methods are trained considering the less memory hungry configuration N_{T_1}. PCA, since its trained offline, is still working with N_{T_3}. Except for the case $n = 30$, the AUC measure reaches the unitary value for each of the three detectors. In addition, these results suggest that larger windows are privileged. As far as PCA is concerned, with the same computational effort, SPE scores are preferred with respect to T^2, since they allow to observe the desired accuracy starting from smaller values of n. Not only PCA allows for a larger separation between the scores assigned to the clean signal with respect to the noisy ones, it also requires far less memory compared to LOF and OCSVM for $n < 500$, while learning on a larger reference set. Working with the same values of n also AE is able to reach AUC $= 1$ with $n \leq 500$ and it shows values of AUC approaching 1 also for lower values of n.

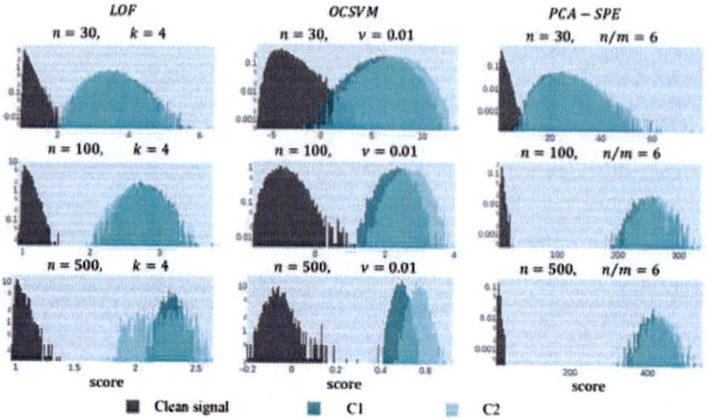

Fig. 6 Distribution of the scores obtained with LOF, OCSVM and PCA in their least expensive setting for windows containing $n = 30, 100, 500$ samples in cases C1 and C2, where noise and clean data have the same energy (see also Table 1)

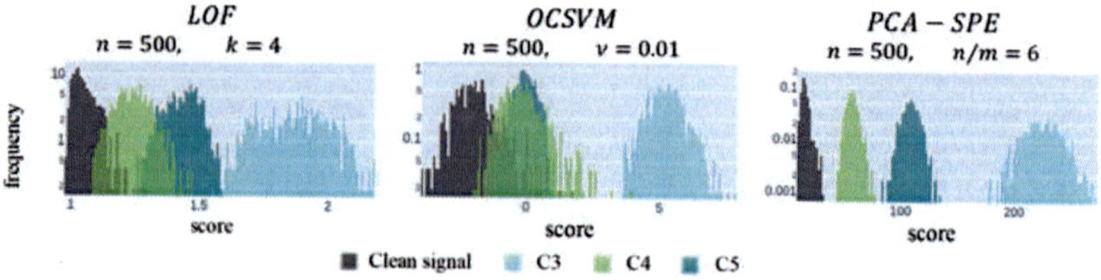

Fig. 7 Distribution of the scores obtained with LOF, OCSVM and PCA in their least expensive settings for window length $n = 500$, in cases C3, C4 and C5 (see Table 1)

In correspondence of $n = 500$, the amount of memory required by LOF shows a drastic drop, OCSVM presents a negligible affection from window size, and both PCA and AE start to exceed LOF. For the same choice of n, however, PCA and AE prove to be more robust for cases C3, C4 and C5, in contrast to LOF, which is still able to detect and distinguish noise signal in case of C3 and C4 only. OCSVM, shows AUC values close to 1 only for C3 (Fig. 7).

Figure 8 depicts the comparison between the performances of PCA and AE when dealing with the aforementioned irregularity cases, proving the models act as proper detectors for $n \gg 500$. They achieve the same detection ability having also a required memory footprint in the same range. Nevertheless, the AE model with convolutional layers is a good candidate for solutions in which multiple time series are monitored.

5 Conclusion

In this work, we have described the work-in-progress about the investigation and application of feature extraction and anomaly detection techniques for analysing

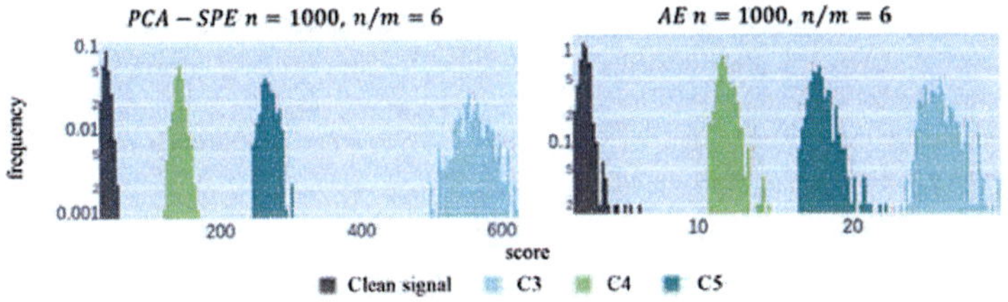

Fig. 8 Distribution of SPE scores for PCA and AE scores, with $n = 1000$ and ratio between the number of input components per window and the principal ones equal to $n/m = 6$, in cases C3, C4 and C5 (see Table 1)

spacecraft telemetries typically generated from LEO satellites. We assessed the performance of four well-know ML methods, properly tuned for the task of anomaly detection for on-board spacecraft health monitoring systems. A key point in our research is the building up of the decision criteria (thresholds, scores) for improving the detection of anomalies while minimizing false alarm conditions. Recorded accuracy results suggest that the ML models are worth deploying for the application of interest.

In the roadmap of our research, further investigations and developments are planned, e.g. in terms of simultaneous monitoring of several telemetry quantities in the same window (multivariate time series analysis), scalability and reliability analysis of developed ML models, focusing on the optimization process needed to implement the ML-based methods in space qualified hardware with limited computational power.

References

1. Ahn, H., Jung, D., Choi, H.L.: Deep generative models-based anomaly detection for spacecraft control systems. Sensors **20**(7) (2020). https://doi.org/10.3390/s20071991
2. Akcay, S., Atapour-Abarghouei, A., Breckon, T.P.: Ganomaly: Semi-supervised Anomaly Detection via Adversarial Training (2018)
3. Amihai, I., Gitzel, R., Kotriwala, A.M., Pareschi, D., Subbiah, S., Sosale, G.: An industrial case study using vibration data and machine learning to predict asset health. In: 2018 IEEE 20th Conference on Business Informatics (CBI), vol. 01, pp. 178–185 (2018). https://doi.org/10.1109/CBI.2018.00028
4. Bishop, C.M.: Pattern Recognition and Machine Learning (Information Science and Statistics), 1st edn. Springer (2007)
5. Breunig, M.M., Kriegel, H.P., Ng, R.T., Sander, J.: Lof: Identifying density-based local outliers. In: Proceedings of the 2000 ACM SIGMOD International Conference on Management of Data, pp. 93–104. SIGMOD '00, Association for Computing Machinery, New York, NY, USA (2000). https://doi.org/10.1145/342009.335388
6. Chandola, V., Banerjee, A., Kumar, V.: Anomaly detection: A survey. ACM Comput. Surv. **41**(3) (Jul 2009). https://doi.org/10.1145/1541880.1541882

7. Di Mascio, S., Menicucci, A., Gill, E., Furano, G., Monteleone, C.: On-board decision making in space with deep neural networks and risc-v vector processors. J. Aerosp. Inf. Syst. **18**(8), 553–570 (2021)
8. Farbiz, F., Miaolong, Y., Yu, Z.: A cognitive analytics based approach for machine health monitoring, anomaly detection, and predictive maintenance. In: 2020 15th IEEE Conference on Industrial Electronics and Applications (ICIEA), pp. 1104–1109 (2020). https://doi.org/10.1109/ICIEA48937.2020.9248409
9. Furano, G., Menicucci, A.: Roadmap for on-board processing and data handling systems in space. In: Dependable Multicore Architectures at Nanoscale, pp. 253–281. Springer, Cham (2018)
10. Jung, D., Kwon, J., Baek, K., Ahn, H.: Attitude Control Simulator for the Korea Pathfinder Lunar Orbiter, pp. 2521–2532 (June 2019)
11. Kingma, D.P., Welling, M.: Auto-encoding variational bayes. ArXiv preprint arXiv:1312.6114 (2013)
12. Lloyd, S.: Least squares quantization in pcm. IEEE Trans. Inf. Theory **28**(2), 129–137 (1982). https://doi.org/10.1109/TIT.1982.1056489
13. MacQueen, J.: Some Methods for Classification and Analysis of Multivariate Observations (1967)
14. Marchioni, A., Enttsel, A., Mangia, M., Rovatti, R., Setti, G.: Anomaly detection based on compressed data: An information theoretic characterization (2021). https://doi.org/10.48550/ARXIV.2110.02579
15. Schölkopf, B., Platt, J., Shawe-Taylor, J., Smola, A., Williamson, R.: Estimating support of a high-dimensional distribution. Neural Comput. **13**, 1443–1471 (2001). https://doi.org/10.1162/089976601750264965
16. Spirkovska, L., Iverson, D., Hall, D., Taylor, W., Patterson-Hine, A., Brown, B., Ferrell, B., Waterman, R.: Anomaly Detection for Next-Generation Space Launch Ground Operations. https://doi.org/10.2514/6.2010-2182
17. Vapnik, V.N.: The Nature of Statistical Learning Theory. Springer (1999)
18. Xiong, L., Ma, H.D., Fang, H.Z., Zou, K.X., Yi, D.W.: Anomaly detection of spacecraft based on least squares support vector machine. In: Prognostics and System Health Managment Confernece, pp. 1–6 (2011)
19. Zetocha, P.: Comparison of AI Technologies for Satellite Anomaly FDIR

Explainable AI with the Information Bottleneck Principle

Gabriele Berardi and Piergiorgio Lanza

Abstract Deep Learning (DL) has proven to be a very powerful technique. Meanwhile, DL achievements have not been match by theoretical understanding. Without firm basis, the design of deep neural networks (DNNs) overlooks profound learning dynamics and remains an empirical process. Developing tools for in-depth understanding could instead speed up design, boost final performance and assure reliability. The information bottleneck principle (IB (Shwartz-Ziv and Tishby in Opening the black box of deep neural networks via information (2017) [1])) is the first framework that offers an explanation of the information flow inside a network. Being a new and unfamiliar framework its results are still actively debated (Geiger in On information plane analyses of neural network classifiers—a review (2020) [2]), but within the IB framework new dynamics have been discovered and performance upper bounds have been set. With a pragmatic eye, inside the IB theory may be the tool we are searching for: the Information Plane, that quantifies how much each part layer knows about the input as a whole and about the desired output. The IB limit is the computational complexity of its main component: mutual information (MI). In this work we adapted the Mutual Information Neural Estimator (MINE (Belghazi in Mutual information neural estimation, pp. 531–540 (2018) [3]) to reproduce the original results on a 12bit dataset and on the smallest prototype of a real-world application (MNIST). The considerable jump in input size has not affected the IB dynamics: the same IB patterns have been observed in the scaled up networks. Hints of new dynamics with train/validation distinction have been observed. This work is a first step towards testing the IB theory on the largest scales and making information plane computation feasible on real-world training applications.

Keywords Information bottleneck · Mutual information · Information plane

G. Berardi (✉) · P. Lanza
Thales Alenia Space (DESI), Torino, Italy
e-mail: gabriele.berardi@thalesaleniaspace.com

P. Lanza
e-mail: piergiorgio.lanza@thalesaleniaspace.com

C. Ieracitano et al. (eds.), *The Use of Artificial Intelligence for Space Applications*,
Studies in Computational Intelligence 1088,
https://doi.org/10.1007/978-3-031-25755-1_15

1 Introduction

Deep Neural Networks (DNNs) have exceeded expectations in all the fields in which they have been introduced. Still, there is not a comprehensive theory on how and why they are so effective. For this reason, building and testing networks is much more a trial and error process than a task with a well-defined approach. This dominant heuristic approach has brought the extraordinary results we have seen, but it is clear that black boxes take much more effort to adapt to new and challenging problems. Moreover, neural networks are not infallible and higher trustworthiness can happen within a reliable theoretical framework.

There are many techniques to design and train a network, but the main metrics used rely on the comparison between desired output and network output. Such simple approach can only scratch the surface of DNNs, that are objects defined by millions or billions of parameters and require resource-intensive sessions to be trained.

This is an opportunity for any comprehensive framework able to explain DNNs behavior. In particular, a more in-depth understanding not only of what comes out, but also of what happens inside DNNs during training could speed up design and boost final performance. Also, it could help justifying the achieved results and assure reliability, that is a very difficult yet required task for any application. This field of study intersects Explainable AI (XAI), but XAI researches are usually focused in making results human-readable more than building predictive frameworks.

The Information Bottleneck Theory applied to DL [1] aims to fill this void and build a theoretical background to deep learning. The starting point is an information theory framework laid out in 2001 by Naftaly Tishby [4], that he has recently applied to DL in a series of articles [1, 5, 6]. Here the IB theory lays out an intuitive interpretation of what DNNs are, starting from a solid information theory base. The main tool used is Mutual Information (MI), that quantifies how much a random variable knows about another. With MI it is possible to explore how much each layer retains about the input as a whole, we can call it context, and how much it has learned about the desired output, that is the semantic content. This two quantities can be used as axis of a 2D plot, the Information Plane (IP). Looking at the trajectories on the information plane, drawn by each layer as the training goes on, opens a much deeper understanding of the DNN internal dynamics than a simple loss approach.

Within the IB framework new discoveries have been made, and this has spark interest inside the DL community since 2017 [7]. The theory and its results are still actively debated with diverse approaches ([2] for a recent literature review), and some of its claims remain with no definite answer. Nevertheless, IB discussion and testing is part of the fundamental effort towards more understandable and efficient deep learning techniques. In this paper we follow IB as it has originally been detailed in [1] and we test some of its results in a controlled environment.

The training process is divided into at least two phases: the initial "ERM" phase and the latter "forgetting phase", that is argued to be fundamental for generalization. It exists a theoretical upper bound to DNNs performance, that must lie under the IB curve. There is a justification for the multi-layer structure, that helps compression exponentially. All these results make DNNs more clear under the IB interpretation.

IB theory has a fundamental limitation: results have been presented only for toy models and have not been applied to normal sized DNNs. The reason for such a limitation is that MI cannot be computed analytically if not in very simple cases: only approximated approaches are possible, and the computational load is significant.

The scope of this work is to test and extend the IB approach, testing its added value to the DL landscape and reducing the gap between IB theoretical results and real-world applications. In particular, we concentrate on the IP, that is the most unique and profound introspective tool for DNNs.

We used approximated MI algorithms (MINE [3, 8]) to reproduce IB results on the 12bit-input case, that is the first case presented in literature where computation can also be done analytically. This both confirmed IB expectations and the reliability of the MINE approximated approach. Then, we adapted MINE to work with grayscale images and we tested again IB on a MNIST [9] classification problem. MNIST is a dataset of grayscale handwritten digits, and it is the classic prototype of image classifications tasks.

The obtained results confirm IB expectations. Moreover, hints of new training phases have been observed, photographing how overfitting appears from an informational flow perspective. Even so, computation remains a problem as some of the presented experiments required hundreds of hours to be generated. The maturated experience with approximated MI computation has laid the basis for further reductions in the IP computational load, that will be explored in a future work.

2 The IB Theory

To present the IB theory some fundamental quantities are here briefly defined. Taking a neural classifier as example, from an information theory point of view it is a data compressor with an input (e.g. an image) and a desired output (e.g. the image semantic content, or label). The image is made by a large amount of pixels, while the label can be completely defined in a much smaller space. Label information is contained inside the image, but it is not stored in any of the single pixels and it is instead highly distributed. The DNN is trained to compress the image and be left with only the useful information to define label. In a more precise definition the DNNs is a Markov chain:

$$X \rightarrow T0 \rightarrow T1 \ldots. \rightarrow Tn \rightarrow Y \tag{1}$$

A Markov chain is a multi-step process where each step state depends only on the step before. Here and following, we indicate with X the dataset distribution, with Ti the DNNs hidden layers activations distributions and with Y the desired output distribution. The output layer distribution is indicated as ~Y, as a remainder that it is the DNN attempt to approximate Y. Markov chains have a fundamental property:

no information can be generated in the process. This means that each DNN layer Ti at best maintains or, more in general, loses information about the relevant semantic content of the input.

To better follow relations between those quantities we can talk about information as entropy. The entropy of a random variable measures the volume of its output space, and corresponds to the mean optimal amount of storage needed to define its state, and it is defined as:

$$H(X) = -\sum_{x \in X} p(x) \log p(x) \tag{2}$$

where X is the random variable distribution, and x is a single outcome. If two random variables are correlated in any way (e.g. image and corresponding label) they share a portion of their entropy. Mutual information (MI) is defined as the amount of shared entropy between two random variables:

$$\mathrm{I(X, Y) = H(X) - H(X|Y) = H(Y) - H(Y|X)} \tag{3}$$

In a more intuitive interpretation, MI represents how much a random variable knows about another. In the contest of IB, MI is the main tool used to explore the learning dynamics.

Two are the relevant quantities that IB concentrate on: the MI between DNN layers and the dataset as a whole, and the MI between the DNN layers and the desired output. The first represents how much the DNN knows about the context, the latter represents how much the DNN knows about semantic content and it is directly correlated with accuracy. Those two quantities are the axis of the Information Plane (IP). Each layer (output layer included) in each epoch is associated with two mutual information values: MI layer-data (context knowledge) and MI layer-labels (semantic knowledge). Those two values are coordinates on the information plane, they evolve during training and through epochs they form a trajectory. Each layer has its own trajectory, that represent its knowledge of context and labels evolving over epochs. Collectively, the layers trajectories define the DNN dynamic on the IP.

Figure 1 shows examples of IPs from [1]. In each panel there are 4 thick lines, the layers trajectories through training. Color gradient indicates the epoch, from the earlier epochs (black) to the latter (yellow). The three panels shows three different IPs, varying training conditions (the size of the training set). IP trajectories are not random and scattered, but a distinct pattern emerges instead. Two phases are visible: a first "fitting phase" (also called ERM), where the network learns quickly about both contest and semantic content, and a second "forgetting phase" (also called diffusion phase), where the DNN decreases context knowledge while still slowly increases semantic knowledge. The phases are distinguished by a sharp turn left of all layers trajectories. In the IB interpretation, the second phase is crucial for generalization: the DNN learns to compress not-useful context features.

This phase distinction arises also looking at the gradient evolution. In the fitting phase the gradient norm is well above its standard deviation, indicating a steady

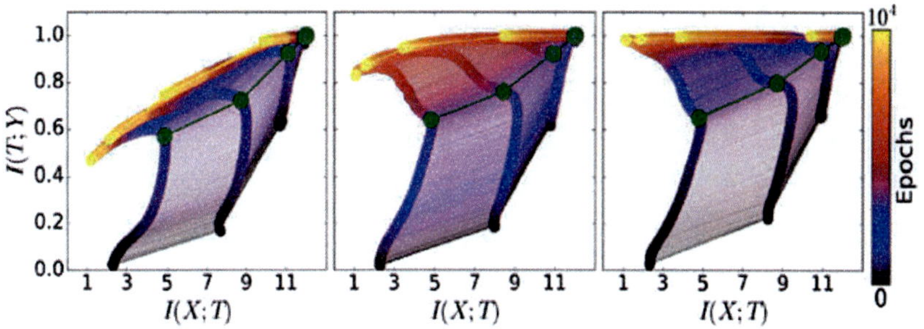

Fig. 1 An example taken from [1] of the emerging training pattern, visible with the means of mutual information. The horizontal axis is the information retained by a layer T about the input as a whole X (context information), the vertical axis is its knowledge of the relevant feature Y (semantic information). Each thick line represents one layer throughout the epochs, while the lighter lines connect layers from the same epoch. The color gradient represents the epoch, from black to yellow different panels show network trained with varying conditions. There is a clear phase distinction along the training process, that manifests as a sharp turn left in the trajectory. This distinction is argued to have profound implications for DNNs generalization ability

gradient descent phase. During the forgetting phase it is the opposite: the gradient standard deviation is greater than its norm, indicating a more diffusion-like movement (Fig. 2).

Another result worth citation from the IB theory is a theoretical upper bound to the DNNs knowledge, represented by a boundary on the information plane. This happens due to the constraints intrinsically introduced by the DNNs encoding/decoding structure, and it is what names the theory (bottleneck of information). The boundary, called information bottleneck curve, is shown in Fig. 3.

We did not test this result, but we cite it as it is relevant to present the IB theory as a first step towards a foundational framework in DL, that can have pragmatic consequences in different DL fields. The interest of part of the DL community about the IB theory is recent and its conclusions are still actively debated. For more in depth explanations and reviews we remand on [1]. The last two years have seen a proliferation of studies on the matter, each with a slightly different approach and interpretation. It is although clear that the IP is a profound inspection instrument to understand what happens under the DNNs surface during training. With a pragmatic eye, a similar tool could become a fundamental resource to monitor and optimize the production process of DNNs. The focus of our work is to reproduce the IP results and evaluate the IP feasibility on real training sessions.

3 Mutual Information and MINE

Mutual information is a fundamental quantity that measures relation between random variables. It has its applications in every branch of data science and, as we have

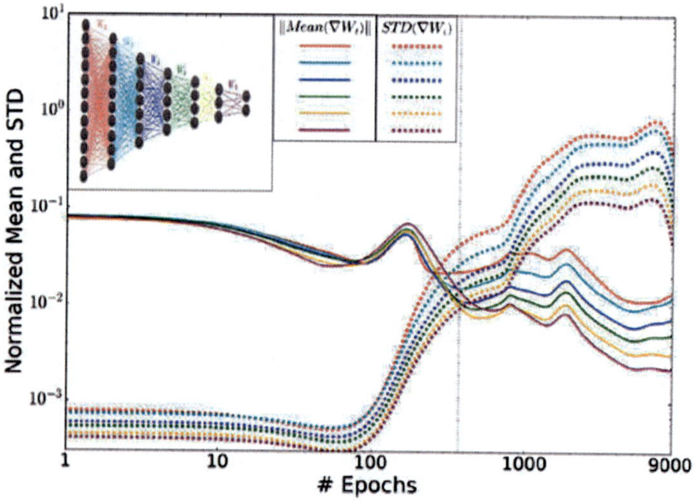

Fig. 2 Gradient norm and standard deviation (std) through epochs from [1]. In the first epochs norm is much higher than std, suggesting a smooth and fast descend towards lower loss. This is the signature of the fitting phase: the network is quickly learning about context with a clean backpropagation signal. After epoch 300 there is change, std becomes greater than norm and the network moves by diffusion in the parameter space. This is the compression phase: the gradient noise makes the DNN move by diffusion in the parameter space and non-relevant information, not being rewarded by backpropagation, begin to slowly disappear. The two distinct behaviors are in agreement with the two phases seen in Fig. 1

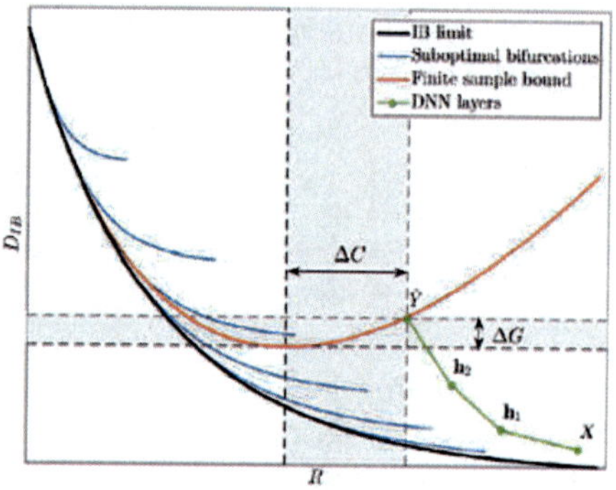

Fig. 3 The information bottleneck curve on the information plane as shown in [1]. Here R = I(X;T) and DIB = − I(Y;T). The black curve delimits what is accessible by any data manipulation. A real DNN will be bounded by a limited sample, that will yield a more strict region (orange line). In green are qualitatively represented the positions of trained DNN layers. The output ~Y of a fully trained network will, in optimal conditions, be a point on the orange line of the finite sample. The minimum of the orange curve represent the maximum semantic information that a network can learn, being trained with a finite sample. This IB result is presented for completeness and to give a more general view of the theory

seen in the previous section, it has profound implications in deep learning. It is in fact a precise measure of how much a network knows about context and semantic content. However, MI computation poses a series of non-trivial problems. From its constructive definition [10]:

$$I(X, Y) = \int_{X*Y} \log\left(\frac{dP_{XY}}{dP_X \otimes dP_Y}\right) dP_{XY} \tag{4}$$

If, for example, the X distribution corresponds to MNIST images, the above integral form would require to sum on the set of all possible images of handwritten digits, that it is not an accessible set and it is not even well-defined. For this reason many different variations of the above formula have been introduced (Poole et al. [11] for a comprehensive review of the most recent ones). In our work we used the Mutual Information Neural Estimator (MINE) that is based on the MI Donsker and Varadhan representation [12]:

$$I(X, Y) = \sup_{\theta} E_{P_{XY}}[T_\theta] - \log\left(E_{P_{XY}}\left[e^{T_\theta}\right]\right) \tag{5}$$

That is, the MI between X and Y is equal to the upper bound of the right hand side varying the generic function T. The main points of this formula are three: it is an approximation by defect, it works also on finite sampled distributions, and it is defined minus a generic parametric function T.

MINE algorithm makes use of this definition implementing the generic function T as a deep neural network. Using the Donsker and Varadhan representation value as loss, MINE algorithm trains the DNN to maximize the MI value. The final MI value will still be, by definition, an approximation by defect (Figs. 4 and 5).

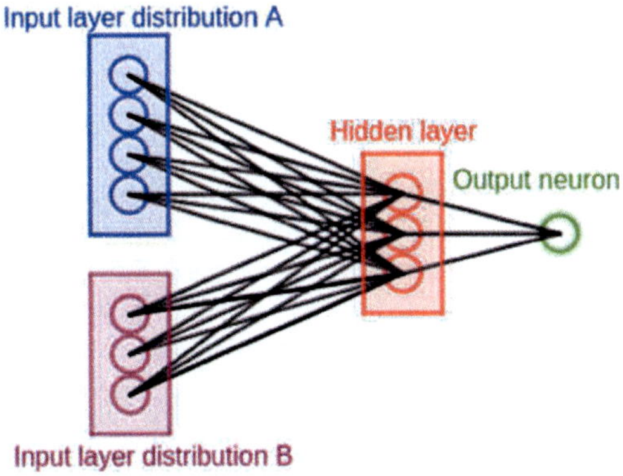

Fig. 4 Depiction of MINE dedicated network to compute MI between two distributions A and B. The Input layer is divided into two parts, each taking one of the two distributions. Each MINE iteration initializes a new MINE network and trains it, up to convergence. The MINE network takes the place of the T function in Eq. (1), and uses the equation value as a loss to be maximized

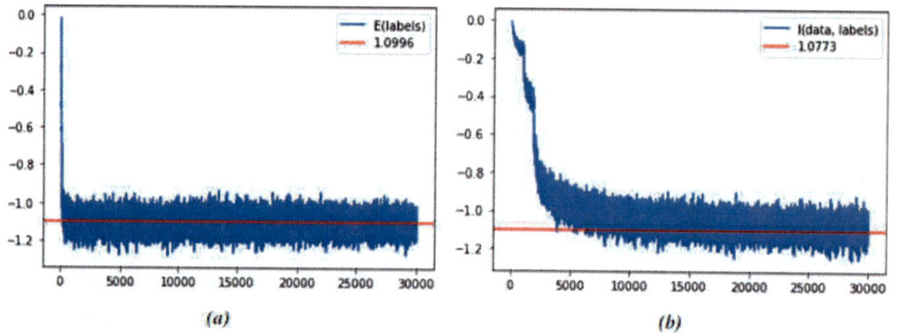

Fig. 5 Our comparison between labels entropy E(labels) computed as **a** H(labels) and **b** I(labels;data) on a set of MNIST images. Both panels show a single MINE iteration: a MINE net is initialized and trained up to convergence. The loss mean over the last few hundreds of iterations is used as MINE final output. There is a good agreement between the two value yield, with a difference below the intrinsic MINE variance

MINE algorithm yields correct results in the cases where the analytical approach is possible. It is however a neural estimator: each MINE iteration requires to define and train a DNN and, being training a stochastic process, the output value is subject to oscillations.

In Fig. 6 we can see that the algorithm is sufficiently stable for our scope. Computational time, on the other hand, is a significant drawback. A complete IP plot that tracks a DNN evolution through training requires a large number of MINE iterations. A five-layer DNN trained for 500 epochs requires 2500 MINE iteration to completely track the IP, where each MINE iteration defines a MINE net and trains it to convergence. This complex setup has brought some of our experiments to last between 100 and 300 h.

4 Experimental Setup

For our experiments we used fully connected networks as DNNs prototypes on two different datasets: 12bit and MNIST. In both cases, each layer has been plotted separately on the IP, using layer-dataset and layer-label MI as coordinates. MI computation has been repeated each epoch or every fixed number of epochs (to reduce computational load) and on the IP are plotted the results of an entire DNN training session. The IP has been differentiated between training and validation set. The reason is that training set is correlated with the network state, while validation set is not. This distinction has not been stressed in literature, but it leads to different results on the IP and it is thus useful to be explored.

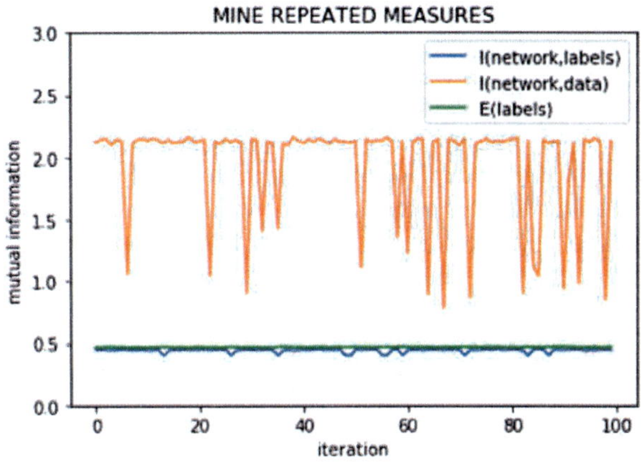

Fig. 6 Repeated MINE iterations that we performed on fixed conditions, to test for numerical stability. Three MI values are computed: E(labels) in green, I(T,labels) in blue and I(T,data) in orange. With a deterministic algorithm, we would expect the same result each iteration. Being instead MINE a neural algorithm, it is subject to variance and occasional convergence on local critical points. Even so, local minima manifests as occasional value drops, and are easily distinguishable from nominal points

4.1 12bit

The 12bit dataset is compatible with the dataset used in [1]. We used MINE, to test the reliability of our approximated approach. The are 4096 possible inputs, and they are binary classified with a simple rule: all numbers with exactly 3 or 4 ones in their binary representation fit in one class, all remaining numbers form the other class. The DNN prototype is a fully connected network with three layers for a sequential structure of 12-8-4-2. The network has been trained with a low learning rate for 20 thousand epochs on a batch of 78% of the possible inputs. Although been trained for many epochs, the network is in a nominal training state and it is not overfitting. Intermediate states are analyzed with the MINE algorithm, that makes use of a dedicated network with one hidden layer and a sequential structure of N-500-1, where the first layer width is adapted to each network layer. Repeated MINE iterations were done in fixed condition to check for numerical stability.

4.2 Resized MNIST

MNIST [9] is a catalogue of 70 thousand 28×28 black and white handwritten digits, and it is a standard benchmark test for small classification networks. The MNIST neural classifier prototype is a fully connected network 40-30-20-10-2, and has been trained for 1500 epochs.

5 Experimental Results

The 12bit IP has confirmed the IB expectations. The same pattern of [1] is visible in Fig. 7, where our training set IP is shown. There is a clear behavior change that separates training in two phases: the fitting phase and the forgetting phase.

A similar pattern, with a less pronounced MI with labels, is visible in the validation set IP (Fig. 8). The network is generalizing enough to have similar behavior both with the training set and the validation set, but being not correlated with the latter it has less shared entropy.

A different scenario arises with the IP relative to an overfitting network. Figure 9 shows training set IP and validation set IP in comparison. We see the usual ERM phase followed by the compression phase but, as soon as the overfitting starts, a different patterns emerges. The overfitting IP dynamics on the training set is similar to a new ERM phase, were both context and semantic knowledge rise quickly. On the contrary, the IP with the validation set shows a rise in context knowledge but significant drop in semantic knowledge. This is coherent with usual overfitting interpretations: the network is losing knowledge about the desired output while learning irrelevant patterns on the training data.

MNIST IP has also confirmed IB expectations. The resulting pattern is very similar to the 12bit cases, Fig. 10 shows the two IP in comparison.

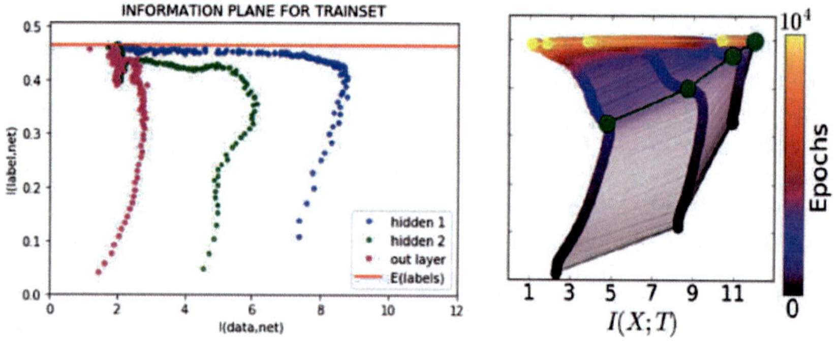

Fig. 7 Information plane comparison between our 12bit results (left panel) and literature results (right panel, figure from [1]). We trained a fully connected network with three layers for a sequential structure of 12-8-4-2, with low learning rate for 20 thousand epochs on a batch of 78% of the possible inputs. Each epoch MINE algorithm has been applied to all layers separately. Our network has shown a first fitting phase (fast movement upper-right) and a latter forgetting phase (slow movement upper right), as expected by the IB theory. The two panels show a very similar pattern

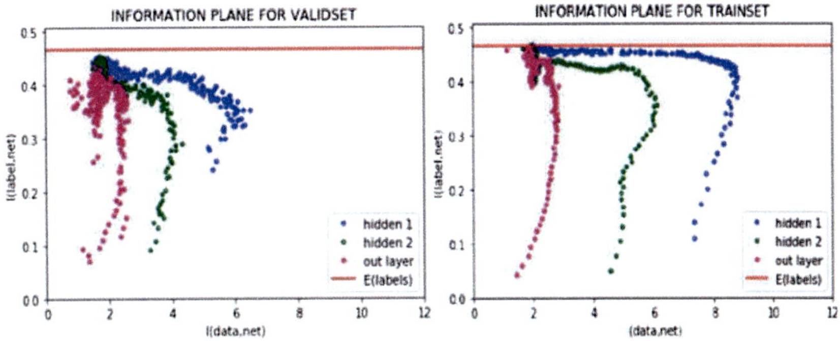

Fig. 8 Comparison between our 12bit IP results, computed with respect to the validation set (left panel) and the training set (right panel). The same two-phase pattern emerges in both panels, with a first fitting phase (fast movement upper-right) and a latter forgetting phase (slow movement upper right), in agreement with IB. The validation IP, however, is more scattered and with less context information, because the validation set is less correlated with the network state. This distinction has not been explored in literature, but it is useful as an agreement between train and validation IP is index of nominal training

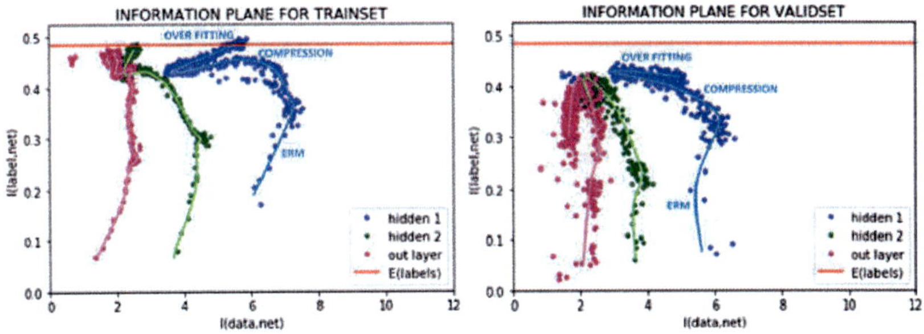

Fig. 9 Comparison of our 12bit IP results, computed with respect to the validation set (left panel) and the training set (right panel) in the case of an overfitting network. The same two phase pattern emerges in both panels, but a change of behavior is observed as overfitting begins. In this third phase, validation and train IP show different patterns. The training IP has what it resembles a new fitting phase, while validation IP begins to lose semantic information. This is in agreement with the usual overfitting interpretation: the DNN focuses on non-relevant features of the training set (increasing both context and semantic knowledge), but fails to generalize to the validation set (losing semantic knowledge). This distinction has not been explored in literature, but IP comparison between training and validation here shows clearly the overfitting state of the network

6 Conclusions

The Information Bottleneck is a profound theory with deep implications for DNNs. It offers a number of theoretical results about DNNs behavior, but also practical instruments for DNNs inspection such as the information plane. The IP could be a very useful tool to understand what happens inside DNNs during training, leading to

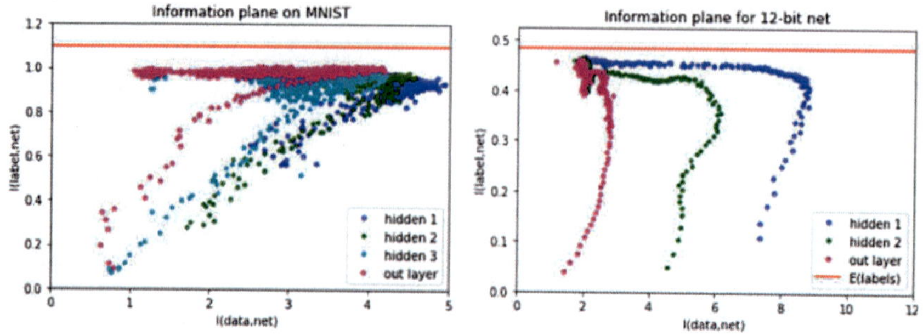

Fig. 10 Comparison between our IP results, on 12bit and MNIST data. The two phase distinction is evident in both cases, confirming IB expectations on MNIST

more conscious design choices and more justified final performances. The main IP drawback is the computation complexity of MI estimation algorithms. We used the Mutual Information Neural Estimator algorithm to review IB expectations on a 12bit dataset and on MNIST. The IB expectations are confirmed as all our experiment showed a clear phase distinction between fitting phase and forgetting phase. We differentiated IP computation between training set and validation set, being two very different sets from a DNN correlation perspective. Nominal training condition have shown the same patterns on training set IP and validation set IP, indicating that the IP trajectory is likely to be a DNN property as the IB would expect. An interesting pattern has been observed on not nominal training condition, namely an overfitting network. Here the DNN behavior is different between training IP and validation IP. In the first case, overfitting presents as a new ERM phase, with an increase of MI with dataset and labels. In the latter, overfitting leads to a significant decrease of MI with labels in all layers, leaving the MI with the dataset mostly unchanged. This behavior is compatible with the usual interpretation of overfitting: the networks begins to focus on non-relevant features on the training dataset that do not generalize to the validation set.

IP has shown a complex and meaningful etiology in this test contest with small networks and dataset. This demonstrates how it would be an useful instrument to better see DNNs state. There is still, however, a fundamental obstacle of computational nature to real-scale application of the IP. MI computation is not yet feasible on large scale, drawing IP through approximated algorithms like MINE takes already a significant number of hours. The simplification of IP computation will be the subject of future work.

References

1. Shwartz-Ziv, R., Tishby, N.: Opening the black box of deep neural networks via information (2017). http://arxiv.org/abs/1703.00810
2. Geiger, B.: On information plane analyses of neural network classifiers—a review (2020). https://doi.org/10.48550/arXiv.2003.09671
3. Belghazi, M.I., et al.: Mutual information neural estimation. In: Dy, J., Krause, A. (eds.) Proceedings of Machine Learning Research vol. 80, pp. 531–540 (2018). http://proceedings.mlr.press/v80/belghazi18a.html
4. Tishby, N., Pereira, F., Bialek, W.: The information bottleneck method. In: Proceedings of the 37th Allerton Conference on Communication, Control and Computation, vol. 49 (2001)
5. Tishby, N., Zaslavsky, N.: Deep learning and the information bottleneck principle. In: 2015 IEEE Information Theory Workshop, ITW 2015 (2015). https://doi.org/10.1109/ITW.2015.7133169
6. Piran, Z., Shwartz-Ziv, R., Tishby, N.: The dual information bottleneck. https://doi.org/10.48550/arXiv.2006.04641
7. Wolchover, N.: New theory cracks open the black box of deep learning. https://www.quantamagazine.org/new-theory-cracks-open-the-black-box-of-deep-learning-20170921/. Last accessed 11 June 2020
8. Yamada, M.: MINE pytorch. https://github.com/MasanoriYamada/Mine_pytorch. Last accessed 11 June 2020
9. LeCun, Y., Cortes, C., Burges, C.J.C.: The MNIST database of handwritten digits. http://yann.lecun.com/exdb/mnist/. Last accessed 11 June 2020
10. Thomas, M.: Cover and Joy A. Thomas. In: Elements of Information Theory (Wiley Series in Telecommunications and Signal Processing). Wiley-Interscience, USA (2006). ISBN: 0471241954
11. Poole, B., et al.: On variational bounds of mutual information. In: Kamalika, C., Salakhutdinov, R. (eds.) Proceedings of Machine Learning Research, vol. 97, pp. 5171–5180 (2019). http://proceedings.mlr.press/v97/poole19a.html
12. Donsker, M., Varadhan, S.: Asymptotic evaluation of certain Markov process expectations for large time (1983). https://doi.org/10.1002/cpa.3160280102

SINAV: An ASI Study of Future AI Applications on Spatial Rovers

Piergiorgio Lanza, Gabriele Berardi, Patrick Roncagliolo, and Giuseppe D'Amore

Abstract In the future Mars or Moon rover scenarios it is always supposed to have a high grade of autonomy on board of the rover. The use of Artificial Intelligence can strongly improve the main goal of an increased rover autonomy navigation. The SINAV (Soluzioni Innovative per la Navigazione Autonoma Veloce) (Agenzia Spaziale Italiana (BANDO DI RICERCA PER TECNOLOGIE ABILITANTI TRASVERSALI Area tematica Tecnologie Spaziali: B) Sistemi Autonomi e Intelligenza Artificiale A.S.I. 2018 [1]) study, co-financed by the Agenzia Spaziale Italiana (ASI), is a consistent step toward this main objective of improving navigation autonomy. In this article we briefly describe the challenging environments of Mars and the Moon that is also included in the main requirements of the SINAV study. Particular care has been devoted in the choice of Deep Learning (DL) for this study. Through Deep Neural Networks (DNN) represent a real and consistent advantage in terms of capability and reliability for rover missions without having to constitute a fancy technical solution where AI is forced in the rover navigational loop. Each DNN explains the problem that we need to solve, how it has been addressed in the past mission and the advantages of the proposed solution. A description of the adopted Machine Learning Operations (MLOps) and the different datasets created is also presented. A final description of the particular hardware and software used during the final tests complete the overall description of the main objectives inherent to Deep Learning of the SINAV study.

P. Lanza (✉) · G. Berardi · P. Roncagliolo
Thales Alenia Space (DESI), Torino, Italy
e-mail: piergiorgio.lanza@thalesaleniaspace.com

G. Berardi
e-mail: gabriele.berardi@thalesaleniaspace.com

P. Roncagliolo
e-mail: patrick.roncagliolo@thalesaleniaspace.com

G. D'Amore
Agenzia Spaziale Italiana, Roma, Italy
e-mail: giuseppe.damore@asi.it

C. Ieracitano et al. (eds.), *The Use of Artificial Intelligence for Space Applications*, Studies in Computational Intelligence 1088,
https://doi.org/10.1007/978-3-031-25755-1_16

Keywords Deep Learning · MLOPs · Mars rover · Space DNN applications

1 Introduction

The uncertainty is a consistent part of any explorative mission. A rover, a lander, or a drone in the case of Mars, shall cope with different uncertainties during their operative life. Until now the human supervisor has always controlled the mission and evaluated, from time to time, different strategies to avoid a possible pitfall or to recover from an unknown state. Typically, during an explorative mission there are two teams controlling the mission: the scientific team and the engineering team. The scientific team with their scientific expectations and the engineering team that support them in their decision as well as an active part during the troubleshooting of the mission. An autonomous rover capable of reaching a chosen destination would be a great relief for any interplanetary mission. A rover able to detect the different types of terrains and determine autonomously its path would be even more advanced. In the evaluation of its minimal path it should be taken into account the geometrical distance, the possible obstacles and the different kinds of terrain. Example: a sandy terrain is typically flat but its traverse is extremely hard. It is much better to choose a different path and avoid as much as possible the sandy terrain. It could lead to a longer solution in terms of elapsed time, but it would be safer and less dispendious in terms of energy.

This AI behavior is what is strictly necessary for future missions. This mission scenario is partially achieved in the Mars mission Perseverance 2020 of NASA. In fact a preliminary terrain exploration is achieved by the first drone used during a Mars mission: Ingenuity helicopter. Its scope is to acquire more information of the terrain and to communicate this information to the Perseverance rover to avoid being stuck as it happened to the Curiosity rover.

Another step forward is to imagine a rover capable of travelling autonomously during the night and use the light day time only for experimentation. Even better, if, the rover is able to catch autonomously interesting rocks on the basis of general description indicated by the Scientific team be it night or day. Multiple rovers, operating concurrently can increase the effectiveness of the exploration in twofold ways: an exhaustive research on specific terrain exploration area and simultaneously reducing the mission time.

ALTEC, as a Prime Contractor, and Thales Alenia Space together AIKO and different Departments of Politecnico di Torino are involved in an innovative study co-financed by ASI started in September 2021 where the new Artificial Intelligence (AI) and autonomous navigation approaches will be studied and applied to the next rover generation. The main goals of SINAV (Soluzioni Innovative per la Navigazione Autonoma Veloce) [1] shall be shown in a final demonstration demo where the rover shall be capable of autonomously reaching a location both in day light and night time.

The SINAV project started in September 2021. The overall project duration is two years. A co-financed budget close to 1.5 Millions of Euro supports this project. It is one of the biggest projects financed by ASI in autonomous navigation and A.I. subjects.

In this article we are mainly focusing on the AI applications instead of autonomous navigation instrumentations and algorithms. Four specific DNN (Deep Neural Networks) will be developed during the SINAV study. For each DNN the advantages of Deep Learning (DL) solution are highlighted with respect to a more consolidated and already used software solution for space applications. During the study a Machine Learning Operations (MLOps) has been adopted for the training and validation phases of the different DNNs.

Most importantly it is also shown how to validate these DNNs in a representative hardware and software environment similar as much as possible to a real Mars mission.

2 Mars and Moon Challenging Environments

Mars and Moon landscape exploration are quite different. This has different impacts on the possible AI applications that are briefly discussed in the previous paragraph.

The Mars to Earth distance is longer than the Moon to Earth distance and this has a dramatic impact on the communications latency time. Radiofrequency signal travels at the speed of the light and in the best case its value is limited to 6.5 min, while in the worst case it can reach 44 min [2]. Besides, communications with Earth would not be possible at all during a 14 day period at every Mars synodic period (approximately 2.1 years) [2]. With these latency times and long periods of time without communications, the control in real time with a human operator in the loop is unrealizable.

As the round trip between the Moon and the Earth is close to 1 s, also with difficulties, in this case it is possible to operate in remote control. Therefore, the autonomy mission level required is higher in the Mars case, because there is no possibility for any fast recovery action by the human operator who supervises the mission. Another influencing factor is the presence of atmosphere on the Mars landscape. At times a turbulent sand storm can occur on Mars. Today, the lack of Mars atmospheric data and a consistent atmospheric model of Mars cannot be used for forecasting. A better approach, using a predictive local model, can be realized on the future Mars mission where, based on multiple data, it is possible to detect the sand storm in a timely manner and to avoid any possible damage of delicate electronic instrumentation. This kind of application is under study at Thales Alenia Space Italia in Turin. Based on many real images of the sky taken under different weather conditions, it is possible to detect the weather condition automatically using DL. This is a typical human task that is easily achieved, but until ten years ago it was very difficult to implement in software prior to the implementation of deep neural networks.

Instead, the absence of atmosphere on the Moon simplifies the image acquisition as there is not any kind of turbulence. Nevertheless, the most interesting places for building a possible Lunar Space Station are located on the South Lunar Pole where illumination is guaranteed through the entire year except for a few days. In this case the solar flux illumination is tangential to the Lunar surface and it creates long shadowed areas and highly contrasted images.

Another critical issue on the Lunar surface is the presence of regolith. The regolith consistency is similar to sand and it is one of NASA's primary concerns because it can get into the mechanical joints of their future rover [3]. Also, in this case, it is fundamental to check the presence of regolith and evaluate in real time if it is safe to travel on it.

To be successful in Space exploration, it is necessary to design the mission specifically according to the environment where it will be operating. The AI approach does not imply any generalization of a unique approach that applies to any context.

3 SINAV Main Requirements

There are also a few considerations due to the Mars and Lunar environments that future exploration missions should take into account. These considerations have been included as a part of the main goals of the SINAV study that are summarized in the Table 1:

A first requirement is inherent to the actual speed of the rover (close to 4 cm/s). This limited average speed is the result of the stop and go approach used in the past mission where the rover takes pictures, evaluates the depth map, the preferred path and finally it moves. Perseverance, the new rover of NASA, uses continuous navigation capabilities which improves the rover moving so it is required the same approach in SINAV study increasing tenfold the rover travel.

Table 1 The six main mission requirements of the SINAV study

O.1 Increase tenfold the rover travel speed from about 10–20 to 100–200 m/h
O.2 Provide continuous navigation capabilities capable of operating even at night
O.3 Use active sensors for autonomous navigation of the rover such as TOF camera and/or laser strips and innovative sensors never used in current missions but in the process of being qualified
O.4 The rover autonomy in the path planning has been extended by means of DNN approach. Furthermore the autonomy is evaluated the information acquired by the on board sensory and integrating them with global information obtained from semantic analysis of images acquired by spacecraft
O.5 Different tests will be run between the rover autonomous functionalities and the same functionalities including spacecraft data
O.6 Common tests will be achieved where autonomous rover capabilities will be incremented with spacecraft data. The test will be executed in an environmental partially representative of Mars terrain

The future exploratory missions in Mars and Moon are also interesting in the cave exploration for understanding better the morphology and also to evaluate possible natural refugee for human basis. This explains why the necessity of a second requirement of a rover's capabilities during the night or in absence of light has been introduced.

Using different new sensor types as Time of Flight (TOF) cameras and employing new DNN to increase the rover autonomy constitute the third and fourth main high level requirements. The fifth requirement is about the consideration that different tests, conducted in different environmental conditions definitively increase the overall confidence in the proposed DNN solutions. A last, sixth high level requirement, is inherent to the inclusion of a drone in the moving rover scenario as it is the case of drone Ingenuity with the Perseverance rover.

In this article we are only focusing on the DL application applied to the rover autonomous navigation.

4 The Value of Deep Learning

In a quite famous and discussed article of Forbes it has been stated that 87% of data science projects failed before introducing them in the production. "One of the biggest [reasons] is sometimes people think, all I need to do is throw money at a problem or put a technology in, and success comes out the other end, and that just doesn't happen," [4]. The second main reason is lack of data and the third one is lack of collaboration as explained in Forbes article. In the book Becoming a Data Head [5] instead is underlined the importance in the correct identification of the customer needs, the outputs and the expectations of the customer about data science solutions. In particular the requirements and the identification of a real application constitute a relevant part of the project that has to be clearly addressed.

In the SINAV study a big effort has been dedicate to drastically reduce the main counterproductive approaches identified above:

- identify pragmatic applications that can be really used in a rover mission. Specifically, for each application, explain how the problem has been addressed in the past and identify the pros and cons of a DNN solution. Furthermore identifying a series of tests that can prove the efficiency of the proposed solution;
- large effort has been taken by the TASI and ALTEC in the generation of different synthetic and real datasets which are distributed between the SINAV partners to alleviate the issue inherent to the lack of data;
- collaboration between all the partners involved in the project towards practical results.

In the next paragraphs are presented different problems inherent to a rover operability and the proposed DNN solutions adopted in the SINAV study.

5 Terrain Traversability Approach

The past Nasa missions have greatly demonstrate how correctly identifying terrain type is crucial for the rover mission. For example in the Fig. 1 the Mars rover *Curiosity* experienced an unusual problem on the of rippled sand [6].

As shown in Fig. 1 the narrow valley has a relatively steep slope on both sides and a floor that is constituted of rippled sand. Initially, the operation team commanded her to drive over the ripples. However, the deep sand was more hazardous than expected, causing high wheel slip and sinkage. As a result, the operations team backed up and chose an alternate path over a harder substrate to continue the traverse toward Mount Sharp [6]. *Curiosity* mission had another critical moment during Sol 450–515 when the control team checked *Curiosity*'s wheels and experienced an unexpectedly high rate of damage. The MSL Wheel Wear Tiger Team identified that the period of highest damage accrual occurred when the rover was driving over angular, embedded rocks [6].

In the past missions the terrain classification in real time, during the rover navigation, was not really feasible. The differentiation between the terrains or rocks was based on a locally analysis of the texture present in the image. The texture analysis was the main approach used in the identification between safe and unsafe terrain. Because the sand bank does not present a consistent large texture this kind of analysis was not really effective to identify this dangerous terrain. An example of texture analysis realized by means of Gray Level Co-occurrence Matrix (GLCM) algorithm is shown in the next picture. The information carried on by the color images was not fully used in this analysis (Fig. 2).

Fig. 1 The hidden valley, imaged by *Curiosity*'s NAVCAM on Sol 717. Image Credit: NASA/JPL [6]

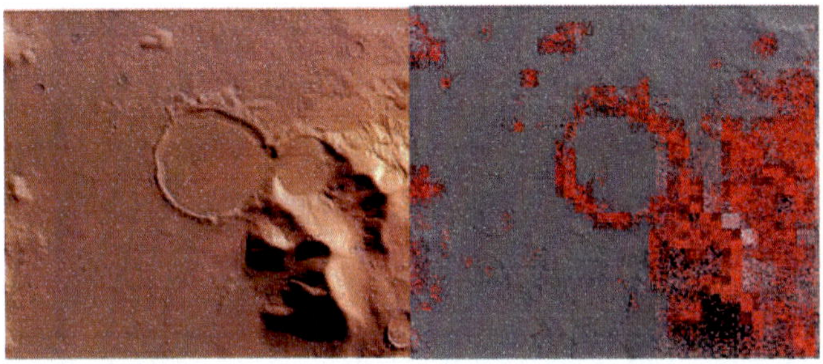

Fig. 2 The GLCM algorithm applied to the Promethei Terra "Hourglass craters". This kind of similar algorithm were used before the introduction of the DNN. They are local algorithm and they are working on local analysis of the image. The dangerous terrain areas for landing are identified by the red color. A specific algorithm version has been developed by Thales

On the basis of these critical experiences and also on the past NASA rover missions NASA decided to create Soil Property and Object Classification (SPOC) team with its own public homepage [7] with the declared goal to classify the different type of terrains present on Mars in supervised way.

The basic idea is to study the terrain type by means of HiRISE (High Resolution Imaging Science Experiment) camera mount on Mars Recognition Orbiter (MRO) which has the resolution of 25 cm/pixel for all eight candidate landing sites, each of which spans over $\sim$100km^2. This analysis helped in the evaluation of the landing site traversability analysis for the Mars 2020 Rover (M2020) mission.

A complete list of the possible Mars terrain has been realized by the SPOC team. It is composed of 17 different categories of Mars terrain where for each terrain a brief description is given. On the basis of this complete analysis a reduced set of terrain categories has been adopted for a real time classification on board of the next rover generation:

Sand corresponds to loose sand without any visible rocks.

Small Rocks is consolidated soil with small rocks.

Large Rocks corresponds to regions with larger rocks on the surface. Bedrock is relatively flat, exposed bedrock.

Outcrop refers to areas generally in the sides of hills with large rock formations that are generally not traversable.

Wheel Tracks refers to visible wheel tracks from the rover. Wheel Tracks were not originally included, but were necessary to prevent misclassifications.

An example of manually labeled terrain in the SPOC context is shown below (Fig. 3).

This approach is also confirm by Treveil [8]. A similar approach will be proposed during the SINAV study. More specifically this kind of processing is named Semantic Segmentation. It is applied only a specific categories of objects presented in the image

Fig. 3 Examples of manually labelled terrains classes in Curiosity Navcam imagery [7]

because a complete segmentation of the image is not strictly necessary during the rover motion. An example of semantic segmentation is reported below (Fig. 4):

6 Depth Map Approach

Usually any rover navigation is based on a Depth Map. The Depth map is typically realized by means of a stereo camera. This approach is consolidated and it is very reliable. A stereo camera system equipped with a 2 × 2K camera is able to resolve the depth map with resolution of a few centimeters at a distance more than 15 m from the stereo cameras. This optical resolution is more than enough in terms of rover travelling distance and rocks avoidance for any moving rovers.

Anyway it requires two cameras. Furthermore if one of the two cameras is malfunctioning or broken it is not possible to operate in this way. Typically a rover has two couple of stereo cameras: one for navigation and one for scientific image acquisition to be safe in this accidental failure. A further consideration is that the usage of stereo camera is not feasible in the case of a micro rover, or drone, because it requires more power and the baseline distance required between the two cameras could be not enough for evaluate a good depth map.

Mono camera systems able to evaluate a depth map are already proposed in the technical literature anyway; they rely on multiple images acquisition with different distance focus: micro mechanical focusing lens systems [9]. These systems are not available for space application because for the low temperature and high level radiation present on Mars terrain environment. Eventually a complete space qualification of these mechanisms is very expensive.

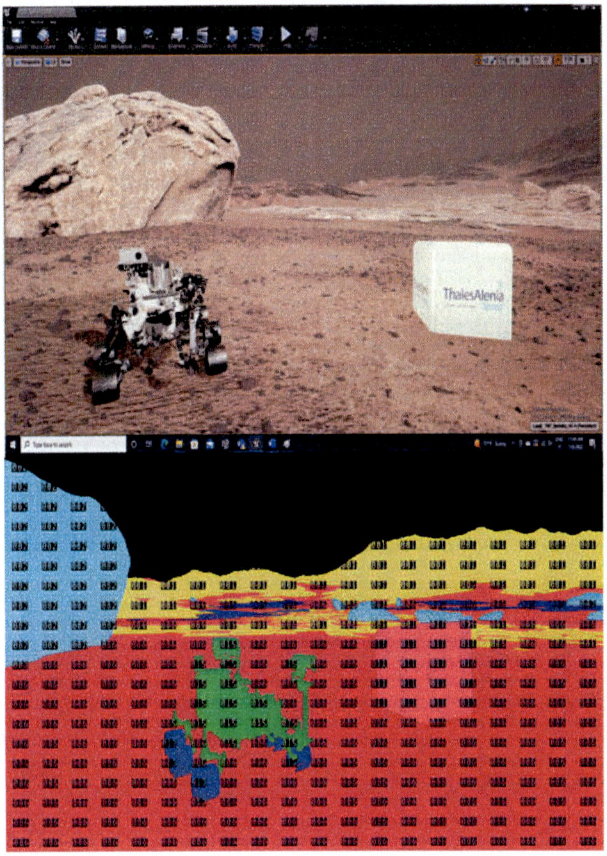

Fig. 4 Example of semantic segmentation from SINAV dataset. Red color is referring to the soil, light blue to the big rock, yellow is referring to the bedrock and dark green is Perseverance rover (NASA model)

As a possible alternative approach in case of a malfunctioning of a stereo camera assembly we can propose to use a DNN able to realize a Depth Map by using only the images acquired by a single camera. It could be loaded as a backup software solution in the case of a rover with stereo camera equipment, or it could be fully operative in the case of a micro rover or a drone because it can capture the depth map [8] (Fig. 5).

7 Object Detection Approach

One of the most known applications in the Computer Vision DL is object detection. There are plenty of datasets with the information inherent to the object detection.

Fig. 5 Examples of synthetic images generated for SINAV dataset. On the left, image with natural texture and colors, in the middle semantic segmentation mask of the previous image and on the right the equivalent image in terms of map depth

They are composed of images where the localization of the specific object is indicated by the coordinates of a rectangular shape surrounding the object. This is one of the most outstanding results obtained by means of the DNN. Previous approaches based on algorithmic image processing were not able to obtain such reliable and accurate results. This task, there is no doubt about it, can be realized only in terms of a DNN for object detection.

In the framework of SINAV study we have identified a specific object that we will use to evaluate the efficiency of our object detection DNN: the Returnable Sample Tube Assembly (RSTA). There are 42 RSTA on board the *Perseverance rover*. Each one will be filled with different Mars sample terrain and left by the rover on an ad-hoc location named depot. RSTAs are rather small, being cylinders of about 150 mm length and 15 mm diameter, without any "active marker" for being easily detected (NASA decided that only "visual means" can be used). Finally the Sample Fetch Rover mission will recover all of them and group them together in a small ball to launch toward the Earth (Fig. 6).

Fig. 6 On the left up corner visualization of the RSTA on black background. In the large image the RSTA is visualized at the distance of 5 m in the ROXY Mars Terrain Facility in Thales Turin

While it is quite easy to detect the RSTA when it is close to the camera it is also necessary to remember that the rover needs to identify the RSTA from a far distance to eventually catch it and put it in the RSTA container. Therefore, as it is possible to see in the image above, the Object Detection algorithm shall be working with a distance between the object (RSTA) and the camera of a few meters. In these operational conditions it is not easy to have good reliability of the object shape especially if this object is partially occluded by a rock or hidden in the sand. A combination of two approaches: digital zoom plus object detection while we are scanning the acquired image can be very beneficial for this specific scientific case.

8 Opportunistic Science Approach

Another useful goal during a rover traversability is to realize an Opportunistic Science approach during the rover mission. In other words, identify a possible application where a DL approach can be useful during the rover exploration. One of the most useful application is the realization of a digital zoom (Super Resolution in the DL terminology). Typically a digital zoom is realized by means of a bilinear or bicubic interpolations on the image. This algorithm is extremely used. Nevertheless the DL alternative is existing and is called Super Resolution. From the different technical survey it seems that DL approach is visually better. Anyway a better understanding in terms of performances and real time execution is necessary for a final choice.

9 SINAV MLOps Approach

SINAV MLOps is defined in accordance with the approach explained in Alhashim and Wonka [10] and hereafter summarized in this block diagram. Every task will be briefly introduced (Fig. 7):

9.1 Scope Project

Scope Project of the MLOps has been detailed described in the previous paragraphs.

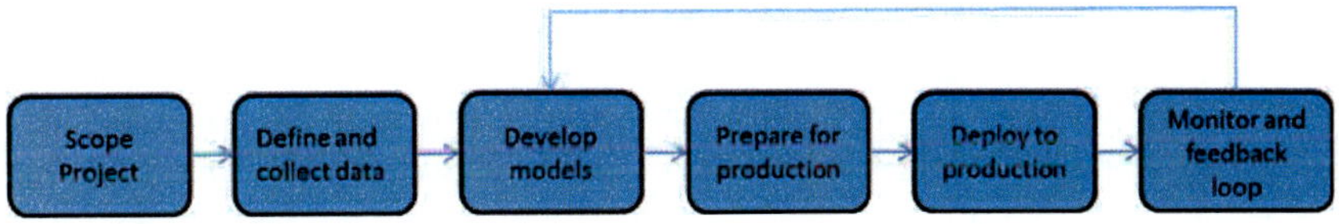

Fig. 7 SINAV MLOPs

9.2 Define and Collect Data

Many datasets will be used along the SINAV study. They can be divided in two categories. Synthetic and real images. The synthetic image dataset are created with Unreal Engine Mars model and a quite sophisticated image processing chain which is in charge to generate images with different illuminance, color tone mapping, image semantic segmentation and automatic verification of the correctness of the image. This approach has allowed the creation of extensive dataset (10000 images or more) in almost automatic way. About the real images three different facilities are used for this scope:

- ROXY Mars terrain outdoor facility environment
- ALTEC Mars terrain indoor facility for ExoMars
- ALTEC Mars terrain indoor facility.

All dataset realized are made available in the SINAV study for the partners.

A last set of dataset is composed by open source public dataset composed by real image on Mars terrain acquired by the NASA rovers. These last datasets are only partially prepared for training of DNN. Mostly are simple images collections.

9.3 Developed Models

Typically the four DNN models will be developed on training and validation on synthetic datasets. In this phase there is an experimentation of different DNN architectures available and their comparison in terms of opportune metrics.

9.4 Prepare for Production

In this phase the DNN architecture are chosen and hyperparameter fine tuning is realized on DNN architecture including an implementation of pre and post processing of the DNN. Preliminary test of feasibility on embedded hardware are realized also in this study phase.

9.5 Deploy for Production

Realization of the software traceability and metric measurement of the chosen DNN in embedded hardware. This is a test phase: the DNN model is working in inference in real time. It is fed by real images acquired from the different facilities. In a last phase testing the model is working with real images of Mars terrain.

9.6 Monitor and Feedback Loop

During this last project phase the different DNN have been implemented into the embedded rover computer and everything is ready for the final demo. During this phase a continuous feedback during the inference of the four DNNs is generated and monitored.

10 SINAV Hardware and Software Solutions

Software developments for space applications is limited by two factors:

- ESA guidelines in the software development
- Limited computational power for space qualified computer.

The ESA software rules are strictly applied as much as possible in the development process. In fact the final code shall be written in C language without any dynamic allocation or pointers as required by ESA guidelines. The idea is to have a software prototype that is not fully space qualified but it is very much close to the ESA main guidelines.

The space qualified computer are strongly limited as computational power because they are designed to withstand to hard rad environment. In particular till now none GPU has been already space qualified.

Until 2015 the space qualified market segment has been dominated by the SPARC architecture. In 2017 Quad Core LEON4 Ft-SoC with 1700 DMIPS and a 5 W power consumption was the last space qualified version realized. Now the modern ARM microprocessor Quad Core ARM R52 SoC can guarantee till 4000 DMIPS with a power dissipation of 10 W. The future space qualified computer mostly will be based on Quad Core ARM architecture [11]. They are also rumors that SpaceX has already used a model of GPU but non official claim has been confirmed.

Therefore the situation is to verify the possibility to realize DNN architecture without GPU involved and based only with ARM architecture.

NVIDIA offers this possibility with the embedded Jetson Nano board which is equipped by 4 GB RAM, Quad Core ARM processor and a GPU which can activated or not. This small board can have four different power modes, consuming from the basic 5 W till a maximum of 20 W.

The SINAV study uses this embedded system because it is possible to experiment with ARM processors and including or not the GPU in the inference. Furthermore the four USB 3.0 connections allows to use the board with different camera setups and models. Hereafter a picture of the Jetson Nano board (Fig. 8):

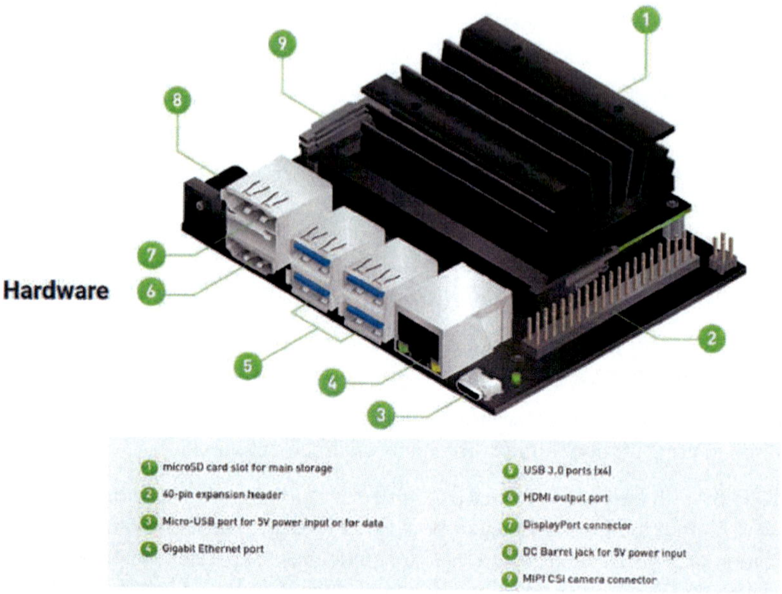

Fig. 8 NVIDIA Jetson Nano board [from NVDIA website]

11 Metrics and Dataset Organization

The adopted metric to evaluate the DNN performances and the dataset organization have been in accordance with the well-known COCO dataset [12]. Briefly the following metrics are adopted:

- The Terrain Traversability DNN performances are evaluated in terms of pixel accuracy. The Pixel Accuracy is perhaps the easiest to understand conceptually for semantic segmentation. *It is the percent of pixels in your image that are classified correctly.*
- The Depth Map DNN performances are evaluated in terms of relative absolute pixel difference. In particular the relative absolute pixel difference is evaluated as the mean relative absolute difference between ground truth and predicted depth.
- The Opportunistic Science realization is a Super Resolution DNN performances which performances are evaluated in terms of pixel-wise loss difference. The Pixel-wise loss is the simplest class of loss functions where each pixel in the generated image is directly compared with each pixel in the ground-truth image.
- The Object Detection DNN performances are evaluated in terms of mAP. The **mAP (mean Average Precision) for Object Detection is based on** AP (Average precision) evaluated in different cases. More information are available in the technical note. This kind of measure performance is quite common for object

detectors like Faster R-CNN, SSD, YOLO. More details are available in the COCO homepage [12].

12 Conclusions

In this article the overall activities and main goals of SINAV study inherent only to Deep Learning has been presented. SINAV study will be completed within next year therefore a few preliminary results are only available till now and they are shown in this article. Other important results will be also provided in the future for the Auto Navigation activities. These two approaches: DNN and auto navigation will be integrated in the final demo to demonstrate with different test setup the overall accuracy and reliability of the new rover navigation. Finally the integration of external data, taken from drone and processed by ALTEC, AIKO and Politecnico di Torino will allow the rover to increment its autonomy in the choice of the trajectory and its execution. The final commitment is to demonstrate that with limited hardware resources and in accordance with ESA software rules it is possible to implement a DL system on board of a future rover mission.

References

1. Agenzia Spaziale Italiana (BANDO DI RICERCA PER TECNOLOGIE ABILITANTI TRASVERSALI Area tematica Tecnologie Spaziali: B) Sistemi Autonomi e Intelligenza Artificiale A.S.I. (2018)
2. Girimonte, D., Izzo, D.: Artificial Intelligence for Space Applications. https://www.esa.int/gsp/ACT/doc/AI/pub/ACT-RPR-AI-2007-ArtificialIntelligenceForSpaceApplications.pdf (2007)
3. Linda Herridege: Commercial CubeRover Test Shows How NASA Investments Mature Space Tech 22th Dec 2020. https://www.nasa.gov/feature/commercial-cuberover-test-shows-how-nasa-investments-mature-space-tech
4. VB Staff: Why Do 87% of Data Science Projects Never Make into Production? https://venturebeat.com/2019/07/19/why-do-87-of-data-science-projects-never-make-it-into-production/ (2019)
5. Treveil, M.,: the Dataiku Team. Introducing MLOps How to Scale Machine Learning in the Enterprise (2020)
6. Rothrock, B., Papon, J., Kennedy, R., Ono, M., Heverly, M.: SPOC: deep learning-based terrain classification for Mars Rover missions. In: AIAA SPACE 2016 13–16 September 2016. Long Beach, California (2020). https://nasa-jpl.github.io/SPOC/. https://doi.org/10.2514/6.2016-5539
7. Gutman, A., Goldmeier, J.: Becoming a Data Head. Wiley & Sons Inc. (2021)
8. Barrett, A.M., Balme, M.R., Woods, M., Karachalios, S., Petrocelli, D., Joudrier, L., Sefton-Nash, E.: NOAH-H, a deep-learning, terrain classification system for Mars: results for the ExoMars Rover candidate landing sites. Icarus **371**, 114701 (2022)
9. de Chanlatte, M., Gadelha, M., Groueix, T., Mech, R.: Leveraging Monocular Disparity Estimation for Single-View Reconstruction (2022)
10. Powel, W.: High-Performance Spaceflight Computing (HPSC) Program Overview. Doc type: 20180003537. https://ntrs.nasa.gov/citations/20180003537 (2018)

11. Alhashim, I., Wonka, P.: High Quality Monocular Depth Estimation via Transfer Learning. https://github.com/ialhashim/DenseDepth (2018)
12. COCO common object in context, https://cocodataset.org/#home

Deep Learning for Navigation of Small Satellites About Asteroids: An Introduction to the DeepNav Project

Carmine Buonagura, Mattia Pugliatti, Vittorio Franzese, Francesco Topputo, Aurel Zeqaj, Marco Zannoni, Mattia Varile, Ilaria Bloise, Federico Fontana, Francesco Rossi, Lorenzo Feruglio, and Mauro Cardone

Abstract CubeSats represent the new frontier of space exploration, as they provide cost savings in terms of production and launch opportunities by being able to be launched as opportunity payloads. In addition, interest in minor bodies is gradually increasing because of the richness and exploitability of the materials present throughout their surface, the scientific return they could yield, and their dangerousness. Moreover, they are characterized by a highly harsh environment. These are the reasons why greater autonomous capabilities are desirable for future space missions. Optical navigation is one of the most promising technique for retrieving spacecraft state, enabling navigation autonomy. Unfortunately, most of these methods cannot be implemented on-board because of their computational burden. This paper presents the "Deep Learning for Navigation of Small Satellites about Asteroids" project, in short "DeepNav", whose aim is to change the current navigation paradigm by exploiting artificial intelligence algorithms for on-board optical navi-

C. Buonagura (✉) · M. Pugliatti · V. Franzese · F. Topputo
Dipartimento di Scienze e Tecnologie Aerospaziali–Politecnico di Milano, Milan, Italy
e-mail: carmine.buonagura@polimi.it

M. Pugliatti
e-mail: mattia.pugliatti@polimi.it

V. Franzese
e-mail: vittorio.franzese@polimi.it

F. Topputo
e-mail: francesco.topputo@polimi.it

A. Zeqaj · M. Zannoni
Centro Interdipartimentale di Ricerca Industriale Aerospaziale, Alma Mater Studiorum–Università di Bologna, Bologna, Italy
e-mail: aurel.zeqaj2@unibo.it

M. Zannoni
e-mail: m.zannoni@unibo.it

M. Varile · I. Bloise · F. Fontana · F. Rossi · L. Feruglio
AIKO, Turin, Italy
e-mail: mattia@aikospace.com

C. Ieracitano et al. (eds.), *The Use of Artificial Intelligence for Space Applications*, Studies in Computational Intelligence 1088,
https://doi.org/10.1007/978-3-031-25755-1_17

gation. As a result, DeepNav will evaluate the performance of fast and light artificial intelligence-based orbit determination for the proximity operations phase around asteroids.

1 Introduction

Over the past decades, governments, space agencies, and private companies have recognized micro- and nano-satellites as appealing platforms for pursuing a wide range of objectives in space, including science, technology demonstration, and Earth observation. With significant cost reductions compared to traditional large satellite missions, these systems offer the opportunity to increase the number of spacecraft launches and enable faster manufacturing development. Among these classes of platforms, the CubeSat standard has become extremely popular. Nano-satellite constellations in low-Earth orbit (LEO) are becoming a reality, while the scientific and industrial community is exploring applications of nano-satellites for interplanetary missions [16]. These, represents new challenges compared to LEO missions, such as surviving in a harsher space environment, communicating with Earth from a greater distance, achieving accurate pointing without exploiting the magnetic fields of celestial bodies, and performing accurate orbital maneuvers. In addition, a paradigm shift is needed when operations are considered: while miniaturization of technology has already enabled a significant reduction in the cost of space segment, the ground segment cost do not scale with satellite size. Ground stations used for deep space missions present high operational costs that are not sustainable with the low costs of nano-satellite missions. These costs are due to the allocation of a flight-dynamic team, big antennas for deep space communications must be scheduled to be operated, and the competition with other missions for access to communications slots [11].

In addition, interest in asteroid exploration is pushing forward for several reasons: they contain an abundance of valuable resources including platinum, gold, iron, nickel, and rare metals, evenly distributed throughout small bodies' mass [1]. Moreover, they could be used as refueling stations to overcome the technological

I. Bloise
e-mail: ilaria@aikospace.com

F. Fontana
e-mail: federico@aikospace.com

F. Rossi
e-mail: francesco@aikospace.com

L. Feruglio
e-mail: lorenzo@aikospace.com

M. Cardone
ASI, Turin, Italy
e-mail: mauro.cardone@asi.it

limits of deep space exploration. Their study could allow to understand the solar system evolution and the accretion process of planets. Moreover, several potential hazardous asteroids (PHAs) have been estimated [8].

Several missions have been performed to study, among others, composition, shape and gravity of these bodies, as measurements from Earth give just approximate information. Therefore, the need to autonomously navigate around these bodies arises. The capability of a navigation system to provide an accurate estimate of spacecraft state strongly depends on the mission scenario, environment, and platform constraints. Optical Navigation (OpNav) represents the most promising autonomous method. It consists of a set of techniques used to obtain an estimate of the spacecraft state from images. The observer state estimation is a process made of different steps. As far as optical navigation is concerned, the first step consists in the image acquisition. The images are processed to extract relevant optical observables, and finally they are further elaborated by an estimation scheme and a navigation filter to retrieve the spacecraft state. These algorithms could be slow and heavy for on-board implementation and so far have been mainly used on regular shaped asteroids and moons. Artificial Intelligence (AI) algorithms can be used to bypass the various steps of traditional methods by directly obtaining the state from the image. In addition, they can be leveraged to increase the achievable accuracy by enhancing generalization capabilities in the proximity of irregularly shaped objects. Furthermore, AI methods enable light and fast computations when facing complex scenarios that would be difficult to process with traditional methods.

The project aim is to evaluate AI-based, Orbit Determination (OD) performance for the proximity operations phase around asteroids. The operative scenario considered is the resolved one, in which the target is a finite-sized object within the camera field of view (FOV). This is the richest and most complex navigation scenario, where surface morphological features, such as craters and boulders, are resolved. For such a scenario, AI and machine learning can enable fast and accurate extraction of relevant information from images that have a high level of detail and a great amount of information.

In this context, the DeepNav (Deep Learning for Navigation of Small Satellites about Asteroids) project was recently selected by the Italian Space Agency, as part of a competitive call focused on the development of new navigation technologies.

The consortium is composed by entities and institutions from Italy. Consortium prime is AIKO,[1] responsible of artificial intelligence research and development for asteroid navigation algorithms, support in integration, testbed execution, and analysis of results. Politecnico di Milano, and in particular the DART[2] team, responsible of research and development of navigation algorithms, testbed development, test execution and analysis of the results. CIRI Aerospace, Universitá di Bologna, and in particular the Radio Science and Planetary Exploration Laboratory,[3] is in charge of the definition of the use cases and requirements, study of simulation environment

[1] https://www.aikospace.com/, last access: 31/05/2022.

[2] https://dart.polimi.it/, last access: 31/05/2022.

[3] https://site.unibo.it/radioscience-and-planetary-exploration-lab/en, last access 31/05/2022.

and navigation filters, implementation of breadboard and related software, support to execution of tests and analysis of results.

The rest of the paper is organized as follows. First, the framework where the DeepNav project is focused is highlighted, with a brief introduction to optical navigation and navigation scales. After such discussion, the goals of the project and the methodology that will be applied to achieve them are described. Expected outcomes are reported, and finally, some conclusions are discussed.

2 Framework

2.1 *Optical Navigation*

Optical navigation consists of a set of techniques used to obtain an estimate of the spacecraft state from images. As far as optical navigation is concerned, diffeent steps are involved, starting with the images acquisition. The images are processed to extract relevant optical observables, and finally they are further elaborated by an estimation scheme and a navigation filter to estimate position and velocity of the observer.

Image Processing The state estimation accuracy is strongly affected by the image processing method used for the optical observable extraction. In deep space, the only available optical observables are the line-of-sight directions of visible objects. Such directions are usually obtained thanks to centroiding algorithms. Regarding the far proximity range, the apparent dimension and the shape of the celestial body are estimated exploiting edge finding and centroiding algorithms. On the contrary, for close proximity operations, there is a need for algorithms able to identify features line-of-sights or to trace their geometry.

Estimation Schemes The estimation schemes take as input the optical observables extracted from image processing. The final objective consists in the preliminary estimation of position and velocity through navigation methods. The most common estimation schemes employ line-of-sight information to use triangulation or least squares estimation techniques. Other traditional navigation techniques elaborate the shape and size of the objects by exploiting perspective geometry rules.

Navigation Filters Estimation schemes do not return sufficiently accurate results about the observer's state. Thus, the last step to obtain an accurate state vector is represented by filtering methods. In doing so, knowledge of the dynamics combined with environmental measurements is exploited to obtain an accurate estimate of the observer's state.

The increasing interest in optical navigation is due to the enhanced request for spacecraft autonomy. The DeepNav project will focus on the close range regime, and will exploit Artificial Intelligence to provide robust solutions regarding feature detection, matching, and to perform the 6D Pose Estimation task. Below, for the seek of completeness, will be a brief survey of all the traditional algorithms, some of them used in previously flown missions to navigate minor bodies by landmarks, and therefore in the close proximity regime.

Navigation Scales Satellite navigation is based on the processing of some measurements, which in the case of optical navigation are acquired from the surrounding environment. This involves the definition of different navigation scales, deeply discussed in [7]: deep space, far range, and close range. Navigation scales can be classified based on the feature pixel ratio (FPR), defined as the ratio between the feature and the pixel angular dimensions (γ and θ, respectively) [7]:

$$FPR = \frac{\gamma}{\theta} \tag{1}$$

Deep Space The distance between the observer and the solar system object is large. In this condition the object is identified as a bright spot in the image plane, as it occupies few pixels. In this condition the $FPR \approx 0$, thus a pixel of the detector contains a large portion of the object, as visible in Fig. 1.

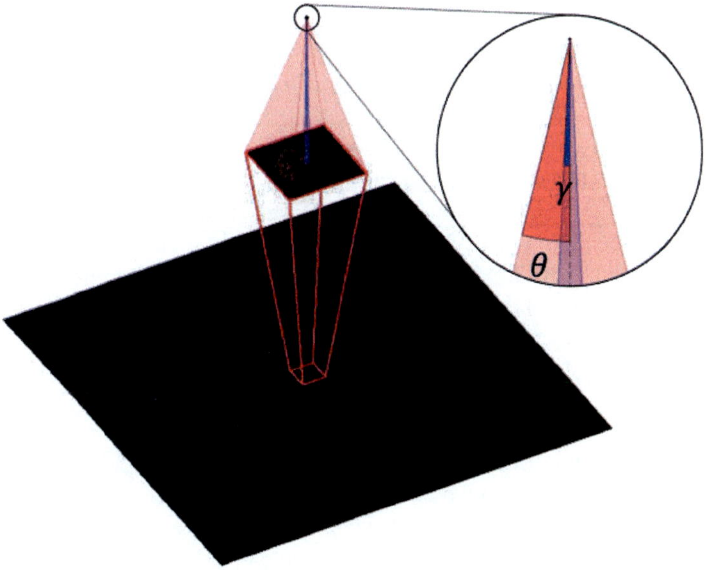

Fig. 1 Visibility of a detector pixel in deep space condition. The angular size of the pixel is depicted in red, while the feature angular size in blue

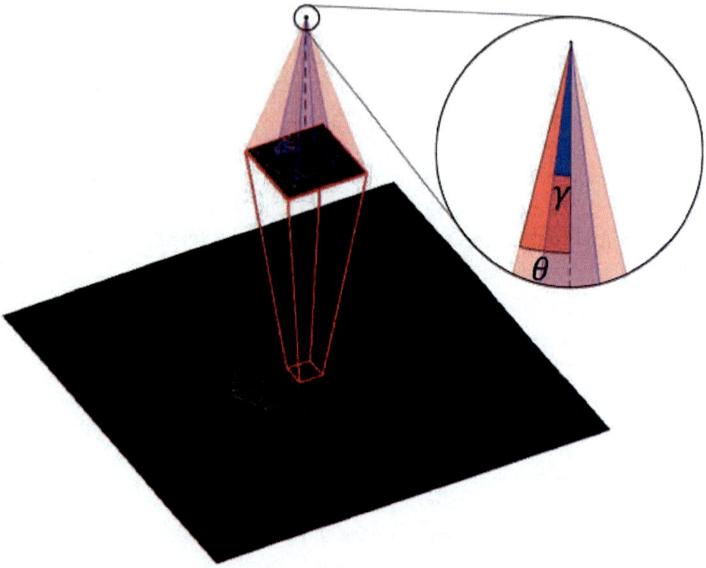

Fig. 2 Visibility of a detector pixel in far range condition. The angular size of the pixel is depicted in red, while the feature angular size in blue

Far Range In the far range scenario, the object occupies a large portion of the camera field of view and the shape is resolved. In Fig. 2, it is possible to see how multiple features are grouped in a pixel resulting in a $FPR \leq 1$.

Close Range The camera is close to the body so that details and surface morphological features are visible and fully resolved. This is the richest optical navigation scale, since many details of the target object such as craters, boulders, and mountains come to light. This scenario will be addressed by the DeepNav project. When a single feature is mapped by several pixels, as shown in Fig. 3, the FPR is greater than one, and the navigation scale is of close range. In this condition the features on the surface of the minor body are fully resolved enabling feature tracking algorithms. Table 1 shows the FPR values in relation to the navigation scale.

Algorithms for Absolute and Relative Navigation Data extracted from cameras return line-of-sight information of interesting points. In this way, the scale ambiguity problem cannot be solved if no maneuvers are performed during the observation of the body.

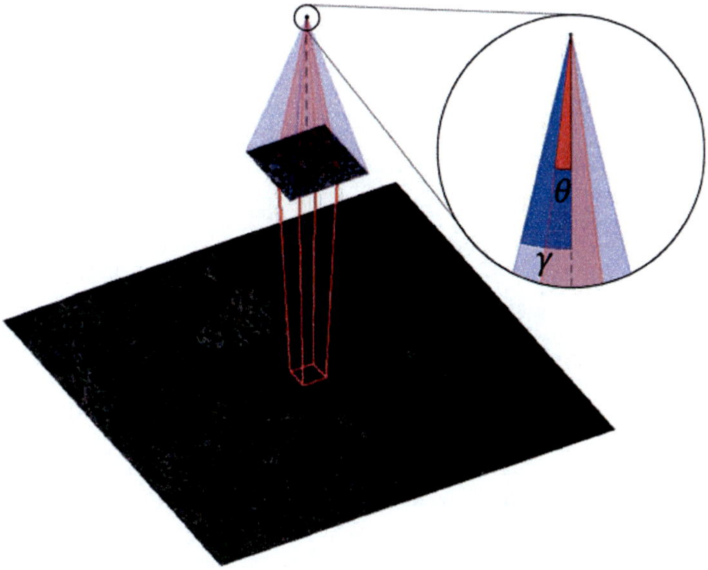

Fig. 3 Visibility of a detector pixel in close range condition. The angular size of the pixel is depicted in red, while the feature angular size in blue

Table 1 Definition of navigation scales

	Deep space	Far range	Close range
FPR	≈ 0	≤ 1	> 1

As described in [9], two types of navigation can be defined: relative and absolute. Relative navigation returns a measure of the camera displacement with respect to an unknown reference frame. Contrary, absolute navigation allows to get the camera position with respect to a known reference system, resulting in a complete determination of its position.

An example of relative navigation algorithm is feature tracking. Since the feature-camera distance is not determined, it is not possible to derive the absolute translation value from the features motion. The motion is performed with respect to an unknown reference as the features are not associated with a known morphological map.

Absolute navigation is based on landmarks, where the observed features are linked to a well known landmark map. The landmarks are expressed in a known reference frame, resulting in the computation of the camera position in this frame. In this way, the motion can be completely reconstructed.

3 Project Development

3.1 Project Objectives

Within this context, the goal of the project is to:

Design and validate up to TRL4 an autonomous orbital determination subsystem for small satellite platforms based on on-board acquired optical imagery for close proximity asteroid navigation scenarios using artificial intelligence techniques.

Three sub-objectives can be defined to achieve this purpose:

- Design and demonstration of an autonomous orbital determination subsystem for small platforms based on on-board acquired optical images.
- Research and implementation of deep learning algorithms for image processing and extraction of information to be used by navigation algorithms.
- Development of a testbed with novel technical characteristics, compatible with small satellite missions, and with relevant computational power.

3.2 Methodology

The DeepNav project is scheduled to last 18 months and can be divided into four distinct phases, ranging from literature study to analysis of results via software and hardware implementation of innovative navigation algorithms. The first phase of the project consists of a literature review of traditional and Artificial Intelligence algorithms in navigation-related scenarios. Vision awareness asks for complex actions, involving for example spatial geometry and feature understanding. More complex needs and critical conditions, such as scale variance and illumination variability, require robust and accurate approaches that Deep Learning (DL) techniques are able to provide. The main topics where DL will be evaluated are: feature extraction and matching [6, 12], Simultaneous Localization and Mapping (SLAM) [3] and 6D Pose estimation [5, 13]. A trade-off in complexity, performance, robustness, and computational burden will be carried out to select the most promising ones. In addition, algorithms for procedural generation of synthetic landmarks on the surface of minor bodies (such as craters and boulders) will be investigated. Finally, the most promising use case scenario with associated navigation requirements will be selected.

The second part of the project consists of the software implementation of the selected algorithms via Python scripting. Regarding artificial intelligence algorithms, the most suitable training strategy (supervised, semi-supervised and unsupervised) will be selected based on the chosen use case. Traditional relative and absolute navigation algorithms will be implemented and their on-board implementation will be investigated. A dataset of high fidelity synthetic images of minor bodies, as shown in Fig. 4, will be made with an in house developed tool [2] for training, validation, and testing purposes. The dataset will be generated by creating a point cloud around the illuminated side of the asteroid in order to capture it at all possible phase angles and poses obtainable during a mission. An example of such a point cloud is shown in Fig. 5. In addition, for testing purposes, feasible trajectories around the body will be simulated.

Two different datasets will be built: the first for software-in-the-loop purposes, the latter to be compatible with the screen resolution of the optical facility. The process for the generation of the datasets is shown in Fig. 6.

Navigation Kalman filters for non-linear scenarios will be implemented and the design of the software and hardware navigation subsystem will be carried out. The performance of the entire navigation algorithm will be evaluated by testing it on synthetic images for the defined use case.

The third phase consists in the optimization of the algorithms for embedded hardware implementation and evaluating their performance in terms of trajectory reconstruction error, run-time performance, and covariance realism. Image processing algorithms will be integrated with filtering ones [14] for the accurate orbit determination and software-in-the-loop tests will be performed for validation. In the meantime, the development and integration of the optical testbed based on the requirements of the project will be carried out.

Finally, in the last phase, the achievement of TRL4 will be demonstrated through a hardware-in-the-loop (HIL) test campaign involving the setup of a dedicated testbed, TINYV3rse [10], designed at the DART Lab and shown in Fig. 7, simulating a feasible mission scenario.

The autonomous navigation technologies developed will be potentially applicable to specific ongoing mission studies employing CubeSats. Among the most recent are the M-ARGO [15] and LUMIO [4] missions. The first is a mission to characterize unknown asteroids in the Solar System, while the second is a mission to study meteoroids in the cis-lunar environment by characterizing the flashes that are produced as a result of impacts on the Moon. Both missions may benefit from the use of autonomous navigation techniques for their cruise or proximity phases.

4 Expected Outcomes

DeepNav will demonstrate autonomous navigation technologies for CubeSats, such as:

- Demonstrate the benefits and evaluate the performance of AI-based autonomous techniques for deep space missions;
- Concrete validation of the applicability of miniaturized sensors and technologies for low-cost and high-throughput scientific and exploration space missions.

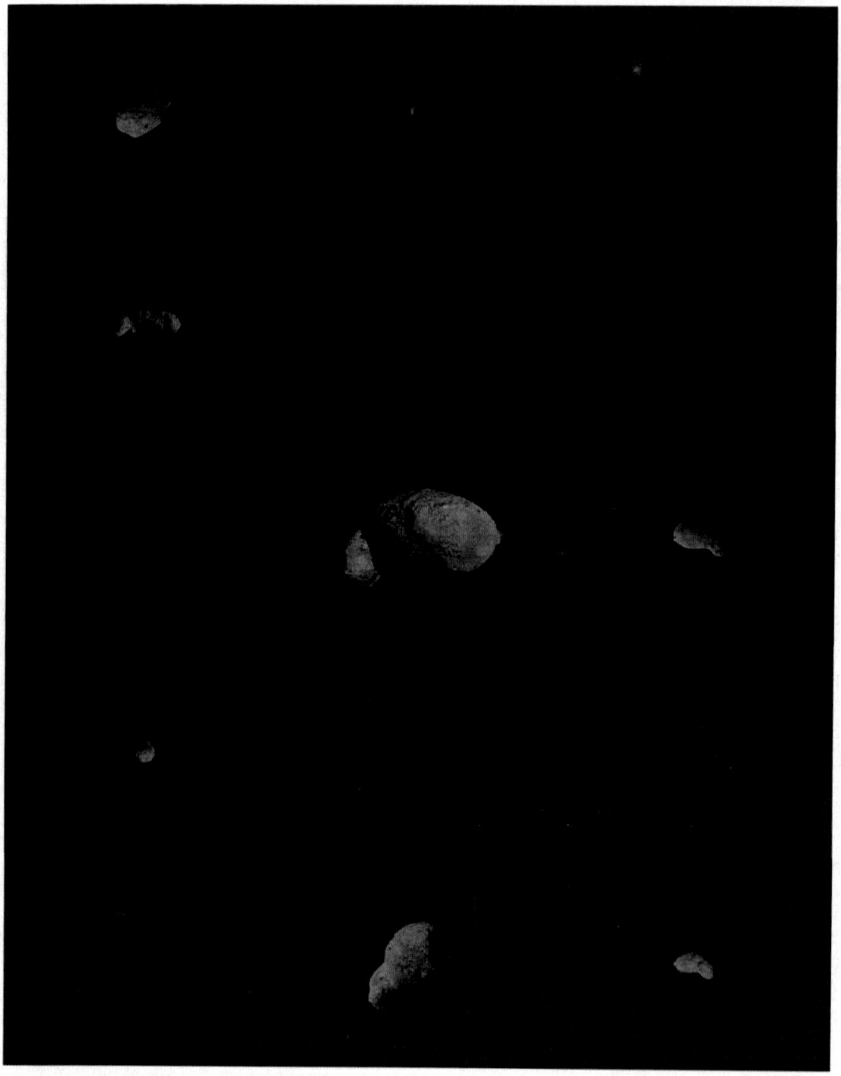

Fig. 4 Sample images of asteroids Itokawa and Bennu at different poses, distances, and lighting conditions

DeepNav will enable significant cost reductions by paving the way for new types of deep space missions, such as:

- Autonomous exploration and characterization of a variety of asteroids and small objects.
- Frequent exploration of well-known Solar System bodies for scientific purposes.

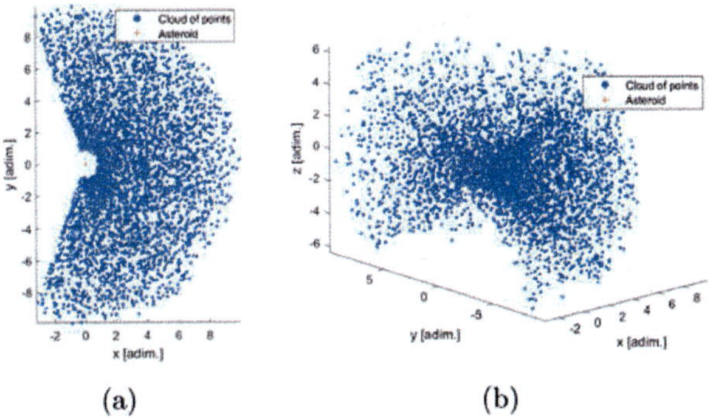

Fig. 5 Example of cloud of points used for training/validation. The Sun is illuminating from the $+x$ direction

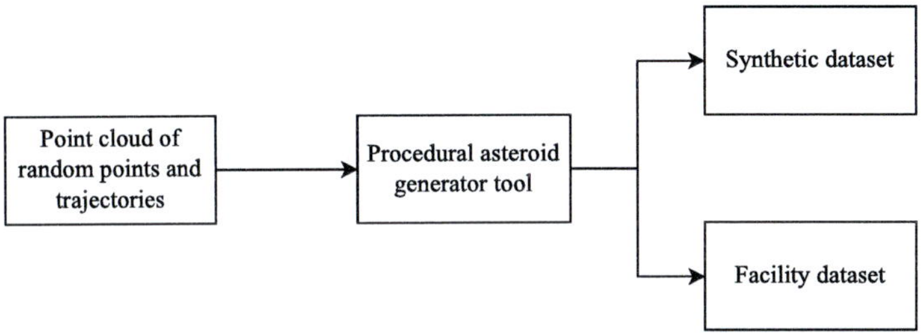

Fig. 6 Logical flow for the construction of the datasets that will be exploited during the project

Fig. 7 TINYV3rse testbed for the DeepNav project

- Low-cost, high-throughput scientific missions to unknown bodies in the Solar System.

5 Conclusions

Deep Learning for Navigation of Small Satellites on Asteroids (DeepNav) is an 18-month project funded by the Italian Space Agency (ASI). The consortium is composed of Italian companies and institutions: AIKO, Politecnico di Milano and Università di Bologna. The project aims to improve the exploration capabilities of small bodies by enabling the autonomy of shoe-boxed sized spacecraft, named CubeSats, exploiting images taken on-board and processed with artificial intelligence algorithms. The operative range of the project is the close range one, where each feature is clearly distinguishable in the camera plane. To achieve the project goal, several steps are involved. Literature review is the first phase, moving through implementation of the algorithms at both software and hardware levels. Finally, simulations will be conducted to validate up to TRL4 an autonomous orbital determination system. This document reflects the project proposal, which is currently at the end of the first phase.

Acknowledgements The DeepNav (Deep Learning for Navigation of Small Satellites about Asteroids) project has been financed under ASI Contract N. 2021-16-E.0. M.P and F.T would also like to acknowledge the funding received from the European Union's Horizon 2020 research and innovation programme under the Marie Skłodowska-Curie grant agreement No 813644.

References

1. Britt, D.T., Yeomans, D., Housen, K., Consolmagno, G.: Asteroid Density, Porosity, and Structure (2003)
2. Buonagura, C., Pugliatti, M., Topputo, F.: Procedural minor body generator tool for data-driven optical navigation methods. In: 6th CEAS Conference on Guidance, Navigation and Control (EuroGNC) (2022)
3. Cadena, C., Carlone, L., Carrillo, H., Latif, Y., Scaramuzza, D., Neira, J., Reid, I., Leonard, J.: Past, present, and future of simultaneous localization and mapping: Toward the robust-perception age. IEEE Trans. Robot. **32**(6), 1309–1332 (2016). https://doi.org/10.1109/TRO.2016.2624754
4. Cervone, A., Topputo, F., Speretta, S., Menicucci, A., Turan, E., Di Lizia, P., Massari, M., Franzese, V., Giordano, C., Merisio, G., Labate, D., Pilato, G., Costa, E., Bertels, E., Thorvaldsen, A., Kukharenka, A., Vennekens, J., Walker, R.: Lumio: A cubesat for observing and characterizing micro-meteoroid impacts on the lunar far side. Acta Astronaut. **195**, 309–317 (2022). https://doi.org/10.1016/j.actaastro.2022.03.032
5. Chen, B., Cao, J., Parra, A., Chin, T.: Satellite pose estimation with deep landmark regression and nonlinear pose refinement. In: Proceedings of the IEEE/CVF International Conference on Computer Vision Workshops, pp. 0–0 (2019)

6. DeTone, D., Malisiewicz, T., Rabinovich, A.: Superpoint: Self-supervised interest point detection and description. In: Proceedings of the IEEE Conference on Computer Vision and Pattern Recognition Workshops, pp. 224–236 (2018). https://doi.org/10.1109/CVPRW.2018.00060
7. Franzese, V.: Autonomous Navigation for Interplanetary CubeSats at Different Scales. Ph.D. Thesis, Politecnico di Milano (2021)
8. Mainzer, A., Grav, T., Masiero, J., Bauer, J., McMillan, R.S.: Characterizing subpopulations within the near-earth objects with neowise: Preliminary results. Astrophys. J. **752**(2), 110 (2012). https://doi.org/10.1088/0004-637x/752/2/110
9. Panicucci, P.: Autonomous Vision-based Navigation and Shape Reconstruction of an Unknown Asteroid During Approach Phase. Ph.D. Thesis, Institut Supérieur de l' Aéronautique et de l'Espace (ISAE) (2021)
10. Panicucci, P., Pugliatti, M., Franzese, V., Topputo, F.: Improvements and applications of the dart vision-based navigation test-bench TinyV3RSE. In: AAS GN&C Conference (2022)
11. Quadrelli, M., Wood, L., Riedel, J., McHenry, M., Aung, M., Cangahuala, L., Volpe, R., Beauchamp, P., Cutts, J.: Guidance, navigation, and control technology assessment for future planetary science missions. J. Guidance Control Dyn. **38**(7), 1165–1186 (2015). https://doi.org/10.2514/1.G000525
12. Sarlin, P., DeTone, D., Malisiewicz, T., Rabinovich, A.: Superglue: Learning feature matching with graph neural networks. In: Proceedings of the IEEE/CVF Conference on Computer Vision and Pattern Recognition, pp. 4938–4947 (2020). https://doi.org/10.1109/CVPR42600.2020.00499
13. Sarlin, P., Unagar, A., Larsson, M., Germain, H., Toft, C., Larsson, V., Pollefeys, M., Lepetit, V., Hammarstrand, L., Kahl, F., et al.: Back to the feature: Learning robust camera localization from pixels to pose. In: Proceedings of the IEEE/CVF Conference on Computer Vision and Pattern Recognition, pp. 3247–3257 (2021)
14. Simon, D.: Optimal State Estimation. Kalman, H Infinity and Nonlinear Approaches (2006)
15. Topputo, F., Wang, Y., Giordano, C., Franzese, V., Goldberg, H., Perez-Lissi, F., Walker, R.: Envelop of reachable asteroids by m-argo cubesat. Adv. Space Res. **67**(12), 4193–4221 (2021). https://doi.org/10.1016/j.asr.2021.02.031
16. Walker, R., Binns, D., Bramanti, C., Casasco, M., Concari, P., Izzo, D., Feili, D., Fernandez, P., Fernandez, J.G., Hager, P., et al.: Deep-space cubesats: thinking inside the box. Astron. Geophys. **59**(5), 5–24 (2018). https://doi.org/10.1093/astrogeo/aty232

Object Recognition Algorithms for the Didymos Binary System

Mattia Pugliatti, Felice Piccolo, and Francesco Topputo

Abstract Optical-based navigation in a binary system such as the Didymos one poses new challenges in terms of image processing capabilities, in particular for what concerns the recognition between the primary and secondary bodies. In this work, the baseline object recognition algorithm used in the Milani mission to distinguish between Didymos and Dimorphos is evaluated against alternative image processing pipelines which use convolutional pooling architectures and machine learning approaches. The tasks of the proposed alternatives is to detect the secondary in the image and to define a bounding box around it. It is shown that these algorithms are capable of robustly predicting the presence of the secondary albeit performing poorly at predicting the components of the bounding box, which is a task that is performed quite robustly by the baseline algorithm. A new paradigm is therefore proposed which merges the strengths of both approaches into a unique pipeline that could be implemented on-board Milani.

Keywords Object recognition · Milani · Didymos · Dimorphos

1 Introduction

Autonomous Optical Navigation (OpNav) is an enabling technology for present and future exploration missions [3]. It exploits an Image Processing (IP) method to extract optical observables and then uses them to generate a state estimate with associated uncertainties. Since images can be inexpensively generated on board with low-cost and low-mass sensors, OpNav is experiencing a growing interest. This is particularly

M. Pugliatti (✉) · F. Piccolo · F. Topputo
Politecnico di Milano, Department of Aerospace Science and Technology, Milan 20156, Italy
e-mail: mattia.pugliatti@polimi.it

F. Piccolo
e-mail: felice.piccolo@polimi.it

F. Topputo
e-mail: francesco.topputo@polimi.it

C. Ieracitano et al. (eds.), *The Use of Artificial Intelligence for Space Applications*, Studies in Computational Intelligence 1088,
https://doi.org/10.1007/978-3-031-25755-1_18

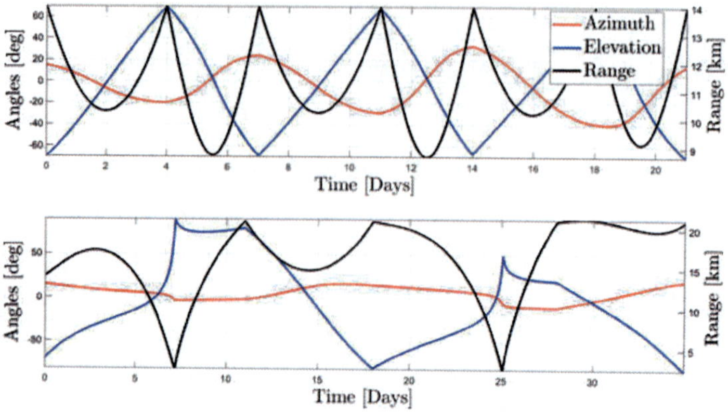

Fig. 1 Azimuth (red), elevation (blue), and range (black) of the FRP (top) and CRP (bottom) phases in the W reference frame [1] of the Didymos binary system

relevant for CubeSat missions, which are often tightly constrained in terms of mass and power. In the case of proximity to small bodies, OpNav can be exploited to reduce costs by enabling an autonomous system and unlocking the capabilities to perform critical operations. By linking OpNav capabilities with guidance and control algorithms, autonomous Guidance Navigation & Control (GNC) systems can be foreseen in the near future in self-exploring missions with reduced or complete absence of humans-in-the-loop. Within this context, the design of an accurate and robust IP algorithm is of paramount importance for a vision-based autonomous GNC system.

Milani is a 6*U* CubeSat that will characterize the Didymos binary system and its close environment [1]. The primary and secondary bodies of the system are called Didymos (D1) and Dimorphos (D2). The first has an estimated diameter of 780 m, the latter of 170 m. Milani's objectives are both scientific and technological [9]: to map and study the composition of D1 and D2, and to demonstrate CubeSat technologies in deep space. Milani is designed with orbital maneuvering capabilities and attitude control and will carry several payloads to characterize the composition of the Didymos system as well as the dust environment.

Milani will be hovering in the Didymos system and will perform the required scientific observations during two main phases: The Far Range Phase (FRP) and the Close Range Phase (CRP). The FRP lasts roughly 21 days while the CRP 35 days. The FRP exhibits symmetrical arcs that develop within 8–14 km from the asteroids. On the other hand, the CRP exhibits asymmetrical arcs within 3–23 km. The characteristics in terms of range and phase angle of these phases are illustrated in Fig. 1.

The IP strategy of Milani [8] roots its foundation on the capability to provide reliable data about D1 for navigation (both on-board and on-ground), but at the same time to be able to distinguish D2 in the image for pointing purposes. This stems from the consideration that D1 being the largest, most visible, and regular body, it would be simpler and more robust to use for navigation. On the other hand, D2 being the focus of the Milani mission, it is important to be able to autonomously distinguish D2 from D1 in the image to eventually be able to point towards it for scientific acquisitions. The object recognition is designed to work in the resolved regime of D1, from 3 to 23 km, which is the applicable range in which both FRP and CRP will mostly occur.

Based on these considerations, the baseline object recognition algorithm for Milani [8] is designed using a simple criterion based on the area of the blobs of pixels detected in the image. The biggest area is assigned to D1 while the second biggest one is assigned to D2, provided that it is located outside an expanded bounding box around D1. This simple algorithm is robust to False Positive (FP) detection of D2 but makes it susceptible to a large number of False Negative (FN), which is somehow acceptable for the purposes of the IP, albeit not ideal. Also, another drawback of the current baseline is that when D2 is the only body in the image, it can be wrongly identified as D1. The latter scenario however is rare and naturally resolves if D2 is tracked by the CubeSat since due to binary dynamics, D1 will eventually reappear in the image.

In this work, the authors investigate different alternatives to the current baseline algorithm which make use of the large amount of data collected for the design of the data-driven IP of Milani. Alternative Machine Learning (ML) methods are considered to understand to what extent the performance of the object recognition algorithm can be improved, especially for what concern the identification of D2 in the images.

The rest of the paper is organized as follows. In Sect. 2 the datasets are explained together with the different IP methods considered in this work. In Sect. 3 a performance comparison between these methods is presented. Finally, in Sect. 4 some final conclusions and considerations are drawn.

2 Methodology

In this section, the methodology used in this work is described in detail. First, in Sect. 2.1 the characteristics of the dataset used to train, validate, and test all IP methods are presented. The baseline IP algorithm currently designed for Milani is described in detail in Sect. 2.2, while new approaches for this task are presented in Sects. 2.3–2.5. Finally in Sect. 2.6 the overall strategy used during training and inference of the IP methods is explained.

2.1 Datasets

In this work, 3 different datasets are used for training, validation, and testing of the various IP algorithms: DB_0, FRP, and CRP. These are illustrated in Fig. 2 in the W reference frame, which is centered on D1 and whose Z-axis is aligned with the spin

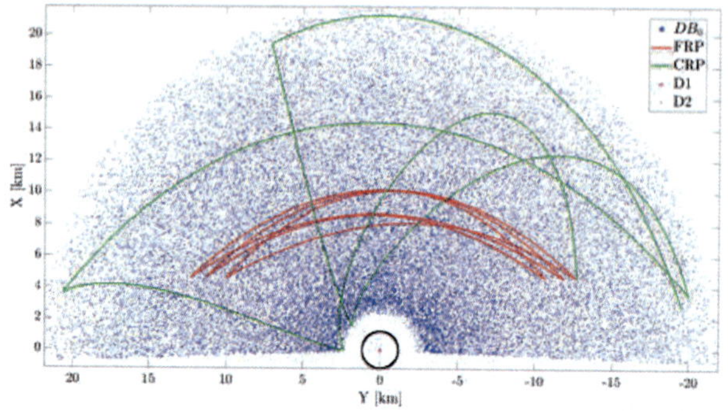

Fig. 2 DB_0, FRP, and CRP datasets used to train, validate, and test the IP algorithms

axis of D1 and the XY-axes are co-planar with the orbital plane of the secondary, with the X-axis following the projection of the Sun in such plane.

Each of these datasets is composed of image-label pairs. Images are represented by synthetic, noisy, grayscale 1536×2048 renderings obtained in Blender[1] out of the different geometric conditions illustrated in Fig. 2. These are splitted into random scattered points used for training and regular ones representative of real trajectories to be flown in the Didymos system. The former are represented by a cloud of points between 3 km and 23 km with azimuth angle between -95 deg and $+95$ deg and elevation between -45 deg and $+45$ deg in the W reference frame in DB_0. The latter are organized observables of the FRP and CRP phases of the Milani mission sampled every 150 s. All images are rendered with an ideal pointing towards D1. Since the focus of the object recognition analysis is mainly to be able to detect D2, this is a reasonable assumption.

For practical purposes, during training different subsets of the DB_0 dataset can be designed. The ones considered in this work are defined as: DS^1_R, DS^3_R, DS^1_{CR}, and DS^2_{CR}. The nomenclature is such that $_{CR}$ and $_R$ stand respectively for "cropped" and "resize" while the apex represents the number of splits in the set. Both DS^1_R and DS^3_R are obtained by resizing the original 1536×2048 px image to a much smaller and easier to handle 96×128 px array. The two subsets differ in terms of the balance of the conditions represented. When considering how D2 appears in the image, 3 of these conditions are identified: one in which D2 is not visible (either because it is in the shadow region, occluded by D1 or outside the region considered in the image), one when D2 is visible but is clearly separated from D1, and finally one when D2 is visible, but positioned close to the bounding box of D1. These three scenarios are clearly represented in Fig. 3. In DS^1_{CR} the entire (unbalanced) DB_0 is represented, while in DS^1_{CR} the dataset is divided into equal splits of these three conditions.

On the other side, both DS^1_{CR} and DS^2_{CR} represent a 128×128 px cropped version of the original 1536×2048 px image around the bounding box of D1. In such images D2 is either not visible or within the bounding box of D1. Because of this, only 2

[1] https://www.blender.org/ last accessed, 4th of July 2022.

Fig. 3 Examples of ground truth labels of D1 (blue) and D2 (red) when D2 is not visible (left), D2 is visible but separated from the bounding box of D1 (center), and D2 is visible but within the bounding box of D1

of the original 3 conditions represented in Fig. 3 are possible. The DS^1_{CR} dataset represents the unbalanced version of such dataset, while the DS^2_{CR} represent the balanced one. The characteristics of these 4 datasets, together with the ones of DB_0, FRP, and CRP are summarized in Table 1.

These four datasets have been designed both in an attempt to identify the best resizing option for the task considered, and to assess the importance of having balanced datasets during training of the IP methods proposed.

Each image is accompanied by a set of labels representing a visibility flag to identify whether or not D2 is in the image and by the 4 components of the bounding box Γ: The first two are the coordinates in *px* on the image plane of the top-left corner of Γ, the latter two represent respectively the width and height of Γ. The ground truth of Γ is obtained in Blender using *Cycles* as the rendering engine and associating different pass indexes to D1 and D2. For representation purposes, a cumulative overlap of the various Γ of D2 in DB_0 is represented in Fig. 4.

2.2 Baseline IP

The baseline IP algorithm used to distinguish between D1 and D2, also referred in this work as IP_b, is now described in its simple steps.

At first, the image is binarized using the Otsu method [2]. Morphological operations such as opening or closing are then applied to the binary image. Thanks to this step, the number of detected blobs in the image is reduced, which is helpful in the following blob analysis. The analysis is performed only on the group of pixels larger than a predefined threshold (this is done to remove small image artifacts) and generates several geometric properties of interest for each blob: e.g., area, bounding box (Γ), centroid coordinates, eccentricity and major axis length of the ellipse fitted to the blob of pixels.

The output of the blob analysis is then used in the object recognition algorithm to distinguish D2 from D1, which is designed as follows:

1. The blobs of pixels are ordered in ascending order based on their areas.
2. The blob with the biggest area is labeled as D1. Its key geometric properties are saved. All other blobs are listed as potential candidates of D2.
3. The bounding box around D1, Γ_{D1}, is expanded by an arbitrary factor in all directions in the image plane. The expanded bounding box Γ^{ex}_{D1} is created.
4. The remaining blobs of pixels that are within Γ^{ex}_{D1} are removed from the list of D2 candidates. These could be false-positive identifications of D2 given by local areas in the terminator region of D1.
5. The biggest blob outside Γ^{ex} is therefore labeled as D2. Its key geometric properties are saved.
6. The number of asteroids detected in the image, and the centroid of D2 are passed as the output of the IP while the other geometrical properties about D1 are passed to the next IP block.

Table 1 Characteristics of the datasets used in this work

	DB_0	FRP	CRP	DS_R^1	DS_R^3	DS_{CR}^1	DS_{CR}^2
Number of images	36400	12102	20164	36400	9225	36400	6150
Image size [px]	1536 × 2048			96 × 128		128 × 128	
D1 is observable [%]	100	100	100	100	100	100	100
D2 is not observable [%]	19.15	11.54	17.21	19.15	33.33	91.55	50
D2 separated from D1 [%]	72.41	76.17	68.75	72.41	33.33	0	0
D2 close to D1 [%]	8.45	12.29	14.04	8.45	33.33	8.45	50

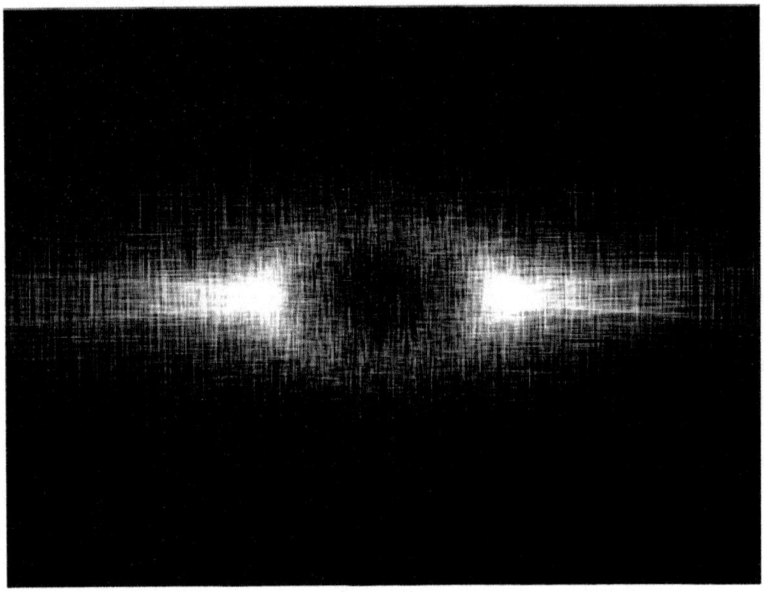

Fig. 4 An overlap image of the bounding boxes of D2 from DB_0 in the image plane

The algorithm described is designed with a conservative philosophy to be robust to FP detection of D2 and instead accommodate a much larger number of FN detection due to its incapability to detect D2 whenever it is located within Γ_{D1}^{ex}. This has been a design choice to keep the algorithm simple yet robust to FP, which could trigger unwanted tracking and pointing behavior in the CubeSat. This work aims to address such issues to design a more robust IP pipeline for the object recognition of D2. To allow for comparison with the other methods, the predicted bounding box of D2 and the number of bodies detected in the image are extracted as output. In the latter case, D2 presence is predicted in the image whenever the number of bodies is equal to 2, being D1 always visible in the image by default. Finally, it is noted that IP_b is designed to work using as input images at their native resolution, while the other IP methods presented requires resized or cropped version of such images.

2.3 Convolutional Extreme Learning Machine

The Convolutional Extreme Learning Machine (CELM) approach consists of a hierarchically organized convolutional pooling architecture in which alternate sequences of convolutions, activation functions, and pooling are executed multiple times until a latent feature vector is generated [6]. Training happens by flowing forward the dataset on the architecture to generate such vectors which are then linked with the target output (in this case the normalized components of Γ) via weighted connections. The weights and biases of the network are set randomly at initialization, while the weights between the latent vector and the output layer are found via a regularized least-square method, as illustrated in [6, 7].

The advantage of using CELMs and the framework supporting them is motivated by the very fast training time, which allows for efficient exploration of the architecture design space [5]. By doing so, the hyper-parameters of the convolutional portion of a Convolutional Neural Network (CNN) architecture can be explored quickly and efficiently. A thorough architecture design search involves the weights and bias initialization strategy (Random, uniform, orthogonal), the type of activation function to be used (none, ReLU, leaky ReLU, tanh, sigmoid), the pooling strategy to be used (mean or max), the number of sequences of convolution, activation function, and pooling to be used in the architecture (from 2 to 6), and the values of the regularization parameter of the least-square problem (from 0.0001 to 10000 in increasing order of magnitudes) is executed each time for the DS_R^1, DS_R^3, and DS_{CR}^2 datasets. A split corresponding to 20% of the dataset considered is used for validation purposes. An example of one of the three winning architectures on these datasets (the one in DS_{CR}^2) is summarized in Table 2.

In inference, the output of the network consists of the 4 components of the normalized bounding box of D2. In this work, the fully connected layer generated by the CELM (such as FC4 in the case of the CELM trained on DS_{CR}^2) is also used by the corresponding Random Forest (RF) method as feature vector to be used both in the classification and regression tasks.

2.4 Convolutional Neural Network

The structure of the CNN method is very similar to the one of the CELM with two major differences: the weights and biases are not set randomly and frozen during training but they are varied via mini-batch gradient descent, and the head of the architecture is extended from the flatten layer of the CELM architecture illustrated in Table 2 with a neural network whose characteristics are searched via a thorough hyper-parameter search. The architecture and initial values of the weights and biases of the same are taken from the corresponding best CELM architectures. From such initialization, CNNs are trained over the DS_R^1, DS_R^3, and DS_{CR}^2 datasets. An example of one of the three winning architectures on these datasets (the one in DS^2_{CR}) is summarized in Table 3.

As for the CELM training, a split corresponding to 20% of the dataset considered is used for validation purposes. The weights and biases of the CNN which achieve the best loss on the validation set are stored during training. These are then used to instantiate the network to be used in inference on the test sets.

Table 2 Architecture of the winning CELM trained on DS^2_{CR}. The kernels are randomly initialized with an *orthogonal* setup, while the *max* pooling strategy and the *leaky ReLU* are used

Layer acronym	Layer type	Output	Number of parameters
I	(InputLayer)	(None, 96, 128, 1)	0
C1	(Conv2D)	(None, 96, 128, 16)	160
A1	(Activation)	(None, 96, 128, 16)	0
P1	(MaxPooling2D)	(None, 48, 64, 16)	0
C2	(Conv2D)	(None, 48, 64, 32)	4640
A2	(Activation)	(None, 48, 64, 32)	0
P2	(MaxPooling2D)	(None, 24, 32, 32)	0
C3	(Conv2D)	(None, 24, 32, 64)	18496
A3	(Activation)	(None, 24, 32, 64)	0
P3	(MaxPooling2D)	(None, 12, 16, 64)	0
C4	(Conv2D)	(None, 12, 16, 128)	73856
A4	(Activation)	(None, 12, 16, 128)	0
P4	(MaxPooling2D)	(None, 6, 8, 128)	0
FC4	(Flatten)	(None, 6144)	0

Table 3 Architecture of the winning CNN trained on DS^2_{CR}. The *max* pooling strategy and the *leaky ReLU* are used

Layer acronym	Layer type	Output	Number of parameters
I	(InputLayer)	(None, 96, 128, 1)	0
CELM	(Functional)	(None, 6144)	97152
D1	(Dense)	(None, 512)	3146240
DO1	(Dropout)	(None, 512)	0
D2	(Dense)	(None, 512)	262656
DO2	(Dropout)	(None, 512)	0
D3	(Dense)	(None, 4)	2052

2.5 Random Forest

A RF is an ensemble of decision trees. By training different trees and combining their output, better performance can be obtained with respect to individual trees [10]. RFs can be used both for classification and regression. In the former, the class is selected by a majority vote among the trees while in the latter the forest prediction is obtained by averaging those of the individual trees.

In this work, two RFs are trained, one for classification and one for regression. The first is used to check for the presence of D2 in the image, while the second to find the coordinates of its bounding box. Differently from the other methods, the RFs take as input the feature vector generated by the CELM discussed before, instead of the entire image. To select the best hyper-parameters settings, a split of 20% of the images is considered for validation purposes, as done for the CELM and CNN.

The training of the RFs is based on the bootstrap aggregating (bagging) technique, which works by splitting the training set into a series of subsets called bootstrap samples that are created randomly and are used to train individual trees. Bootstrap samples are drawn with replacement so they are independent of each other. For the decision trees training, the following parameters are set: the number of predictors sampled for each decision split is equal to the square root of the total number for classification and one-third for regression; the minimum number of observations per tree leaf is 1 for classification and 5 for regression; node splitting is based on the Gini diversity index for classification and the mean squared error for regression. The trees in the RF architectures are not pruned. The number of trees is set to 50 for classification and 100 for regression.

2.6 Overall IP Strategy

To overcome the limitation of IP_b, different training and inference pipelines are put in place combining the different IP techniques presented in the previous sections. Figure 5 summarizes such strategy, which is now described in details.

First, different CELMs are trained on DS_R^1, DS_R^3, and DS_{CR}^2 with the same procedure illustrated in Sect. 2.3. The 3 winning architectures obtained in this step are used as basis for the other methods. An equal number of CNNs are trained with these architectures using gradient-descent methods on the same datasets, as illustrated in Sect. 2.4, while the fully connected layers generated by the CELMs in inference over all datasets are used as input by RFs in place of images. Moreover, while CELMs and CNNs are designed exclusively for regression, RFs are designed both for regression and classification. First, RFs are trained using the fully connected layers of the CELMs trained on the DS_R^1, DS_R^3, and DS_{CR}^2 datasets for classification. Three of these RFs are trained over such datasets as well. These are referred to as $RF_{(DS_R^1)}^{FC(DS_R^1)}$, $RF_{(DS_{CR}^2)}^{FC(DS_{CR}^2)}$, and $RF_{(DS_R^3)}^{FC(DS_R^3)}$ where in the notation $RF_{(i)}^{FC(j)}$ the i reflects the dataset

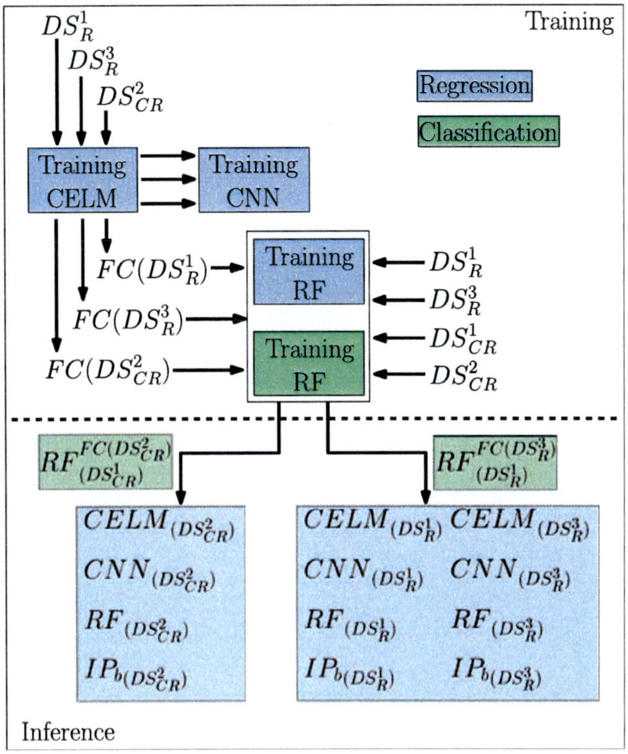

Fig. 5 Summary of the training and inference strategy, as well as the relationships between the various IP methods

used for training the RF and the j the dataset used by the CELM to generate the fully connected layer used by the RF. Moreover, two additional RFs are trained over DS^1_{CR} and DS^1_R but using the fully connected layers of the CELMs trained from DS^2_{CR} and DS^3_R. Out of these five classifiers, two are selected for $_R$ and $_{CR}$ images, respectively $\mathrm{RF}^{FC(\mathrm{DS}^2_{CR})}_{(\mathrm{DS}^1_{CR})}$ and $\mathrm{RF}^{FC(\mathrm{DS}^3_R)}_{(\mathrm{DS}^1_R)}$.

After this training for classification, the RFs are trained over the DS^1_R, DS^3_R, and DS^2_{CR} datasets (the same ones considered with CELM and CNN) for regression.

At inference time, the performance of all methods introduced (CELM, CNN, and RF), are illustrated on the FRP and CRP for the images which provided a positive identification of D2 by the corresponding classifier.

3 Results

To evaluate and compare the classification performance of different IP algorithms to perform the detection of D2 in the images, the Accuracy (A), Precision (P), and Recall (R) metrics are defined considering the True Positive (TP), True Negative (TN), FP, and FN detection of D2 in the image:

$$\mathrm{A} = \frac{\mathrm{TP} + \mathrm{TN}}{\mathrm{TP} + \mathrm{FP} + \mathrm{TN} + \mathrm{FN}} \tag{1}$$

$$\mathrm{P} = \frac{\mathrm{TP}}{\mathrm{TP} + \mathrm{FP}} \tag{2}$$

$$\mathrm{R} = \frac{\mathrm{TP}}{\mathrm{TP} + \mathrm{FN}} \tag{3}$$

Additionally, the Intersection over Union (IoU) metric is defined to characterize the performance of the predicted bounding box to overlap with the true one:

$$\mathrm{IoU} = \frac{\text{Area of Overlap}}{\text{Area of Union}} =$$

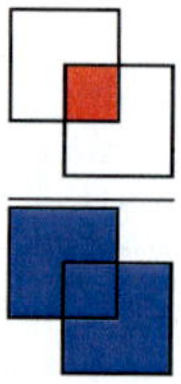

(4)

where a value of 100% would signify a perfect overlap and 0% a completely wrong prediction of the bounding box characteristics.

The classification performances achieved with the various RFs are compared in Table 4 with the ones previously achieved with the IP_b, both on FRP and CRP datasets.

From Table 4 it is possible to see that as discussed in Sect. 2.2, IP_b suffers from a high number of FN while achieving very small numbers of FP. It is immediately possible to appreciate that the RF classifiers dramatically reduce such a large number of FN, albeit somehow losing performance in terms of FP. This is reflected in the P and R metrics of the different methods. Comparing the performances of the RF with the one of IP_b, it can be concluded that the former would be outperforming the latter across all metrics considered but R, which is particularly high for IP_b due to

Table 4 Performances of the classifiers on the FRP and CRP datasets

Metric	IP_b	$RF_{(DS^1_R)}^{FC(DS^1_R)}$	$RF_{(DS^2_{CR})}^{FC(DS^2_{CR})}$	$RF_{(DS^3_R)}^{FC(DS^3_R)}$	$RF_{(DS^1_{CR})}^{FC(DS^2_{CR})}$	$RF_{(DS^1_R)}^{FC(DS^3_R)}$
			FRP			
TP	9179	10480	1629	10652	1577	10464
TN	1396	1038	9748	266	10173	1076
FP	1	359	431	1131	6	321
FN	1526	225	294	53	346	241
A [%]	87.38	95.17	94.01	90.22	**97.09**	95.36
P [%]	**99.99**	96.69	79.08	90.40	99.62	97.02
R [%]	85.75	97.90	84.71	**99.50**	82.01	97.75
			CRP			
TP	13373	16167	3056	16534	2720	15949
TN	3465	2369	15818	800	16624	2404
FP	6	1102	902	2671	96	1067
FN	3320	526	388	159	724	744
A [%]	83.51	91.93	93.60	85.97	**95.93**	91.02
P [%]	**99.96**	93.62	77.21	86.09	96.59	93.73
R [%]	80.11	96.85	88.73	**99.05**	78.98	95.54

its over-conservative design. Based on these results and considering separately the performance of the classifier working with $_{CR}$ and $_R$ images, it has been decided to use the $RF_{(DS^1_{CR})}^{FC(DS^2_{CR})}$ and $RF_{(DS^1_R)}^{FC(DS^3_R)}$ classifiers respectively during the preprocessing step before regression.

On the other side, when considering the regression performance in quantifying the properties of the bounding box, it has been observed that all ML methods considered performed poorly when compared to the simpler IP_b, which remains the only one to generate reliable estimates. These results are summarized in Table 5. Of all possible combinations between IP method and dataset, it is possible to see that CNNs are the ones achieving similar, albeit smaller, value of the IoU, immediately followed by RFs and then very poorly by CELMs.

The higher performance of IP_b in identifying the bounding box is driven by two factors. First, when D2 is completely separated from D1 it is relatively easy after binarization to determine its bounding box via the *regionprops* algorithm. Second, the relative frequency on the test datasets in which D2 is either absent or completely separated from D1 is very high (respectively in 11.54% and 76.17% of the cases in FRP and 17.21% and 68.75% of the cases in CRP) while the cases in which D2 is found within the bounding box of D1 (which are the ones in which IP_b is not even capable to determine the presence of D2) are relatively low (only 12.29% and 14.04% respectively in FRP and CRP). The coupling of these two phenomena makes it possible for the IP_b to achieve better performance in identifying the properties of Γ, albeit these are shadowed by the incapability of the same to detect D2 when imaged

Table 5 Values of IoU achieved over the FRP and CRP datasets by the various IP methods trained on different datasets

Dataset	IP_b	CELM	CNN	RF
		FRP		
DS_R^1	70.14	14.13	62.94	51.47
DS_R^3	70.14	10.81	56.79	27.58
DS_{CR}^2	17.73	9.25	47.34	31.48
		CRP		
DS_R^1	64.97	16.64	61.83	48.32
DS_R^3	64.97	13.54	52.43	24.52
DS_{CR}^2	13.10	10.58	48.56	29.89

above D1. This same reasoning is also reflected by the trend observed on the performance when considering the same IP method trained over different datasets. It is observed that methods trained over unbalanced datasets tend to perform better on the FRP and CRP datasets, which by default are also unbalanced. Looking only at global performance metrics, one would be encouraged to train the data-driven ML methods over unbalanced datasets, which would however be incapable to overcome the same limitations already existing in the design of IP_b. To conclude, the results show that CNNs are capable to outline the bounding box of D2 even when D2 is appearing within the bounding box of D1 (when the proper dataset is considering during training) and that such capability comes at the price of poorer outlining performances even in the simpler cases in which D2 is completely separated from D1.

4 Conclusions

In this work, different IP algorithms are designed and tested for the specific scenario of object recognition of Dimorphos, the secondary body of the Didymos binary system. Their performance in terms of accuracy, precision, recall, and IoU are evaluated over two datasets representative of the two nominal phases of the Milani mission: FRP and CRP.

It is found that ML methods may be very beneficial in the classification task to determine the presence of D2 in the image and that a simple RF taking as input a fully connected layer generated by a CELM can improve detection performances when compared to the baseline IP algorithm of Milani.

On the other hand, it is found that the ML methods considered performed poorly at predicting the bounding box characteristics when compared to the baseline algorithm. This is explained by the low relative frequency of the cases in which these algorithms excel compared to the high relative frequency at which the baseline algorithm performs well.

Based on this observation, a hybrid approach is proposed for Milani which combines the strengths of the baseline algorithm with the ones of the ML ones. In inference, whenever the detected number of bodies by IP_b is equal to 2, the nominal IP_b pipeline is used to determine the bounding box using the *regionprops* function, since in this case D2 will most probably be largely separated from D1 in the image. On the contrary, if the number of detected bodies in the image is equal to 1, a classifier based on RF can be run on the resized or cropped image to verify the presence of D2. In the cases in which the presence of D2 is predicted by the classifier, a more sophisticated CNN can be run on the image to predict the bounding box.

Future work could be focused on the usage of more advanced fine-tuned pre-trained Mask-R CNN architectures, which represent the state of the art for object recognition, in place of the more simple architectures considered in this work.

Acknowledgements M.P and F.T would like to acknowledge the funding received from the European Union's Horizon 2020 research and innovation programme under the Marie Skłodowska-Curie grant agreement No 813644.

References

1. Ferrari, F., et al.: Preliminary mission profile of Hera's Milani CubeSat. Adv. Space Res. **67**(6), 2010–2029 (2021)
2. Nobuyuki, O.: A threshold selection method from gray-level histograms. IEEE Trans. Syst. Man Cybern. **9**(1), 62–66 (1979)
3. Quadrelli, M.B., et al.: Guidance, navigation, and control technology assessment for future planetary science missions. J. Guidance Control Dyn. **38**(7), 1165–1186 (2015)
4. Szeliski, R.: Computer Vision, 2nd edn. Springer International Publishing, New York (2022)
5. Saxe, A.M. et al.: On random weights and unsupervised feature learning. In: Proceedings of the 28th International Conference on Machine Learning, pp. 1089–1096. Omnipress, Bellevue, Washington, USA (2011)
6. Huang, G.B., et al.: Local receptive fields based extreme learning machine. IEEE Comput. Intell. Mag. **10**(2), 18–29 (2015)
7. Rodrigues, I.R., et al.: Convolutional extreme learning machines: A systematic review. Informatics **8**(2), 11–33 (2021)
8. Pugliatti, M. et al.: Design of the on-board image processing of the Milani mission. In: 44th AAS GN&C Conference, pp. 1–22. Univelt, Breckenridge (2022)
9. Pugliatti, M. et al.: The milani mission: Overview and architecture of the optical-based GNC system. In: SciTech Forum, pp. 1–20. Univelt, San Diego (2022)
10. Breiman, L.: Random forests. Mach. Learn. **45**(1), 5–32 (2001)

Towards an Explainable Artificial Intelligence Approach for Ships Detection from Satellite Imagery

Cosimo Ieracitano, Nadia Mammone, and Francesco Carlo Morabito

Abstract In recent years, ship detection and classification from remote satellite imagery using advanced Artificial Intelligence (AI) algorithms raises great attention in military and civilian applications, such as marine resources monitoring, maritime rescue assistance, protection in territorial sea areas and so on. In this contest, here, the "Ships in Satellite Imagery" Kaggle dataset is taken into account and a customised Convolutional Neural Network (CNN) is proposed to perform the 2-way classification task: presence of ship (denoted as *ship class*) versus no ship presence (denoted as *no-ship class*), reporting an accuracy of 98.83%. Furthermore, the explainability of the developed neural network is also investigated by means of an occlusion sensitivity analysis in order to explore which part of the satellite images was involved mostly in the discrimination procedure. The promising outcomes together with the xAI based methodology encourage the use of the proposed ship detection model in space sector.

Keywords Earth observation · Satellite images · Ship detection · Explainable artificial intelligence · Convolutional neural network

1 Introduction

The analysis of remote sensing imagery has become of great interest in several fields, such as energy, agriculture, intelligence and defense. To this end, commercial imagery providers, e.g., Planet [1], employs constellations composed of small satellites to acquire images of the Earth daily. In this context, marine ship detection attracted significant attention in marine security. In recent years, ship detection

C. Ieracitano (✉) · N. Mammone · F. C. Morabito
DICEAM, University Mediterranea of Reggio Calabria, Via Graziella Feo di Vito, 89122 Reggio Calabria, Italy
e-mail: cosimo.ieracitano@unirc.it

C. Ieracitano et al. (eds.), *The Use of Artificial Intelligence for Space Applications*, Studies in Computational Intelligence 1088,
https://doi.org/10.1007/978-3-031-25755-1_19

approaches based on Artificial Intelligence (AI) have been emerging. For example, Kang et al. [12] proposed a contextual region-based Convolutional Neural Network (CNN) with multilayer fusion for synthetic aperture radar (SAR) ship identification; while Guo et al. [5] proposed CenterNet++, a single-stage detector with high accuracy for small ships recognition. Hong et al [7] developed an improved-YOLOv3 ("you only look once" version 3) for automatic ship detection by means of SAR and optical ship datasets, achieving higher performance (2–3% more) when compared with the original YOLOv3. Xu et al. [16] presented an on-board ship detection framework based on the standard Constant False Alarm Rate (CFAR) method and deep learning (DL) techniques, reporting precision of 85.9% for the experimental data. Li et al. [14] proposed a multidimensional domain DL strategy based on a Faster RCNN and on the analysis of spatial and frequency-domain features. Jiang et al. [11] introduced a multi-channel fusion SAR image processing approach based on YOLO-V4 and reporting an average precision of 90.37%. Gadamsetty et al. [4] applied hashing concept (by using SHA-256) and YOLOv3 employed to extract features from deep CNN, for satellite imagery detection using the "air bus ship detection" Kaggle dataset, achieving an accuracy of 96.4%. Yan et al. [18] developed a novel ship detector by proposing a lightweight network module able to mitigate jamming and complex backgrounds while improving the features of objects. Alghazo et al. [3] evaluated tow different CNN on the Airbus satellite images dataset for maritime ship detection reporting a accuracy of 89.7%. Recently, Zhao et al. [20] proposed an Adversarial Learning Attention (ALA) and Compensation Loss Module system for ship recognition; whereas, Xie et al. [15] developed YOLO-CASS (YOLO Coordinate Attention SAR Ship) based deep learning framework tested on the SAR Ship Detection Dataset (SSDD) synthetic aperture radar. Xu et al. [17] presented a low-resolution marine object (LMO) detection YOLO system, called LMO-YOLO.

Most of the aforementioned works are based on YOLO method. In contrast, this paper proposes a novel detection approach based on explainable Artificial Intelligence (xAI) to recognise ships from satellite imagery. Specifically, first, a CNN is implemented to classify images with ships and no-ships by using the "Ships in Satellite Imagery" Kaggle dataset [2]. Next, in order to study the behavior of the trained CNN and explain which input area is more relevant to the classification process, xAI is employed. In particular, an Occlusion Sensitivity Analysis (oSA) is carried out to investigate the highly importance regions. It is to be noted that the developed CNN was able to achieve accuracy rate up to 98.93% and that the xAI was able not only to explain the behavior of the proposed CNN but also detect the position of the ship in the satellite image.

The paper is organized as follows: Sect. 2 describes the dataset; Sect. 3 presents the methodology, including the developed CNN and xAI approach; Sect. 4 reports the achieved results; while, Sect. 5 concludes the paper.

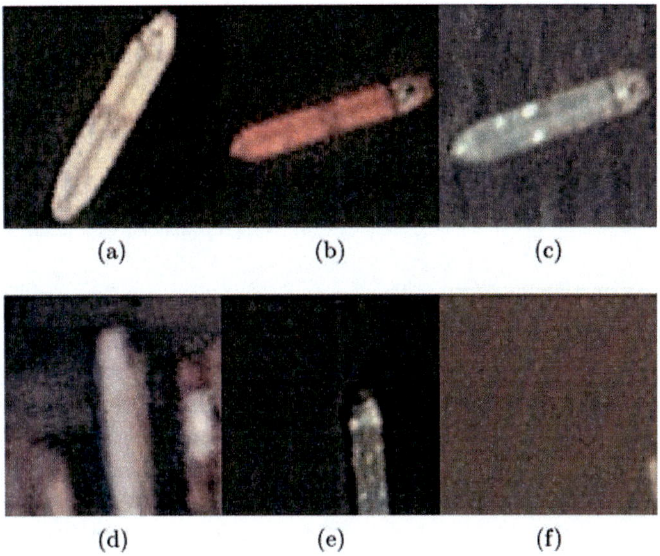

Fig. 1 Examples of images belonging to "ship" (**a–c**) and "no-ship" class (**d–e**)

2 Data Description

In this paper, the Kaggle open source ("Ships in Satellite Imagery") dataset is employed [2]. It consists of images acquired from Planet satellite imagery placed at the San Francisco Bay and San Pedro Bay areas of California (extracted from PlanetScope full-frame visual scene products). Specifically, 4000 RGB images of 80 X 80 are included: 3000 with "no-ship" and 1000 with "ship" label. The "ship" class includes images with full appearance of a single ship, with different dimensions, orientations, and environmental conditions (e.g., atmospheric noises). In contrast, the "no-ship" category has (1) images of different random land cover features (e.g., building, water, vegetation etc.) with no portion of ships; (2) partial ships; (3) mislabeled images due to noise caused by bright pixels or strong linear properties. Figure 1, shows images from "ship" and "no-ship" extracted from the Kaggle "Ships in Satellite Imagery" dataset [2].

3 Methodology

3.1 Convolutional Neural Network Architecture

A CNN is a deep learning model, commonly used for image processing [8–10] and allows to estimate the most significant features automatically by means of a set processing layers: convolution, activation and pooling [13]. The developed CNN is

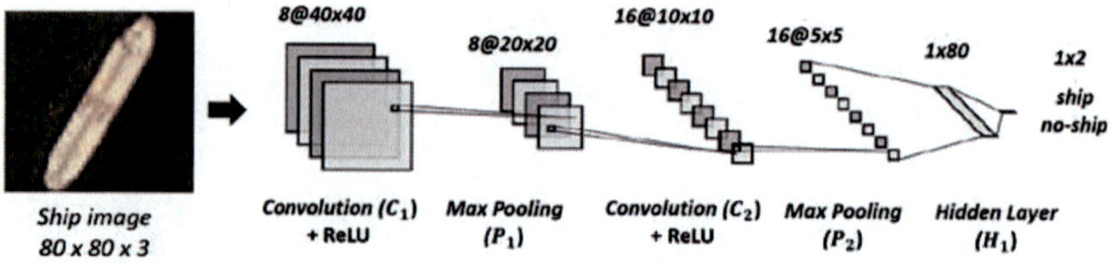

Fig. 2 The proposed Convolutional Neural Network for performing the ship versus no-ship classification task

shown in Fig. 2. It consists of two convolutional layers, two *Rectified Linear Unit* (ReLU) activation layers and one max-pooling layer. A standard neural network with one hidden layer ends the classification model. More specifically, the first convolutional layer is composed of 8 filters 4 x 4 x 3 that move along the satellite image with a step $s = 2$ and padding $p = 1$. Hence, eight activation maps 40 x 40 are extracted; whereas, the max pooling reduces the dimension of the extracted features maps from 40 x 40 to 20 x 20, by applying a filter 2 x 2 and stride $\tilde{s} = 2$. Similarly, the second convolutional layer, uses 16 filters of size 4 x 4 x 8 with $s = 2$ and $p = 1$, resulting in 16 features maps 10 x 10; while, the second max pooling layer, has a filter sized 2 x 2 with $\tilde{s} = 2$ that downsamples the previous output from 10 x 10 to 5 x 5. These are reshaped into a single array si1 x A with A = 5 x 5 x 16 = 400. The latter feds a 1-hidden layer NN of 80 hidden units and a softmax output used for the binary discrimination task: ship versus no-ship. It is to be noted that the network was set up according to a trial and error strategy and estimating the performance of models with different number of layers and filter size, as reported in Table 1. The proposed CNN was implemented using Matlab R2021b and trained with the optimized Adaptive Moment method (learning and exponential decay parameters of 0.0001, 0.9 (β_1) and 0.999 (β_2), respectively) for 100 epochs for about 1 min on a high performance computer with 64 GB RAM installed and one graphics processing unit NVIDIA GeForce RTX 2080 Ti .

3.2 *Explainable Ship Detection Model*

Explainability of decision models is one of the main challenge of AI that aims to "open" the *black box* and investigate which area is mostly involved in the classification chain [6]. In this context, methodologies of explainable Artificial Intelligence (xAI) are indeed emerging. In this paper, the Occlusion Sensitivity Analysis (OSA [19]) technique is exploited. OSA occludes parts of the input image using a moving filter. The occluded image inputs a pre-trained CNN for estimating changes of the classification rate. The result is referred to as *occlusion map* or *heatmap* where the most significant input regions for the classification are highlighted with coloration ranged between blue (low significance) and red (high significance). The occlusion

Table 1 Architectures of different CNN and computational cost

Model	C_1 +ReLU	P_1	C_2+ReLU	P_2	HL_1	HL_2	Output	Time cost (s)
CNN_1	Filters =8@ 4x4x3 s = 2	Filters = 2x2 $\tilde{s} = 2$	–	–	1000	–	2	~71
CNN_2	Filters = 8@ 4x4x3 s = 2	Filters = 2x2 $\tilde{s} = 2$	–	–	1000	500	2	~79
CNN_3	Filters = 8@ 4x4x3 s = 2	Filters = 2x2 $\tilde{s} = 2$	Filters = 16@4x4x8 s = 2	Filters = 2x2 $\tilde{s} = 2$	200	–	2	~86
CNN_4	Filters= 8@ 4x4x3 s = 2	Filters = 2x2 $\tilde{s} = 2$	Filters = 16@4x4x8 s = 2	Filters = 2x2 $\tilde{s} = 2$	80	20	2	~86
CNN_5	Filters= 8@ 4x4x3 s = 2	Filters = 2x2 $\tilde{s} = 2$	Filters = 16@4x4x8 s = 2	Filters = 2x2 $\tilde{s} = 2$	80	–	2	~84
CNN_6	Filters= 8@ 8x8x3 s = 2	Filters = 4x4 $\tilde{s} = 2$	–	–	1000	–	2	~74

map is usually overlapped to the original input image to derive information about the importance of the input areas. Here, OSA was applied by using the images, correctly identified as ship/no-ship as input to the pre-trained CNN (i.e., CNN reported in Sect. 3.1, Fig. 2).

3.3 Performance Metrics

In order to assess the effectiveness of the developed CNN the following common parameters are estimated:

$$Specificity\ (Sp) = \frac{TN}{TN + FP} \tag{1}$$

$$Sensitivity\ (Ss) = \frac{TP}{TP + FN} \tag{2}$$

$$Positive\ Predicted\ Value\ (PPV) = \frac{TP}{TP + FP} \tag{3}$$

$$Negative\ Predicted\ Value\ (NPV) = \frac{TN}{TN + FN} \tag{4}$$

$$F1 - score\ (FS) = \frac{2 * TP}{2 * TP + FP + FN} \tag{5}$$

$$Accuracy\ (Acc) = \frac{TP + TN}{TP + TN + FP + FN} \tag{6}$$

with TP = True Positive, i.e., images properly identified as "ship"; TN = True Negative, i.e., images properly identified as "no-ship"; FP = False Positive, i.e., "no-ship" images wrongly recognized as "ship"; FN = False Negative, i.e., "ship" images misclassified as "no-ship".

4 Results

Classification performance Table 2 shows the achieved results for different CNN architectures. It is worth noting that high and comparable performance were observed with all the configurations developed. Indeed, the lowest performance was achieved by CNN_1 composed of one convolutional, one pooling layer and 1-hidden layer NN with 1000 units, reporting *Sp* of 97.30±1.57%, *Ss* of 99.00±0.70%,

Table 2 Results of CNN models with different architectures

CNN	Sp [%]	Ss [%]	PPV [%]	NPV [%]	FS [%]	Acc [%]
CNN_1	97.30±1.57	99.00±0.70	99.10±0.52	97.04±2.05	99.05±0.52	98.57±0.77
CNN_2	96.90±2.64	99.17±0.42	98.97±0.87	97.49±1.24	99.07±0.52	98.60±0.79
CNN_3	97.90±1.52	99.20±0.36	99.30±0.50	97.62±1.06	99.25±0.31	98.87±0.48
CNN_4	97.60±2.01	99.33±0.42	99.20±0.66	98.00±1.23	99.27±0.42	98.90±0.63
CNN_5	98.20±1.32	99.17±0.32	99.40±0.44	97.52±0.95	99.28±0.33	98.93±0.50
CNN_6	97.90±1.20	99.27±0.26	99.30±0.40	97.81±0.76	99.28±0.14	98.82±0.21

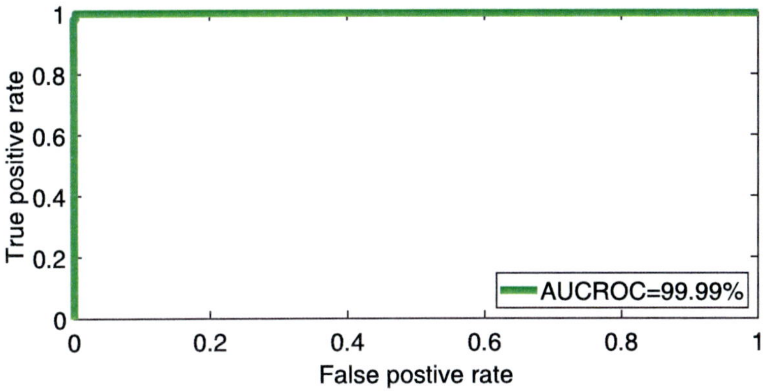

Fig. 3 ROC curve and AUC score of the proposed CNN

PPV of 99.10±0.52%, NPV of 97.04±2.05%, FS of 99.05±0.52% and Acc of 98.57±0.77%; while, the highest result was reported by CNN_4 composed of two convolutional layers, two max pooling layers and 1-hidden layer NN with 80 hidden units, achieving an accuracy score up to 98.93±0.50%. Furthermore, the Area Under the Receiver Operating Characteristic (ROC) curve (AUROC) was also estimated as shown in Fig. 3, reporting AUROC of 99.99%.

Explainability for ship detection OSA was carried out to explore the input region that mostly contributed in the ship detection. Figure 4, shows examples of importance maps of ship (Fig. 4a–f) and no-ship (Fig. 4g–l) images of the test set. Red colour represents the regions mostly relevant in the classification; vice-versa, blue colour denotes areas not relevant in the discrimination task. The developed CNN was capable of detecting effectively regions of ships in satellite images that indeed were related to the input portions thoroughly significant for discriminating ship images. In particular, it is worth noting that only three specific areas were particularly relevant: midship, stern and bow.

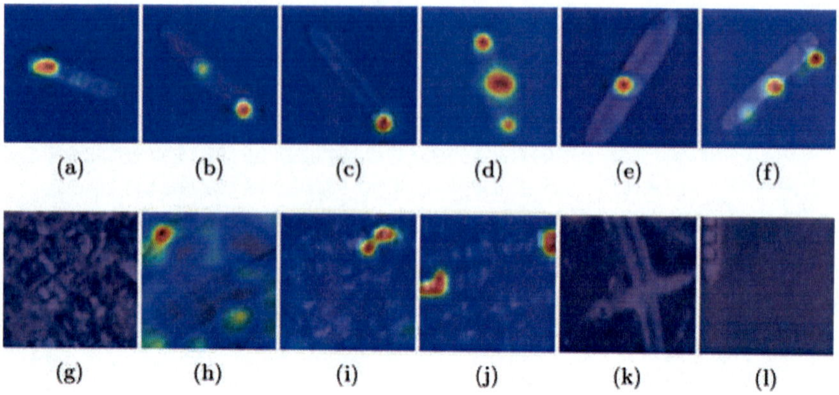

Fig. 4 Occlusion sensitivity maps of ship (**a–f**) and no-ship (**g–l**). Coloration is ranged between blue (low significance) and red (high significance)

5 Conclusion

In the present work, an explainable Artificial Intelligence methodology for detecting ships from satellite imagery is proposed. Specifically, the Kaggle open source ("Ships in Satellite Imagery") dataset was employed [2]. To this end, a costomised CNN consisted of two convolutional and max pooling layers followed by 1-hidden layer NN was developed to perform the binary classification task: ship versus no-ship, reporting and accuracy rate up to 98.93%. Moreover, to explore which region of the satellite images contributed highly in the discrimination, a well-known xAI technique was also used. In particular, the occlusion sensitivity analysis reported that the proposed CNN allowed to successfully detect ships in satellite images and no-ships (such as buildings, bare earth etc.) as shown in Fig. 4. It is to be noted that this paper is supposed to be a preliminary study, indeed, the xAI strategy can be used for developing a more precise ship detection system. However, the present work has some drawbacks. The main limitation is that the ship class contains only full appearance of single ships, while partial ships are in the no-ship class. This means that the system is specialized only to detect full shapes of ships. In the future, the dataset will be organised including in the "ship" class also partial ships. In addition, the system will be also tested using other satellite dataset for further comparisons and validation.

Acknowledgements This work was granted by the Programma Operativo Nazionale (PON) "Ricerca e Innovazione" 2014-2020 CCI2014IT16M2OP005 (CUP C35F21001220009 code: I05).

References

1. Planet. https://www.planet.com/
2. Ships in Satellite Imagery. https://www.kaggle.com/datasets/rhammell/ships-in-satellite-imagery
3. Alghazo, J., Bashar, A., Latif, G., Zikria, M.: Maritime ship detection using convolutional neural networks from satellite images. In: 2021 10th IEEE International Conference on Communication Systems and Network Technologies (CSNT), pp. 432–437. IEEE (2021)
4. Gadamsetty, S., Ch, R., Ch, A., Iwendi, C., Gadekallu, T.R.: Hash-based deep learning approach for remote sensing satellite imagery detection. Water **14**(5), 707 (2022)
5. Guo, H., Yang, X., Wang, N., Gao, X.: A centernet++ model for ship detection in sar images. Pattern Recognit. **112**, 107787 (2021)
6. Holzinger, A., Malle, B., Saranti, A., Pfeifer, B.: Towards multi-modal causability with graph neural networks enabling information fusion for explainable ai. Inf. Fusion **71**, 28–37 (2021)
7. Hong, Z., Yang, T., Tong, X., Zhang, Y., Jiang, S., Zhou, R., Han, Y., Wang, J., Yang, S., Liu, S.: Multi-scale ship detection from sar and optical imagery via a more accurate yolov3. IEEE J. Sel. Top. Appl. Earth Obs. Remote Sens. **14**, 6083–6101 (2021)
8. Ieracitano, C., Mammone, N., Paviglianiti, A., Morabito, F.C.: A conditional generative adversarial network and transfer learning-oriented anomaly classification system for electrospun nanofibers. Int. J, Neural Syst (2022)
9. Ieracitano, C., Mammone, N., Versaci, M., Varone, G., Ali, A.R., Armentano, A., Calabrese, G., Ferrarelli, A., Turano, L., Tebala, C., et al.: A fuzzy-enhanced deep learning approach for early detection of covid-19 pneumonia from portable chest x-ray images. Neurocomputing (2022)
10. Ieracitano, C., Morabito, F.C., Hussain, A., Mammone, N.: A hybrid-domain deep learning-based bci for discriminating hand motion planning from eeg sources. Int. J. Neural Syst. **31**(09), 2150038 (2021)
11. Jiang, J., Fu, X., Qin, R., Wang, X., Ma, Z.: High-speed lightweight ship detection algorithm based on yolo-v4 for three-channels rgb sar image. Remote Sens. **13**(10), 1909 (2021)
12. Kang, M., Ji, K., Leng, X., Lin, Z.: Contextual region-based convolutional neural network with multilayer fusion for sar ship detection. Remote Sens. **9**(8), 860 (2017)
13. Krizhevsky, A., Sutskever, I., Hinton, G.E.: Imagenet classification with deep convolutional neural networks. Commun. ACM **60**(6), 84–90 (2017)
14. Li, D., Liang, Q., Liu, H., Liu, Q., Liu, H., Liao, G.: A novel multidimensional domain deep learning network for sar ship detection. IEEE Trans. Geosci. Remote Sens. **60**, 1–13 (2021)
15. Xie, F., Lin, B., Liu, Y.: Research on the coordinate attention mechanism fuse in a yolov5 deep learning detector for the sar ship detection task. Sensors **22**(9), 3370 (2022)
16. Xu, P., Li, Q., Zhang, B., Wu, F., Zhao, K., Du, X., Yang, C., Zhong, R.: On-board real-time ship detection in hisea-1 sar images based on cfar and lightweight deep learning. Remote Sens. **13**(10), 1995 (2021)
17. Xu, Q., Li, Y., Shi, Z.: Lmo-yolo: A ship detection model for low-resolution optical satellite imagery. IEEE J. Sel. Top. Appl. Earth Obs. Remote Sens (2022)
18. Yan, H., Li, B., Zhang, H., Wei, X.: An antijamming and lightweight ship detector designed for spaceborne optical images. IEEE J. Sel. Top. Appl. Earth Obs. Remote Sens. **15**, 4468–4481 (2022)
19. Zeiler, M.D., Fergus, R.: Visualizing and understanding convolutional networks. In: European Conference on Computer Vision, pp. 818–833. Springer (2014)
20. Zhao, S., Zhang, Z., Guo, W., Luo, Y.: An automatic ship detection method adapting to different satellites sar images with feature alignment and compensation loss. IEEE Trans. Geosci. Remote Sens. **60**, 1–17 (2022)

Investigating Vision Transformers for Bridging Domain Gap in Satellite Pose Estimation

Alessandro Lotti, Dario Modenini, and Paolo Tortora

Abstract Autonomous onboard estimation of the pose from a two dimensional image is a key technology for many space missions requiring an active chaser to service an uncooperative target. While neural networks have shown superior performances with respect to classical image processing algorithms, their adoption is still facing two main limitations: poor robustness on real pictures when trained on synthetic images and scarce accuracy-latency trade off. Recently, vision Transformers have emerged as a promising approach to domain shift in computer vision owing to their ability to model long range dependencies. In this work we therefore provide a study on vision Transformers as a solution to bridging domain gap in the framework of satellite pose estimation. We first present an algorithm leveraging Swin Transformers and adversarial domain adaptation which achieved the fourth and fifth places at the 2021 edition of the ESA's Satellite Pose Estimation Competition, challenging researchers to develop solutions capable of bridging domain gap. We provide a summary of the main steps we followed showing how larger models and data augmentations contributed to the final accuracy. We then illustrate the results of a subsequent development which tackles the limitations of our first solution, proposing a lightweight variant of our algorithm, not requiring access to test images. Our results show that vision Transformers can be a suitable tool for bridging domain gap in satellite pose estimation, although with limited scaling capabilities.

Keywords Pose estimation · Machine vision · Vision Transformers

A. Lotti (✉) · D. Modenini · P. Tortora
Department of Industrial Engineering, Alma Mater Studiorum Università di Bologna, 47121 Forlì, Italy
e-mail: alessandro.lotti4@unibo.it

D. Modenini
e-mail: dario.modenini@unibo.it

P. Tortora
e-mail: paolo.tortora@unibo.it

D. Modenini · P. Tortora
Interdepartmental Centre for Industrial Research in Aerospace, Alma Mater Studiorum Università di Bologna, 47121 Forlì, Italy

C. Ieracitano et al. (eds.), *The Use of Artificial Intelligence for Space Applications*, Studies in Computational Intelligence 1088,
https://doi.org/10.1007/978-3-031-25755-1_20

1 Introduction

On-Orbit Servicing and Active Debris Removal have recently attracted a lot of attention from the space community. Those will allow to repair, inspect, refuel, upgrade, or de-orbit nonfunctional satellites or satellites near the end-of-life, impacting both design philosophy and business models. This new class of missions will require to maneuver two or more spacecraft in close proximity up to docking, an operation with high risk of catastrophic failures. The development of technologies to enable autonomous close proximity operations is therefore a necessary step towards a sustainable space economy.

In this framework the relative pose, i.e. position and orientation, between the involved probes should be known as accurately as possible, since such information must be fed into guidance and control laws. In the worst case scenario, where an active chaser is required to service an uncooperative target, the pose shall be estimated autonomously by the chaser through electro-optical sensors. To this end, monocular cameras are an attractive solution because of their low power consumption and compact size. These advantages, however, come at the expense of a significant algorithmic complexity. In recent years many solutions have been proposed who leverage Neural Networks (NN) to infer the three dimensional pose of a known satellite from a two dimensional image. However, their adoption is still facing two main limitations.

1. Poor robustness against domain shifts: models are typically trained on rendered satellite images and their accuracy does not transfer to an operational scenario.
2. Poor accuracy-latency trade off: current methods are often too computational expensive for real time onboard execution or not enough accurate especially under Domain Shift (DS).

A recent effort to push research forward in bridging DS is represented by the 2021 edition of the Satellite Pose Estimation Competition[1] (SPEC) hosted by the European Space Agency and Stanford's Space Rendezvous Lab. The competition however, focused on the accuracy only, with no constraints in terms of computational burden. Therefore, how to bridge domain gap while keeping computational complexity compatible with a deployment onboard typical S/C hardware is an open research question.

In this work we thus investigate Visual Transformers (VT) as a possible answer as they have recently emerged as a promising alternative to Convolutional Neural Networks (CNN).[2]

In the first part of the paper we focus on domain gap only, by proposing an algorithm which leverages adversarial training. It consists of two NNs tasked to perform object detection and to regress a set of landmarks coordinates. The pose is then estimated through a Perspective-n-Point (PnP) solver. This approach, equipped with Swin Transformers, placed fourth and fifth on the sunlamp and lightbox leaderboards

[1] https://kelvins.esa.int/pose-estimation-2021/.

[2] https://github.com/Microsatellites-and-Space-Microsystems/pose_estimation_domain_gap.

of SPEC 2021 reaching 3.20 % (6.24°) and 5.48% (9.69°) normalized position (orientation) errors respectively. Despite the good accuracies achieved, this solution had some limitations which guided subsequent development: the need to access unlabeled real images, which are typically not available prior to the mission, and the high computational burden. In an attempt to overcome both drawbacks, in the second part of the paper a new algorithm is presented: after removing the object detection step, we develop a hybrid model consisting of a CNN equipped with a customized Vision Transformer (ViT) which leverages a stronger data augmentation pipeline to promote domain generalization. We show how this solution is light enough to be executed on a Coral Dev Board Mini 1.5 GHz quadcore CPU at 0.56 fps.

Summarizing, our main contributions are:

- A pose estimation pipeline based on adversarial training and Swin Transformers robust to domain gap.
- A lightweight algorithm leveraging the strength of both CNNs and ViTs towards low power real-time pose estimation across domain gap.

All the models discussed in this work have been developed in TensorFlow Keras.

2 Related Work

2.1 Monocular Pose Estimation in Space

Conventional monocular pose estimation algorithms [1–4] rely on matching handcrafted features extracted from the image to the target's 3D model. These methods often lack scalability to different satellites and showcase poor robustness to image aberrations, which are ubiquitous in space. Recently, NNs have emerged as a promising solution as they can learn complex representations from raw data, such as pixel values, without extensive engineering. Some authors [5, 6] proposed end-to-end solutions by learning a mapping from the images to the pose space. Others [7–10] employed NNs to identify on the image a set of keypoints, defined a priori, while pose estimation is handled by a dedicated solver. As these methods have shown superior accuracies [11], the same approach is adopted in this paper.

2.2 Domain Generalization

We refer to Domain Generalization (DG) as the ability to learn a model which generalizes well to unseen domains. Wang et al. [12] presented a taxonomy of DG. Domain Adversarial Neural Network (DANN) [13, 14] is a popular representation learning method where the feature extractor tries to confuse a discriminator encouraging the learning of domain-invariant features. Data manipulation through augmentation [15] instead promotes generalization without the need to access test images.

In the framework of satellite pose estimation, previous works have been tested on severely limited datasets collected in a laboratory environment [11, 16]. Only recently, Park et al. [17] released SPEED+ [18], the first comprehensive dataset focusing on domain gap in space, and proposed a CNN [19] leveraging augmentations, multi-task learning, and online refinement.

2.3 *Transformers in Vision*

Transformers [20] have dominated the field of Natural Language Processing (NLP) owing to their ability to model long range dependencies in data. Vision Transformers [21] are the first direct application of this architecture to images which are splitted into patches that are treated the same as tokens in NLP. ViTs achieve state-of-the art performances on image classification but require large scale datasets. Nonetheless, Zhang et al. [22] showed how ViTs typically generalize better than CNNs under domain shift, because of their stronger inductive bias towards shapes. This property is crucial, since synthetic and real images are characterized by different visual features. Liu et al. [23] proposed a general purpose transformer denoted as Swin, leveraging shifted windows to limit self-attention. This method not only decreases the computational complexity from quadratic to linear with respect to the image size but also achieves good performances for dense prediction tasks.

3 Pose Estimation Competition and SPEED+

From October 2021 to March 2022, a new edition of the Pose Estimation Competition was held. The contest was based on the SPEED+ [18] dataset, depicting the Tango spacecraft in a variety of poses. More in detail, participants have been provided 59960 synthetic images for training (80%) and validating (20%) their algorithms and two unlabeled test sets of hardware-in-the-loop pictures, denoted as *lightbox* (6740) and *sunlamp* (2791). The former is obtained by emulating realistic Earth albedo while the latter is aimed at simulating direct sunlight. Sample images are reported in Fig. 1.

Submissions were ranked according to an average score E accounting for rotation and normalized position errors, further corrected for machine precision:

$$\bar{e}_{ti} = \frac{\left|\boldsymbol{t}_{BCi} - \hat{\boldsymbol{t}}_{BCi}\right|}{|\boldsymbol{t}_{BCi}|_2},\ e_{ti} = \begin{cases} 0 \ \ if \ \bar{e}_{ti} < 0.002173 \\ \bar{e}_{ti} \qquad else \end{cases}$$

$$\bar{e}_{qi} = 2\arccos\left(|\langle \boldsymbol{q}_i, \hat{\boldsymbol{q}}_i \rangle|\right),\ e_{qi} = \begin{cases} 0 \ \ if \ \bar{e}_{qi} < 0.169^\circ \\ \bar{e}_{qi} \qquad else \end{cases}$$

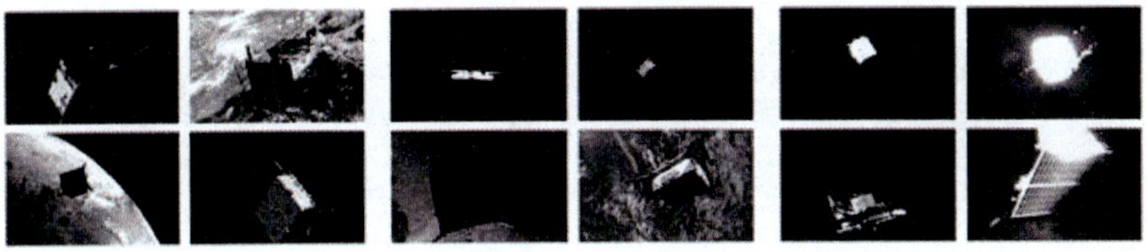

Fig. 1 Sample images from the SPEED+ dataset [18]: synthetic (left), lightbox (center), sunlamp (right) domains. Images have resolution 1200 × 1920 px

$$E = \frac{1}{N}\sum_{i=1}^{N}\left(e_{ti} + e_{qi}\right)$$

where $\boldsymbol{t}_{BCi}$, $\boldsymbol{q}_i$ and $\hat{\boldsymbol{t}}_{BCi}$, $\hat{\boldsymbol{q}}_i$ the ground truth and estimated position vectors and attitude quaternions associated to the i-th image while N is the number of images. The pose is referred to a camera-fixed reference frame.

Two separate leaderboards for lightbox and sunlamp were foreseen. A maximum of two submissions every 24 h applied and the server returned only best results (i.e. if no progresses had been made with respect to the previous submissions, no feedback was returned). At the time of this writing, ground truth labels for the test sets have not been released yet and the scoring server is working in post-mortem mode under the same rules. While this ensures the fairness of the results, it limits the possibility to assess the performance of intermediate processing steps and to benchmark individual solutions. For these reasons, which are out of the control of the authors, the accuracies reported from 5.4 on will be referenced to our best solutions and not to the ground truth, for the purpose of comparing models and variants.

4 Dataset Preprocessing

Both our methods leverage the knowledge of the satellite three dimensional model. This latter had been obtained by manually selecting a set of landmarks on a group of images from SPEED [24] and by exploiting Multiview triangulation.[3] Figure 2 highlights the position of the 11 keypoints which we employed throughout this work. Those have been then projected onto the image plane exploiting the known satellite pose of synthetic images to retrieve 2D landmarks coordinates and ground truth bounding boxes.

[3] https://it.mathworks.com/help/vision/ref/triangulatemultiview.html.

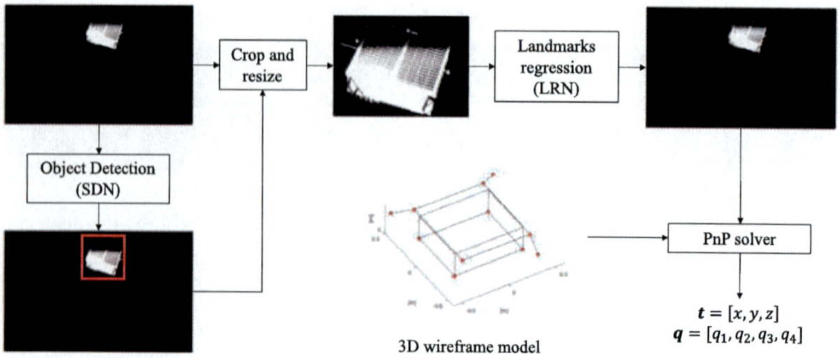

Fig. 2 Three-stage pose estimation pipeline

5 Three Stage Domain Adversarial Approach

The first algorithm we propose leverages adversarial learning. Since this approach has been developed as part of our participation to SPEC 2021, we also did not pose any constraint in terms of computational complexity. In this section we summarize the main steps which contributed to the final accuracy. A comparison with the same pipeline equipped with a CNN of similar size is also provided.

5.1 Architecture and Training Setup

Our first solution consists of a three-stage pose estimation pipeline as depicted in Fig. 2. In this framework, a first NN is trained to identify the satellite on the image plane. The purpose of this step is to reduce the loss of information due to direct image resizing. During inference the predicted bounding box is enlarged by the 15%. A second network is then tasked to regress a set of landmark coordinates on the resulting Region of Interest (RoI). Those are then fed into a Perspective-n-Point (PnP) solver together with the three dimensional model of the satellite, to estimate the pose. To this end we adopted an EPnP [25] with RANSAC, setting the maximum reprojection error to 5 px, followed by a Levenberg–Marquardt (LM) refinement step.

Networks' Structure.
We refer to the two networks as Satellite Detection Network (SDN) and Landmark Regression Network (LRN). These share the same structure which consists of a backbone Encoder (E), a Regression head (R), and a Domain discriminator (D). Since the dataset comprises only one target, object detection is treated as a simple regression task on the bounding box vertices, with no classification.

As a backbone, we adopt Swin Transformers.[4,5] Average pooling is applied to the output of last NN layer to reduce the feature map dimensions. The two heads consist of Fully Connected Networks (FCN) with 3 layers and Gaussian Error Linear Unit (GELU) activation, except for the output where no activation is present.

Adversarial Training.

Adversarial learning is exploited for both SDN and LRN to promote domain generalization, under the same setup. During training, the backbone encourages the emergence of domain-invariant features by trying to fool a discriminator. The latter shall classify the images as either synthetic or real, i.e. with no distinction made between sunlamp and lightbox. To this end, we defined domain discrimination and regression losses respectively as:

$$L_{ADV} = \frac{1}{N_B} \sum_{i=1}^{N_B} H\left(\hat{y}_{Di}, y_{Di}\right)$$

$$L_{REG} = \frac{1}{N_S} \sum_{i=1}^{N_S} mae\left(\hat{\boldsymbol{r}}_i, \boldsymbol{r}_i\right)$$

where H is the cross-entropy between the regressed $\left(\hat{y}_D\right)$ and ground truth (y_D) binary domain labels and N_B is the batch size. Regression is performed on synthetic images only (N_S) through mean absolute error (mae). Vector $\hat{\boldsymbol{r}}$ represents the regressed bounding box and landmarks coordinates for SDN and LRN respectively, normalized between 0 and 1.

The training objectives are:

$$\hat{\theta}_D = \arg\min_{\theta_D}(L_{ADV})$$

$$\hat{\theta}_R = \arg\min_{\theta_R}(L_{REG})$$

$$\hat{\theta}_E = \arg\min_{\theta_E}(L_{REG} - \lambda L_{ADV})$$

The hyperparameter λ is updated at the beginning of each epoch according to a scheduling adapted from [13]:

$$\lambda = \frac{1}{50}\left(\frac{1}{1 + e - \frac{10 \cdot \text{epoch}}{total\ epochs}}\right) - 0.01$$

A schematization of the NNs structure and training regime is depicted in Fig. 3.

[4] https://github.com/shkarupa-alex/tfswin.

[5] https://github.com/rishigami/Swin-Transformer-TF.

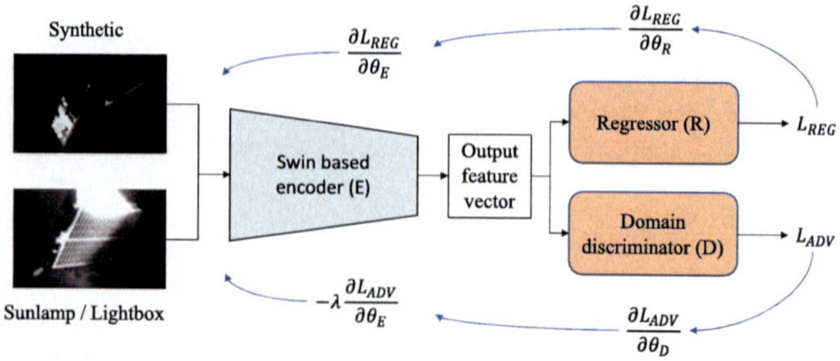

Fig. 3 Schematization of the Swin Transformers based model adopted for SPEC 2021

Common Training Setup and Geometrical Augmentations.
The SDN and LRN for all model variants described in 5.2 have been trained for 40 epochs, with a batch size of 48 and AdamW optimizer with weight decay 1e^{-8}. The learning rate follows a cosine decay scheduling starting from 5e^{-5}. Random rotations are applied as a baseline augmentation.

All trainings are performed with squared images as input. For the SDN, this is achieved simply through resizing. On the other hand, LRN variants are trained on ground truth, squared bounding boxes, which are randomly enlarged (up to 15%) and shifted in both horizontal and vertical directions. To enable adversarial training in this case, we leveraged the predictions of the first SDN model on the real images. This allowed to have a zoom similar to synthetic ones and to apply them the same augmentations.

5.2 *Main Experiments*

We tested backbones of different size, resolutions, and augmentations, while keeping fixed the overall structure of the pose estimation and training algorithms.

In our experiments we leveraged two variants, out of the four available, of Swin Transformers, namely the "Tiny" and "Base" ones. The former is pretrained on the Imagenet dataset while the latter on Imagenet-22 k. Tables 1 and 2 report the architecture details we adopted in our main submissions to SPEC 2021 as well as the accuracy achieved. During the competition, submissions were ranked on a random portion of the test sets.

Solutions A and B share the same SDN, built on top of a Swin Base model at 384 px. Random brightness and contrast are applied during both SDN and LRN training to all images. Enlarging both LRN backbone size and resolution, from 224 (A) to 384 px (B), allowed to cut the pose estimation error by the 33% and 48% on the lightbox and sunlamp domains respectively, confirming that larger Transformers tend to generalize better.

Resolution was further increased to 448 px in variants C and D, where we explored potential benefits brought by pixel-level data augmentations on the synthetic images.

Table 1 Backbones and resolutions employed in our main submissions

Submission code	SDN		LRN	
	Swin model	Image resolution [px]	Swin model	Image resolution [px]
A	Base	384	Tiny	224
B	Base	384	Base	384
C	Base	448	Base	448
D	Base	448	Base	448

Table 2 Performances achieved by our main submissions at SPEC 2021

Submission code	Lightbox			Sunlamp		
	e_t	e_q [rad]	E	e_t	e_q [rad]	E
A	0.1099	0.4895	0.5993	0.0542	0.2740	0.3282
B	0.0964	0.3080	0.4045	0.0438	0.1274	0.1712
C	0.0760	0.2465	0.3226	–	–	–
D	**0.0530**	**0.1673**	**0.2202**	**0.0383**	**0.1206**	**0.1590**

Fig. 4 Comparison between the original image (left) and sunflare augmented (right)

To this end, training images for D were transformed by adding sunflare[6] (offline) (Fig. 4) and by applying random brightness, contrast, gaussian blurring, and gaussian noise, each with 50% probability. Similar to [19] we experienced significant improvements on the lightbox domain with a further 32% error reduction.

Despite the simplicity of our object detection approach, our SDN performed well in terms of Intersection-over-Union (IoU), reaching a satisfactory mean value of 0.9653 on the synthetic validation set even with the simplest variant A. Overall, the number of networks' parameters is about 140 million for variant A and 207 million for the others.

5.3 *Best Performances*

At the end of the competition, all submissions have been re-evaluated on the full test sets. In Table 3 we report the partial sunlamp and lightbox leaderboards in which our method ranked fourth and fifth respectively.

[6] https://albumentations.ai/docs/api_reference/augmentations/transforms/.

Table 3 SPEC 2021 final leaderboards (partial); all values are multiplied by 10^2

Lightbox leaderboard				Sunlamp leaderboard			
Team name	e_t	e_q [rad]	E	Team name	e_t	e_q [rad]	E
TangoUnchained	1.79	5.56	7.35	lava1302	1.13	4.76	5.89
VPU	2.15	7.99	10.14	VPU	1.18	4.93	6.12
lava1302	4.83	11.63	16.46	TangoUnchained	1.50	7.50	9.00
haoranhuang_njust	3.15	14.19	17.34	**u3s_lab**	**3.20**	**10.9**	**14.11**
u3s_lab	**5.48**	**16.92**	**22.40**	haoranhuang_njust	2.84	14.7	17.51
chusunhao	3.28	28.59	31.87	bbnc	8.19	38.3	46.51

Table 4 Comparison of the relative accuracies achieved by our domain-adversarial approach with a CNN and a Swin model having similar size

SDN backbone	LRN backbone	$\tilde{E}_{sunlamp}$	$\tilde{E}_{lightbox}$	Parameters (SDN + LRN)
Swin_Tiny	Swin_Tiny	0.240	0.305	73.95 M
EffNet-B5	EffNet-B5	0.429	0.399	79.09 M

Starting from next section, all scorings will be referred to our best solutions, this is acknowledged by adding a $\sim$ mark.

5.4 Comparison with a CNN Based Pipeline

To demonstrate the effectiveness of vision Transformers at bridging domain gap, we compared the accuracy achieved by our method when equipped with a Swin model and a CNN of similar size. In Table 4 we report the results of an experiment where we built the SDN and LRN on top of Swin Tiny and EfficientNet-B5 [26] backbones. The training setup is the same described in 5.1 and all augmentations have been applied. The input size is set to 224 px for both networks. Since the training batch is small and we train our models on Google's Tensor Processing Units (TPU), in the EfficientNet model we replaced the original Keras BatchNormalization layers with SyncBatchNormalization[7] which synchronizes the global batch statistics across replicas. We also applied a 1×1 convolution to the CNN feature map to reduce the number of output channels to 768, the same of the Swin Tiny model. The result is then fed into an average pooling layer. Both backbones are initialized with Imagenet pretrained weights. The results, reported in Table 4, highlight the effectiveness of VTs over a traditional CNN under this configuration.

[7] www.tensorflow.org/api_docs/python/tf/keras/layers/experimental/SyncBatchNormalization.

Table 5 On device inference runtimes with TFlite converted models

SDN backbone	LRN backbone	GFLOPS[8]	Pipeline fps
Swin_Tiny	Swin_Tiny	9.04	0.15
EffNet-B5	EffNet-B5	9.92	0.22

5.5 On Device Inference

While our domain adversarial approach achieved good accuracy, its computational burden is too high for a low power CPU in both the Swin Tiny and EfficientNet variants. Indeed, when deploying the software on an embedded device such as the 3W Coral Dev Board Mini, the achieved frames per second is not compatible with real-time processing. We report the performances achieved by our models upon TFLite conversion in Table 5. Notably the CNN variant is about the 47% faster.

6 Lightweight Dual Stage Domain Agnostic Approach

In this section we propose a lightweight algorithm based on a hybrid CNN + ViT model that achieves competitive accuracy-computational burden trade off when trained solely on synthetic images.

6.1 Model Structure

The most natural way of reducing both computations and memory footprint is to shrink the image resolution and move from a three-stage pose estimation pipeline to a dual stage one. The scheme is therefore the same reported in Fig. 2, except for the object detection step, which is removed in favor of a direct resizing of the image. Although such an approach implies working with less informative input data, when the target occupies a fairly large portion of the image, as for Tango within SPEED+ [18] dataset (average size ≈650 px), the expected impairment is limited. The model we propose consists of a hybrid architecture between a CNN and a ViT, followed by a Fully Connected (FC) head with three layers and GELU activations, except for the last, performing direct regression of landmarks coordinates. A 0.1 dropout is added before the first and second layer and to the ViT encoder.

The CNN we opted for is an EfficientNetLite-4[9] which is obtained by optimizing the original EfficientNet model for mobile CPUs removing squeeze-and-excitation blocks and replacing swish activations with RELU6. The input size is set to 320×512

[8] The FLOPS account for the NNs only.

[9] https://blog.tensorflow.org/2020/03/higher-accuracy-on-vision-models-with-efficientnet-lite.html.

Table 6 Structure of the ViT encoder

Layers	Hidden size	MLP	Heads
1	192	576	3

Table 7 Accuracy achieved by our CNN + ViT method

Lightbox			Sunlamp		
$\tilde{e}_t$	$\tilde{e}_q$ [rad]	$\tilde{E}$	$\tilde{e}_t$	$\tilde{e}_q$ [rad]	$\tilde{E}$
0.109	0.339	0.448	0.082	0.324	0.406

px, which preserves the original image aspect ratio. The input to the ViT encoder consists of 1×1 patches obtained from the CNN feature map. Even in this case we replaced the standard BatchNormalization layers with SyncBatchNormalization. We also made some adjustments to the Vision Transformer. In particular we did not prepend any learnable "classification token" to the sequence of patches whose state, at the output of the last Transformer encoder, is employed as the image representation in the original paper [21]. We instead apply an average pooling to the encoder output and take the resulting feature vector as the image representation. Table 6 summarizes the details of the Transformer encoder we adopted, which is inspired by ViT-Tiny [27]. We employed a single encoder layer to keep the computational complexity as low as possible.

6.2 Training Setup

The model is trained for 80 epochs with a batch size of 64 and AdamW optimizer. A weight decay of $1e^{-8}$ is applied, while the learning rate follows a cosine decay scheduling starting from $1e^{-4}$. All details related to data augmentations are reported in 6.3. The backbone is initialized from Imagenet pretrained weights.

As in the previous approach, the landmarks coordinates regressed by the network are then fed into a RANSAC EPnP solver followed by a LM refinement step. Table 7 summarizes the accuracies achieved.

6.3 Effects of Data Augmentations

In this context we exploited data augmentations only to enhance domain generalization. The idea is to expose the model to images with different appearances to promote the learning of domain invariant features related to the satellite shape rather than its semblance.

We first divided the train dataset in three even groups. Then, we processed two groups through sunflare addition and style randomization [28]. During training, each

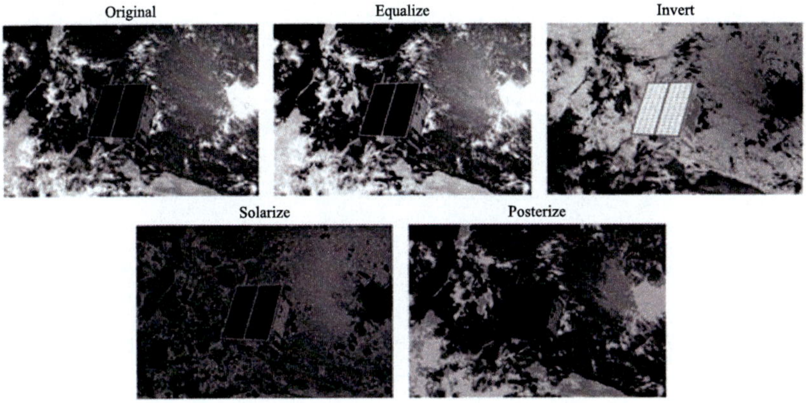

Fig. 5 Visualization of the main augmentations

Table 8 Comparison of the relative accuracy achieved by our CNN + ViT model under different data augmentations setup. Random rotation is always applied

Augmentations applied	Lightbox			Sunlamp		
	$\tilde{e}_t$	$\tilde{e}_q$ [rad]	$\tilde{E}$	$\tilde{e}_t$	$\tilde{e}_q$ [rad]	$\tilde{E}$
Synthetic images with baseline transformations	0.178	0.489	0.667	0.327	0.920	1.247
Addition of sunflare and style randomization (offline)	0.135	0.417	0.552	0.113	0.455	0.568
Addition of equalize, invert, posterize, solarize	0.109	0.339	0.448	0.082	0.324	0.406

input image is randomly rotated and further processed through a data augmentation pipeline which consists of applying one strong transformation among equalize, invert, posterize, and solarize (Fig. 5) and a sequence of baseline transformations. Those include random brightness, contrast, and addition of blurring and gaussian noise, each applied with 50% probability.

Table 8 displays the results of three experiments which highlight the contribution of the main augmentation steps towards the final accuracy.

6.4 *Comparison Between Pure Transformer Based and Hybrid Solution*

To validate our architecture we replaced the NN with a Swin Tiny model equipped with the same regression head. We added a 1×1 convolution to the transformer's feature map to reduce the number of channels to 192, as in our ViT. The results,

Table 9 Comparison with the relative accuracy achieved by pure Transformer based solution

Model	Lightbox			Sunlamp			Parameters
	$\tilde{e}_t$	$\tilde{e}_q$ [rad]	$\tilde{E}$	$\tilde{e}_t$	$\tilde{e}_q$ [rad]	$\tilde{E}$	
SwinTiny + FC head	0.176	0.523	0.699	0.123	0.469	0.592	27.7 M
CNN + ViT	0.109	0.339	0.448	0.082	0.324	0.406	12.5 M

Table 10 CNN + ViT variant details and performances, fps are measured on the Coral Dev Board Mini CPU upon TFLite conversion

GFLOPS	Pipeline fps	Average $\tilde{E}$
4.53	0.56	0.427

reported in Table 9 show how, under this setup, the generalization of the Transformer-only solution is worse than that of our hybrid model. We ascribe this behavior, at least partly, to the lack of adversarial signal which brings forth the data-hungry nature of Transformers, thereby limiting their scalability to small datasets.

6.5 *On Device Inference*

Our domain-agnostic, hybrid (CNN + ViT) approach is about 3.73 × times faster than our first algorithm when equipped with Swin Tiny (Table 5), with a reduction of relative accuracy (averaged between lightbox and sunlamp) from 0.273 to 0.427, showcasing a competitive accuracy-runtime trade-off. This is illustrated in Table 10.

7 Conclusions

In this work we propose two alternative approaches towards robust spacecraft pose estimation across domain gap, leveraging vision Transformers and either adversarial training or data augmentation. A first algorithm exploiting Swin Transformers and adversarial domain adaptation was developed. Tested against SPEED+ [18] dataset, it was found to outperform a state-of-the art CNN of similar size. However, vision Transformers involve operations not well supported by embedded devices, which may undermine their deployment in an operational scenario. In addition, while achieving good accuracy, adversarial training requires images from both domains, which is possible only if one has access to a facility that faithfully reproduces on-orbit conditions.

In the second part of the paper we tackle those limitations by investigating a hybrid architecture free from adversarial training, which leverages the advantages of both

CNNs and vision Transformers: respectively, the ability to learn from limited data and to capture long range correlations. The proposed solution runs at 0.56 fps on the Coral Dev Board Mini CPU, by exploiting a single NN. When compared with our first method equipped with Swin Tiny (Table 4), this translates into a 273% runtime advantage. The latter comes at the expense of a limited 56% average relative accuracy reduction, thus resulting in an improved accuracy-latency trade-off.

References

1. D'Amico, S., Benn, M., Jørgensen, J.L.: Pose estimation of an uncooperative spacecraft from actual space imagery. Int. J. Space Sci. Eng. **2**, 171 (2014). https://doi.org/10.1504/IJSPACESE.2014.060600
2. Liu, C., Hu, W.: Relative pose estimation for cylinder-shaped spacecrafts using single image. IEEE Trans. Aerosp. Electron. Syst. **50**, 3036–3056 (2014). https://doi.org/10.1109/TAES.2014.120757
3. Capuano, V., Kim, K., Hu, J., Harvard, A., Chung, S.-J: Monocular-based pose determination of uncooperative known and unknown space objects. In: 69th International Astronautical Congress (2018)
4. Capuano, V., Alimo, S.R., Ho, A.Q., Chung, S.-J.: Robust features extraction for on-board monocular-based spacecraft pose acquisition. In: AIAA Scitech 2019 Forum, San Diego, California (2019). https://doi.org/10.2514/6.2019-2005
5. Sharma, S., Beierle, C., D'Amico, S.: Pose estimation for non-cooperative spacecraft rendezvous using convolutional neural networks. In: 2018 IEEE Aerospace Conference, pp. 1–12 (2018). https://doi.org/10.1109/AERO.2018.8396425
6. Proenca, P.F., Gao, Y.: Deep Learning for Spacecraft Pose Estimation from Photorealistic Rendering (2019). https://arxiv.org/abs/1907.04298
7. Chen, B., Cao, J., Parra, A., Chin, T.-J.: Satellite pose estimation with deep landmark regression and nonlinear pose refinement. In: 2019 IEEE/CVF International Conference on Computer Vision Workshop (ICCVW), pp. 2816–2824 (2019). https://doi.org/10.1109/ICCVW.2019.00343
8. Park, T.H., Sharma S., D'Amico S.: Towards robust learning based pose estimation of noncooperative spacecraft. In: 2019 AAS/AIAA Astrodynamics Specialist Conference. https://arxiv.org/abs/1909.00392
9. Black, K., Shankar, S., Fonseka, D., Deutsch, J., Dhir, A., Akella, M. R.: Real-time, flight-ready, non-cooperative spacecraft pose estimation using monocular imagery. In: 31st AAS/AIAA Space Flight Mechanics Meeting (2021). https://arxiv.org/abs/2101.09553
10. Hu, Y., Speierer, S., Jakob, W., Fua, P., Salzmann, M.: Wide-depth-range 6D object pose estimation in space. In: 2021 IEEE/CVF Conference on Computer Vision and Pattern Recognition (CVPR), pp. 15865–15874. https://doi.org/10.1109/CVPR46437.2021.01561
11. Kisantal, M., Sharma, S., Park, T.H., Izzo, D., Märtens, M., D'Amico, S.: Satellite pose estimation challenge: dataset, competition design and results. IEEE Trans. Aerosp. Electron. Syst. **56**, 4083–4098 (2020). https://doi.org/10.1109/TAES.2020.2989063
12. Wang, J., Lan, C., Liu, C., Ouyang, Y., Qin, T., Lu, W., Chen, Y., Zeng, W., Yu, P.S.: Generalizing to Unseen Domains: A Survey on Domain Generalization (2022). https://arxiv.org/abs/2103.03097
13. Ganin Y., Lempitsky, V.: Unsupervised domain adaptation by backpropagation. In: 32nd International Conference on Machine Learning, pp. 1180–1189 (2015). https://arxiv.org/abs/1409.7495
14. Ganin, Y., Ustinova, E., Ajakan, H., Germain, P., Larochelle, H., Laviolette, F., Marchand, M., Lempitsky, V.: Domain-adversarial training of neural networks. J. Mach. Learn. Res. **17**, 59:1–59:35 (2016). https://arxiv.org/abs/1505.07818

15. Cubuk, E.D., Zoph, B., Shlens, J., Le, Q.: RandAugment: practical automated data augmentation with a reduced search space. In: Larochelle, H., Ranzato, M., Hadsell, R., Balcan, M.F., and Lin, H. (eds.) Advances in Neural Information Processing Systems, pp. 18613–18624. Curran Associates, Inc. (2020)
16. Pasqualetto Cassinis, L., Menicucci, A., Gill, E., Ahrns, I., Sanchez-Gestido, M.: On-ground validation of a CNN-based monocular pose estimation system for uncooperative spacecraft: Bridging domain shift in rendezvous scenarios. Acta Astronaut. **196**, 123–138 (2022). https://doi.org/10.1016/j.actaastro.2022.04.002
17. Park, T.H., Märtens, M., Lecuyer, G., Izzo, D., D'Amico, S.: SPEED+: Next-Generation Dataset for Spacecraft Pose Estimation across Domain Gap (2021). https://arxiv.org/abs/2110.03101
18. Park, T.H., Märtens, M., Lecuyer, G., Izzo, D., D'Amico, S.: Next Generation Spacecraft Pose Estimation Dataset (SPEED+). Stanford Digital Repository (2021). https://doi.org/10.25740/wv398fc4383
19. Park, T.H., D'Amico, S.: Robust Multi-Task Learning and Online Refinement for Spacecraft Pose Estimation across Domain Gap. https://arxiv.org/abs/2203.04275
20. Vaswani, A., Shazeer, N., Parmar, N., Uszkoreit, J., Jones, L., Gomez, A.N., Kaiser, Ł., Polosukhin, I.: Attention is all you need. In: Guyon, I., Luxburg, U.V., Bengio, S., Wallach, H., Fergus, R., Vishwanathan, S., and Garnett, R. (eds.) Advances in Neural Information Processing Systems. Curran Associates, Inc. (2017)
21. Dosovitskiy, A., Beyer, L., Kolesnikov, A., Weissenborn, D., Zhai, X., Unterthiner, T., Dehghani, M., Minderer, M., Heigold, G., Gelly, S., Uszkoreit, J., Houlsby, N.: An image is worth 16×16 words: transformers for image recognition at scale. In: 9th International Conference on Learning Representations, ICLR 2021, Virtual Event, Austria, May 3–7 (2021)
22. Zhang, C., Zhang, M., Zhang, S., Jin, D., Zhou, Q., Cai, Z., Zhao, H., Liu, X., Liu, Z.: Delving Deep into the Generalization of Vision Transformers under Distribution Shifts (2022). https://arxiv.org/abs/2106.07617
23. Liu, Z., Lin, Y., Cao, Y., Hu, H., Wei, Y., Zhang, Z., Lin, S., Guo, B.: Swin transformer: hierarchical vision transformer using shifted windows. In: 2021 IEEE/CVF International Conference on Computer Vision (ICCV). IEEE, Montreal, QC, Canada, pp. 9992–10002 (2021). https://doi.org/10.1109/ICCV48922.2021.00986
24. Sharma S., D'Amico S.: Pose estimation for non-cooperative rendezvous using neural networks. In: 2019 AAS/AIAA Astrodynamics Specialist Conference, Ka'anapali, Maui, HI. https://arxiv.org/abs/1906.09868
25. Lepetit, V., Moreno-Noguer, F., Fua, P.: EP*n*P: an accurate O(*n*) solution to the P*n*P problem. Int. J. Comput. Vis. **81**, 155–166 (2009). https://doi.org/10.1007/s11263-008-0152-6
26. Tan, M., Le, Q.V.: EfficientNet: Rethinking Model Scaling for Convolutional Neural Networks (2019). https://arxiv.org/abs/1905.11946
27. Steiner, A., Kolesnikov, A., Zhai, X., Wightman, R., Uszkoreit, J., Beyer, L.: How to Train Your ViT? Data, Augmentation, and Regularization in Vision Transformers (2021). https://arxiv.org/abs/2106.10270
28. Jackson, P.T., Atapour-Abarghouei, A., Bonner, S., Breckon, T.P., Obara, B.: Style augmentation: data augmentation via style randomization. In: Proceedings of the IEEE Conference on Computer Vision and Pattern Recognition Workshops, pp. 83–92 (2019)

Detection of Clouds and Cloud Shadows on Sentinel-2 Data Using an Adapted Version of the Cloud-Net Model

Bram Eijgenraam and Simone Mancon

Abstract How to deal with the presence of weather affected satellite data is an unavoidable topic in the processing of optical imagery. Clouds and cloud shadows significantly alter the obtained spectral signatures. To accurately detect these phenomena is a key component for scientific analysis and commercial applications based optical satellites. Most existing operational cloud detection algorithms are so-called rule-based. Their performance is highly variable from one region to another. New deep learning techniques have proven to give significant improvements. Their main strength lies in their capability to adjust the weights to any kind of training set. When one's interest, regarding optical satellite images, is limited to a few geographical locations, a deep learning technique can secure an accurate cloud and cloud shadow detection algorithm. Our research is build on a previous study, where a convolutional neural network (CNN) named 'Cloud-Net' was developed and tested on Landsat data. In our study we have elaborated on this CNN, by analyzing our results on Sentinel-2 data and by applying modifications on the model setup. The results have been compared to the SEN2COR algorithm. It was found that for the detection of clouds the overall CNN accuracy outperforms SEN2COR (95.6 versus 92.0% respectively), and the F1 score was also higher (89.5 versus 78.1%). For the detection of cloud shadows, the modified Cloud-Net model also gave better results regarding overall accuracy (90.4 versus 84.4%) and F1 score (76.1 versus 41.9%).

Keywords Convolutional neural network · Clouds · Cloud shadows · Sentinel-2

B. Eijgenraam (✉) · S. Mancon
Aresys s.r.l. Milano, Milano, Italy
e-mail: abraham.eijgenraam@aresys.it

S. Mancon
e-mail: simone.mancon@aresys.it

C. Ieracitano et al. (eds.), *The Use of Artificial Intelligence for Space Applications*, Studies in Computational Intelligence 1088,
https://doi.org/10.1007/978-3-031-25755-1_21

1 Introduction

The first step in the processing of optical satellite imagery is to remove clouds and cloud shadows from the dataset. This then provides clean data fit for all sort of applications. At Aresys for example optical satellite imagery has been investigated to monitor stockpiles. These stockpiles like coal and iron ores are stored at harbours all over the world and can be tracked by Sentinel-2 data. For the detection of clouds and clouds shadows we used to rely on rule-based algorithms like SEN2COR, MAJA and FMASK [9, 13, 14]. These models use specific intensity thresholds in order to determine to which specific class a pixel belongs to. The rule-based algorithms are optimized to achieve the best overall global performance. This means that from one region to another the results can vary a lot. On the other hand, a CNN is able to adjust its weights to a particular set of geographical features. For most applications one's interest only lie in a limited amount of images, and a CNN can therefore be the solution in these cases. In recent years, deep learning models have shown great potential to outperform these rule-based algorithms [2, 4]. A particular promising research was done by Mohajerani & Saeedi [6], where a convolutional neural network called 'Cloud-Net' was developed. An overall accuracy of 96.48% was achieved. This surpassed the FMASK algorithm, which achieved an accuracy of 94.89% on the same dataset.

In this contribution, we present a new version of the Cloud-Net model, and the results are compared to the SEN2COR algorithm. The first difference is that in our study, the CNN has been used for the training of two binary classification models. One for thick clouds and one for cloud shadows. This instead of only one model for cloud classification only. The second difference is that the training and testing has been done on Sentinel-2 data instead of Landsat, which allows for a higher resolution analysis. Thirdly, different input bands have been used compared to the original model. In the study of Mohajerani & Saeedi [6], only the RGB-NIR band combination was utilized. In this project, two different band combinations are used: RGB-NIR for cloud detection and RGB-NIR-SWIR for cloud shadow detection. Lastly, some main network hyperparameters have been changed like loss function, number of iterations and learning rate.

The goal of our research is to let an adapted version of Cloud-Net locally outperform the SEN2COR algorithm. In our study six different areas have been selected. This also means that any conclusions coming forth apply to these areas only. For future work similar experiments need to be done on different geographical locations to generalize the conclusions.

Table 1 Literature comparison of different overall cloud and cloud shadow detection accuracies on Sentinel-2 data in %

Algorithm	Tarrio et al.	Sanchez et al.	Domnich et al.
FMASK	81.2	90	81
SEN2COR	77.8	79	82
MAJA	82.0	69	71
KappaMask	–	–	91

Table 2 Cloud-Net and FMASK comparison of cloud and cloud shadow detection performances on Landsat data in %

Algorithm	Overall accuracy	Precision	Recall
FMASK	94.89	77.71	97.22
Cloud-Net	96.48	91.23	84.85

2 Related Work

Different researches have been done on the detection of clouds and cloud shadows, with a focus on both rule-based and deep learning techniques. The findings from four different studies are presented in this section, three on Sentinel-2 data [3, 9, 13] and one on Landsat data [6]. Table 1 shows the overall accuracies of cloud and cloud shadow detection models on Sentinel-2 data.

The three studies all include the classical rule-based algorithms, namely FMASK, SEN2COR, and MAJA. In the work of Domnich et al. [3] a new model named KappaMask was proposed, which is a CNN following a U-Net structure. In their research this machine learning model outperformed the classical classification models. Comparing these classical models to each other regarding the three studies, we see a large difference in their overall accuracy scores. This is most likely related to the fact that they were tested on different datasets, and the performance can highly vary from one region to another. Tarrio et al. [13] used a total of 6 Sentinel-2 scenes , distributed across the Eastern Hemisphere. Sanchez et al. [9] investigated 5 areas in the Amazon region, so an approach from a more regional point of view. Domnich et al. [3] selected 21 scenes spread across Northern Europe.

Another U-Net model was proposed in the work of Mohajerani & Saeedi [6]. It was trained and tested on Landsat data and compared to the performance of the FMASK algorithm. Their results are summarised in Table 2.

Also in this study the overall accuracy of the proposed machine learning model outperformed the classical rule-based algorithm. However, regarding precision and recall scores we see a big difference between the two. Cloud-Net scores 13.5 % higher on the precision, while FMASK scores 12.5 % higher on recall. The overall performance of Cloud-Net is higher but one's model preference may differ based on certain performance requirements.

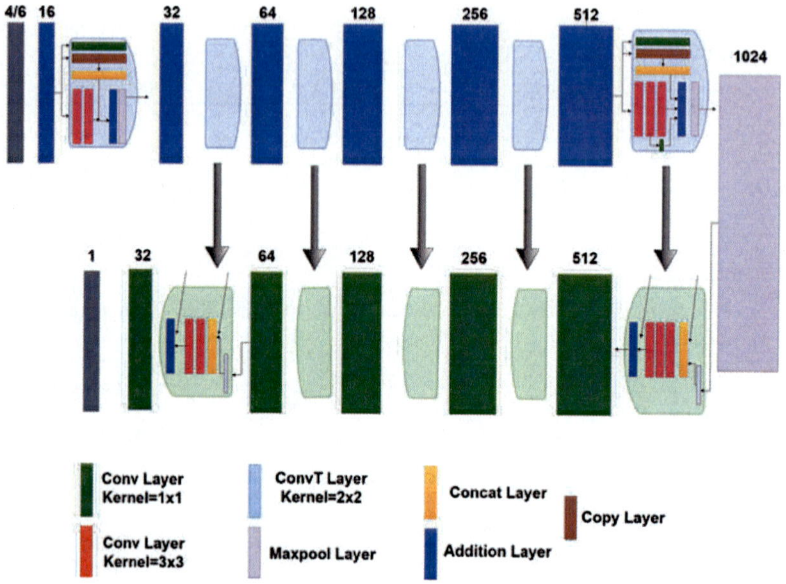

Fig. 1 Architecture of the Cloud-Net model. Expanding arm displayed on top, contracting arm at the bottom. Numbers displayed on top of each block indicate the number of channels corresponding to each hidden layer. The large arrows represent the skip connections

3 Method

In our research we elaborate on the Cloud-Net model by Mohajerani & Saeedi [6]. Cloud-Net is able to perform end-to-end pixel classification on satellite imagery. The CNN follows a so-called U-Net architecture. U-Net has proven to give high performance results [7, 8], and over the years a lot of variations have been proposed for a wide range of applications. One of the advantages of using a U-Net structure is that it works with very few training samples, and it gives high performance for segmentation tasks [1]. The main architecture of the original Cloud-Net model has been preserved. A schematic overview is presented in Fig. 1.

The Cloud-Net model contains a total of 12 hidden layers, with five skip connections. These skip connections help the network to preserve and utilize the learned contexts from the earlier layers. As a result, the network is capable of capturing more features. A drop-out function is also applied on the output of hidden layer 7, the step moving from the expanding arm to the contracting arm. The idea behind this dropout function is to prevent the network from over-fitting. Over-fitting is a common issue in networks that have many hidden layers and therefore a lot of parameters. This technique was successfully applied with several types of neural networks and it shows significant improvements [10]. A 2D dropout function is used, which means that instead of single elements, entire channels are set to zero. This value is kept the

Table 3 Neural network parameter settings

Parameter	Adapted Cloud-Net	Original Cloud-Net
batch size	5	12
image size	256x256	192x192
Loss function	Binary cross entropy loss	Jaccard loss function
Number of iterations	7000	2000
Learning rate	0.0001	0.0001 with decay rate of 0.7
Optimizer	Adam	Adam

same as in the original model, namely 0.15. Therefore 15% of all channels consists entirely of zero values.

Regarding the hyperparameters some changes have been made compared to the work of Mohajerani & Saeedi [6]. Table 3 gives an overview of the used settings.

The batch size, number of iterations and learning rate have been selected by performing a grid-search. The numbers presented in Table 3 are the optimal settings found.

Since we changed from one multi-class to two binary classification models, we changed the loss function to binary cross entropy loss. A smaller batch size compared to the original Cloud-Net model was selected in order to save computation power. As a result a higher number of iterations was needed. A difference with the original proposed learning rate is that we don't use a decay rate. Our motivation for this is that the Adam optimizer already handles learning rate optimization, as is explained by the authors who introduced Adam [5]. The training of the network has been done on a supercomputer named Lisa [11]. The system is installed and maintained by SURFsara. The trainings have been done on a single CPU node. The total of 7000 iterations took 7 hours for the training of the cloud detection model (4 channels) and approximately 9.5 hours for the cloud shadow detection (6 channels).

4 Experimental Setup

4.1 Dataset

The Sentinel-2 images used for training and testing have been manually labelled by us with the "Computer Vision Annotation Tool" (CVAT). A total of 6 locations have been selected: Rotterdam Port (Netherlands), Geonoa Port (Italy), Negev Desert (Israel), Qinhuangdao Port (China), Aoshan Bay (China) and Lanshan Rizhao (China).[1] The

[1] These locations are chosen because they are of interest for the Earth Observation Department of Aresys, a company specialized in Remote Sensing where this project has been carried out. No further details regarding this project will be discussed in this paper.

Table 4 Label distribution of each class in %

	Clouds	Cloud shadows
Training	27	21
Testing	21	21

images have been divided into smaller tiles of 256 by 256 pixels. This gave a total of 236 images. We then made a selection on the images to be used for the training and testing phase. We aimed that the majority of the selected images contained at least more than 20% of one of the two classes. Regarding the clouds, the amount of images used for training and testing is 85 and 20 respectively. For the shadows it is 91 and 22. An overview of the label distribution is presented in Table 4.

4.2 *Band Selection*

In the original version of Cloud-Net four bands were used: The blue, green, red and near infrared (NIR) bands. However, which input bands to use is arbitrary. A high resolution is preferred. Band 2, 3, 4 and 8 (RGB + NIR) are all 10 m resolution bands, which are therefore good to consider. Band 5, 6, and 7 have a 20 m resolution, but are focused on vegetation monitoring. Bands 9 and 10 have a resolution of 60 m, which is considered too low. Bands 11 and 12 are two short wave infrared (SWIR) bands with a resolution of also 20 m, and are therefore also considered to be potentially useful. This gives us a total of 6 bands to consider: B02, B03, B04, B08, B11 and B12.

It is well known that both clouds and cloud shadows are clearly distinguishable in the visible spectrum, therefore the RGB channels will be included in both models. Regarding the NIR band, it is argued in the work of Tan & Tong [12] that cloud regions usually have higher NIR values than that of non-cloud regions. Therefore the B08 band will be included in the cloud detection model. In the literature there was no motivation found to use the SWIR bands for cloud detection. Regarding the cloud shadow detection, in the study of Zhu & Helmer [15] it is stated that for cloud shadows, direct solar radiation is blocked by clouds, so the shadow pixels are illuminated by scattered light. Because the atmospheric scattering is weaker at longer wavelengths, the NIR and SWIR bands of shadow pixels are much darker than surrounding clear pixels. Therefore it is assumed, that the NIR and SWIR bands are useful for the detection of cloud shadows. The final band selections are stated in Table 5.

Table 5 Sentinel-2 band selection

Model	Band selection
Clouds	RGB+NIR
Cloud shadows	RGB+NIR+SWIR

4.3 Performance Evaluation

During the testing phase, the predictions of the trained model are compared to the labelled ground truth values. Four evaluation metrics are used:

$$\begin{aligned}\text{overall accuracy} &= \frac{\text{TP} + \text{TN}}{\text{TP} + \text{TN} + \text{FP} + \text{FN}},\\ \text{precision} &= \frac{\text{TP}}{\text{TP} + \text{FP}},\\ \text{recall} &= \frac{\text{TP}}{\text{TP} + \text{FN}},\\ \text{F1 score} &= \frac{2 * (\text{recall} * \text{precision})}{\text{recall} + \text{precision}}.\end{aligned} \tag{1}$$

Where TP is the number of true positive pixels, TN is the number of true negative pixels, FN is the number of false negative pixels and FP corresponds to the number of false positive pixels. These classes will also be represented in a confusion matrix.

Precision and recall are inversely proportional to each other. This means that there is a performance trade-off between these two metrics. Depending on the application, one's preference can be a classification model with a focus on either high precision or recall. Regarding the cloud and cloud shadow classification, the most common application is to discard all the pixels that are classified as cloud or cloud shadow. A model where the focus is on a high precision score will give you an underestimation of the pixels that need to be detected, but most of the pixels that are detected are correct. When the focus is on a high recall score this will give you the opposite: Most of the pixels that need to be detected are detected, but among the detected pixels a lot of them are wrongly classified.

5 Results

In this paragraph the results of the comparison between our modified Cloud-Net and the SEN2COR algorithm are presented. A summary is given in Table 6. The reference images have been manually labelled by us in CVAT. For the cloud detection, 20 images (256x256 pixels) were used. For the cloud shadow detection we analyzed 22 images. The corresponding confusion matrices are given in Fig. 2.

The results indicate that the modified Cloud-Net model outperforms the SEN2COR algorithm. The overall accuracy, recall and F1 score of the CNN are all significantly higher. Only the precision score of SEN2COR is higher, but the difference is relatively small compared to the difference in recall performance.

In our study the SEN2COR algorithm gives a slightly better precision score (3.1 and 5.7% difference regarding cloud and cloud shadow detection respectively), while the modified Cloud-Net model gives a significantly higher recall score (21.8 and

Table 6 Test results of the modified Cloud-Net model and the SEN2COR algorithm, for the classification of clouds and cloud shadows in %

Classification model	Overall acc.	Precision	Recall	F1 score
Mod. Cloud-Net (clouds)	95.6	90.0	89.1	89.5
SEN2COR (clouds)	92.0	93.1	67.3	78.1
Mod. Cloud-Net (shadows)	90.4	75.9	76.3	76.1
SEN2COR (shadows)	84.4	81.6	28.2	41.9

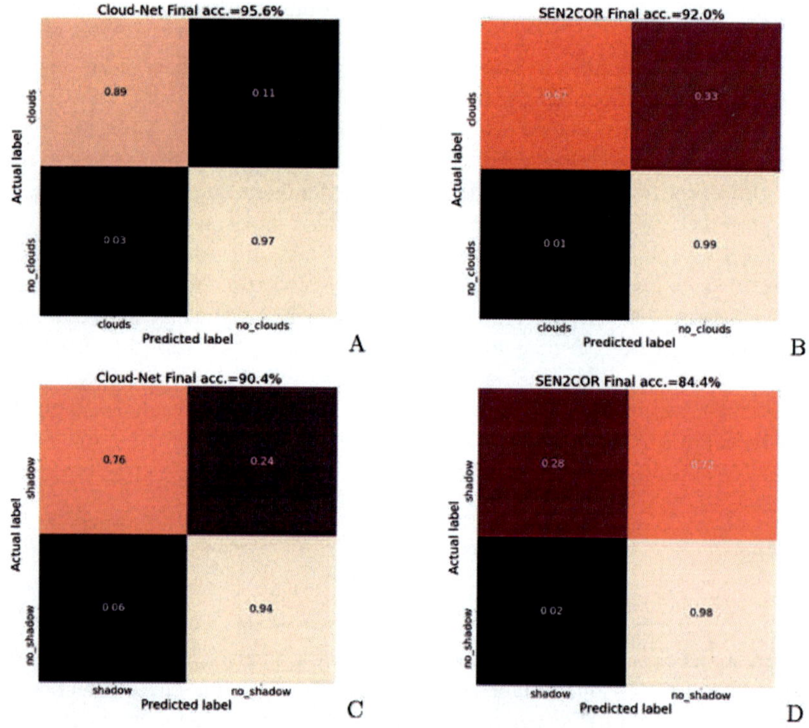

Fig. 2 Confusion matrices for the modified Cloud-Net model and SEN2COR results. Left top: CNN cloud detection. Right top: SEN2COR cloud detection. Left bottom: CNN cloud shadow detection. Right bottom: SEN2COR cloud shadow detection

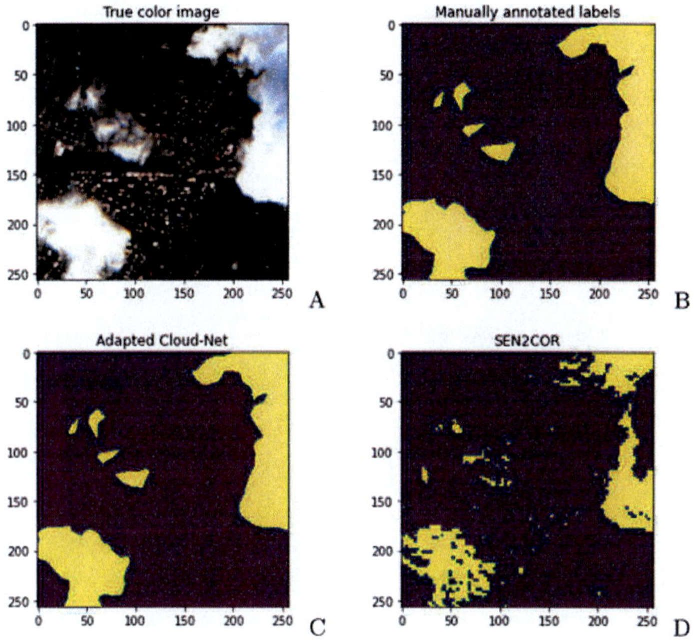

Fig. 3 Thick clouds classification example of one of the testing images. Left top: True color image. Right top: Manually annotated labels (ground truth). Left bottom: Adapted Cloud-Net results. Right bottom: SEN2COR results

48.1% difference). Considering this large difference it can be stated that the higher precision of the SEN2COR algorithm does not outweigh the difference in recall performance. The F1 score gives the harmonic mean between precision and recall. Table 6 tells us that Cloud-Net clearly outperforms SEN2COR on this score.

Figure 3 shows one of the examples where the modified Cloud-Net model clearly performs better compared to SEN2COR, with an overall accuracy of 96.6 versus 90.7% respectively.

It can be seen that the modified Cloud-Net predictions match the manually annotated labels well. The SEN2COR algorithm was able to detect most of the cloud pixels, but some spots were missed. The CNN shows more continuity in the results without having missed individual pixels in between like the SEN2COR algorithm. This is most likely related to the fact that the CNN takes the spatial context into account, while SEN2COR analyzes each pixel individually. Most cloud pixels are surrounded by other cloud pixels. This information is included in the CNN and it is a fundamental difference between the two techniques. Since there is a strong coherence between the individual pixels it gives the CNN a major advantage compared to the rule-based algorithms.

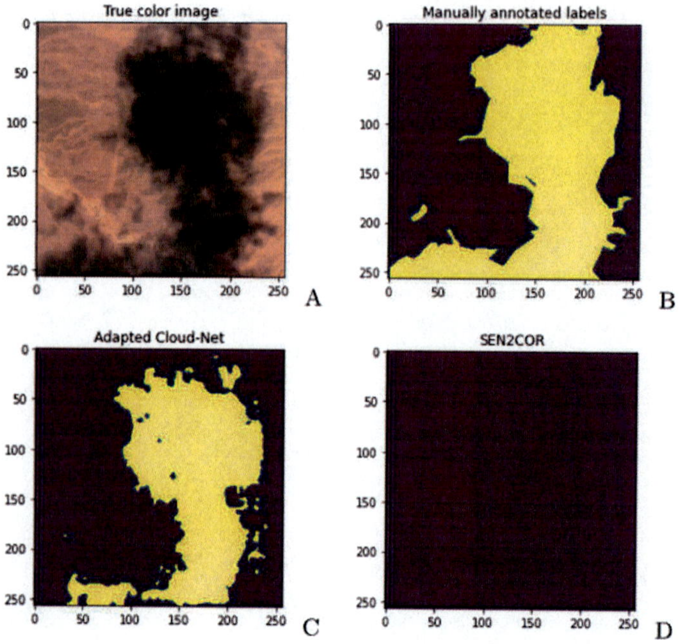

Fig. 4 Cloud shadow classification example of one of the testing images. Left top: True color image. Right top: Manually annotated labels (ground truth). Left bottom: Adapted Cloud-Net results. Right bottom: SEN2COR results

Regarding the cloud shadow detection, we notice a strong performance difference for the SEN2COR algorithm for desert and non-desert areas. In total 9 out of 22 test images are located in the Negev Desert. For many of these images SEN2COR was not able to capture any cloud shadow pixels. An example is provided in Fig. 4.

The CNN is able to capture most cloud shadow pixels, although there are some spots the model has missed. It is striking to see that the SEN2COR results don't return any cloud shadow labels. From this we derive another major advantage that a CNN provides compared to a rule-based algorithm: The ability to locally adapt to particular situations. The SEN2COR parameters are optimized in order to give the best global results. This means that the performance can vary from one region to another. This may result in very inaccurate predictions, as can be seen in Fig. 4. It is not very convenient to dive into the source code of SEN2COR and tweak the threshold settings, in order to make it more accurate. The CNN is more flexible in order to adapt from one region to another. Once the model setup is completed, you can change the training data to make your model fit any area of interest.

6 Conclusion

Our adapted version of the Cloud-Net model is able to outperform the SEN2COR algorithm on the 6 regions that were selected for training and testing. Regarding the cloud detection, the overall accuracy score of the modified Cloud-Net was higher (95.6 versus 92.0%) and also the F1 score was better (89.5 versus 78.1%). A major difference between the two models is that the SEN2COR algorithm missed individual pixels or small clusters of pixels within a cloud area. The CNN did not miss these individual pixels and the predictions matched the manually annotated labels better. The cloud shadow detection results of the modified Cloud-Net model also outperformed SEN2COR. The overall accuracy was better (90.4 versus 84.4%) and the F1 score was better (76.1 versus 41.9%). The SEN2COR algorithm missed the vast majority of the cloud shadow pixels on the Negev Desert images. The modified Cloud-Net model detected detected most of the shadow pixels, which supports the hypothesis that a CNN is able to locally adapt to specific geographical features.

References

1. Alom, M., Hasan, M., Yakopcic, C., Taha, T., Asari, V.: Recurrent Residual Convolutional Neural Network based on U-Net (R2U-Net) for Medical Image Segmentation. CoRR,abs/1802.06955. (2018)
2. Bai, T., Li, D., Sun, K., Chen, Y., Li, W.: Cloud detection for high-resolution satellite imagery using machine learning and multi-feature fusion. Remote Sens. **8** (2016)
3. Domnich, M., Sünter, I., Trofimov, H., Wold, O., Harun, F., Kostiukhin, A., Järveoja, M., Veske, M., Tamm, T., Voormansik, K., Olesk, A., Cadau, E., Piciarelli, C., Lee, H., Eum, S., Longépé, N., Boccia, V.: KappaMask: AI-based cloudmask processor for sentinel-2. Remote Sens. **13** (2021)
4. Ghasemian, N., Akhoondzadeh, M.: Introducing two Random Forest based methods for cloud detection in remote sensing images. Adv. Space Res. **62** (2018)
5. Kingma, D.P., Ba, J.: Adam: A method for stochastic optimization. Int. Conf. Learn. Represent. (2014)
6. Mohajerani, S., Parvaneh, S.: Cloud-Net: An end-to-end cloud detection algorithm for landsat 8 imagery. In: IGARSS 2019–2019 IEEE International Geoscience and Remote Sensing Symposium, pp. 1029–1032. Yokohama, Japan (2019)
7. Pan, Z., Xu, J., Guo, Y., Hu, Y., Wang, G.: Deep learning segmentation and classification for urban village using a worldview satellite image based on U-Net. Remote Sens. **12** (2020)
8. Ronneberger, O., Fischer, P., Brox, T.: U-Net: Convolutional networks for biomedical image segmentation. In: LNCS, p. 9351 (2015)
9. Sanchez, A.H., Picoli, M., Câmara, G., Andrade, P., Chaves, M., Lechler, S., Soares, A., Marujo, R., Simoes, R., Ferreira, K., Queiroz, G.: Remote sensing comparison of cloud cover detection algorithms on sentinel-2 images of the amazon tropical forest. Remote Sens. **12** (2020)
10. Srivastava, N., Hinton, G., Krizhevsky, A., Sutskever, I., Salakhutdinov, R.: Dropout: A simple way to prevent neural networks from overfitting. J. Mach. Learn. Res. **15**, 1929–1958 (2014)
11. SURFsara. Lisa cluster computer (2021). Retrieved from: https://userinfo.surfsara.nl/systems/lisa
12. Tan, K., Tong, X.: Cloud extraction from chinese high resolution satellite imagery by probabilistic latent semantic analysis and object-based machine learning. Remote Sens. **8** (2016)

13. Tarrio, K., Tang, X., Masek, J.G., Claverie, M., Ju, J., Qiu, S., Zhu, Z., Woodcock, C.E.: Comparison of cloud detection algorithms for Sentinel-2 imagery. Sci. Remote Sens. **2** (2020)
14. Zekoll, V., Main-Knorn, M., Alonso, K., Louis, J., Frantz, D., Richter, R., Pflug, B.: Comparison of masking algorithms for sentinel-2 imagery. Remote Sens. **13** (2021)
15. Zhu, X., Helmer, E.: An automatic method for screening clouds and cloud shadows in optical satellite image time series in cloudy regions. Remote Sens. Environ. **214** (2018)

PRISMA Hyperspectral Image Segmentation with U-Net Convolutional Neural Network Using Singular Value Decomposition for Mapping Mining Areas: Preliminary Results

Andrea Dosi, Michele Pesce, Anna Di Nardo, Vincenzo Pafundi, Michele Delli Veneri, Rita Chirico, Lorenzo Ammirati, Nicola Mondillo, and Giuseppe Longo

Abstract This work is focused on a deep learning model–U-Net convolutional neural network–with the purpose of segmenting relevant imagery classes, for detecting mining areas using hyperspectral images of the PRISMA Earth Observation mission, funded by the Italian Space Agency (ASI). To avoid the typical problem of hyperspectral data redundancy and to improve the computational performances without losing accuracy, the Singular Value Decomposition (SVD) is applied to the hyperspectral *data cube*, taking only the first three singular values, thus projecting the

Project carried out using PRISMA Products, © of the Italian Space Agency (ASI), delivered under an ASI License to use.

A. Dosi (✉) · M. Pesce · A. Di Nardo · V. Pafundi · G. Longo
University of Naples Federico II, Department of Physics E. Pancini and University of Naples Federico II, Naples, Italy
e-mail: andrea.dosi@unina.it

M. Pesce
e-mail: michele1.pesce@telecomitalia.it

M. Delli Veneri
Department of Electrical Engineering and Information Technology (DIETI), Naples, Italy
e-mail: michele.delliveneri@unina.it

R. Chirico · L. Ammirati · N. Mondillo
University of Naples Federico II, Department of Earth, Environmental and Resources Sciences, Naples, Italy
e-mail: rita.chirico@unina.it

L. Ammirati
e-mail: lorenzo.ammirati@unina.it

N. Mondillo
e-mail: nicola.mondillo@unina.it

A. Dosi
Accenture Technology Solutions S.r.l, Industry X Area, Milano, Italy

C. Ieracitano et al. (eds.), *The Use of Artificial Intelligence for Space Applications*, Studies in Computational Intelligence 1088,
https://doi.org/10.1007/978-3-031-25755-1_22

multi-dimensional *data cube* to a three channels image. The method is applied to a PRISMA surface reflectance scene of South-West Sardinia, one of the oldest mining districts in the world. The Quadrilátero Ferrífero mining district (Minas Gerais, Brazil) will also be analyzed to test the transferability of the model to other mining areas worldwide.

Keywords Deep learning · Remote sensing · Hyperspectral imaging · PRISMA

1 Introduction

Hyperspectral imaging (HSI) is a widely used technique in remote sensing applications since hyperspectral sensors can rapidly collect images of the Earth's surface, providing compositional information about superficial materials in many, narrow and contiguous spectral bands. HSI has found application in Earth observation and environmental monitoring due to the wide range of spatial and spectral information that can be obtained. Surface-exposed materials show spectral absorption features over the Visible Near Infrared (VNIR) and the Shortwave Infrared (SWIR) regions of the electromagnetic spectrum (400–2500 nm), which allow to accurately recognize and map them over a specific area of interest. Supervised semantic segmentation of hyperspectral images can prove very challenging due to the high dimensionality of the data, redundant information, and limited availability of training samples.

Several works using satellite hyperspectral images focused on Deep Learning-based architecture have been recently developed [8]. More specifically, HSI and Deep Learning methods have been used for geological remote sensing applications, focusing on mineral detection and mapping for exploration purposes, as well as soil composition evaluation [14].

In the present work, a Deep Learning pipeline capable of performing automatic segmentation of PRISMA hyperspectral satellite cubes is created for the purpose of mapping both the active and abandoned mining areas. The automatic detection of mining areas within large areas may provide valuable information useful in planning reclamation and restoration activities. PRISMA, a space-borne hyperspectral sensor developed by the Italian Space Agency (ASI), can capture images in a continuum of 240 spectral bands ranging between 400 and 2500 nm, at a spatial resolution of 30 m [9]. The method was applied to a PRISMA Bottom-Of-Atmosphere (BOA) reflectance scene of the South-West Sardinia and The Quadrilátero Ferrífero area (Minas Gerias State, Brazil). The SW Sardinia mining district (see Fig. 1a) was chosen since it is one of the oldest mining districts in the world, known since pre-Roman times for the exploitation of lead-silver-copper-zinc and barium deposits [2, 5].

The Quadrilátero Ferrífero area, instead, is located in the central portion of the Minas Gerais State (northern Brazil) (see Fig. 1b). The area comprises several iron ore deposits (e.g., the Córrego do Feijão mine, the West-, Central-, and East-Mine of the Usiminas mining complex, the Esperança, Jangada, and the now-exhausted Águas Claras ore deposits) [3], and is well known because of the collapse of the tailing "Dam B1" of the Córrego do Feijão Mine (Brumadinho), a large socio-environmental flood-disaster occurred in January 2019, recently studied [1] by using satellite multispectral images.

2 The Main Study Area

The study area is in the historic Italian mining district located in the southwestern part of Sardinia, within the Carbonia-Iglesias province, known as "Sulcis carboniferous basin" (see Fig. 1a). The Sulcis mining activity started around 1850 in Bacu Abis municipality where the coal outcroppings were located. Thereafter considering the economic necessities and the discovery of new techniques, the underground mining activity started. Nowadays, in the south-Sardinia there are more than fifty mining sites that characterize the territory. Geologically the area is in Sulcis basin characterized by the extensive orogenic volcanic sequences referable to the Oligo-Miocene magma cycle [12] and below the continental environment deposits on a Paleozoic metamorphic basement. The sedimentary rocks occurred in this area from the Paleozoic to the Quaternary. The basement is featured by Paleozoic and Mesozoic rocks, which crop out on the eastern and northern borders of the Sulcis basin [15]. The sedimentary succession filling the basin, uncomfortably covering the basement, can be subdivided from the bottom to the top, into four Formations: Calcari a Macro-

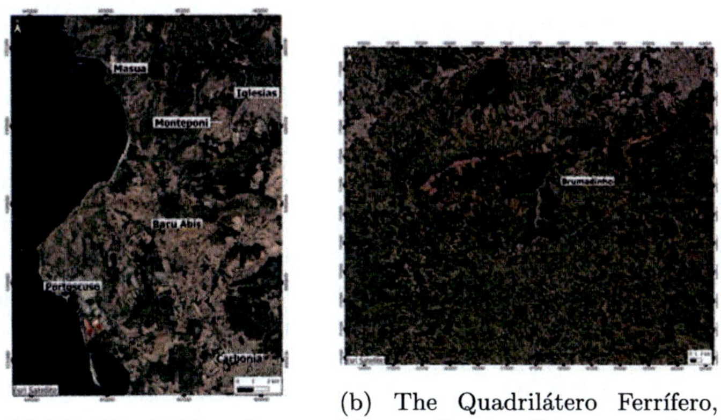

(a) The South West Sardinia mining district

(b) The Quadrilátero Ferrífero, Minas Gerais State, northern Brazil

Fig. 1 Mining area

foraminiferi Fm. (limestones); Miliolitico Fm., consisting of sandstones, marls and limestones (20–70 m thick); Lignitifero Fm., an association, 70 to 150 m thick, of clays, marly limestones, bituminous limestones, marls and conglomerates, interbedded with coal seams, and the Cixerri Fm., consisting of sandstones, conglomerates and marls (average thickness = 300 m) [15]. The Lignitifero Fm. coal seams have a thickness comprised between 1 and 10 m and consist of pure coal layers commonly 10 cm thick, rarely reaching 30–50 cm of thickness, interbedded with clays. In the Seruci-Nuraxi Figus area, the mined horizon is located at 350–450 m below the surface. Sedimentary rocks are covered by 100–300 m of volcano-pyroclastic rocks and ignimbrites [6, 15].

3 Pre-Processing

The method was applied to PRISMA Bottom-Of-Atmosphere (BOA) reflectance scenes ("PRS_L2C_STD") of the South-West Sardinia area (acquired at 10:25:24 on 15 July 2021–UTC) and of the Quadrilátero Ferrífero mining district (acquired at 13:05:52 on 21 May 2022–UTC). Level 2C PRISMA hyperspectral imagery was accessed from the mission website http://prisma.asi.it/. The VNIR and SWIR datacubes were pre-processed using tools available in ENVI 5.6.1 (L3Harris Technologies, USA). Errors in absolute geolocation were corrected using the Refine RPCs (HDF-EOS5) task, included within the PRISMA Toolkit in ENVI.

The PRISMA hyperspectral bands greatly affected by atmospheric absorption were excluded from further processing. The final data cube includes 181 bands out of 234.

To improve accuracy and validate the results, two different land cover data, respectively for the Sardinia area and Brazil area, were used to build the ground truth.

Regarding the Sardinia area, land cover data were downloaded from *Corine Land Cover (CLC) 2006*, Version 2020_2021 datasets (European Environment Agency–EEA) produced within the frame *Copernicus Land Monitoring Service* (https://land.copernicus.eu/pan-european/corine-land-cover/clc-2006).

This ground truth land cover was used as labelling for the training and validation set.

However the *Esa WorldCover–Map* (https://esa-worldcover.org/en) together with the *Global-scale mining polygons* (Version 1) [11] were used for the Brazil area (Quadrilatero Ferrífero mining district) to test the accuracy and the Intersection over Union (IoU) of the transfer learning. To avoid redundant and noisy information, only the vegetation (including dense, sparse and very sparse vegetation together with crops), water areas (sea, rivers and lakes), mining areas and background (all the remaining including urban areas) areas were selected. Then, concerning only the Sardinia area, the size of the image was reduced: redundant sea areas and San Pietro Island (that does not contain mining areas) were cut off (see Fig. 2).

Cutting off the redundant information has the advantage to reduce the noise, but on the other hand it has the disadvantage of a further reduction of the training and

(a) Full image with relevant areas (b) Reduced image with relevant areas

Fig. 2 Images with relevant areas

validation dataset (70% of the initial number of pixels) used by the deep learning model. This problem will be overcome by applying *ad hoc* data augmentation as explained in the next sections.

4 Methodology

In this section the methodology used to map the mine sites in the SW Sardinia area is briefly explained.

4.1 Singular Value Decomposition

To deal with the problem of redundant information in the hyperspectral *data cube*, a feature extraction method based on Singular Value Decomposition (SVD) is proposed [17, 20]. To best represent the HSI, the first three singular values were selected for the following reasons: first of all the *data cube* explained variance decrease as the number of singular components increase. In particular, the first three singular values retain more than 99% of the total variance (see Fig. 3a). Secondly, by plotting the mean pixels information per class over singular values, a general decreasing and then flattening trend can be appreciated approaching the third singular value. (see Fig. 3b). For the previous reasons the first three singular values were then selected as feature component of the original data. This component effectively disentangles vegetation, sea and mining areas distributions while beginning from the third component, the distributions become overlapped, degrading the information (see Fig. 4). Consequently, taking only the first three singular values, the 3-dimensional multichannel *data cube* was reduced to a 3-channels image.

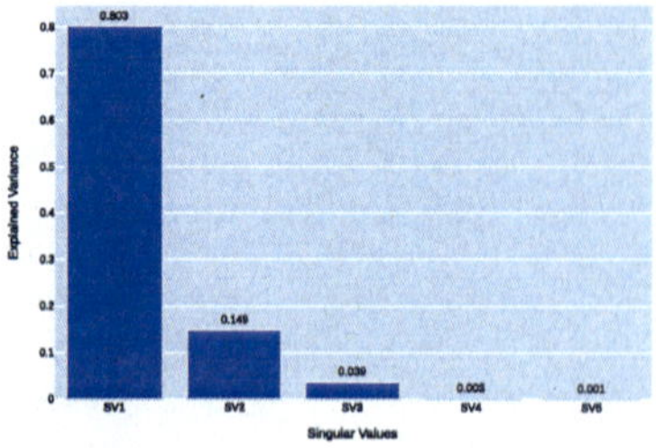

(a) Explained Variance over SV

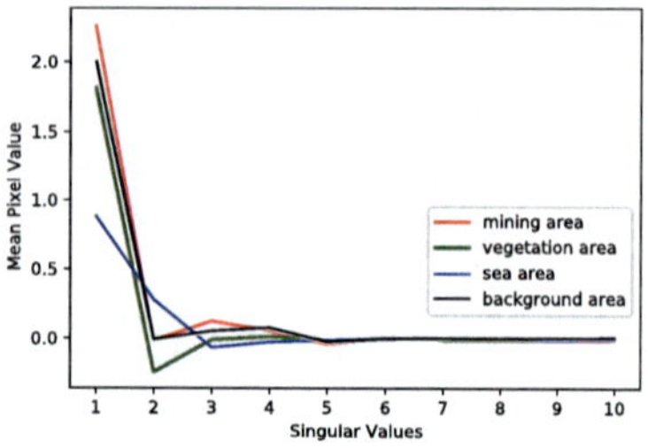

(b) Classes Information Contribution

Fig. 3 Singular Value Decomposition graphs

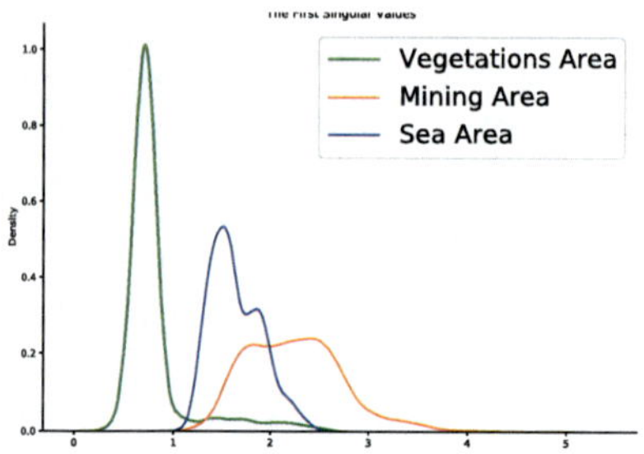

(a) The distributions of the first Singular Value

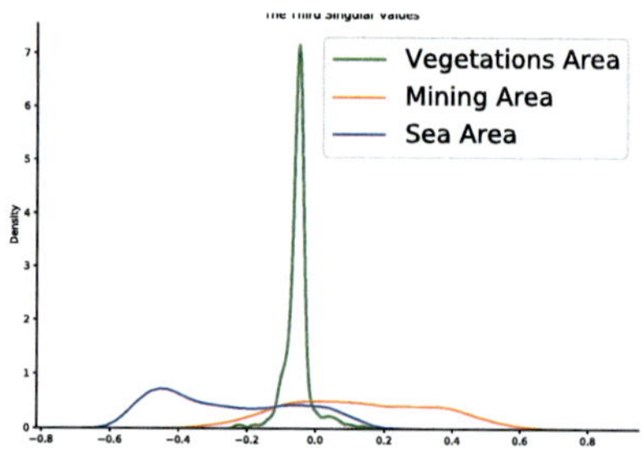

(b) The distributions of the third Singular Value

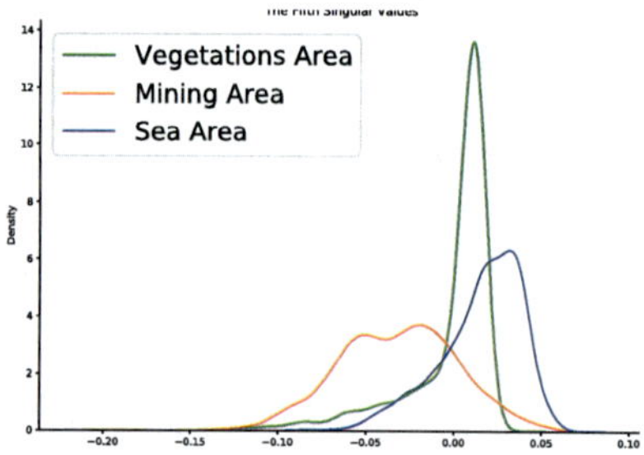

(c) The distributions of the fifth Singular Value

Fig. 4 Singular Value Decomposition distribution per class

4.2 *Data Augmentation*

The image resulting data cube was then normalized between 0 and 1 and divided into overlapping patches of 128 × 128 pixels. The overall number of patches is 243. As previously mentioned, after performing SVD, the entire *data cube* has been reduced to a 3-channels 2-dimensional image containing most of the information.

To increase the size of the dataset, to reduce the risk of overfitting and to deal with the imbalance dataset (see Table 1: the percentage of patches with mining pixels is 21, 5% circa) an *ad hoc* data augmentation was performed through the following steps: firstly, a random *rotation, translation* and *flipping* [4, 19] were applied on the full dataset. Secondly, to increase the number of patches with mining pixels random *rotation, translation* and *flipping*, together with the addition of a *gaussian random noise* were performed, taking only the patches with mining pixels. Finally, to further increase the number of patches, a third random *rotation, translation* and *flipping* were done again only on the mining pixels. After the third step of data augmentation, the number of patches increases to 5226 and the percentage of patches with at least one pixel belonging to the mining areas increases to 77, 6% (see Table 2).

4.3 Model Selection: U-Net Convolutional Network

Semantic segmentation of the three channels image was performed using a U-Net convolutional neural network architecture [16]. This architecture consists of a contracting path and an expansive path, which gives it the U-shape. The contracting path (or *encoding*) is a typical convolutional network that consists of repeated application of convolutions, each followed by a rectified linear unit (ReLU) and a max pooling operation. During the contraction, the spatial information is reduced while feature information is increased. The expansive pathway (or *decoding*) combines the feature and spatial information through a sequence of up-convolutions and concatenations with high-resolution features from the contracting path (see Fig. 5). This particular

Table 1 Before the data augmentation: percentage of patches with at least one pixel belongs to the following four classes

Class	Percentage of patches
Background	79.972
Water	79.561
Vegetation	92.249
Mines	21.536

Table 2 After the data augmentation : percentage of patches with at least one pixel belongs to the following four classes

Class	Percentage of patches
Background	82.357
Water	79.257
Vegetation	97.837
Mines	77.573

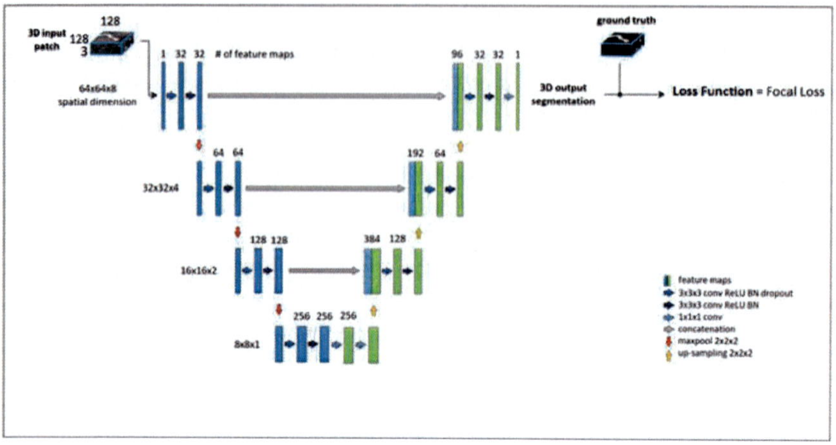

Fig. 5 The U-Net Architecture

CNN can be trained end-to-end from very few images, thus allowing to avoid the problem of the small number of the annotated hyperspectral datasets [18].

The adopted model is able to detect placer mining disturbance from high-resolution multispectral imagery minimizing misclassification errors [10]. It is also effective in environmental mapping as the model architecture enables the learning of the textures and of other spatially discriminative features [7].

5 Experimental Preliminary Results

5.1 Sardinia Area Results

First Experiment: The "Classic" Training and Validation Split The resulting Bok was then split in a *training set* (80% of the Bok) and a *validation set* (20% of the Bok). The model was then trained for a total of 200 epochs. Using the Google Colab Standard GPU, the neural network was trained in approximately 40 minutes. In order to deal with the segmentation problem, the focal loss was used as loss a function. The model achieved a validation accuracy of more than 99%. The validation loss function is approximately zero (see Fig. 6) and the Mean IoU is more than 96% (see Table 1). Table 3 shows the resulting IoU for each class calculated on the validation set.

Figure 7 shows the ground truth of the Sardinia area of interest and the prediction on the corresponding full Sardinia image (both training and validation set).

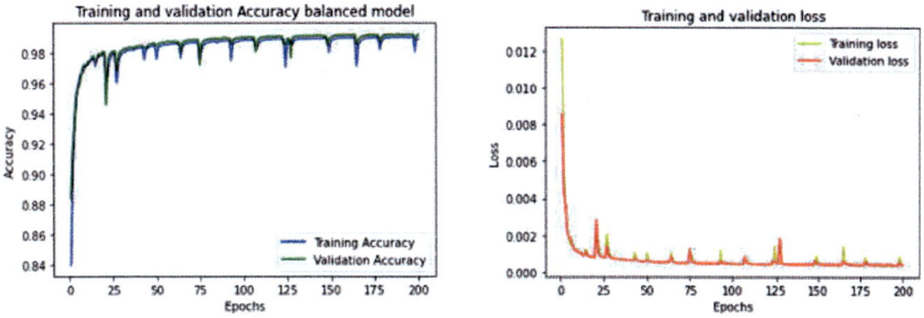

Fig. 6 Training and validation Accuracy and Loss Function

Table 3 IoU obtained for each class

Class	IoU
Background	0.938
Water	0.991
Vegetation	0.987
Mines	0.955

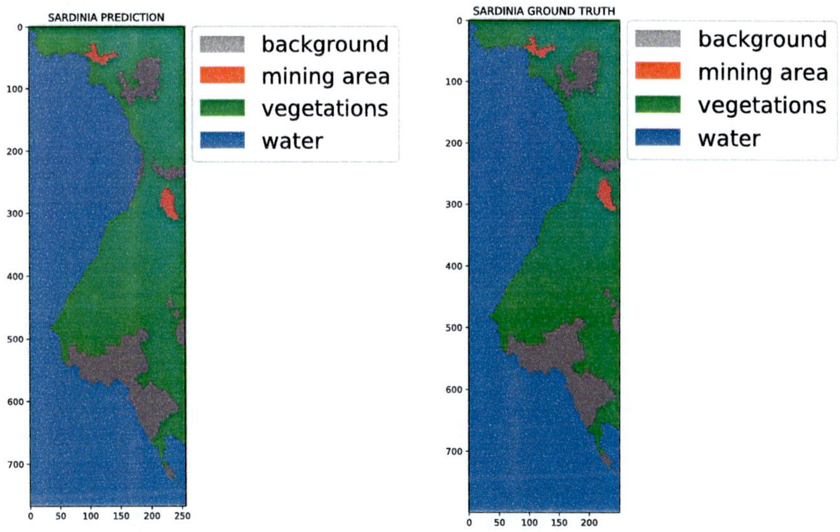

Fig. 7 The Sardinia ground truth and the corresponding prediction

Table 4 One versus Rest Confusion Matrix

	Positive	Negative
Positive	0.970 ± 0.0257	0.00113 ± 0.000257
Negative	0.0444 ± 0.0147	0.998 ± 0.000806

Table 5 One versus Rest Precision and Recall

Metrics	Results
Precision	0.956 ± 0.0253
Recall	0.998 ± 0.0272

Second Experiment: k-fold Validation and Callback To further reduce the risk of overfitting and to improve the time performance of the deep learning model, a k-fold validation (taking k = 5) and a callback with patience = 5 (that is, stop the training when the loss function does not improve for 5 consecutive epochs anymore, and save the best result of the training) were implemented. Different loss function was also used: the categorical cross entropy fits well to multiclass semantic segmentation problems. The overall time of the $k = 5$ different training was approximately 22 minutes. The use of the callback, therefore, improves the speed of training almost twice, despite five different training were made.

To evaluate the performance of this model to correctly map the mining area, the *One versus Rest* metric was used. More in detail, the confusion matrix, precision, recall were presented in the form of *mean* ± *standard deviation* of the k training results (see Tables 4 and 5).

5.2 *Brazil Area Results*

In order to test the generality and transferability of the model, the Quadrilátero Ferrífero mining district (Minas Gerais, Brasil) was also analyzed. As previously mentioned, the *Esa WorldCover–Map* together with *Global-scale mining polygons* (Version 1) were used to build the Brazil ground truth.

To avoid the introduction of bias, the SVD and the normalization previously fitted on the Sardinia dataset were used and the atmosphere absorption bands were excluded. The Brazil image was divided into non-overlapped patches of 128×128 pixels. The two different IoUs, corresponding to the predictions of the two models previously trained on the Sardinia dataset, are shown in Table 6. The two images predicted by the two experiments compared with the ground truth are in Fig. 8.

After that, some considerations can be made: firstly, the very low mining and background IoU can be explained by the low resolution of the *Corine Land Cover (CLC)* used to build the mask useful to train the model, with respect to the images

Table 6 First and second experiment: IoU for each class

Class	First exp IoU	Second exp IoU
Background	0.138	0.0971
Water	0.558	0.465
Vegetation	0.912	0.897
Mines	0.0576	0.0447

Fig. 8 PRISMA ground truth of the Quadrilátero Ferrífero Area and the corresponding first and second experiment prediction

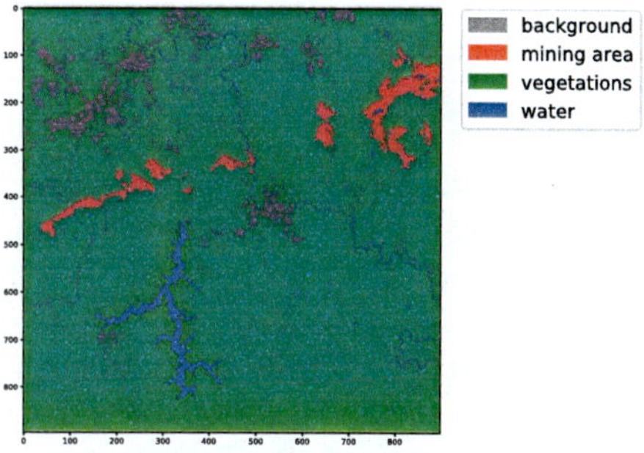

(a) The Brazilian Ground Truth Image

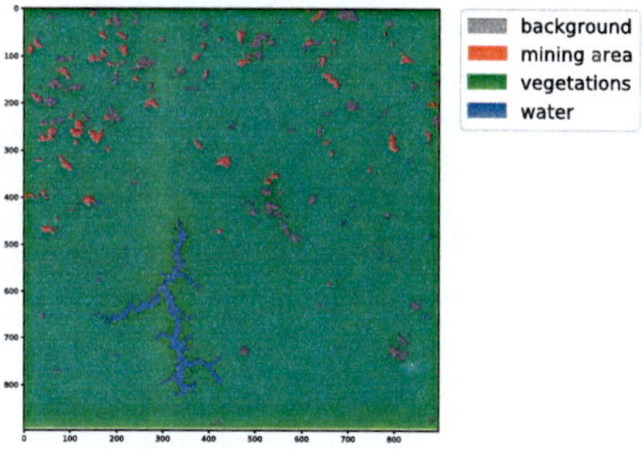

(b) The Brazilian First Experiment Prediction

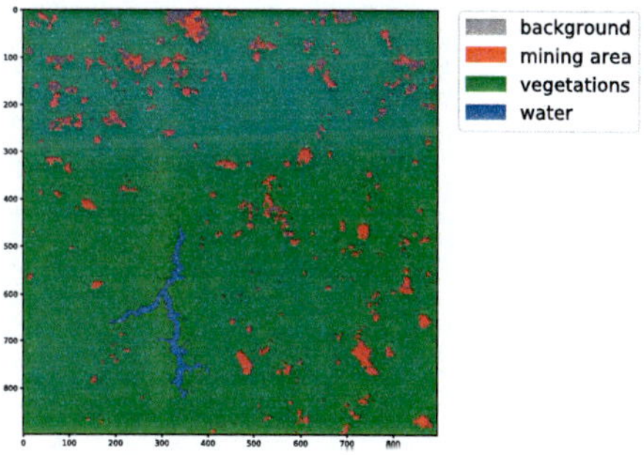

(c) The Brazilian Second Experiment Prediction

acquired by *PRISMA*. Secondly, the Quadrilátero Ferrífero mining district has a different geological setting and material compositions compared to the SW-Sardinia area. The experiments are then able to recognize the background area and the mining area only partially.

6 Discussion and Conclusions

The preliminary results reveal that the proposed model exhibits competitive advantages with regards to state-of-the-art HSI semantic classification methods, demonstrating higher computational performance (relatively few training epochs and competitive run time compared with 3d convolutional neural network [13]) and good accuracy when dealing with limited amounts of training data. This makes the proposed approach a powerful tool for the analysis of large and complex hyperspectral datasets with a good generalization attitude. Two large surface-exposed mining areas of SW-Sardinia (i.e., Masua and Monteponi mines; (see Fig. 1a) were accurately identified by applying the SVD method to the PRISMA image. There are, however, also some unsolved challenges in recognizing minor mines, mainly because of their smaller extension compared to the PRISMA 30m-spatial resolution and, since they are abandoned mining areas, are now covered by vegetation. Even if the Quadrilátero Ferrífero mining district has a different geological setting and material compositions compared to the SW-Sardinia area, the method is able to recognize larger mines occurring in the area (even if only partially). Moreover, the *Corine Land Cover (CLC)* used to build the mapping of the different classes, seems to bring underestimated predictions due to the low resolution of the land cover. So, the integration of the SVD with the U-Net convolutional neural network method can be transferred to other environments worldwide but both more spectral data and a more accurate land cover are needed to retrain the model, improving the predictions accuracy and the reliable discrimination of different mines in other mining areas. As evidence of this, the water bodies and the full range of vegetated areas were identified in both the investigated areas (see Figs. 7 and 8). The preliminary results show that the SVD approach using PRISMA hyperspectral images represents a feasible tool for mapping mining areas. Moreover, it offers the potential for obtaining significant information potentially useful for monitoring environments strongly affected by intense mining activity and planning post activities management. We plan to further improve the method by combining the SVD approach with the extraction of specific spectral features for mineral exploration purposes.

References

1. Ammirati, L., Chirico, R., Di Martire, D., Mondillo, N.: Application of multispectral remote sensing for mapping flood-affected zones in the brumadinho mining district (minas gerais, brasil). Remote Sens. **14**(6) (2022). https://doi.org/10.3390/rs14061501, https://www.mdpi.com/2072-4292/14/6/1501
2. Boni, M., Balassone, G., Iannace, A.: Base metal ores in the lower paleozoic of southwestern sardinia. In: Carbonate-Hosted Lead-Zinc Deposits: 75th Anniversary Volume. Society of Economic Geologists (Jan 1996). https://doi.org/10.5382/SP.04.03
3. Chemale, F., Rosière, C.A., Endo, I.: The tectonic evolution of the quadrilátero ferrífero, minas gerais, brazil. Precambr. Res. **65**(1), 25–54 (1994). https://doi.org/10.1016/0301-9268(94)90098-1, https://www.sciencedirect.com/science/article/pii/0301926894900981
4. Chen, L.C., Papandreou, G., Kokkinos, I., Murphy, K., Yuille, A.L.: Deeplab: Semantic image segmentation with deep convolutional nets, atrous convolution, and fully connected crfs. IEEE Trans. Pattern Anal. Mach. Intell. **40**(4), 834–848 (2017)
5. Cidu, R., Fanfani, L.: Overview of the environmental geochemistry of mining districts in southwestern sardinia. Italy. Geochem. Explor. Environ. Anal. **2**, 243–251 (2002). https://doi.org/10.1144/1467-787302-028
6. Fadda, A.O.: The Sulcis Carboniferous Basin-geology, Hydrogeology. Mines-Carbosulcis s.p.a. Cagliari, Italy (1994)
7. Flood, N., Watson, F., Collett, L.: Using a u-net convolutional neural network to map woody vegetation extent from high resolution satellite imagery across queensland, australia. Int. J. Appl. Earth Obs. Geoinf. **82**, 101897 (2019)
8. Jia, S., Jiang, S., Lin, Z., Li, N., Xu, M., Yu, S.: A survey: Deep learning for hyperspectral image classification with few labeled samples. Neurocomputing **448**, 179–204 (2021). https://doi.org/10.1016/j.neucom.2021.03.035, www.sciencedirect.com/science/article/pii/S0925231221004033
9. Loizzo, R., Daraio, M., Guarini, R., Longo, F., Lorusso, R., Dini, L., Lopinto, E.: Prisma mission status and perspective. In: IGARSS 2019–2019 IEEE International Geoscience and Remote Sensing Symposium, pp. 4503–4506 (2019). https://doi.org/10.1109/IGARSS.2019.8899272
10. Malik, K., Robertson, C., Braun, D., Greig, C.: U-net convolutional neural network models for detecting and quantifying placer mining disturbances at watershed scales. Int. J. Appl. Earth Obs. Geoinf. **104**, 102510 (2021). https://doi.org/10.1016/j.jag.2021.102510, www.sciencedirect.com/science/article/pii/S0303243421002178
11. Maus, V., Giljum, S., Gutschlhofer, J., da Silva, D.M., Probst, M., Gass, S.L.B., Luckeneder, S., Lieber, M., McCallum, I.: Global-scale Mining Polygons (Version 1) (2020). https://doi.org/10.1594/PANGAEA.910894
12. Morra, V., Secchi, F.A., Assorgia, A.: Petrogenetic significance of peralkaline rocks from cenozoic calc-alkaline volcanism from sw sardinia. Italy. Chem. Geol. **118**(1), 109–142 (1994). https://doi.org/10.1016/0009-2541(94)90172-4, www.sciencedirect.com/science/article/pii/0009254194901724
13. Paoletti, M., Haut, J., Plaza, J., Plaza, A.: A new deep convolutional neural network for fast hyperspectral image classification. ISPRS J. Photogram. Remote Sens. **145**, 120–147 (2018). https://doi.org/10.1016/j.isprsjprs.2017.11.021, www.sciencedirect.com/science/article/pii/S0924271617303660, deep Learning RS Data
14. Paoletti, M., Haut, J., Plaza, J., Plaza, A.: Deep learning classifiers for hyperspectral imaging: A review. ISPRS J. Photogram. Remote Sens. **158**, 279–317 (2019). https://doi.org/10.1016/j.isprsjprs.2019.09.006, www.sciencedirect.com/science/article/pii/S0924271619302187
15. Pasci, S.C.: Notes to 1:50.000 Geological Map of Italy, Sheet 564, Carbonia. Servizio Geologico d'italia–ispra and Regione Autonoma della sardegna (2012)
16. Ronneberger, O., Fischer, P., Brox, T.: U-net: Convolutional networks for biomedical image segmentation. CoRR **abs/1505.04597** (2015). http://arxiv.org/abs/1505.04597

17. Sarker, Y., Fahim, S.R., Hosen, M.S., Sarker, S., Mondal, M., Das, S.: Regularized singular value decomposition based multidimensional convolutional neural network for hyperspectral image classification (2020). https://doi.org/10.1109/TENSYMP50017.2020.9230701
18. Wambugu, N., Chen, Y., Xiao, Z., Tan, K., Wei, M., Liu, X., Li, J.: Hyperspectral image classification on insufficient-sample and feature learning using deep neural networks: A review. Int. J. Appl. Earth Obs. Geoinf. **105**, 102603 (2021). https://doi.org/10.1016/j.jag.2021.102603, www.sciencedirect.com/science/article/pii/S030324342100310X
19. Wong, S.C., Gatt, A., Stamatescu, V., McDonnell, M.D.: Understanding data augmentation for classification: When to warp? In: 2016 International Conference on Digital Image Computing: Techniques and Applications (DICTA), pp. 1–6. IEEE (2016)
20. Wu, J.z., Yan, W.d., Ni, W.p., Bian, H.: Feature extraction for hyperspectral data based on mnf and singular value decomposition. In: 2013 IEEE International Geoscience and Remote Sensing Symposium–IGARSS, pp. 1430–1433 (2013). https://doi.org/10.1109/IGARSS.2013.6723053

Earth Observation Big Data Exploitation for Water Reservoirs Continuous Monitoring: The Potential of Sentinel-2 Data and HPC

Roberta Ravanelli, Paolo Mazzucchelli, Valeria Belloni, Filippo Bocchino, Laura Morselli, Andrea Fiorino, Fabio Gerace, and Mattia Crespi

Abstract The impact of climate change on freshwater availability has been widely demonstrated to be severe. The capacity to timely and accurately detect, measure, monitor, and model volumetric changes in water reservoirs is therefore becoming more and more important for governments and citizens. In fact, monitoring over time the water volumes stored in reservoirs is mandatory to predict water availability for irrigation, civil and industrial uses, and hydroelectric power generation; this information is also useful to predict water depletion time with respect to various scenarios. Nowadays, water levels are usually monitored locally through traditional ground methods by a variety of administrations or companies managing the reservoirs, which are still not completely aware of the advantages of remote sensing applications. The continuous monitoring of water reservoirs, which can be performed by satellite data without the need for direct access to reservoir sites and with an overall cost that is independent of the actual extent of the reservoir, can be a valuable asset nowadays: water shortage and perduring periods of droughts interspersed with extreme weather events (as it has been experienced across all Europe in the latest years) make the correct management of water resources a critical issue in any European country (and especially in Southern Europe). The goal of this work is therefore to provide a methodology and to assess the feasibility of a service to routinely monitor and

R. Ravanelli (✉) · V. Belloni · F. Bocchino · M. Crespi
Sapienza University of Rome, DICEA, Rome, Italy
e-mail: roberta.ravanelli@uniroma1.it

P. Mazzucchelli · F. Gerace
Aresys S.r.l, Milan, Italy

L. Morselli
CINECA, Casalecchio Di Reno, Italy

A. Fiorino
Società Risorse Idriche Calabresi S.p.a, Catanzaro, Italy

M. Crespi
Sapienza School for Advanced Studies, Rome, Italy

C. Ieracitano et al. (eds.), *The Use of Artificial Intelligence for Space Applications*, Studies in Computational Intelligence 1088,
https://doi.org/10.1007/978-3-031-25755-1_23

measure 3D (volumetric) changes in water reservoirs, exploiting Artificial Intelligence (AI) to improve the geometrical resolution of the available Sentinel-2 imagery (10 m).

Keywords Optical imagery · Sentinel-2 · Super-resolution · Deep image prior

1 Introduction

Water reservoirs are essential for freshwater supply and hydroelectric power production. Sustainable management of water reservoirs is thus indispensable to monitor hydrological stresses and to predict water availability for irrigation, civil and industrial uses. Nowadays Remote Sensing, driven by the unprecedented technological development of Earth Observation sensors, can represent a viable and low-cost solution for a continuous and long-term monitoring of fundamental reservoir parameters such as water volume/area/level.

At present, water levels are usually monitored locally through traditional ground methods (e.g. gauge stations) by a variety of administrations or companies managing the reservoirs, which are still not completely aware of the advantages of remote sensing applications. The water level measurements collected by gauge stations are generally used to compute the area and volume of the reservoir under investigation from the volume-area-elevation curves derived from topographic and bathymetric information of the reservoir itself [1]. However these data are not always easily available. Moreover, this traditional monitoring methodology presents limitations connected to the need for direct access to reservoirs, for the installation and maintenance of the gauge stations (difficult especially in remote areas), and the issues related to the spatial continuity of the altimetric/bathymetric data [1].

On the other hand, the use of Remote Sensing technologies can limit the monitoring costs (independent of the actual extent of the reservoir) and provide frequently updated spatially continuous data that facilitate mapping and analysis of water bodies [1].

In this context, our work proposes a methodology to remotely monitor the surface extent of artificial reservoirs through the analysis of AI super-resolved Sentinel-2 imagery, with the final aim of monitoring the volumetric variation of the water stored in the reservoir itself, through the integration of the remote sensed planimetric water extent with the local information on water level. This goal is strictly related to United Nations (UN) Sustainable Development Goals (SDGs) [2] related to water availability (SDG 6) and climate change effect monitoring (SDG 13) and with the Recovery Plan Next Generation EU [3].

2 Proposed Approach

The goal of this work is to provide a methodology and to assess the feasibility of a service to routinely monitor and measure 3D (volumetric) changes in water reservoirs, by exploiting AI to improve the original geometrical resolution of the available Sentinel-2 imagery (10 m). The service will be run on a dedicated High Performance Computing (HPC) platform.

Specifically, the proposed methodology implements the following main steps:

1. AI powered super resolution of Sentinel-2 imagery;
2. detection of the extent of the reservoirs;
3. computation of 3D volumetric changes by integrating the previously computed water surface extent with local information on the water level.

2.1 AI Powered Super Resolution

The goal of super-resolution is to take a low resolution (LR) image I0 and upsampling factor t, and generate a corresponding high resolution (HR) version I.

State-of-the-art approaches to single-image super-resolution are currently based on deep convolutional neural networks (ConvNets), as in Ledig et al. [4] and Tai et al. [5]. Popular approaches for image generation such as generative adversarial networks, variational autoencoders and direct pixel-wise error minimization also use ConvNets.

It must be noted that ConvNets excellent performances are due to the fact that they learn realistic data priors from the training phase performed on a large dataset of image prototypes. Thus, the ability to perform a specific super-resolution task is related to the availability of high-quality training datasets. However, in Lempitsky et al. [6] it is demonstrated that, in fact, not all image priors must be learned from data; instead, a great deal of image statistics is captured by the structure of generator ConvNets, in-dependent of learning. We follow the deep image prior approach (Lempitsky et al. [6]) to provide a super-resolution algorithm that does not depend on the availability of training datasets.

The super resolution problem is hence formulated as an optimization task:

$$\min xE(x; x_0) + R(x)$$

where R(x) is an image prior, and it can be represented by a ConvNet by training it using a large number of examples. However, the optimization problem above can be re-written as:

$$\min \theta E(g(\theta); x_0) + R(g(\theta))$$

Fig. 1 Original Sentinel-2 image acquired on 23rd June 2022 over the Menta dam lake, RGB bands

with g(θ) defined as a mapping from subspace θ to subspace x. If such a mapping is chosen to be a deep ConvNet with parameter space θ, g(θ) = fθ(z) (where z is a fixed random input), the optimization problem is recast as a minimization in the space of neural network's parameters with respect to the initial image space x.

Within this new formulation, no training dataset is still needed, but just the initial reference image x0 is used in the restoration process.

The same approach has been proposed for satellite optical imagery in Ma et al. [7] and Sun et al. [8]. The main disadvantage of the proposed approach is that, while avoiding any training phase, the actual super resolution step can be regarded as a training phase itself, with a similar (high) computational cost. Thus, the availability of HPC premises (in particular, GPU HW), is mandatory to allow any practical implementation of the deep image prior super resolution approach on the Sentinel-2 imagery. In particular, we are currently testing an upsampling factor t equal to 4 for all the Sentinel-2 bands. An example is reported in Figs. 1, 2, 3 and 4, where a Sentinel-2 image is upsampled using both deep image prior and the standard bilinear and bicubic algorithms.

2.2 *Detection of the Horizontal Extent of the Reservoirs*

Two different methods to detect the horizontal extent of the reservoirs, originally developed for Landsat imagery, are currently under investigation.

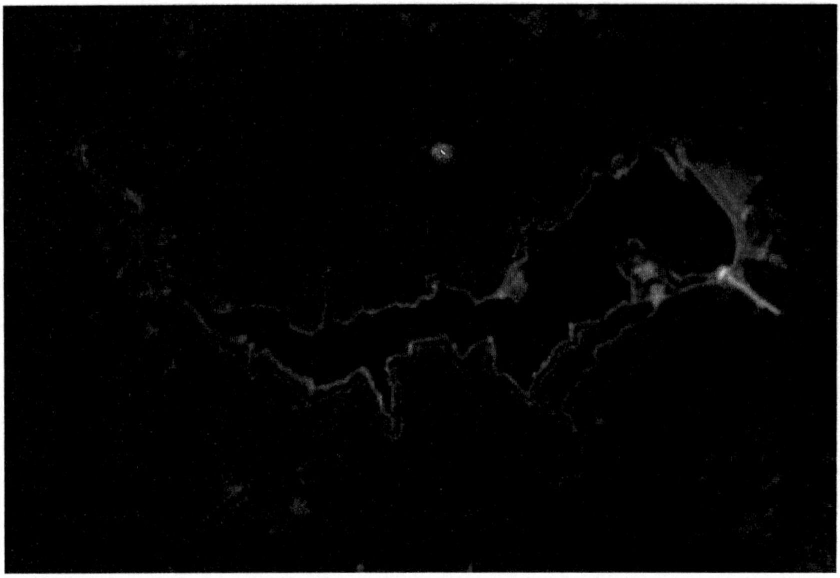

Fig. 2 4x upsampled Sentinel-2 image with bilinear interpolation

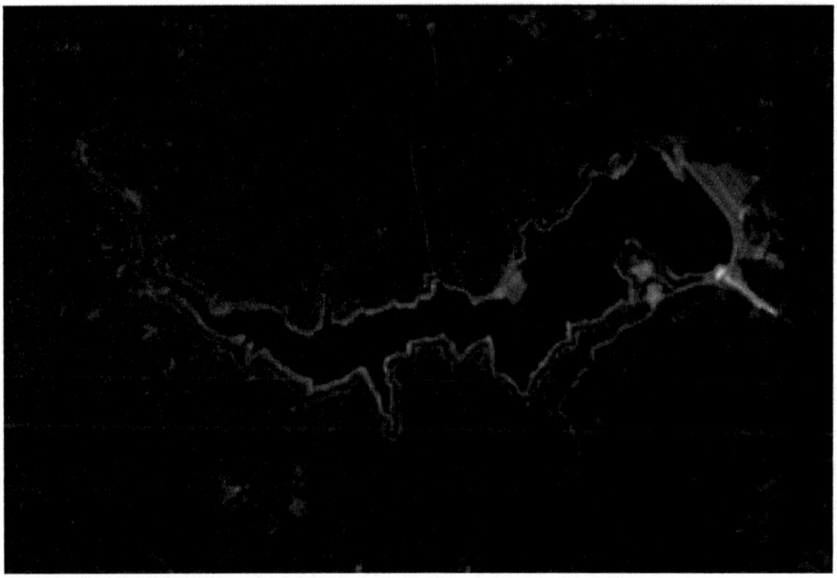

Fig. 3 4x upsampled Sentinel-2 image with bicubic interpolation

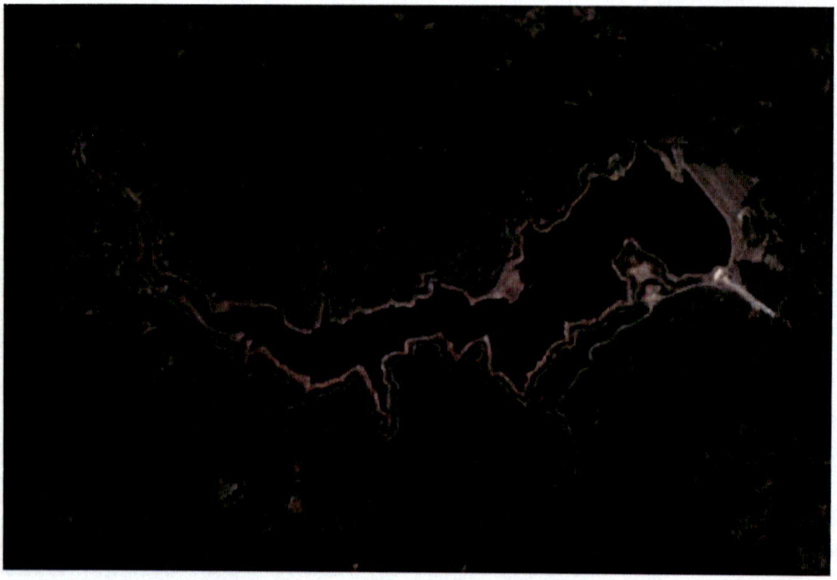

Fig.4 4x upsampled Sentinel-2 image with deep image prior

The first method [9] relies on the use of spectral indexes since it is well known that water significantly absorbs most radiation at near-infrared (NIR) wavelengths and beyond. In particular, the authors used the Automatic Water Extraction Index (AWEI) on the atmospherically corrected surface reflectance values of Landsat 5, 7 and 8 imagery, and implemented a non-parametric unsupervised method based on the Canny edge filter and Otsu thresholding to detect water pixels [9, 10].

The second method [1] generates false color composites starting from the Short-wave infrared (SWIR1)/NIR/Red (R) bands of Landsat images and then applies a HSV transformation to classify the pixels that correspond to water areas.

2.3 *Computation of the 3D Volumetric Changes*

The 3D volumetric changes will be computed by integrating the information of in-situ measured water level with the corresponding water surface estimated on the basis of the previous step of the procedure.

The seasonal variations of the reservoir water storage allow defining a volumetric 3D model of the reservoir itself (i.e., the varying water level acts as a layer stripping procedure, as shown in Fig. 5). Indeed, the ability to map the time variations of the shallower areas of the reservoir (whose extent may vary over time by the depositional process of the debris carried by river tributaries) plays a key role in reservoir and water resources management.

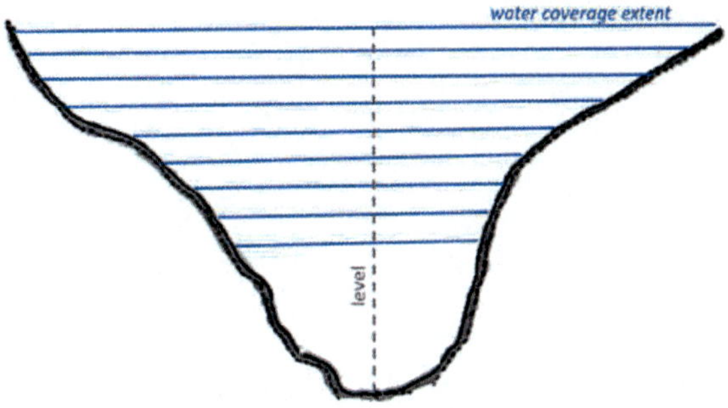

Fig. 5 Relation between estimated water coverage exent and reservoir filling level (shown for a selected reservoir cross-section)

3 Testing Sites

The experiment is driven by the actual needs of the end-user, Società Risorse Idriche Calabresi S.p.a. (So.Ri.Cal.), that operates two different reservoirs in the south of Italy (Calabria region) providing the freshwater supply for nearly two million people. Therefore, the availability of a dynamic 3D mapping could greatly improve its resource usage planning. The two selected reservoirs (Menta dam lake, with a total volume of 17.9 hm^3, and Alaco dam lake, with a total volume of 32 hm^3), although just 100 km apart, show quite different environments and average shore steepness.

4 Preliminary Results

The proposed methodology is currently under development. The first preliminary results were obtained for the year 2021 on the Sentinel-2 images at original resolution available for the two areas of interest. At present, only Alaco dam lake has been investigated.

Firstly, all the images of Sentinel-2 available for the considered period and for the specific area of interest were filtered by selecting the ones with the lowest cloud coverage and masking out the remaining cloudy pixels.

A total of 4 images, one for every three months of the considered period, were then obtained by computing the median values of the pixels from the filtered set of images. The two methods for the water extent detection were hence applied to each of the four images. The results for the Alaco dam lake are shown in Figs. 6 and 7.

The validation of the two methods is currently ongoing.

5 Concluding Remarks

The continuous monitoring of water reservoirs can be performed by satellite data without the need for direct access to reservoir sites and with an overall cost that is

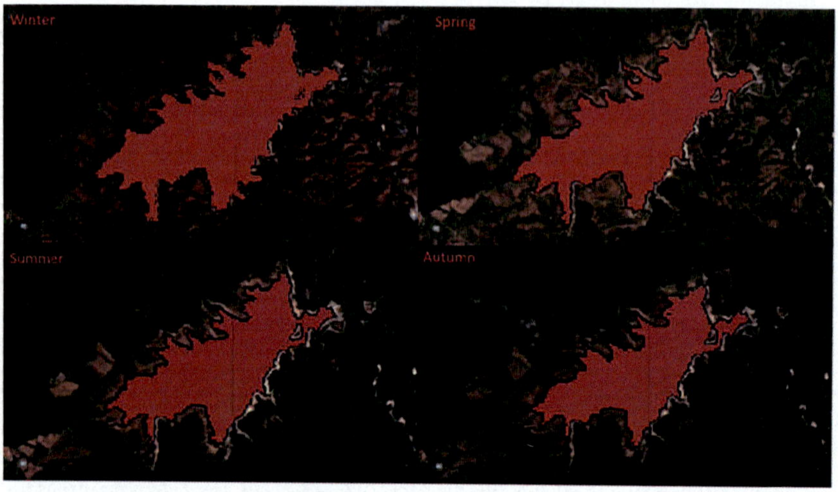

Fig. 6 Results of the detection of the water extent for the first (spectral index based) method for the 2021 year over the Alaco dam lake

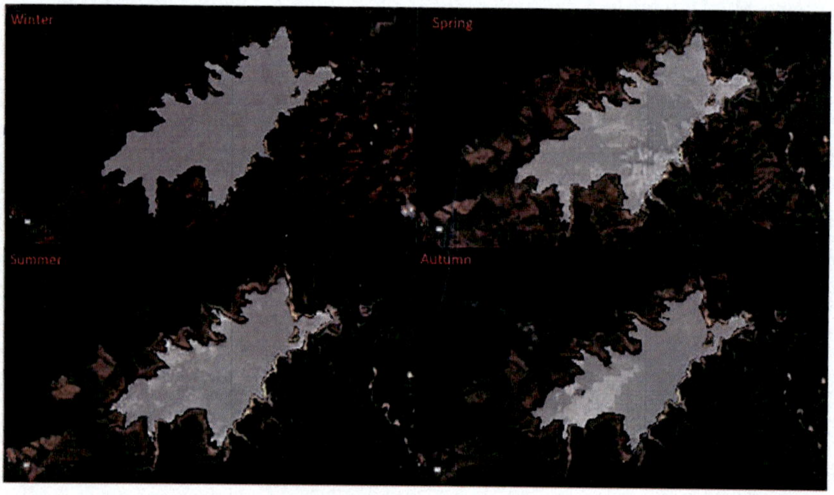

Fig. 7 Results of the detection of the water extent for the second (HSV based) method for the 2021 year over the Alaco dam lake

independent of the actual extent of the reservoir. It can be a valuable asset nowadays: water shortage and long periods of droughts interspersed with extreme weather events make the correct management of water resources a critical issue.

A complete data processing chain is currently under development, starting from satellite raw data up to water body detection and surface estimation. The use of

AI-powered super-resolution techniques will allow achieving the required resolution starting from the ESA Sentinel-2 constellation (belonging to EU Copernicus Programme).

Funding This project [11] has received funding from the European High-Performance Computing Joint Undertaking (JU) under grant agreement No 951745. The JU receives support from the European Union's Horizon 2020 research and innovation programme and Germany, Italy, Slovenia, France, Spain.

References

1. Valadão, L.V., Cicerelli, R.E., de Almeida, T., Ma, J.B.C., Garnier, J.: Reservoir metrics estimated by remote sensors based on the Google Earth Engine platform. Remote Sens. Appl. Soc. Environ. **24**, 100652 (2021)
2. UN sustainable development goals. www.un.org/sustainabledevelopment/
3. Recovery plan for Europe-next generation EU. https://www.ec.europa.eu/info/strategy/recovery-plan-europe_en#nextgenerationeu
4. Ledig, C., Theis, L., Huszar, F., Caballero, J., Cunningham, A., Acosta, A., Aitken, A., Tejani, A., Totz, J., Wang, Z., Shi, W.: Photo-realistic single image superresolution using a generative adversarial network. IEEE Computer Society, CVPR (2017)
5. Tai, Y., Yang, J., Liu, X.: Image super-resolution via deep recursive residual network. IEEE Computer Society, CVPR (2017)
6. Lempitsky, V., Vedaldi, A., Ulyanov, D.: Deep image prior. In: 2018 IEEE/CVF conference on computer vision and pattern recognition, pp. 9446–9454 (2018). https://doi.org/10.1109/CVPR.2018.00984
7. Ma, X., Hong, Y., Song, Y.: Super resolution land cover mapping of hyperspectral images using the deep image prior-based approach. Int. J. Remote Sens. **41**(7), 2818–2834 (2020). https://doi.org/10.1080/01431161.2019.1698079
8. Sun, Y., Liu, J., Yang, J., Xiao, Z., Wu, Z.: A deep image prior-based interpretable network for hyperspectral image fusion. Remote Sens. Lett. **12**(12), 1250–1259 (2021). https://doi.org/10.1080/2150704X.2021.1979270
9. Sengupta, D., Chen, R., Meadows, M.E., Banerjee, A.: Gaining or losing ground? Tracking Asia's hunger for 'new' coastal land in the era of sea level rise. Sci. Total Environ. **732**, 139290 (2020)
10. Donchyts, G., Schellekens, J., Winsemius, H., Eisemann, E., Van de Giesen, N.: A 30 m resolution surface water mask including estimation of positional and thematic differences using landsat 8, srtm and openstreetmap: a case study in the Murray-Darling Basin Australia. Remote Sens. **8**(5), 386 (2016)
11. FF4EuroHPC project. HPC for reservoir monitoring. www.ff4eurohpc.eu/en/experiments/2022031511555138/hpc_for_reservoir_monitoring

Retrieval of Marine Parameters from Hyperspectral Satellite Data and Machine Learning Methods

Federico Serva, Luigi Ansalone, and Pierre-Philippe Mathieu

Abstract The PRISMA hyperspectral mission of the Italian Space Agency, operational since 2019, is providing high spectral resolution data in the range 400–2500 nm, in support of multiple environmental applications, such as water quality and ecosystem monitoring. In this work we discuss how hyperspectral data can be used to simultaneously retrieve aerosol and marine properties, including sediment properties and chlorophyll, by using a coupled radiative transfer model (RTM). As physics-based methods are computationally expensive, we investigate the use of machine learning methods for emulation and hybrid retrievals, combining physics with machine learning. We find that assumptions on the covariance matrices strongly affect the retrieval convergence, which is poor in the coastal waters we considered. We also show that RTM emulation provides substantial speed-up and good results for AOD and sediment variables, however further parameter tuning seems necessary.

Keywords Hyperspectral data · Retrieval methods · Machine learning

1 Introduction

Continuous, global and high resolution monitoring of water quality and marine ecosystems are important steps to achieve the Sustainable Development Goals targets [22]. Ongoing satellite missions, equipped with multi and hyperspectral sensors operating in the optical range, allow us to observe inland and marine waters globally and in near-real time, depending on favorable weather conditions. In the next few years the extension of the current Earth observation (EO) records as well as the launch of dedicated satellites will allow us to better understand natural variability and human impacts [15]. As discussed by [2], spectral bands in the visible (VIS,

F. Serva (✉) · L. Ansalone
Agenzia Spaziale Italiana, Via del Politecnico snc, 00133 Rome, Italy
e-mail: federico.serva@terrarum.eu

F. Serva · P.-P. Mathieu
European Space Agency, Via Galileo Galilei 1, 00044 Frascati, Italy

C. Ieracitano et al. (eds.), *The Use of Artificial Intelligence for Space Applications*,
Studies in Computational Intelligence 1088,
https://doi.org/10.1007/978-3-031-25755-1_24

such as blue, green, red) and in the shortwave infrared (SWIR) are well suited for detecting ecosystem status and changes in the oceans. Operational monitoring can be achieved already with multispectral missions such as Sentinel 2A/B [5], with spatial resolution up to 10 m and 13 spectral bands in total. One major issue when retrieving marine parameters is the small relative contribution of the ocean signal in the overall radiance measured at the top-of-atmosphere (TOA), which is affected by atmospheric conditions and thus needs to be interpreted after atmospheric correction [12].

In this work we discuss the use of hyperspectral data provided by the innovative PRISMA (Precursore Iperspettrale della Missione Operativa) mission of the Italian Space Agency to pursue these objectives, which is operational since 2019 [10]. For an overview of the mission characteristics and data availability also see [19] and the dedicated ASI website.[1] The mission has near-global observing capabilities and provides hyperspectral data (400–2500 nm) with high spectral resolution (240 bands), which allows a more accurate characterization of the observed scenes. The potential of the mission for water applications has been demonstrated in preliminary analyses [13, 20], but to our knowledge one-step approaches as ours were not employed so far. While hyperspectral data from airborne platforms and space-borne satellites was relatively limited in the last decades [24], multiple hyperspectral missions from both public and private entities are entering in operations, and therefore the interest is high in a timely data exploitation for marine applications [11]. Retrieval of atmospheric and marine parameters will also be the focus upcoming missions both in Europe (with CHIME[2]) and the United States (with SBG[3] and PACE[4]). Processing of the data provided by these missions will benefit from simultaneous atmosphere-ocean retrievals [12].

In order to retrieve relevant marine parameters related with algal and sediment matters, we describe two possible approaches, both relying on radiative transfer calculations performed with an open-source coupled atmosphere-ocean radiative transfer model (RTM). In principle the outlined methodology is applicable to any kind of scenario, assuming that ancillary information on the area of interest is available to define suitable first guesses and set boundary conditions. In a more standard approach, we use a Gauss-Newton iterative method to iteratively solve the nonlinear retrieval problem, with a state vector including both atmospheric and marine parameters. Furthermore, following the large number of works using machine learning (ML) for retrieval of vegetation traits, we discuss how machine learning can be applied to our marine case study. Preliminary results highlight the challenge posed in the definition of a suitable retrieval algorithm, due to the limited amount of in-situ data available for most regions, as well as the higher computational efficiency offered by ML approaches. We argue that data pre-processing and band selection based on physical considerations can be helpful to improve the preliminary results presented.

---

[1] https://www.asi.it/en/earth-science/prisma/.

[2] https://directory.eoportal.org/web/eoportal/satellite-missions/c-missions/chime-copernicus.

[3] https://sbg.jpl.nasa.gov/.

[4] https://pace.gsfc.nasa.gov/.

This paper is organized as follows: Sect. 2 provides information on PRISMA and the retrieval methods, both standard and ML-based; Sect. 3 provides results obtained with the two approaches; concluding remarks and perspectives are given in Sect. 4.

2 Methods and Data

2.1 The Coupled Atmosphere-Ocean RTM

In order to model how a certain combination of atmospheric and marine parameters influences the observed radiation, we use the Ocean Successive Orders with Atmosphere–Advanced (OSOAA) RTM [7], version 1.6. The software is freely distributed[5] by its developers and it gives users many customization options. The model solves the radiative transfer equation with the successive orders of scattering method [9, 16], where radiance intensity is computed by adding the contribution of photons scattered once, twice or multiple times, and it is useful for inhomogeneous media. Note that the model works with monochromatic inputs, thus several runs are needed to compute the whole spectral interval of interest. For our case, this means computing scattering and absorption in the atmosphere and the ocean, as well as transmission and reflection processes at their interface. OSOAA calculate all components of the Stokes vector [7], but since PRISMA does not provide information on polarization, we do not consider these terms. The normalized OSOAA intensity is converted into physical units considering the observed solar irradiance spectra.[6]

In our idealized setup, we are interested in the TOA radiance and the effects of aerosol particles in the atmosphere, algal and sediment particles in the ocean. We focus on aerosol optical depth (AOD), chlorophyll and suspended sediments concentrations, absorption coefficient for yellow substance and mineral particles. Several boundary conditions are obtained from ancillary data, as specified below, and for simplicity we adopt the available oceanic (i.e. hydrated) aerosol model and assume a light sand ocean bottom. Absorption from atmospheric gases is not considered and default values are assumed for the remaining parameters used in the simulations.

2.2 PRISMA and Ancillary Data

Acquisitions from the PRISMA satellite can be downloaded by users through the dedicated portal, either requesting a new acquisition or extracted from the operational archive. In our case we consider a hyperspectral image acquired near the city of Ravenna, Italy, which is located at the mouth of the Po river, and influenced by the

[5] https://github.com/CNES/RadiativeTransferCode-OSOAA.

[6] Irradiance data is freely available from https://www.nrel.gov/grid/solar-resource/spectra.html.

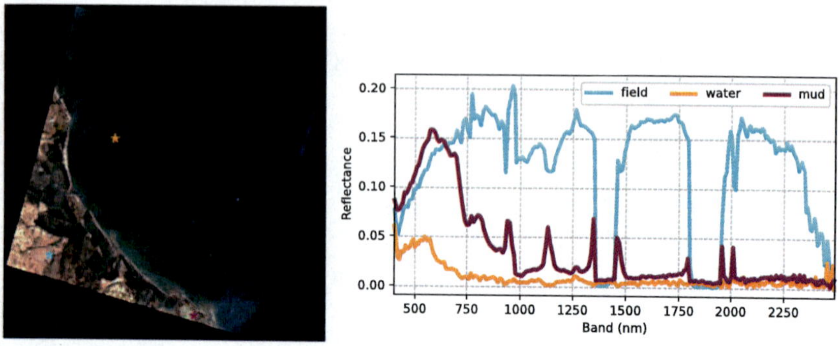

Fig. 1 The PRISMA scene considered for the retrieval exercise, acquired near the city of Ravenna, Italy. Left: red-green-blue composite of the data cube; marks indicate coastal waters (orange), agricultural field (cyan), and muddy waters (purple). Right: reflectance values along the full spectrum for the three points indicated in the composite image. Note that these plots are based on L2D (geolocated and geocoded reflectance at surface) data

proximity to the Po Valley, a major pollution hotspot [23]. The image acquisition time was 20211015101102-20211015101107, with identifier 8291. For the retrieval we consider the L1 file, providing calibrated radiance at TOA as a function of latitude, longitude and band. No specific processing of the data is performed, except for the conversion from digital number to physical units with the available metadata and selection of the spectral interval between 400 and 700 nm, similar to what was done in recent works using the OSOAA RTM [8].

In Fig. 1 (left) we report a false colour view of the scene (note this is based on L2D data), from which agricultural fields, optically complex and cleaner water off-coast can be seen. To illustrate the potential of PRISMA data, we select three points (coastal water, water with heavy sediment loads and land) and show the respective spectra in Fig. 1 (right). The spectral signature of the land point suggests this is characterized by bare soil or dry grass, as reflectance values remain relatively high throughout [21]. The reflectance of the water pixel is substantially smaller, as we already noted the signal from the ocean is less intense than that coming from land pixels. In the sediment-rich areas near the coast, reflectance is larger than for cleaner water, and a complex spectral signature can be observed. Note that near 1370 and 1900 nm absorption from atmospheric water vapor is very strong and reflectance values reach zero in these bands [29].

For the atmosphere, we use data obtained from the reanalysis system EAC4 [17] produced by the Copernicus Atmosphere Monitoring Service (CAMS), providing analyzed dynamical and composition variables for almost two decades, globally, with a spatial resolution of 80 km and 3 hours temporally. For the closest grid point and time available, we consider surface wind speed, pressure and aerosol optical depth (AOD) at 550 nm. AOD is the only atmospheric parameter currently included into the state vector for retrieval, but surface pressure could be optimized too (as it acts as scale factor for the radiance spectra, not shown).

For the first guess, we need to specify the absorption coefficients of both yellow substance (organic) and detritus (inorganic). Since absorption coefficients can be combined to obtain the total absorption coefficient [31], we assume half of the total absorption obtained from the CCI ocean colour dataset [28]. Marine depth is also an ancillary information, downloaded from the Copernicus Marine Service (CMEMS) archive.

2.3 Solving the Nonlinear Inverse Problem

The (nonlinear) problem we are trying to solve is to find the optimal state vector x, composed by the parameters to be retrieved, such that the difference between the observed radiance y and the modeled radiance $F(x)$ is minimized, where F is the forward (RTM) model. Of course errors in the formulation of the forward model F would be reflected into the modeled radiances we are trying to match with PRISMA observations. In the *a posteriori* approach, the problem can be formulated assuming Gaussian probability distributions as discussed in [26]:

$$-2\ln(P(x|y)) = [y - F(x)]^T S_e^{-1}[y - F(x)] + [x - x_a]^T S_a^{-1}[x - x_a] + c, \quad (1)$$

with the aim of finding a best estimate x and error characteristic that describe this distribution. In this equation, x is the state vector, S are covariance matrices for the first guess (subscript a) and the observation error (subscript e).

To solve the nonlinear problem of finding the minimum of the gradient of cost function, one can use the Newtonian iteration method [26, Par. 5.3], where at the $i - th$ iteration the state vector can be written as:

$$x_{i+1} = x_i + (S_a^{-1} + K_i^T S_e^{-1} K_i)^{-1}[K_i^T S_e^{-1}(y - F(x_i)) - S_a^{-1}(x_i - x_a)] \quad (2)$$

where K is the Jacobian matrix, the first derivative of the forward model. In our setup, we limit the number of iterations to a maximum of ten, or we require convergence when the modeled spectrum is within 5% of observations. Note that this physics-based method requires multiple RTM runs to estimate the Jacobian, thus making the procedure computationally intensive (taking several seconds for each spectrum with standard hardware).

2.4 Machine Learning Methods

ML methods are useful for the retrieval problem as they are capable of capturing the associations between a set of input (e.g., the biogeophysical parameters) and output parameters (e.g., radiances). In the last few years many groups applied these methods to the retrieval of vegetation properties from multispectral satellites such as Sentinel

2, since as explained by [6], ML methods can help both in the forward modeling step (from state vector to radiance) as well as for solving the inverse problem (from radiance to state vector).

De Sa et al. [27] explored different techniques for a hybrid retrieval, and here we follow their methodology to explore the applicability of ML for our tasks, by using Random Forest (RF) regression and Gaussian processes (GP). Here we follow their methodology, which is shared by several recent works dealing with multi-spectral data, using the *scikit-learn* implementation.[7]

RF regression [3] is an ensemble classifier combining classification and regression trees, in which data are binned to different sets to maximize the purity of each. For a candidate split s at a node t, the purity variation (related with the gain in information) can be expressed as:

$$\Delta i_{(s,t)=i_t-p_L i_{(t_L)}-p_R i_{(t_R)}}, \tag{3}$$

where p_R and p_L are the data proportions, i are measures of purity before and after splitting. With an ensemble of K decision trees DT and data x, the regression is

$$RF_{(x)} = \frac{1}{K}\sum_{k=1}^{K} DT_x. \tag{4}$$

We use 850 trees and 10^{-3} for minimum number of samples for split and internal node and to be at a leaf node.

Gaussian process (GP) regression [25] is a non-parametric regression method, based on the Bayes method and a set of covariance kernel functions to be optimized. A GP function $f(x)$ is a set of Gaussianly distributed random variables in a multivariate space, then for a mean value μ, a function k is the covariance between data and the Gaussian noise with standard deviation σ

$$f(x) \sim GP(\mu(x), k(x, x') + I\sigma_y^2), \tag{5}$$

where x are observations and k is the prior covariance function, and $I\sigma_y^2$ is noise. Therefore the resulting function is a combination of Gaussian processes, here estimated using the rational Quadratic kernel for covariance (one of those available), and 80 restarts of the optimizer.

While we are not considering these approaches here, previous works showed the potential for GP methods for RTM emulation, especially in light of substantial computational gains. Among them, [14] demonstrated that emulation performs well with widely used RTM in the context of land surface retrieval, while recently [4] developed surrogate models for RTMs used for retrieval of atmospheric trace gases from Sentinel-5P EO data.

[7] The toolbox developed for the recently launched EnMAP mission is also using this library for regression.

3 Results and Discussion

3.1 RTM Sensitivity to Inputs

To understand how the modeled radiance responds to changes in the inputs, we run the RTM in the same configurations changing one parameter at the time, as reported in the panels of Fig. 2. Note that the standard OSOAA output is normalized radiance, which is shown in these plots.

We can see that the effect of higher AOD values (dimensionless) is to increase radiance values throughout the spectrum, particularly going to the red and SWIR bands. While values above unity are not common in the area of interest, they can occur in association with dust transport from Northern Africa [1]. Due to the model formulation, the sensitivity to chlorophyll concentration (units mg m^{-3}) can be appreciated below 600 nm, and differently from AOD, it reduces the radiance at TOA due to absorption. It is worth noting that in eutrophic waters, as in coastal areas or inland water bodies, larger concentrations are frequently observed [18]. For sediment par-

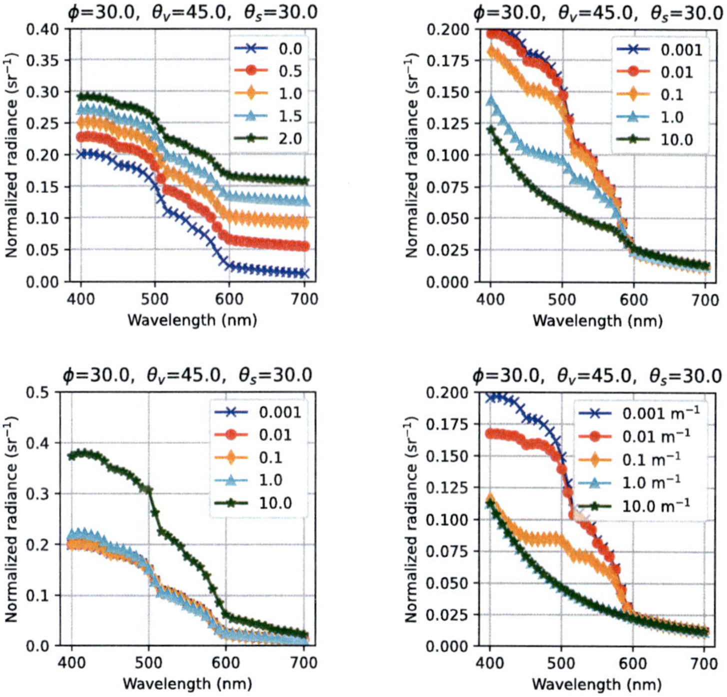

Fig. 2 Sensitivity of the OSOAA modeled radiance to changes in the biogeophysical parameter values: aerosol optical depth (top left), chlorophyll (top right), sediments (bottom left), and particle absorption (bottom right). Results for detritus and yellow substance absorption are qualitatively similar (not shown)

ticles (units mg l^{-1}), the qualitative spectral change is more similar to AOD, as they increase scattering in the visible range, consistent with what was found in Fig. 1. Finally, the effect of larger absorption coefficient of particles (units m^{-1}) is to reduce the radiation reaching the TOA, with a spectral response which depends on the model assumptions on their characteristics.

These results highlight the spectrally complex response of the modeled radiation to changes in the state vector parameters, which makes the retrieval in the general case a difficult task already with a small number of unknowns, as in our case.

3.2 Variational Retrieval

In order to retrieve the state vector that best matches observed radiance when used with the forward model, we use the variational methodology laid out above. The procedure iteratively updates the entries of the state vector as long as the overall difference of the modeled spectrum is larger than 5% or up to a maximum of ten iterations. Here and in the next sections we consider one single pixel in coastal waters.

The first guess and final spectra after the maximum number of iterations are reported in Fig. 3 (right). By comparing the initial and final spectrum, we can see a tendency for lower radiance values going towards the last iteration. While the overall difference from the observed spectrum is reduced (not shown), the agreement with the observed spectrum is poor. The lack of convergence indicates that the retrieval procedure was not successful for one or more target parameters.

To gain further insight on the individual parameter evalution across the iterations, we report their values in Fig. 4. We can see that AOD, chlorophyll and detritus absorption values oscillate between the various iterations, while interestingly both

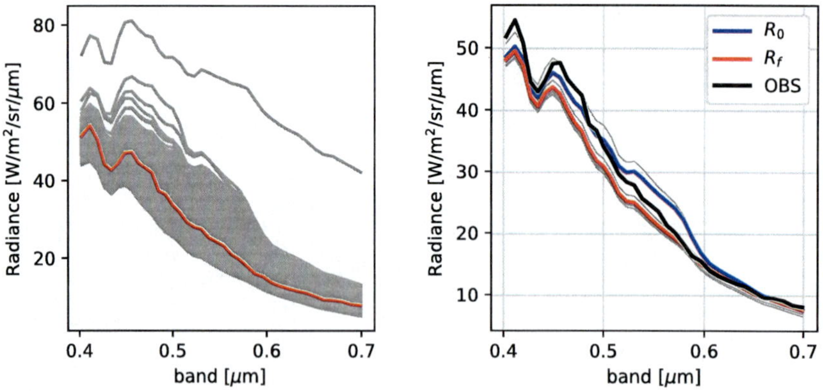

Fig. 3 Left: examples of spectra generated with OSOAA (grey lines) and a random sampling of the retrieval parameters ranges, also used for the machine learning retrieval, compared against the PRISMA spectrum for a pixel of interest (red). Right: comparison of the PRISMA spectrum (black) with the spectrum obtained with the first guess (blue) and the one obtained after the iterations (red)

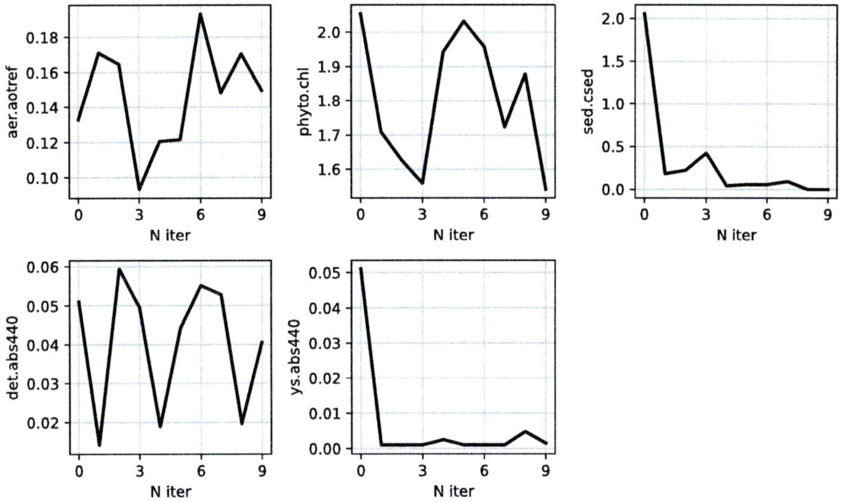

Fig. 4 Evolution of the retrieval parameters through the various iterations, 0 being the first guess values. A maximum of ten iterations was imposed, and convergence was not reached in this case

sediment concentrations and yellow substance absorption at 440 nm go almost to zero already after the first few iterations.

These results suggest that the adoption of diagonal covariance matrices may not suitable be for the retrieval in coastal waters, therefore additional in-situ data should be used to explore the relationships between the different parameters and estimate off-diagonal terms. Additionally, the sensitivity of each band to changes in each of the parameters to be retrieved could be used to select a number of more sensitive bands and thus improve the algorithm convergence.

3.3 Machine Learning Retrieval

An alternative method to retrieve the optimal combination of parameters for use with the forward model is to learn the association between them and the spectral response. Following [27], here we illustrate the results obtained for the validation set with RF and GP regression algorithms. In order to perform this, we generated 1500 spectra based on Latin hypercube sampling of the the parameters space, as shown in Fig. 3 (left). Note that the number of pixels in a hyperspectral scene is around one million. Ten percent of the data is taken for training, the rest is used for validation.

With this approach we can estimate the optimal state vector given a radiance spectrum, thus not needing an iterational procedure as in the physics-based case. The results for the training fold are shown in Fig. 5 as scatter plots, with predicted against target values. It is evident that the best results are achieved for AOD (cor-

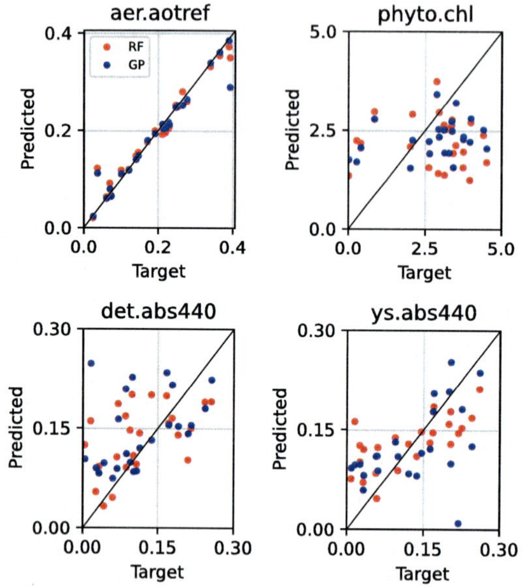

Fig. 5 Scatter plots of target values (x-axis) and emulated values (y-axis) for the RF (red) and GP (blue) methods. In the best case, points should fall on the 1:1 line (in black). Only a subset of randomly chosen points is shown for better readability

relation 0.992 and 0.979 for RF and GP, respectively) and sediment concentrations (0.961 and 0.951), as points lie close to the 1:1 line. As was seen in the RTM sensitivity analysis, these parameters tend to have an overall scaling effect on the whole spectrum, therefore it is likely they are easier for the models to learn. Conversely, systematic biases affect the remaining three parameters, which appear to be more dispersed (chlorophyll 0.252 and 0.262, detritus 0.550 and 0.474, and yellow substance 0.632 and 0.616). It is interesting to notice that while the sensitivity of the radiance to these parameters is qualitatively similar (see Fig. 2), the degree of nonlinearity may be larger for chlorophyll in the range of interest. While the results are quantitatively not satisfactory, the qualitative behaviour is acceptable, so we would expect to obtain better results after a more accurate hyperparameter tuning and using larger training datasets.

4 Conclusions

The availability of new satellite missions providing hyperspectral data is raising interest in developing new methodologies for data exploitaition. Here we consider data from the PRISMA hyperspectral mission for an acquisition taken over a complex coastal scenario, near the mouth of the Po river in Italy. We explore different retrieval methods all relying on the OSOAA RTM to perform a simultaneous estimation of both atmospheric and marine parameters, a task which has recently received attention in the literature [30] especially for computational efficiency considerations.

In the remote sensing context, a distinction between waters where phytoplankton and their derivatives play a dominant role in determing the optical characteristics and waters where resuspended sediments and land discharge are important is often used [18]. While algorithms are often tuned to perform well in one case or the other, thus reducing the flexibility of the retrieval, in our case the algorithm is designed to deal with different regimes, based on the ancillary data.

We performed a physics-based retrieval based on the OSOAA RTM and an iterative Newtonian method, finding strong dependency on the specification of the covariance matrices. The lack of convergence indicates that insight from available in-situ data should be used to better define parameter cross-correlations. We also followed the methodology of [27] to perform a hybrid retrieval with Random Forests and Gaussian process regression, but without performing an ad-hoc tuning of these models so far. While results are quantitavely acceptable only for two out of five parameters, these preliminary results indicate that we may be able to obtain good results with less computational cost compared to the previous method. Moreover an investigation on the impact of noise in the case of hyperspectral data need to be made, since hybrid methods are very sensitive to it [27], and further ML methods (e.g., gradient boosting) can be explored for application in our retrieval task.

Acknowledgements The work of the first author has been supported by a joint ASI/ESA postdoctoral fellowship. Useful discussions with colleagues from the ESA Φ-lab and with G. L. Liberti (CNR) are kindly acknowledged. All data sources are freely available as mentioned in the text (PRISMA data are accessible only upon registration on the mission portal); for the PRISMA acquisitions, the information was generated by the authors under an ASI License to Use; Original PRISMA Product–©ASI–(2021).

References

1. Antoine, D., Nobileau, D.: Recent increase of Saharan dust transport over the Mediterranean Sea, as revealed from ocean color satellite (SeaWiFS) observations. J. Geophys. Res. **111**(D12), D12214 (2006). https://doi.org/10.1029/2005JD006795, https://doi.wiley.com/10.1029/2005JD006795
2. Blondeau-Patissier, D., Gower, J.F.R., Dekker, A.G., Phinn, S.R., Brando, V.E.: A review of ocean color remote sensing methods and statistical techniques for the detection, mapping and analysis of phytoplankton blooms in coastal and open oceans. Prog. Oceanogr. **123**, 123–144 (2014). https://doi.org/10.1016/j.pocean.2013.12.008, www.sciencedirect.com/science/article/pii/S0079661114000020
3. Breiman, L.: Random forests. Mach. Learn. **45**(1), 5–32 (2001). https://doi.org/10.1023/A:1010933404324
4. Brence, J., Tanevski, J., Adams, J., Malina, E., Džeroski, S.: Surrogate models of radiative transfer codes for atmospheric trace gas retrievals from satellite observations. Mach. Learn. (2022). https://doi.org/10.1007/s10994-022-06155-2

5. Caballero, I., Steinmetz, F., Navarro, G.: Evaluation of the first year of operational sentinel-2A data for retrieval of suspended solids in medium- to high-turbidity waters. Remote Sens. **10**(7), 982 (2018). https://doi.org/10.3390/rs10070982, www.mdpi.com/2072-4292/10/7/982
6. Camps-Valls, G., Sejdinovic, D., Runge, J., Reichstein, M.: A perspective on gaussian processes for earth observation. Nat. Sci. Rev. **6**(4), 616–618 (2019). https://doi.org/10.1093/nsr/nwz028, www.academic.oup.com/nsr/article/6/4/616/5369430
7. Chami, M., Lafrance, B., Fougnie, B., Chowdhary, J., Harmel, T., Waquet, F.: OSOAA: a vector radiative transfer model of coupled atmosphere-ocean system for a rough sea surface application to the estimates of the directional variations of the water leaving reflectance to better process multi-angular satellite sensors data over the ocean. Opt. Express **23**(21), 27829 (2015). https://doi.org/10.1364/OE.23.027829, www.opg.optica.org/abstract.cfm?URI=oe-23-21-27829
8. Chami, M., Larnicol, M., Minghelli, A., Migeon, S.: Influence of the suspended particulate matter on the satellite radiance in the sunglint observation geometry in coastal waters. Remote Sens. **12**(9), 1445 (2020). https://doi.org/10.3390/rs12091445, www.mdpi.com/2072-4292/12/9/1445
9. Chami, M., Santer, R., Dilligeard, E.: Radiative transfer model for the computation of radiance and polarization in an ocean-atmosphere system: polarization properties of suspended matter for remote sensing. Appl. Opt. **40**(15), 2398 (2001). https://doi.org/10.1364/AO.40.002398, www.opg.optica.org/abstract.cfm?URI=ao-40-15-2398
10. Cogliati, S., Sarti, F., Chiarantini, L., Cosi, M., Lorusso, R., Lopinto, E., Miglietta, F., Genesio, L., Guanter, L., Damm, A., Pérez-López, S., Scheffler, D., Tagliabue, G., Panigada, C., Rascher, U., Dowling, T.P.F., Giardino, C., Colombo, R.: The PRISMA imaging spectroscopy mission: Overview and first performance analysis. Remote Sens. Environ. **262**, 112499 (2021). https://doi.org/10.1016/j.rse.2021.112499, www.sciencedirect.com/science/article/pii/S0034425721002170
11. Dierssen, H.M., Ackleson, S.G., Joyce, K.E., Hestir, E.L., Castagna, A., Lavender, S., McManus, M.A.: Living up to the hype of hyperspectral aquatic remote sensing: Science, resources and outlook. Front. Environ. Sci. **9** (2021). https://www.frontiersin.org/article/10.3389/fenvs.2021.649528
12. Frouin, R.J., Franz, B.A., Ibrahim, A., Knobelspiesse, K., Ahmad, Z., Cairns, B., Chowdhary, J., Dierssen, H.M., Tan, J., Dubovik, O., Huang, X., Davis, A.B., Kalashnikova, O., Thompson, D.R., Remer, L.A., Boss, E., Coddington, O., Deschamps, P.Y., Gao, B.C., Gross, L., Hasekamp, O., Omar, A., Pelletier, B., Ramon, D., Steinmetz, F., Zhai, P.W.: Atmospheric correction of satellite ocean-color imagery during the PACE era. Front. Earth Sci. **7** (2019). https://www.frontiersin.org/article/10.3389/feart.2019.00145
13. Giardino, C., Bresciani, M., Braga, F., Fabbretto, A., Ghirardi, N., Pepe, M., Gianinetto, M., Colombo, R., Cogliati, S., Ghebrehiwot, S., Laanen, M., Peters, S., Schroeder, T., Concha, J.A., Brando, V.E.: First evaluation of PRISMA level 1 data for water applications. Sensors **20**(16), 4553 (2020). https://doi.org/10.3390/s20164553, www.mdpi.com/1424-8220/20/16/4553
14. Gómez-Dans, J.L., Lewis, P.E., Disney, M.: Efficient emulation of radiative transfer codes using gaussian processes and application to land surface parameter inferences. Remote Sens. **8**(2), 119 (2016). https://doi.org/10.3390/rs8020119, www.mdpi.com/2072-4292/8/2/119
15. Groom, S., Sathyendranath, S., Ban, Y., Bernard, S., Brewin, R., Brotas, V., Brockmann, C., Chauhan, P., Choi, J.k., Chuprin, A., Ciavatta, S., Cipollini, P., Donlon, C., Franz, B., He, X., Hirata, T., Jackson, T., Kampel, M., Krasemann, H., Lavender, S., Pardo-Martinez, S., Mélin, F., Platt, T., Santoleri, R., Skakala, J., Schaeffer, B., Smith, M., Steinmetz, F., Valente, A., Wang, M.: Satellite ocean colour: Current status and future perspective. Front. Mar. Sci. **6** (2019). https://www.frontiersin.org/article/10.3389/fmars.2019.00485
16. Hansen, J.E., Travis, L.D.: Light scattering in planetary atmospheres. Space Sci. Rev. **16**(4), 527–610 (1974). https://doi.org/10.1007/BF00168069
17. Inness, A., Ades, M., Agustí-Panareda, A., Barré, J., Benedictow, A., Blechschmidt, A.M., Dominguez, J.J., Engelen, R., Eskes, H., Flemming, J., Huijnen, V., Jones, L., Kipling, Z., Massart, S., Parrington, M., Peuch, V.H., Razinger, M., Remy, S., Schulz, M., Suttie, M.: The

CAMS reanalysis of atmospheric composition. Atmos. Chem. Phys. **19**(6), 3515–3556 (2019). https://doi.org/10.5194/acp-19-3515-2019, www.acp.copernicus.org/articles/19/3515/2019/
18. Kirk, J.T.O.: Light and Photosynthesis in Aquatic Ecosystems, 3rd edn. Cambridge University Press, Cambridge (2010). http://ebooks.cambridge.org/ref/id/CBO9781139168212
19. Lopinto, E., Ananasso, C.: The PRISMA hyperspectral mission. In: European Association of Remote Sensing Laboratories Symposium Proceedings. Matera, Italy (2014)
20. Niroumand-Jadidi, M., Bovolo, F., Bruzzone, L.: Water quality retrieval from PRISMA hyperspectral images: First experience in a Turbid Lake and comparison with sentinel-2. Remote Sens. **12**(23), 3984 (2020). www.mdpi.com/2072-4292/12/23/3984
21. Pepe, M., Pompilio, L., Gioli, B., Busetto, L., Boschetti, M.: Detection and classification of non-photosynthetic vegetation from PRISMA hyperspectral data in croplands. Remote Sens. **12**(23), 3903 (2020). www.mdpi.com/2072-4292/12/23/3903
22. Politi, E., Paterson, S.K., Scarrott, R., Tuohy, E., O'Mahony, C., Cámaro-García, W.C.A.: Earth observation applications for coastal sustainability: Potential and challenges for implementation 1. Anthropocene Coasts (2019). https://doi.org/10.1139/anc-2018-0015
23. Pozzer, A., Bacer, S., Sappadina, S.D.Z., Predicatori, F., Caleffi, A.: Long-term concentrations of fine particulate matter and impact on human health in Verona. Italy. Atmos. Pollut. Res. **10**(3), 731–738 (2019). https://doi.org/10.1016/j.apr.2018.11.012, www.sciencedirect.com/science/article/pii/S1309104218303465
24. Qian, S.E.: Hyperspectral satellites, evolution, and development history. IEEE J. Sel. Top. App. Earth Obs. Remote Sens. **14**, 7032–7056 (2021). https://doi.org/10.1109/JSTARS.2021.3090256
25. Rasmussen, C.E., Williams, C.K.I.: Gaussian Processes for Machine Learning. MIT Press, Cambridge, Mass, Adaptive Computation and Machine Learning (2006)
26. Rodgers, C.D.: Inverse Methods for Atmospheric Sounding: Theory and Practice, Series on Atmospheric, Oceanic and Planetary Physics, vol. 2. World Scientific (2000). https://www.worldscientific.com/worldscibooks/10.1142/3171
27. de Sa, N.C., Baratchi, M., Hauser, L.T., van Bodegom, P.: Exploring the impact of noise on hybrid inversion of PROSAIL RTM on sentinel-2 data. Remote Sens. **13**(4), 648 (2021). https://doi.org/10.3390/rs13040648, www.mdpi.com/2072-4292/13/4/648
28. Sathyendranath, S., Brewin, R.J.W., Brockmann, C., Brotas, V., Calton, B., Chuprin, A., Cipollini, P., Couto, A.B., Dingle, J., Doerffer, R., Donlon, C., Dowell, M., Farman, A., Grant, M., Groom, S., Horseman, A., Jackson, T., Krasemann, H., Lavender, S., Martinez-Vicente, V., Mazeran, C., Mélin, F., Moore, T.S., Müller, D., Regner, P., Roy, S., Steele, C.J., Steinmetz, F., Swinton, J., Taberner, M., Thompson, A., Valente, A., Zühlke, M., Brando, V.E., Feng, H., Feldman, G., Franz, B.A., Frouin, R., Gould, R.W., Hooker, S.B., Kahru, M., Kratzer, S., Mitchell, B.G., Muller-Karger, F.E., Sosik, H.M., Voss, K.J., Werdell, J., Platt, T.: An ocean-colour time series for use in climate studies: The experience of the ocean-colour climate change initiative (OC-CCI). Sensors **19**(19), 4285 (2019). https://doi.org/10.3390/s19194285, www.mdpi.com/1424-8220/19/19/4285
29. Shaw, G., Burke, H.h.K.: Spectral imaging for remote sensing. Lincoln Lab. J. **14**(1) (2003)
30. Ssenyonga, T., Frette, O., Hamre, B., Stamnes, K., Muyimbwa, D., Ssebiyonga, N., Stamnes, J.J.: A new algorithm for simultaneous retrieval of aerosols and marine parameters. Algorithms **15**(1), 4 (2022). https://doi.org/10.3390/a15010004, www.mdpi.com/1999-4893/15/1/4
31. Wang, G., Lee, Z., Mishra, D.R., Ma, R.: Retrieving absorption coefficients of multiple phytoplankton pigments from hyperspectral remote sensing reflectance measured over cyanobacteria bloom waters: Retrieval of absorption coefficients of multiple pigments. Limnol. Oceanogr. Methods **14**(7), 432–447 (2016). https://doi.org/10.1002/lom3.10102, www.onlinelibrary.wiley.com/doi/10.1002/lom3.10102

Lunar Site Preparation and Open Pit Resource Extraction Using Neuromorphic Robot Swarms

Jekan Thangavelautham

Abstract Decentralized multirobot systems offer a promising approach to automate labor-intensive dull and dangerous tasks in remote environments. Use of multiple robots provide inherent advantages including parallelism, fault-tolerance, reliability and scalability in design. However, this is offset by significant challenges from antagonism, when multiple robots trying to perform the same task interfere and undo the work of others causing unreliable performance or worse, gridlock. Conventional approaches rely on domain knowledge of a task at hand. In the absence of domain knowledge, human designers learn the required multirobot coordination strategies through trial and error. Using an artificial Darwinian approach, we automate the controller design process. This approach uses the Artificial Neural Tissues (ANT) that combines a typical feed-forward wired neural-network with a wireless chemical signaling scheme that facilitates self-organized task decomposition. ANT overcomes limitation with conventional and variable topology neural networks to find efficient solutions to a resource-collection task intended for unstructured environments. In this resource collection task, teams of robots need to forage for resources, collect and dump at a designated location. They need to learn to interpret a series of unlabeled cues to identify the dump location and may choose to exploit assets such as a light beacon to home in on these locations. The key to ANT's advantage is its ability to perform trial and error exploration more efficiently enabling it to acquire creative solutions such as bucket brigades that enable effective cooperation with increased number of robots.

Keywords Multirobot · Neural-networks · Evolutionary algorithms

J. Thangavelautham (✉)
Space and Terrestrial Robotic Exploration (SpaceTREx) Laboratory, University of Arizona, Tucson AZ 85721, USA
e-mail: jekan@arizona.edu

C. Ieracitano et al. (eds.), *The Use of Artificial Intelligence for Space Applications*, Studies in Computational Intelligence 1088,
https://doi.org/10.1007/978-3-031-25755-1_25

1 Introduction

Use of multiple robots to autonomously perform labor intensive tasks has many useful applications. Decentralized control methods offers some inherent advantages, including fault tolerance, parallelism, reliability, scalability and simplicity in agent design [1]. This approach is loosely inspired by behavior of social insects such as ants, termites and bees. These eusocial insects without any centralized coordination produce scalable group behaviors to build elaborate structures including tunnels, hives and cathedral mounds containing internal heating and cooling systems [2].

Conventional methods and behavior-based control requires domain knowledge of a task at hand. It is often unclear how best to organize and control the individuals to perform the required group behaviors. Work by [3–6] rely on task-specific human knowledge to develop simple "if-then" rules and coordination behaviors to solve multirobot tasks. Domain knowledge and a priori human experience is used to determine whether the robots are assembled into formation or are scattered throughout the work area. Decentralized methods use an automated algorithm to perform the partitioning but require imposing physical boundaries or requiring mapping and localization capability in an unstructured setting. Wrong choice of behaviors limits overall system performance and result in *antagonism* when multiple individuals trying to perform the same task interfere or undo each other, resulting in unreliable group performance or worse, gridlock.

Our innovative approach uses an artificial Darwinian approach to automate the controller design process through trial and error learning. In this work, we use an artificially evolvable neural network architecture that draws heavily from nature called the "Artificial Neural Tissue" framework [7, 8]. The Artificial Neural Tissue (ANT) is an adaptive approach that learns to solves tasks through trial and error. ANT superimposes on a typical feed forward or recurrent neural-network a mapping scheme that can activate and inhibit segments of the genome (genotype) and the neural network (phenotype).

Here we focus on the resource-collection task, which is motivated by plans for open pit mining and processing of raw material such as water (ice) on the lunar surface. In this paper the ANT framework is expanded from [13, 14]. The innovation presented include methods for producing scalable controllers including using motor primitives that produce increasingly diverse toolbox of behaviors. We show that although the search space maybe increased from introducing these methods, the diverse behaviors introduced can better tackle antagonism than previous methods. It is shown that even though each individual robot has no "blueprint" of the overall task and operate with only local sensor data. The desired global behavior appears through emergence from the local interactions of robots.

Importantly, conventional fully connected neural network approaches are shown unable to find solutions to the resource collection task. Fixed topology partially connected neural networks rarely find solutions and this requires judicious design by an experimenter. ANT controllers can automatically find solutions to the task without human input in the design of the network topology and choice of output behaviors. Importantly ANT controllers discover mechanisms used by social insects to solve tasks. This includes *templates* [15] (unlabeled environmental cues), *stigmergy* [16] (indirect communication mediated through the environment), and self-organization [17]. Interestingly, ANT shows improved performance when provided less constraints in terms of output behaviors. Using this approach, ANT is shown to tackle the problem of controller scalability and antagonism.

The paper is organized as follows. Section 2 presents background and related work. Section 3 presents the Artificial Neural Tissue approach. Section 4 presents the resource-collection task used to demonstrate ANT's capabilities. Results and discussion follow in Sect. 5.

2 Background

The design and control of multirobot systems have been heavily inspired by evidence of cooperation in social insect societies. Some of the identified mechanisms facilitate cooperative behaviors such as templates, stigmergy, and self-organization. *Templates* are features in the environment perceptible to the individuals within the collective [15]. *Stigmergy* is a form of coordination mediated through the environment [16]. *Self-organization* describes emergence, where local behaviors give rise to a globally coordinated structure in systems which are not in equilibrium [17]. These are advantageous traits of natural systems which would worthwhile implementing on robotic systems. However, many multiagent systems can also suffer from negative traits such as *antagonism*. Antagonism occurs when multiple individuals trying to perform the same task interfere or undo each other and reduce the efficiency of the group or worse lead to gridlock.

In insect nests, templates may be features or events or they may be created by the insect colonies. Examples include light gradients, temperature gradients, chemical gradients or humidity gradients. In robotics, template-based methods include use of light fields to direct the creation of circular [18] and linear walls [?] and planar annulus structures [5]. Spatiotemporally varying templates allow the creation of complex structures [19].

Stigmergy describes exploitation of changes in the environment as a means to perform communication or coordination between agents. Ants and termites perform stigmergy through the use of pheromone trails. Stigmergy has been modelled and used extensively in collective-robotic construction tasks, including blind bull dozing [4], box pushing [3], heap formation [20] and tiling pattern formation [21].

The works above rely on using a well studied, identified mechanism for collective coordination by using user-defined deterministic "if-then" rules or probabilistic behaviors. Our focus is to design a simple controller for agents that have access only to local information, but are able to work together to achieve a global objective. This is difficult to perform manually, since it is hard to predict the consequences of local behaviors gobally. A simple method with limited effectiveness is to design a controller for a single robot to complete the task and have it treat any other robots as obstacles to be avoided. Using multiple robots, the workspace is partitioned or allocated for each robot. This often requires additional sensing and accurate localization which is challenging in unstructured environments. Foraging benefits from this category of controllers, with the robots have minimal inter-robot interaction and being dispersed and working in parallel to maximize area coverage. A more complete solution includes an extra set of rules to handle the interaction with other robots gracefully. This usually includes communication between the robots or use of sensing methods to maintain a minimal distance. It is rarer to find scenarios where the controllers are designed ground-up with cooperation and interaction. Multirobot formations fall into this category and as will shown here could provide significant performance improvements for specific tasks. Environmental templates can also be used to govern multirobot cooperation behaviors, and seed and/or trigger flocking formations [22].

In [20], a group of robots make use of stigmergy in performing a clustering task. In [4], robots perform the opposite task, area clearing, and [18] uses templates to direct the construction of a wall. For these tasks, a single robot is capable of performing the entire task. The controllers use a series of "if-then" rules that treat other robots as obstacles to be avoided. In [20], we note that for more than three robots, the task completion efficiency begins to decrease. In [4], although increasing robot numbers increases efficiency, there is less control over the final shape and size of the area cleared.

For the box-pushing task [3], a controller for a two-robot system was designed with cooperation in mind. This controller is shown to exhibit improved performance compared to two non-cooperating robots trying to perform the same task. However, these controllers communicate at each step to share complete state information with one another. Unfortunately, this does not scale up to larger groups of robots.

Most conventional methods are designed first with a single agent in mind, and then are scaled up (based on human knowledge) with ad-hoc arbitration rules that enable multiple robots to interact. The result is that the interactions between robots are much less cooperative and more antagonistic. It is more challenging to manually design controllers with cooperation in mind, because it is challenging to predict or control global behaviors that will self-organize from even known local interactions. Designing globally coordinated controllers manually can turn into a process of trial and error, especially in the case of the first two examples described above.

A method to simplify manual design of cooperating robot controllers is to encode the controllers as look-up tables and allow a GA (Genetic Algorithm) to evolve the table entries. This method solves a heap formation task in [24] and a 2×2 tiling formation task in [21]. This method offers a simple yet elegant strategy to obtaining group behaviors required to solve a task.

Unfortunately, lookup tables lack dimensional scalability as the size of the look-up table grows exponentially with linear increase in sensory input. Look-up tables are also poor at generalization. Neural network controllers perform better generalization since they can encode a compressed representation of lookup tables. Neural controllers have been used to solve the 3×3 tiling tasks [25] and to build walls, corridors, and briar patches [26], and have been used for multirobot communication and coordination [27]. When using fixed neural networks networks, the size of the network must be specified ahead of time. Choosing the wrong topology may lead to a network that is either unable to solve the problem or is difficult to train [7]. Similarly variable topology neural networks such as NEAT [28] face the same problem, wrong topologies can make the problem difficult or impossible to solve. The problem occurs when there is spatial crosstalk between parallel interconnected neurons [23]. One or more spurious neurons can drown out important signals coming from rest of the network. A larger network faces more challenges as there is a greater chance spurious neurons can drown out important signals, making training difficult or intractable.

The ANT controllers presented here is a bio-inspired approach that simultaneously addresses both the problems in designing rule-based systems manually and the problem of spatial crosstalk inherent in conventional fixed and variable topology neural networks. ANT uses a coarse coding regulation mechanism to dynamically activate and inhibit networks of neurons within a larger pool. This reduces the number of active interconnected neurons needed thus reducing spatial crosstalk. These observations will be presented later in the paper provides good scalability and generalization of sensory input [7]. ANT does not rely on detailed task-specific knowledge. It starts with a blank slate and evolves controllers to optimize a user-specified global fitness function. Evolutionary training results in novel solutions that learn to exploit template, stigmergy and self-organization. and better handle the negative effects of antagonism.

3 Artificial Neural Tissue

The ANT architecture presented in this paper consists of a developmental program encoded in a software defined 'genome,' that constructs an artificial neural tissue and regulatory functions. The tissue consists of two types of neurons, *decision neurons* and *motor-control neurons*, or motor neurons placed in three-dimensional space. Assume a tissue containing randomly generated motor neurons connected electrically by wires (Fig. 1a). Chances are most neurons will produce incoherent output, while a few may produce desired functions by random chance. If the signal from all of these neurons are combined, then these "noisy" neurons would drown out the output signal (Fig. 1b) due to spatial crosstalk [23]. Within ANT, decision neurons emit chemicals that diffuse omnidirectionally (shown shaded) (Fig. 1d). By coarse-coding multiple diffusion fields, the desired motor neurons are selected and "noisy" neurons inhibited, referred to as neural regulation. With multiple diffusion fields superpositioned (Fig. 1d ii), there is some redundancy and when one decision neuron is damaged the desired motor neurons are still selected.

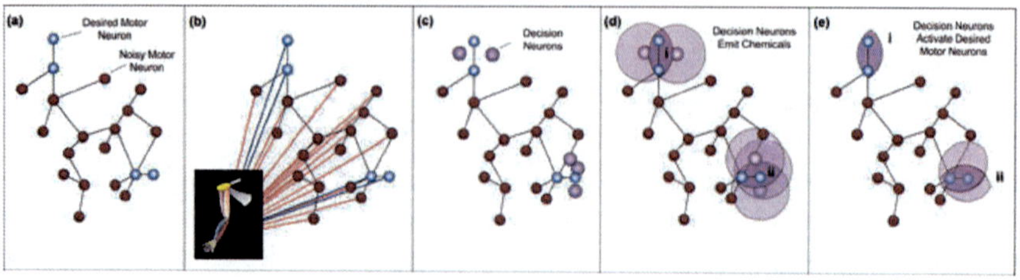

Fig. 1 In a randomly generated tissue, most motor neurons would produce spurious/incoherent output **a** that would 'drown out' signals from a few desired motor neurons due to spatial crosstalk [23] **b**. This can make training intractable for difficult tasks. Neurotransmitter (chemicals) emitted by decision neurons **c** selectively activate networks of desired motor neurons in shaded regions (i) and (ii) by coarse-coding overlapping diffusion fields as shown **d**. This inhibits noisy motor neurons and eliminates spatial crosstalk **e**

3.1 Motor Neurons

We imagine the motor neurons within the tissue to be arranged in a regular rectangular lattice in which the neuron N_λ occupies the position $\lambda = (l, m, n) \in \mathbb{I}^3$ (Fig. 2). Depending on the activation functions used, the state $s_\lambda \in \mathbb{S}$ of the neuron is either binary, i.e., $\mathbb{S}_b = \{0, 1\}$ or can be real, $\mathbb{S}_p = [0, 1]$ or $\mathbb{S}_r = [-1, 1]$.

Each neuron N_λ nominally receives inputs from neurons N_κ where $\kappa \in \Uparrow(\lambda)$, the nominal input set. Here we shall assume that these nominal inputs are the 3×3 neurons centered one layer below N_λ; in other terms, $\Uparrow(\lambda) = \{(i, j, k) \mid i = l-1, l, l+1;\ j = m-1, m, m+1;\ k = n-1\}$. (As will be explained presently, however, we shall not assume that all the motor neurons are active all the time.)

The sensor data are represented by the activation of the sensor input neurons $N_{\alpha_i}, i = 1 \ldots m$, summarized as $A = \{s_{\alpha_1}, s_{\alpha_2} \ldots s_{\alpha_m}\}$. Similarly, the output of the network is represented by the activation of the output neurons $N_{\omega_j}, j = 1 \ldots n$, summarized as $\Omega = \{s_{\omega_1^1}, s_{\omega_2^1}, s_{\omega_3^2} \ldots s_{\omega_n^b}\}$, where $k = 1 \ldots b$ specifies the output behavior. Each output neuron commands one behavior of the agent. (In the case of a robot, a typical behavior may be to move forward a given distance. This may require the coor-

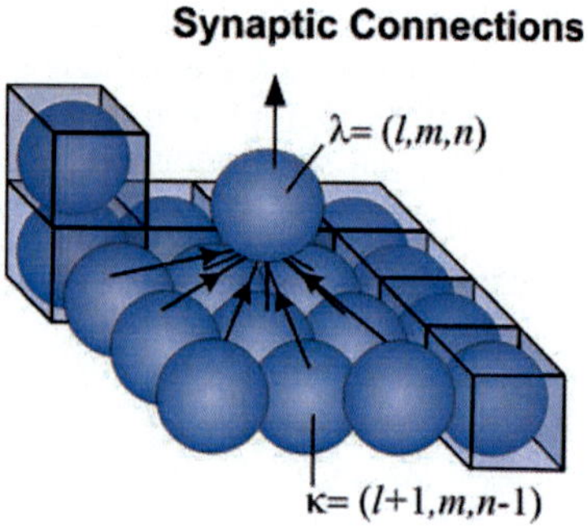

Fig. 2 Synaptic connections between ANT motor neurons from layer $l + 1$ to l

dinated action of several actuators. Alternatively, the behavior may be more primitive such as augmenting the current of a given actuator.) If $s_{\omega_j^k} = 1$, output neuron ω_j votes to activate behavior k; if $s_{\omega_j^k} = 0$, it does not. Since multiple neurons can have access to a behavior pathway, an arbitration scheme is imposed to ensure the controller is deterministic where $p(k) = \sum_{j=1}^{n} \gamma(s_{\omega_j^i}, k) s_{\omega_j^i} / n_k$ and $n_k = \sum_{j=1}^{n} \gamma(s_{\omega_j^i}, k)$ is the number of output neurons connected to output behavior k where $\gamma(s_{\omega_j^i}, k)$ is evaluated as follows:

$$\gamma(s_{\omega_j^i}, k) = \begin{cases} 1, & \text{if } i = k \\ 0, & \text{otherwise} \end{cases} \tag{1}$$

and resulting in behavior k being activated if $p(k) \geq 0.5$.

As implied by the set notation of Ω, the outputs are not ordered. In this embodiment, the order of activation is selected randomly. We are primarily interested here in the statistical characteristics of relatively large populations but such an approach would likely not be desirable in a practical robotic application. However, this can be remedied by simply assigning a sequence *a priori* to the activations. We moreover note that the output neurons can be redundant; that is, more than one neuron can command the same behavior, in which case for a given time step one behavior may be 'emphasized' by being voted multiple times.

3.2 The Decision Neuron

The coarse coding nature of the artificial neural tissue is performed by the decision neurons. Coarse coding is a distributed representation that uses multiple overlapping coarse fields to encode a finer field [9, 10]. For example, coarse receptive fields can be overlapping circles and the area represented by the 'finer' field more accurately encodes for a position than the individual coarse receptive fields.

Decision neurons occupy nodes in the lattice as established by the evolutionary process (Fig. 3). The effect of these neurons is to excite into operation or inhibit the motor control neurons (shown as spheres). Once a motor control neuron is excited into operation, the computation outlined in (3) is performed. Motivated as we are to seek biological support for ANT, we may look to the phenomenon of chemical communication among neurons. In addition to communicating electrically along axons, some neurons release chemicals that are read by other neurons, in essence serving as a 'wireless' communication system to complement the 'wired' one.

Each decision neuron can be in one of two states, one in which it diffuses a neurotransmitter chemical or remains dormant. The state of a decision neuron T_μ, μ is binary and determined by one of the activation functions (see Sect. 3.3). Assuming the decision neurons use the modular activation function described in Sect. 3.3, the inputs to T_μ are all the input sensor neurons N_α; *i.e.*, $s_\mu = \psi_\mu(s_{\alpha_1} \ldots s_{\alpha_m})$ where $\sigma_\mu = \sum_\alpha v_\alpha^\mu s_\alpha / \sum_\alpha s_\alpha$ and v_α^μ are the weights. The decision neuron is dormant if

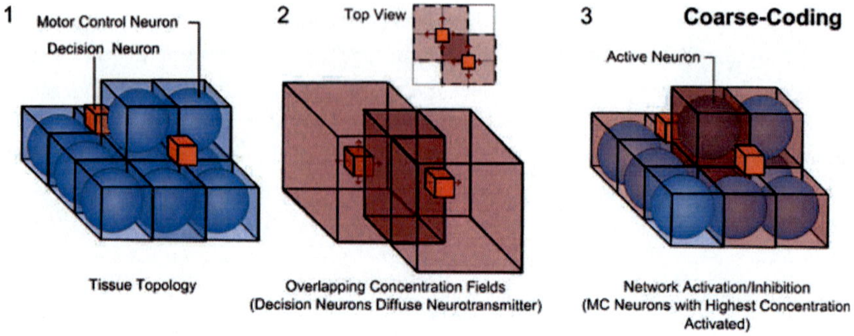

Fig. 3 Coarse coding regulation being performed by two decision neurons (shown as squares) that diffuse a chemical, in turn activating a motor neuron column located at the center (right)

$s_\mu = 0$ and releases a virtual neurotransmitter chemical of uniform concentration c_μ over a prescribed field of influence if $s_\mu = 1$.

Motor control neurons within the highest chemical concentration field are excited into operation. Only those neurons that are so activated will establish the functioning network for the given set of input sensor data. Owing to the coarse coding effect, the sums used in the weighted input of (2) are over only the set $\overline{\Uparrow}(\lambda) \subseteq \Uparrow(\lambda)$ of active inputs to N_λ. Likewise the output of ANT is in general $\overline{\Omega} \subseteq \Omega$. The decision neuron's field of influence is taken to be a rectangular box extending $\pm d^r_\mu$, where $r = 1, 2, 3, ...,$ from μ in the three perpendicular directions. These three dimensions along with μ and c_μ, the concentration level of the virtual chemical emitted by T_μ, are encoded in the genome.

Decision neurons emit chemicals that are used to selectively activate and inhibit motor control neurons. We label this component of ANT as the *neural regulatory system*. This is akin to genes being able to activate or inhibit other genes in a gene regulatory system.

3.3 Activation Function

The modular activation function allows selection among four possible threshold functions of the weighted input σ. The use of two threshold parameters, allows for a single neuron to compute the XOR function, in addition to the AND and OR functions. For this version of the modular activation function,

$$
\begin{aligned}
(\sigma) &= \begin{cases} 0, & \text{if } \sigma \geq \theta_1 \\ 1, & \text{otherwise} \end{cases} \\
(\sigma) &= \begin{cases} 0, & \text{if } \sigma \leq \theta_2 \\ 1, & \text{otherwise} \end{cases} \\
(\sigma) &= \begin{cases} 0, & \min(\theta_1, \theta_2) \leq \sigma < \max(\theta_1, \theta_2) \\ 1, & \text{otherwise} \end{cases} \\
(\sigma) &= \begin{cases} 0, & \sigma \leq \min(\theta_1, \theta_2) \text{ or } \sigma > \max(\theta_1, \theta_2) \\ 1, & \text{otherwise} \end{cases}
\end{aligned}
\tag{2}
$$

and the weighted input σ_λ for neuron N_λ is nominally taken as

$$
\sigma_\lambda = \frac{\sum_{\kappa \in \Uparrow(\lambda)} w_\lambda^\kappa s_\kappa}{\sum_{\kappa \in \Uparrow(\lambda)} s_\kappa} \tag{3}
$$

with the proviso that $\sigma = 0$ if the numerator and denominator are zero. Also, $w_\lambda^\kappa \in \mathbb{R}$ is the weight connecting N_κ to N_λ. We may summarize these threshold functions in a single analytical expression as

$$
\psi = (1 - k_1)[(1 - k_2) + k_2] + k_1[(1 - k_2) + k_2] \tag{4}
$$

where k_1 and k_2 can take on the value 0 or 1. The activation function is thus encoded in the genome by k_1, k_2 and the threshold parameters $\theta_1, \theta_2 \in \mathbb{R}$.

4 Resource-Collection Task

We utilize the resource-collection task to test the effectiveness of the ANT controller. A team of robots fetch resource material dispersed within the workspace and deposits it in a designated dumping area for further processing by a larger robot (Fig. 4). The workspace is a two-dimensional grid space with one robot occupying four grid squares. Each robot controller needs to acquire a number of well-tuned behaviors

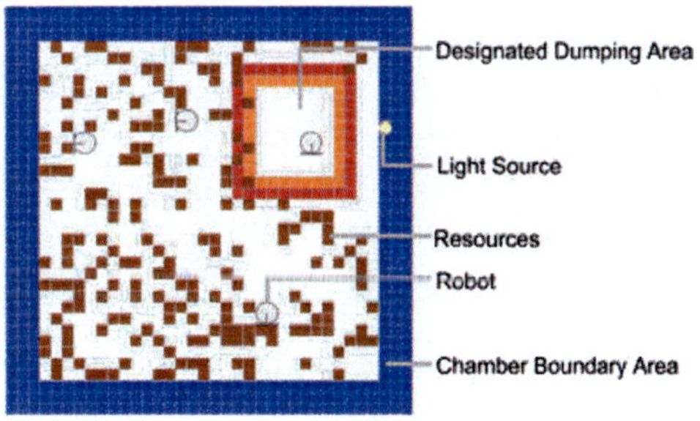

Fig. 4 2D grid world model of experiment workspace

Table 1 Sensor Inputs

Sensor Variables	Function	Description
$V_1 \ldots V_4$	Resource detection	Resource, No Resource
$C_1 \ldots C_4$	Template detection	Blue, Red, Orange, Floor
S_1, S_2	Obstacle detection	Obstacle, No obstacle
LP_1	Light position	Left, Right, Center
LD_1	Light range	0–10 (distance to light)

including gathering resource material dispersed within the workspace, avoiding the perimeter, avoiding robot collisions, and forming resources into mounds at the designated dumping location. The dumping location has perimeter markings on the floor and a light beacon mounted nearby. The two colored border are intended to allow the controller to determine whether the robot is inside or outside the dumping area. Solutions can be found without utilizing the light beacon. However, its presence improves the efficiency of the solutions found, as it allows the robots to efficiently move to the dumping area instead of randomly perusing the workspace in search of the perimeter. The fitness function for the task measures the amount of resource material accumulated in the dumping location after a finite number of time steps averaged over 100 different initial conditions.

Sensory inputs to the ANT controller are shown in Table 1. These inputs are used to detect resources, templates, obstacles and a light beacon. The robots are modeled on a fleet of Argo V1.0 rovers designed and built at the University of Toronto (see Fig. 5). A layout of the sensor inputs is shown in Fig. 6. The robots come equipped

Fig. 5 One of four Argo robots use to validate the ANT controllers for the resource-collection task. Each robot is equipped with a bulldozer blade, front and back sonars and a pair of Logitech® Quickcams™ affixed to pan-tilt units

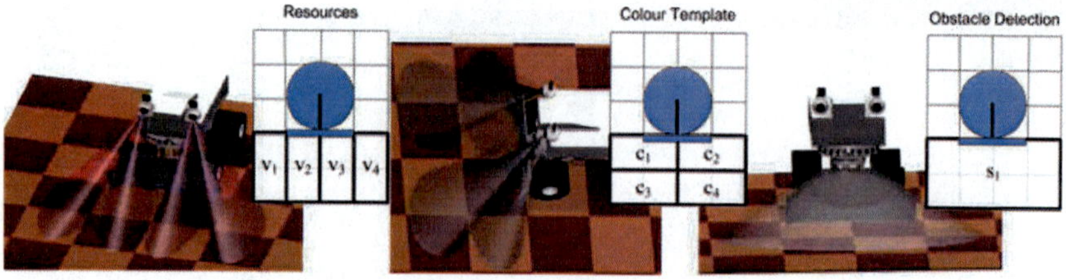

Fig. 6 Input sensor mapping, with simulation model inset

with a pair of webcams that can independently tilt and pan. In-addition the rovers contain body-mounted arrays of sonars. All raw input data are discretized. The sonar sensors are used to compute the values of S_1 and S_2. One of the cameras is tilted downwards and used to detect resource material and colored floor templates. The other camera is used to tilt up and track the light beacon. To identify colored floor templates and resources, a naïve Bayes filter is used to perform color recognition [29]. Simple vision-based heuristics are used to determine the values of $V_1 \ldots V_4$ in-addition to $C_1 \ldots C_4$ based on the grid locations shown. To detect the light beacon, the electronic shutter speed is adjusted to show only the light beacon and mask out all other surrounding objects. The position of the light LP_1 is estimated using the camera pan angle. The light source distance, LD_1 is estimated based on size of the light within the captured image. The robots have access to four internal memory bits, that can be modified using some of the basis behaviors/behavior primitives. Together there are $2^4 \times 4^4 \times 2^2 \times 3 \times 11 \times 2^4 = 8.7 \times 10^6$ possible combination of sensor inputs.

Table 2 lists a pre-ordered set of basis behaviors the robots can perform. These behaviors are activated based on ANT controller output, and all occur within a single time-step. Noting that each behavior in Table 2 can be triggered or not for any one of 8.7×10^6 possible combination of sensor inputs, there is a total of $2^{12\times 8.7\times 10^6} \approx 10^{3\times 10^7}$ possible states in the search space! In some experiments, ANT uses a set of behavior primitives shown in Table 3, where the order of execution of the behavior primitives is an evolved parameter.

Task decomposition is required strategy to tackle very large search spaces and find desired solutions. ANT using its coarse-coding scheme described earlier is shown to perform task decomposition [8], by segmenting the search space. ANT controllers are first evolved in a gridworld simulation environment. The Evolutionary Algorithm population size for training is $P = 100$, crossover probability $p_c = 0.7$, mutation probability $p_m = 0.025$ and a tournament size of $0.06P$. The tissue is initialized with a 3×6 seed culture consisting of motor control neurons in one layer and pre-grown to include between 10–110 neurons (selected randomly) before starting training. Initializing with random number of neurons was found to produce a diverse initial population irrespective of the task being trained for.

Table 2 Basis Behaviors

Order	Behavior	Description
1	Dump resource	Move one grid square back; turn left
2	Move forward	Move one grid square forward
3	Turn right	Turn 90° right
4	Turn left	Turn 90° left
5, 7, 9, 11	Bit set	Set memory bit i to 1, $i = 1 \ldots 4$
6, 8, 10, 12	Bit clear	Set memory bit i to 0, $i = 1 \ldots 4$

Table 3 Motor Primitives

Neuron ID	Behavior	Coupled Motor Signals
1	Move forward	Left motor 1 || Right motor 1
2	Move backward	Left motor −1 || Right motor −1
3	Turn right 90°	Left motor 1 || Right motor −1
4	Turn left 90°	Left motor −1 || Right motor 1
5	Pivot right	Left motor 0 || Right motor −1
6	Pivot left	Left motor 0 || Right motor 1
7	Pivot right	Left motor 1 || Right motor 0
8	Pivot left	Left motor −1 || Right motor 0
9	Slide right	Turn right, fwd, turn left
10	Slide left	Turn left, fwd, turn right
11	Diagonal right	Fwd, turn right, fwd, turn left
12	Diagonal left	Fwd, turn left, fwd, turn right
13, 15, 17, 19	Bit set	Set memory bit i to 1, $i = 1 \cdots 4$
14, 16, 18, 20	Bit clear	Set memory bit i to 0, $i = 1 \cdots 4$

5 Results and Discussion

Figure 7 shows the fitness of ANT controllers evaluated at each generation of the artificial evolutionary process. The fitness is simply a measure of the fitness of the population best at each generation. The evolutionary runs show an increase in performance with more robots. With more robots, there is a decrease in work efficiency, with each robot having a smaller area to cover in trying to gather and dump resources. The simulation runs identify a point of diminishing returns is reached, where beyond this point, additional robots provide minimal contribution to the collective. The evolutionary

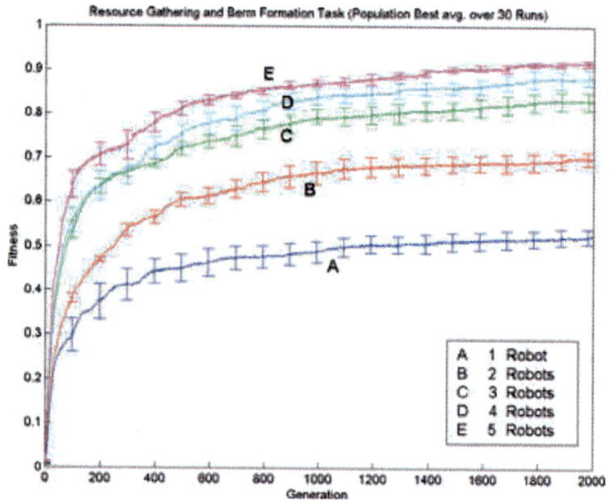

Fig. 7 Evolutionary performance comparison of ANT-based solutions for one to five robots. Error bars indicate standard deviation

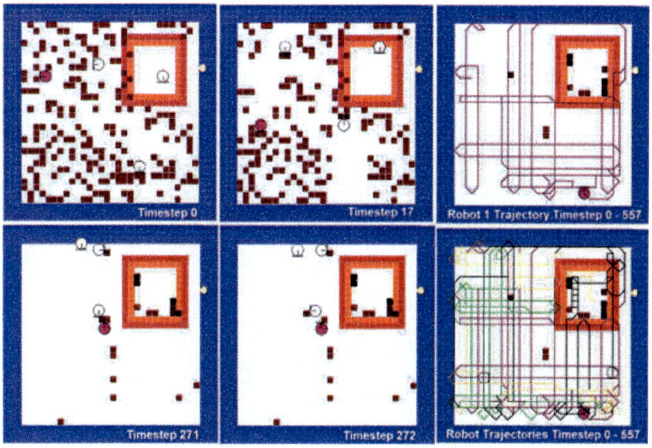

Fig. 8 Workspace snapshots of robots and trajectories for the resource gathering task

process enables self-organized decomposition of a goal task based on global fitness, and the tuning of behaviors depending on the robot density. Snapshots of an ANT controller solution completing the resource collection task is shown Fig. 8.

5.1 Behavior Scalability

Next we compare ANT to other methods for the resource gathering task, including fixed topology feed-forward neural networks and variable topology NEAT [28] in Fig. 9. For this comparison, topologies for NEAT and fixed neural networks are randomly generated and contain between 10 and 110 neurons to make the comparison meaningful with ANT. For the fixed feed-forward neural network, we compare two variants, ones that are fully connected (FC) and partially connected (PC), with a maximum of nine connections feeding into one neuron. As expected, the worst results are obtained with the fully connected neural networks, where fitness performance is lower than all other cases and finds no solutions (fitness > 0.9). Fully connected neural networks are faced with the problem of spatial crosstalk, where noisy signals can drown out signals from important neurons making training slow, difficult or intractable. Partially connected fixed and variable topology networks tend to have more 'active' synaptic connections and thus takes longer for each neuron to tune these connections. ANT is an improvement as the decision neurons learn to actively mask out/quickly shut off spurious neurons producing fitter solutions compared to conventional feed-forward neural networks (Fig. 10).

Further, we analyze the effect of segmenting the output basis behaviors into motor primitives. Motor primitives are taken as discrete voltage signals to a motor. This enables many combinations of motor actuation possible and thus increases the search

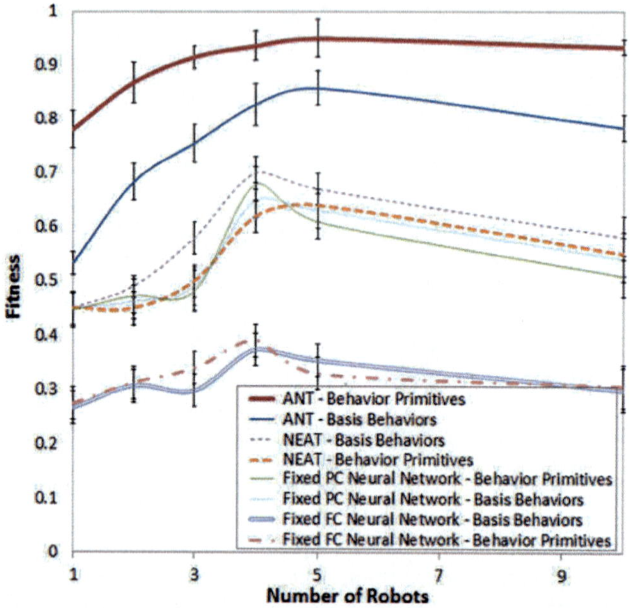

Fig. 9 Fitness Comparison for the resource-collection task using various neural network control approaches. ANT based controllers, particularly using behavior primitives show the best performance. Conventional approaches including fixed and variable network topology, with partially connected networks show comparable performance, while fully connected neural networks show the worst performance

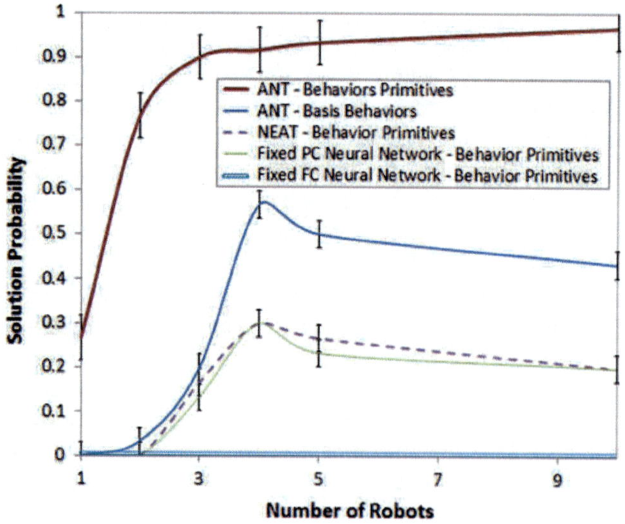

Fig. 10 Probability of obtaining a full solution (fitness >0.9) for the resource gathering task using various neural network control approaches. Fixed fully connected (fc) neural networks are unable to find solutions to the task

space of possible controller output. Further, we allow the order of execution of the behavior primitives to be genetically evolved instead of using a predetermined order presented previously. ANT shows increased fitness and improved scalability, as will be seen later using motor primitives. However we find that conventional approaches, including standard fixed topology neural networks and variable topology NEAT are unable to exploit motor primitives to obtain a substantial fitness advantage.

There exists a trade-off, increased search space decreases performances because it takes longer to find good solutions, however more combinations of outputs can find diverse or fitter solutions. Analyzing the results, it is found that controllers evolve to traverse the experimental area diagonally and in zig-zag patterns. Traveling diagonally in combination with phototaxis permits the ANT controllers to travel shorter distances (and in shorter time) to and from the dumping area. Zig-zag traversal improves area coverage when collecting resources. After most resources are collected, the robots settle into a series of periodic motions of following a rectangular trail inside the perimeter (Fig. 8). However one or more robots may encounter an obstacle and thus a make a left turn or right turn breaking away from the trail thus traversing an unexplored area. Zig-zag traversal allows the robots settle into multiple trails that overlap better enabling them to traversing unexplored areas and forage for resources.

5.2 *Controller Scalability*

Having found that ANT can find fitter solution to the resource gathering task than conventional methods, we analyze the fittest ANT solutions from the simulation runs for scalability in robot number while holding the amount of resources constant (Fig. 11). The results for ANT using basis behaviors is presented in [13]. The results for behavior primitives show an improvement in fitness compared to those with basis behaviors due to the behavior primitives providing many more possible solution strategies for slight increase in search space. A controller evolved for a single robot and running it on a multirobot system shows poor performance as the controller is unable to adjust to the shock of a multirobot setting which requires social coordination. These observations suggest the training conditions need to be carefully weighed and modified to match representative target conditions. We find when more than four robots are utilized, that results in a gradual *decrease* in performance due to the increased effects of antagonism.

The scalability of the evolved solution is found to heavily depend on the number of robots used during training. The single-robot controller lacks the social behavior needed to function well within a multirobot setting. For example, such controllers fail to develop collision avoidance or bucket brigade behaviors. On the other hand, the robot controllers evolved with two or more robots perform considerably worse when scaled down showing the solutions are dependent on social interactions to succeed.

With ANT using behavior primitives, we can easily evolve the controller under multiple sets of initial conditions. While this is theoretically possible with basis

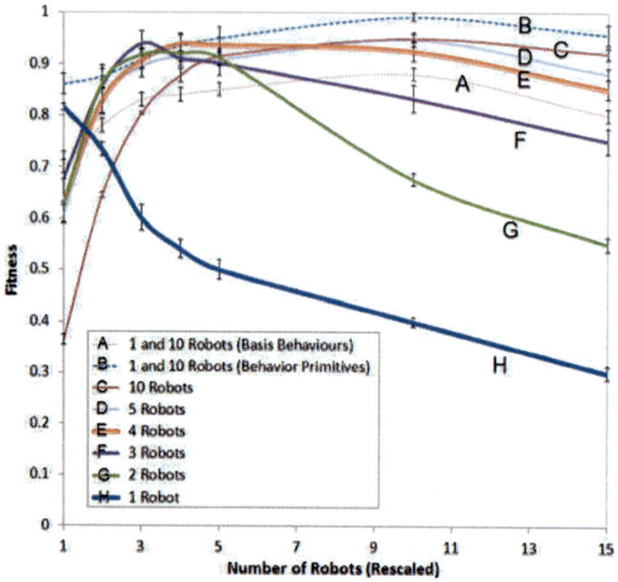

Fig. 11 Comparison of scalability of ANT using behavior primitives evolved using multiple robot scenarios and one to five robots. The results are shown compared with ANT using basis behaviors using multiple robot scenarios

behaviors, we don't see a improvement in performance (Fig. 11). With this approach, we can choose a mixture of training runs consisting of the one and ten robot scenarios. Analyzing the scalability of this controller, the fitness performance on average is higher than all others, showing substantially improved scalability, with excellent performance for both single robot and multirobot scenarios. This modified training approach using behavior primitives overcomes the scalability problem encountered with ANT using basis behaviors. So why does the preselected basis behaviors perform worse? The behavior primitives allow for a bigger combination of solutions, hence even with this additional training constraint of one and ten robot scenarios, ANT using behavior primitives can still find satisfactory solutions. However with the preselected basis behaviors, the number of solutions possible is restricted and hence in this scenario, the evolutionary algorithm is unable to find fitter solution under the training constraints.

We further analyze to get understanding of how the ANT using behavior primitives obtains its advantage. Figure 12 shows fitness performance of the ANT using behavior primitives evolved for 1 and 10 robots evaluated for scenarios range from 1 and 50 robots. The results a show performance plateau for substantial range of robot densities compared to the other ANT variants which then steadily decreases due to the effects of antagonism. This suggest that the controller has successfully evolved a general method that is adaptive to the number of robots, over robot density that varies by 25-fold (i.e from 1 robot to 25 robots). As will be shown later, this

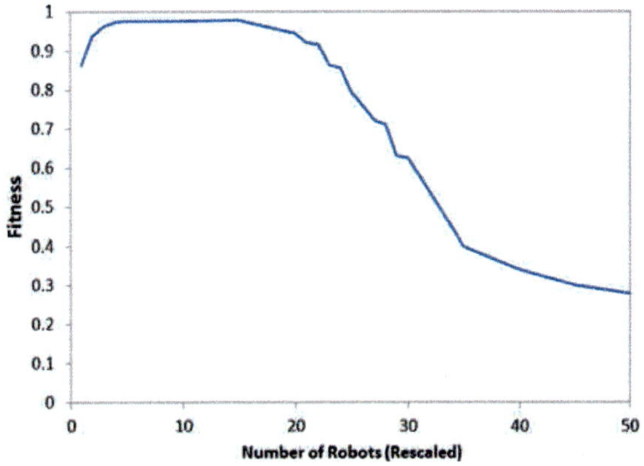

Fig. 12 Scalability performance of ANT controller evolved for one and tens robots using motor primitives evaluated for between one and 50 robots. For between 1 and 20 robots, the controller show excellent scalability performance, but with further increased in number of robots, the performance gradually decreases due to antagonism

further supports who this method discovers and exploits a bucket brigade formation to efficient search, collect and dump the resources at the designated area. A bucket brigade formation is quite adaptive to number of individuals presents. With increased number individual, the distance between each individual in the brigade is shortened up to the point where the individuals are side by side. Beyond that point, there is no advantage to having any more individuals. A further increase in individual results antagonism, which would decrease the overall efficiency of the group.

Bucket Brigade Behavior Analysis of the evolved solutions indicate that the robots learn to dump resources into the dumping area, but interestingly not all robots deliver resource all the way to the dumping area every time. This makes sense as handing off resources to nearby robots is more energy and time efficient than having to carry the resources all the way to the dumping area. The robots evolve to pass the resource material from one to another forming a mobile "bucket brigade" (see Fig. 8). This technique improves the overall efficiency of the group as less time is spent transporting resources to and from the dumping area. Since the robots cannot directly communicate with one another, these encounters happen by chance and varies by location.

Figure 13 shows the probability of obtaining a bucket brigade solution for the resource collection task. Increased number of robots in the workspace increases the probability ANT finds a bucket brigade solution. This intuitively makes sense, because increased number of robots would require robots to act in a coordinated fashion to assemble a bucket brigade. Too few robots decreases the chance of robot-robot encounters making it unlikely for the behavior to be evolved. A bucket brigade approach is overall beneficial because it increases the efficiency of the group. The robots are designated to a limited area, where they are more effective at gather all of the resources and transport it to the neighboring robot, until the resources are gathered to the dumping area. ANT manages to exploit the bucket brigade strategy while fixed neural networks are found unable to exploit this strategy for increased of number of robots. This shows a major weakness in conventional neural networks

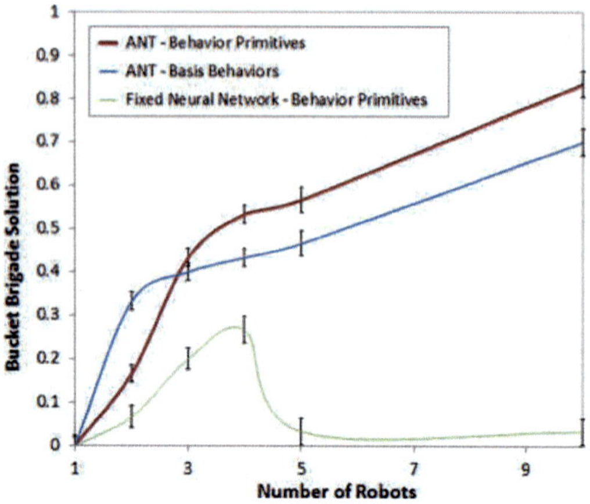

Fig. 13 The probability of solutions using a bucket brigade strategy versus number of robots. ANT controllers learn to use the bucket brigade strategy with increased number of robots because it increases group performanc

in facilitating creative solution strategies. Fixed feed-forward neural networks show limited scalability for increased number of robots. ANT with behavior primitives is found to be more effective than ANT with basis behaviors at using a bucket brigade solution. Behavior primitives provide a larger toolbox of behaviors that enable the robots to select the most effective ones through trial and error to complete the task. It should be noted that even with behavior primitives, ANT is not totally immune to antagonism, rather the controller solutions are made more robust to greater range of robot densities.

6 Conclusions

A neural network architecture called the "Artificial Neural Tissue" (ANT) has been successfully applied to a multirobot resource gathering task with applications to open pit mining. In comparison, conventional fixed and variable topology feed-forward neural networks tested were unable to find robust or creative solutions. ANT is provided a global fitness function, a simulation environment, behavior primitives and sensor input. This facilitates ANT discovering creative solutions that might otherwise be overlooked by a human supervisor. ANT controllers exhibit improved scalability over conventional methods and can reduce the effects of antagonism caused by increasing number of robots. ANT evolved controller are found to be adaptive by producing a spectrum of effective group behaviors such as bucket brigades for varying robot densities. Future work will proceed on applying the technology towards end-to-end robotic site preparation, resource collection and construction tasks on the lunar surface.

References

1. Cao, Y., Fukunaga, A., Kahng, A.: Cooperative mobile robotics?: Antecedents and directions. Auton. Robots **4**, 1–23 (1997)
2. Bristow, K., Holt, J.: Can termites create local energy sinks to regulate mound temperature? J. Therm. Biol. **12**, 19–21 (1997)
3. Matarić, M.J., Nilsson, M., Simsarian, K.T.: Cooperative multi-robot box-pushing. In: IEEE/RSJ IROS, pp. 556–561 (1995)
4. Parker, C.A., Zhang, H., Kube, C.R.: Blind bulldozing: Multiple robot nest construction. In: IEEE/RSJ Int. Conference on Intelligent Robots and Systems, pp. 2010–2015 (2003)
5. Wilson, M., Melhuish, C., Sendova-Franks, A.B., Scholes, S.: Algorithms for building annular structures with minimalist robots inspired by brood sorting in ant colonies. Auton. Robots **17**, 115–136 (2004)
6. Werfel, L., Y. Y., Nagpal, R.: Building patterned structures with robot swarms. In: IJCAI, pp. 1495–1502 (2005)
7. Thangavelautham, J., D'Eleuterio, G.M.T.: A coarse-coding framework for a gene-regulatory-based artificial neural tissue. In: Advances In Artificial Life: Proc. of the 8th European Conference on ALife, pp. 67–77 (2005)
8. Thangavelautham, J., D'Eleuterio, G.: Tackling learning intractability through topological organization and regulation of cortical networks. IEEE Trans. Neural Networks Learn. Syst. **23**, 552–564 (2012)
9. Albus, J.S.: Brains. Behavior and Robotics. BYTE Books, McGraw-Hill (1981)
10. Hinton, G.: Shape representation in parallel systems. In: Proceedings of the 7th International Joint Conference on Artificial Intelligence, pp. 1088–1096 (1981)
11. Garthwaite, J., Charles, S.L., Chess-Williams, R.: Endothelium-derived relaxing factor release on activation of nmda receptors suggests role as intercellular messenger in the brain. Nature **336**(6197), 385–388 (1988)
12. Hölscher, C.: Nitric oxide, the enigmatic neuronal messenger: Its role in synaptic plasticity. Trends Neurosci. **20**(7), 298–303 (1997)
13. Thangavelautham, J., Smith, A., Boucher, D., Richard, J., D'Eleuterio, G.M.T.: Evolving a scalable multirobot controller using an artificial neural tissue paradigm. In: Proceedings of the IEEE International Conference on Robotics and Automation (2007)
14. Thangavelautham J., Xu, Y.: Co-evolution of multi-robot controllers and task cues for off-world open pit miningm. In: Proceedings of the International Symposium on Artificial Intelligence, Robotics and Automation in Space (2020)
15. Bonabeau, E., Dorigo, M., Theraulaz, G.: Swarm Intelligence: From Natural to Artificial Systems. Oxford University Press, New York (1999)
16. Grassé, P.: La reconstruction du nid les coordinations interindividuelles; la theorie de stigmergie. Insectes Sociaux **35**, 41–84 (1959)
17. Bonabeau, E., Theraulaz, G., Deneubourg, J.-L., Aron, S., Camazine, S.: Self-organization in social insects. Trends Ecol. Evol. **12**, 188–193 (1997)
18. Stewart, R., Russell, A.: Emergent structures built by a minimalist autonomous robot using a swarm-inspired template mechanism. In: The First Australian Conference on ALife (ACAL2003), pp. 216–230 (2003)
19. Stewart, R., Russell, A.: Building a loose wall structure with a robotic swarm using a spatio-temporal varying template. In: IEEE/RSJ Int. Conference on Intelligent Robots and Systems, pp. 712–716 (2004)
20. Beckers, R., Holland, O.E., Deneubourg, J.L.: From local actions to global tasks: Stigmergy and collective robots. In: Fourth Int. Workshop on the Syntheses and Simulation of Living Systems, pp. 181–189. MIT Press (1994)
21. Thangavelautham, J., Barfoot, T., D'Eleuterio, G.M.T.: Coevolving communication and cooperation for lattice formation tasks (updated). In: Advances In Artificial Life: Proc. of the 7th European Conference on ALife (ECAL), pp. 857–864 (2003)

22. Labella, T., Dorigo, M., Deneubourg, J.-L.: Self-organised task allocation in a group of robots. In: Alami, R., Chatila, R., Asama, H (eds.) Distributed Autonomous Robotic Systems 6, pp. 389–398. Springer Japan (2007)
23. Jacobs, R., Jordan, M., Barto, A.: Task decomposition through competition in a modular connectionist architecture: The what and where vision tasks. Cogn. Sci. **15**(2), 219–250 (1991)
24. Barfoot, T.D., D'Eleuterio, G.M.T.: An evolutionary approach to multiagent heap formation. Cong. Evol. Comput. (CEC) **1999**, 427–435 (1999)
25. Thangavelautham, J., D'Eleuterio, G.M.T.: A neuroevolutionary approach to emergent task decomposition. In: 8th Parallel Problem Solving from Nature Conference, vol. 1, pp. 991–1000 (2004)
26. Crabbe, F.L., Dyer, M.G.: Second-order networks for wall-building agents. Int. Joint Conf. Neural Networks **3**, 2178–2183 (1999)
27. Trianni, V., Dorigo, M.: Self-organisation and communication in groups of simulated and physical robots. In: Biological Cybernetics, vol. 95, pp. 213–231. Springer, Berlin (2006)
28. Stanley, K., Miikkulainen, R.: Continual coevolution through complexification. In: Proceedings of the Genetic and Evolutionary Computation Conference (GECCO-2002). Kaufmann, pp. 113–120 (2002)
29. Hastie, T., Tibshirani, R., Friedman, R.: The Elements of Statistical Learning. Springer, New York (2001)

Artificial Intelligence for SAR Focusing

Oreste Trematerra, Quirino Morante, and Federica Biancucci

Abstract The article proposes a method of focusing Synthetic Aperture Radar (SAR) RAW data using a deep learning technique. The approach is based on the implementation of a suitable Convolutional Neural Network (CNN), a Deep Neural Network model used to learn efficient representation of input data. Pairs of images acquired by Sentinel-1 mission are downloaded from the Copernicus Open Access Hub and different pre-processing stages are performed before training the Neural Network model to reconstruct the focused image from the RAW data. A regression problem needs to be addressed, in this case the aim is to find and train a suitable Neural Network architecture that is able to process the RAW SAR decompressed data and predict the corresponding focused version to extract useful information directly on-board the satellite.

Keywords Synthetic Aperture Radar (SAR) · Sentinel-1 · CNN · Deep neural network · RAW data

1 Introduction

Before the space age, remote sensing was done exclusively with photographic cameras (aerial photo). Aerial photography was extensively employed during both World Wars for military reconnaissance [1].

Data from Earth observation satellites enables to see how the Earth is changing as a result of human activity. According to the Union of Concerned Scientists (UCS), over 700 Earth observation satellites are in orbit in 2019. They are an important scientific tool with a wide range of applications in areas such as:

- weather forecasting;
- wildlife conservation;

O. Trematerra (✉) · Q. Morante · F. Biancucci
Thales Alenia Space Italia, Via Saccomuro, 24, Rome, Italy
e-mail: Oreste.Trematerra@thalesaleniaspace.com

C. Ieracitano et al. (eds.), *The Use of Artificial Intelligence for Space Applications*, Studies in Computational Intelligence 1088,
https://doi.org/10.1007/978-3-031-25755-1_26

- agriculture;
- resource management;
- natural disaster response;
- climate science.

Today we are moving towards EO missions where response time is increasingly challenging. These missions require near-real time data analysis and ground processing is one of the principal problems of delays in delivering the product on time.

Synthetic Aperture Radar (SAR), is a remote sensing-imaging technique used to form images starting from high frequency radar signals emitted in the direction of a specific Earth's surface region in order to detect its physical properties. These signals reflect off the target region and are received by an antenna on the same imaging device. Once a sufficient number of signal echoes have been received, they are used to provide high-resolution two-dimensional images independent from daylight, cloud coverage and weather conditions with the purpose of monitoring dynamic processes on the Earth surface in a reliable, continuous and global way [2].

Data acquired by a SAR on-board a satellite (Raw Data), must be processed to produce a visible image (at least the L1 Single-Look Complex SLC product). For this reason, data have to be transferred to the Ground Segment (GS) and processed through a time/computing-consuming focusing algorithm mainly due to the use of FFT and IFFT operations. This process (RAW data exchange GS and satellite) is the mainly limit to use SAR mission for application where the time performance is the key. Solutions to move the focus processing directly on-board had failed due limited resources on board of satellite.

This article proposes a Deep Learning based approach to reconstruct the L1 SLC Sentinel-1 product which consists in focused SAR data by means of an Artificial Neural Network (ANN) architecture that is able to simulate the entire multistep SAR processing pipeline that nowadays is used for image focusing. In this way all computational operations within the pipeline are performed through the forward propagation of a single Neural Network (Fig. 1).

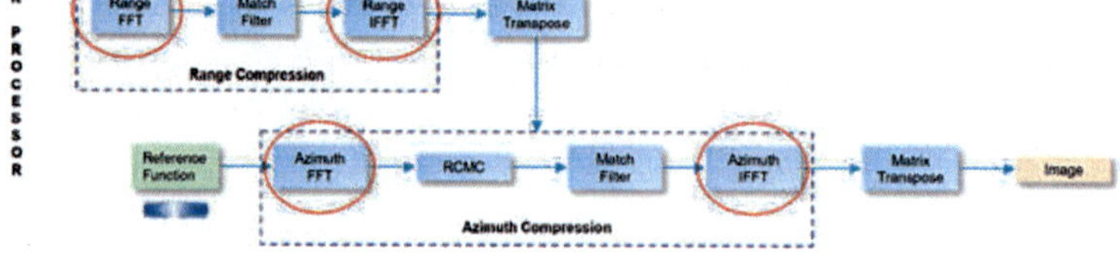

Fig. 1 Example of ground segment algorithm and FFT\IFTT operations

2 Method

SAR processing needs High Performance Computing (HPC) algorithms to process data quickly. Considering the limited computing hardware on-board, data must be transmitted to GS for further processing where processors at state of art are available. The entire end-to-end pipeline from user request to final product delivery takes place in two different physique locations: the Ground Segment (GS) and the Space Segment (SS). The user's request is first assessed to check if its feasibility is in line with the mission plan. Then the request is up-linked on the Space Segment to perform data acquisition. This is followed by the data download, its processing and delivery to the end user.

Computing hardware platforms are mainly used in space missions for data transfer that involves sending RAW data to GS and data processing is done in centralized servers on Earth. Large data transfers could cause network bottlenecks and increases latency, thus increasing the system response time (Fig. 2).

The computational complexity related with the processing steps is another key point to start thinking about a new solution to be applied directly on-board satellite. The growing need for immediate feedback is leading to the use of new technologies that can be integrated directly on-board to allow the RAW data to be analysed directly and provide feedback without going through processing in the GS.

One possible solution requires the use of Artificial Intelligence (AI) algorithms. Deep learning inference can be effectively executed very close to the data sources (e.g., imaging sensors) at the edge device. This may improve the spacecraft autonomy, mitigate costly requirements of transmitting large volumes of data to a remote processing site, accelerate decision-making, and increase reliability in communication-denied environments (i.e., in space) [3].

These considerations lead to think about a new setup where the GS is used to train the Artificial Intelligence architecture exploiting existing appropriate hardware for high computational effort and to handle large amounts of data. Once the AI model is ready, weights are placed on-board satellite to perform near/real time inference over new acquisitions and guarantee a low computational effort. The proposed approach aims to build a neural network architecture to be trained in a supervised way considering L0 data as input and L1 data as labels.

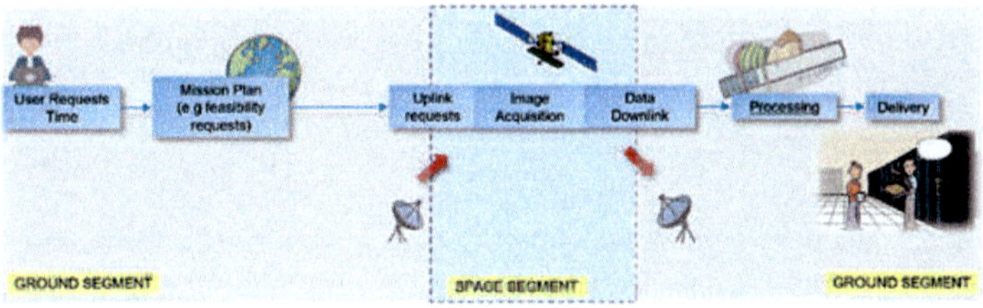

Fig. 2 EO legacy approach

2.1 Dataset

Data are picked-up from the Copernicus Open Access Hub website [4]. The Hub requires an initial authentication to retrieve products and the user credentials must be provided to set up the query. For this purpose, Sentinelsat, a powerful Python API and a command line interface to search, download and retrieve the metadata for Sentinel products, must be downloaded to perform the API connection in Python. Sentinelsat provides different methods for retrieving product metadata from the server. Among them, the query() method is selected for OpenSearch, which supports filtering products by their attributes and returns metadata for all matched products at once. It includes different kind of information that the user can set up in order to search for specific data. For the project purpose each couple of RAW (L0) and SLC (L1) data must be downloaded setting the StripMap (SM) operational mode and the same sensing period in order to be sure that the specific SLC data corresponds to the processed version of the RAW one.

2.2 Dataset Preparation

The packet format used within the Sentinel mission is largely based on the Consultative Committee for Space Data Systems (CCSDS) standard [5]. Sentinel-1 is a set of spaceborne radars operated by the European Space Agency (ESA). The C-band (centre on 5.405 GHz) radiofrequency [I,Q] data streams are relayed by the European Data Relay Satellites (EDRS) before being received by ESA listening stations, processed and stored for being disseminated to users [6]. While ESA will provide free of charge L1 processed synthetic aperture radar data, RAW/L0 original data are not supported. Nevertheless, ESA is providing access to these data and the user can decode and decompress them.

Flexible Dynamic Block Adaptive Quantization (FDBAQ) decompression must be performed to obtain original [I,Q] data stream. It is based on Block Adaptive Quantization (BAQ) compression which uses a scalar quantizer depending on the statistics of the RAW data, in order to quantize them with fewer bits then required by the Shannon entropy. At the end of this process, RAW data are in the same format as the ones acquired on-board. In this way, once the neural network model is trained, it is possible to perform future inferences over new acquisition on-board the satellite (Fig. 3).

The dataset is completely ready once each pair of L0/RAW and L1/SLC data are one by one assigned:

1. Totally 79 data pairs were downloaded, decompressed and de-formatted. Considering that L0 and L1 data files sizes range from a minimum of 2.12 GB up to 6.5 GB, the entire dataset is quite large (~850 GB).
2. Each L0 file is stored in a double channel array, where the dual channel is reserved for the real and imaginary component of the radar echo and each individual value

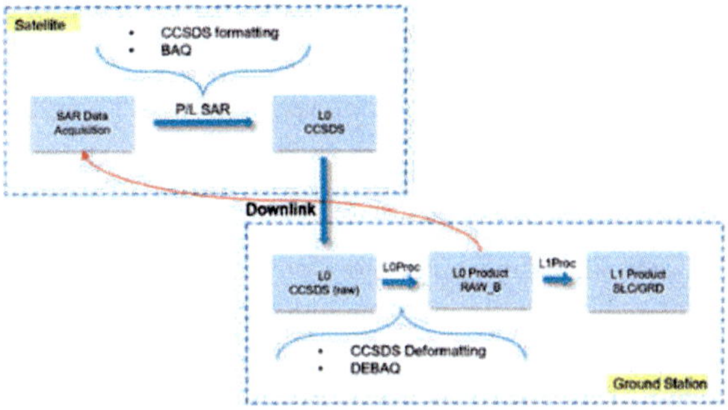

Fig. 3 SAR data flow from SAR RAW up to L1 product

has an accuracy ~~of~~ 8 decimal places. For example, a decompressed L0 data has a size of [48570 × 22,760 × 2].

3. Also, L1 data maintain real and imaginary component. To simplify the problem and to not require additional storage space, the absolute value of L1 data is stored as labels.
4. For each couple of data, L1 azimuth and range dimensions are always smaller than the corresponding L0. The reason is that the SAR processor excludes some values in azimuth and range direction during RAW data processing due to filter matching. To achieve the same size and to ensure a more accurate match between RAW and SLC image, excessive last rows and columns of the RAW array are removed. For example, if the original L0 size is equal to [48570 × 22,760 × 2] and the corresponding L1 size is equal to [46780 × 20860], this means that RAW final 1790 rows and 1900 columns are deleted (Fig. 4).

Fig. 4 Example of RAW (left) and corresponding SLC (right) data

Once excessive rows and columns are removed, trimming of the patches is performed. In the RAW data, the signal energy from a point target is spread in range and azimuth and the purpose of SAR focussing is to collect this dispersed energy into a single pixel in the output image. Accordingly with this energy spread, the size of a single tile is set equal to [916 $\times$ 3034 $\times$ 2] for L0 arrays and [916 $\times$ 3034] for L1 images (respectively azimuth $\times$ range direction). Moreover, the sliding window that is used to cut the tiles performs a 10% overlap to further increase the dataset dimension and to include transmission signal information (chirp) that can go beyond the predefined tile length. At this point, a total number of 48780 patches pairs are collected and this implies that the final dataset size is equal to 1 TB.

2.3 *Normalization*

Rarely, neural networks, as well as statistical methods in general, are applied directly to the raw data of a dataset. The first normalisation is carried out considering the processed SLC data (the labels of the dataset). By analysing each array, L1 pixel values can range from a minimum of 0 up to a maximum of 65000 (16-bit coding). To limit the range, all pixels with a value greater than or equal to 255 are set equal to 255. Then, min–max normalisation is performed to re-scale entire SLC dataset between [0, 1] with respect to minimum and maximum value of the whole dataset.

With regard to RAW data, it is necessary to refer to the adaptability of the neural network which, once trained, must carry out inference on board the satellite. As RAW data is not normalised on-board, some trainings will be carried out using the original data. On the other hand, other trainings have been done by normalising the RAW data between [-1, 1] (min–max normalisation) in order to maintain also the negative components of the original file, since the global minimum of RAW data is equal to -3.1327314 and the maximum is equal to 2.925825.

2.4 *Artificial Neural Networks*

The entire dataset consists of 22,897 couple of images. In order to enable the learning process, these images are randomly split in train, validation and test set, that represent respectively the 70, 25 and 5% of the entire dataset.

Two different Artificial Neural Network (ANN) structures are provided to understand which is the best architecture that led to learning a mapping from RAW L0 input data to visible focused L1 image. The procedure adopted consists in setting two neural network models, a Dual Convolutional Neural Network (Dual-CNN) architecture and a CNN architecture that includes residual layers. The primary goal is to obtain an understandable image so that some kind of information can be retrieved from them.

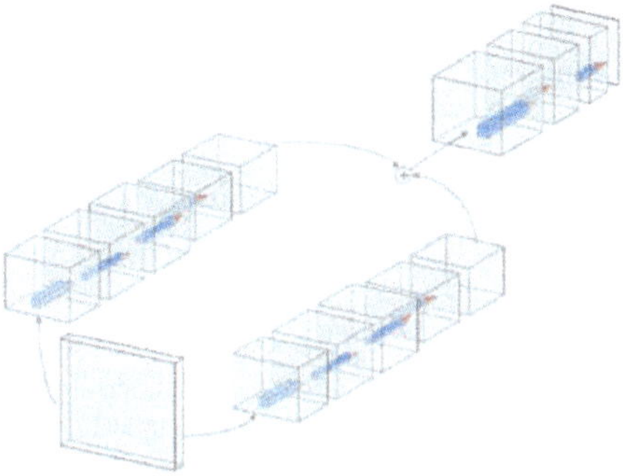

Fig. 5 First network architecture[1]

Setting the parameters of deep neural networks is a heuristic and complicated procedure, especially on atypical data without previous experiments. There is no golden rule for the depth of the network, filter size, the number of filters, stride size, and other hyperparameters. Hyperparameters should be set according to the size of the dataset and the type of data. Due to the limited amount of time, computing power and data, very few combinations of hyperparameters could be tested. From now on, the framework used to implement neural network architectures and train them is Pytorch [7], an optimized Deep Learning tensor library based on Python and Torch and is mainly used for applications using GPUs and CPUs.

Regardless of the model structures that will be described, the mainly used layers are:

- Convolutional layers (Conv);
- Batch-Normalisation layers (BN);
- Dropout layers (Drop);
- Residual layers (Res);
- Max-Pooling2D layers (MaxPool).

The first proposed method consists of a Dual-CNN model design. There is a first section composed by two networks used to extract diverse reliable features; then these features are combined by some final convolutional layers to enhance the extracted features by fusing the features from previous two sub-networks. This technique is particularly useful for images corrupted by unknown types of noise, such as many real-world corrupted image and blind noise [8].

Obviously, the output size of the two networks is the same so that the final result of both can be summed up. Final layer sizes are respectively designed with Conv + ReLU, they are characterised by a gradually decreasing number of feature maps and a filter size of 3×3. Usually, this kind of structure is implemented to recover image structure (super-resolution applications), for noise/artifacts removal and edge-preserving filtering [9] (Fig. 5).

[1] The 3D representation is realised with the https://alexlenail.me/NN-SVG/LeNet.html framework which allows to visualise the structure of the neural network including the convolutional layers

The second neural network architecture is inspired by [10] whose purpose is similar to that proposed by this article, but there are substantial differences in several points. The authors focus only on one type of scenario (mountains) and apply various pre-processing operations to the RAW data. These operations are computationally expensive to be performed on-board the satellite, for this reason, more attention is paid only to their network structure. To map the RAW SAR echo data to the SAR image, the authors used a convolutional encoder similar to VGG11 [11]. The encoder consists of a set of convolutional blocks with an increasing number of filters, from 64 up to 512, each with kernel dimensions 3 × 3. Every convolutional layer is connected to the next via a Leaky ReLU() activation function. Following some convolutional block, MaxPool2D with stride 2 was used, to reduce the dimensionality of the encoded data by two in each dimension.

A similar structure is implemented in this case by modifying the number of feature maps (because of memory problems), the filter sizes and the number of layers (e.g., fully connected layers are not used). In addition, spatial dropout [12] (Drop2D()) layers are included to prevent overfitting, save memory and allow regularization.

The ResNet is born to solve the problem of vanishing gradients using a shortcut link between blocks of layers. In this way, the output of these external layers is added to the output of stacked layers and it is possible to achieve better results by combining information processed in different ways. This network portion is characterized by the presence of residual blocks containing three convolutional layers, connected through ReLU() activation function.

The residual block output is added to the output of a Conv layer of the main network, checking that the dimensions are the same.

For this work, the same network architecture was not used, being too deep for the available hardware. It was an inspiration to combine convolutional layers and residual layers (Fig. 6).

Fig. 6 Second network architecture

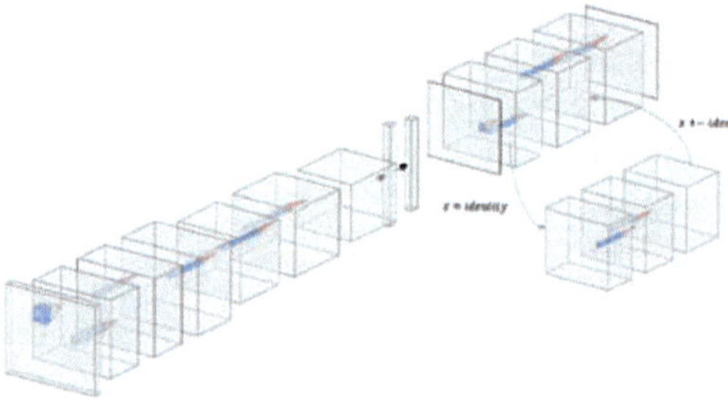

used; it does not give the possibility to represent also Batch Normalisation and MaxPooling2D layers but it is very useful to get a visual idea of the general structure of the network.

2.5 *Loss Function, Optimizer and Hyperparameters*

The objective function (also named loss function) used to check correct model parameters and to measure the discrepancy between the predicted reconstructed *Net*(*Yj*) and the corresponding ground-truth *Yj* is the Mean Absolute Error (*MAE*) [13].

$$MAE = \frac{1}{N}\sum_{j=1}^{N}\left|Y_j - Net\left(Y_j\right)\right|$$

The reason for choosing MAE as a loss function is that in this circumstance it makes sense to give more weight to points further away from the dataset mean.

The optimizer that will be used to train both neural networks is Adam [14]. It is a first-order optimization algorithm whose specific name refers to a minor extension of gradient descent.

Hyperparameters are model parameters whose values are set before training, often by trial and error. Among these values are: batch size, parameters that are part of the learning algorithm like momentum and the Learning Rate (LR), regularization (e.g., L1, L2), the number of layers and neurons. The selection of true hyperparameters combination is very important and will directly affect the convergence of the neural network. Since the model has several hyperparameters, it is necessary to find the best combination of these values searching in a multi-dimensional space. That's why hyperparameter tuning, which is the process of finding the right values of the hyperparameters, is a very complex and time-expensive task.

3 Results

3.1 *Settings*

The results presented in this section were obtained using the hardware technologies listed in the following tables. Each workstation differs each other for RAM size, type of Processors and storage capacity; some activities have been executed in parallel way to speed up pre-processing and training activities. For simplicity they will be identified as workstation n°1, workstation n°2 and workstation n°3 (Tables 1, 2 and 3).

Table 1 Workstation n°1

TASI Dell workstation n°1
Specifics:
Intel® Core™i7-6700 K CPU@4.00 GHz; 16 GB of RAM; NVIDIA GeForce RTX 2080 SUPER 8 GB of VRAM

Table 2 Workstation n°2

TASI Dell workstation n°2
Specifics:
Intel® Xeon(R) CPU E5-2640 v4@2.40Ghz; 64 GB of RAM;

Table 3 Workstation n°3

TASI Dell workstation n.3
Specifics:
Intel Xeon CPU 32 core at 3,7 GHz; 404 GB of RAM; Dual NVIDIA Quadro RTX 600 with 24 GB of VRAM each

Table 4 Dataset normalization

Input (RAW data)	Labels (SLC data)
No normalization	No normalization
No normalization	[0, 255]
No normalization	[0, 1]
[−1, 1]	[0, 1]

At this point data are stored to the workstation n°1 where they are normalised and a smaller portion of the original dataset is prepared to test some trainings on the current workstation.

Three different configurations of the dataset were used to check how normalisation would affect training performance (Table 4):

The first three settings respect the principle of leaving the RAW data unaltered, as if them had just been acquired on-board the satellite. Instead, the normalisation of labels is indifferent in relation to the final purpose. The fourth setting respects the type of normalisation that is carried out at the state of the art before training a Neural Network model but involves the normalisation of RAW data.

The choice of not normalising the labels is ruled out almost immediately: the model is not able to simulate the dynamics of the data (from RAW to SLC values) since the range of original SLC data values is very wide (minimum of 0 and maximum of 65000) while RAW data range from a minimum of −3.1327314 and a maximum equal to 2.925825.

At this point all simulations are performed considering 3 and 4th configurations and using the two Neural Network architectures presented in the previous section.

Final results are obtained considering a smaller portion of the dataset randomly sampling few pairs of images from the original one and dividing them into training and validation set. Workstation n°1 allowed the use of a very small batch size without resizing tiles (e.g., equal to 2, maximum equal to 4 by decreasing the number of parameters of the neural network).

Table 5 Hyperparameters values

Hyperparameters	Values
Learning rate	$1 * e^{-4}$
Weight, bias initializer	Normal
Weight decay	$1 * e^{-5}$ patience = 5
Leaky ReLU slope	0.1

For this reason, the Pytorch Dataset Class is used to resize input RAW tiles up to [100 × 300 × 2] and SLC label tiles to [100 × 300]. In this way it is possible to test several Neural Network structures and a variable value of the batch size.

The trainings carried out showed that best hyperparameters configuration is (Table 5):

Setting ADAM optimizer and MAE as loss function.

At this point, with this set of hyperparameters (except for the Learning Rate that is initially set equal to $1 * e^{-4}$ or $1 * e^{-3}$), the training is moved to workstation n°3 in order to use higher amount of data.

There is only one consideration to be made about this workstation, it does not have sufficient storage space. This means that the training is carried out while keeping the data saved on an external hard disk; thus, the data transfer slows down the simulations. For this reason, trainings with more than 40/50 epochs were not performed, considering that a simulation of 40 epochs takes about one week.

In the next section, the results obtained with the listed set of hyperparameters, MAE loss, ADAM optimizer, smaller portion of the entire dataset and using the two described network architectures will be presented. Only few changes were made to the kernel size of some hidden layers and alternative tests concerning the use of ReLU() or Leaky ReLU() as activation functions.

3.2 Models' Comparisons

Considering the first Neural Network structure, training is carried out twice setting a number of epochs equal to 30. The initial optimizer learning rate was set to $1 * e^{-3}$ in the first case and equal to $1 * e^{-4}$ for the second one, the batch size is set equal to 16 and all other parameters were left as default.

Once the training is complete, the model is tested by making some inferences using the data from the test set.

Better results were obtained considering the second Neural Network architecture. In this case training is carried out three different times changing learning rate and batch size values and setting a number of epochs equal to 40. The MAE loss demonstrates the training stability checking the training and validation loss andament (Fig. 7).

The best configuration is characterised by:

- initial optimizer learning rate equal to $1 * e^{-4}$;

Fig. 7 MAE loss

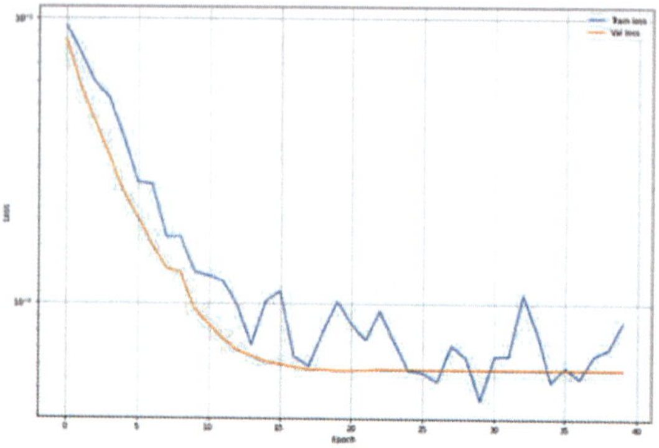

- weight decay value equal to $1 * e^{-5}$ (i.e., L2 penalty);
- batch size equal to 16.

At the end of the training phase, the model is tested over the test set.

The reconstruction of the images is not precise, but there are visible improvements compared to the previous case. If the tile in question represents only sea, the image reconstructed by the model will tend to be only black. In mountainous areas, reconstruction begins to pick up differences in colour between lighter and greyer areas. Instead, when there are more details, the final reconstruction tends to homogenise the pixel values, failing to output a crisp and clean image (Figs. 8, 9 and 10).

In any case, an important result is obtained by performing inference on several successive tiles and then attaching them together.

By increasing the contrast in both the original and predicted image the end result is much more useful and satisfying. From the predicted tiles it is possible to understand which is the observed context, the output inference allows to realize that a portion

Fig. 8 CNN residual layers inference, mountains

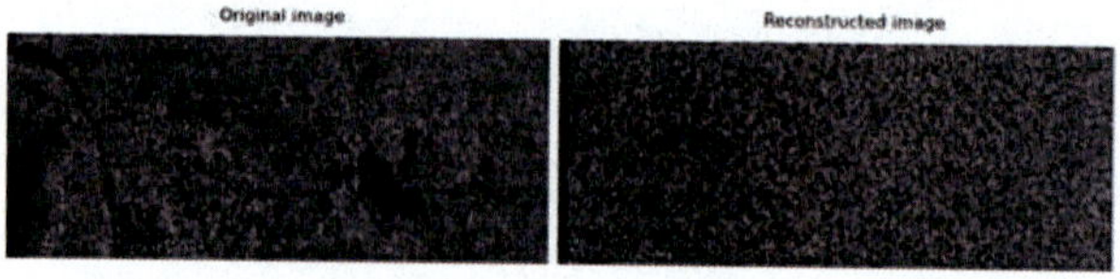

Fig. 9 CNN residual layers inference, country and rivers

Fig. 10 CNN residual layers inference, sea

Fig. 11 True labels, portion of a coastline

of a coast is being captured. These first results are in line with the purposes of this work (Figs. 11, 12, 13 and 14).

Good results can be observed performing inference over the total number of tiles that compose a single image and then connecting them together (Fig. 15).

Fig. 12 Predicted output, portion of a coastline

Fig. 13 True labels, portion of a coastline with higher contrast

Fig. 14 Predicted output, portion of a coastline with higher contrast

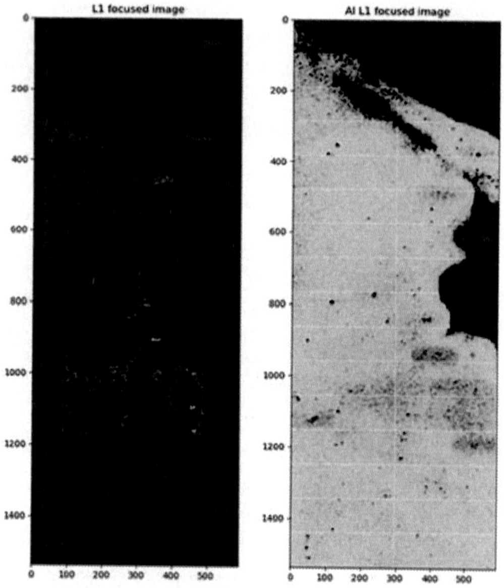

Fig. 15 L1 processed image on GS (left) and focused version by using AI (right)

4 Conclusions

The article presents an approach to reconstructing focused SLC images starting from RAW Sentinel-1 data by training a suitable neural network. The general objective, of the Earth Observation mission purposes, is not to obtain the sharpest possible output images but to be able to interpret the reconstructed image in such a way as to give near-time feedback to the end user.

An important point to consider concerns the optimal timing for inference on-board. Inference over a single tile (using a simple laptop) requires a mean time of 0.3 s. while the inference time to focus all tiles that compose the entire image is equal to 7 s. These timings were calculated by performing inferences in a laptop computer and using a graphical interface that also displays current hardware characteristics.

Once the model is trained in Ground Station (GS) infrastructure, the inference phase can be even executed on Space Segment (SS) using only Neural Network weights placed on-board satellite.

Moreover, the model update can be easily performed by sending a small file containing the new weights and the same model could be used in missions with different work bands by tuning the network parameters, adapting them to the specific case, or performing Transfer Learning to re-use weights of the model and training only new final layers to adapt it on the new task.

4.1 Time and Available Resources

Due to time and resource constraints, the results presented in this work are preliminary; a full dataset and optimization activities are the next steps foreseen in a coming phase at Thales Alenia Space Italy to have more accurate results. In principle, 79 scenarios (Both RAW and SLC data) of the entire planet were downloaded. Considering that the final dataset is in any case large and diverse, more accurate results could have been obtained with a higher amount of data (more storage space).

The training and parameters tuning phase was carried out over two months and each training stage required approximately one week of training by setting a number of epochs equal to 50, by using the entire dataset and not reshaping tiles. This can also be considered a limiting factor, as it cannot be ruled out that by increasing the number of epochs the final result could have improved.

Another thing to consider is the hardware technology that has been used. The workstation n°2 was used to perform BAQ decoding and tile cutting to use the available RAM. Once the dataset was ready, first trainings were carried out on workstation n°1 using a small percentage of the total data and a maximum batch size of 4. Workstation n°3 is the reference workstation to train the model using the complete dataset. This one is equipped with a GPU of Nvidia, but in this case the problem is related to the storage space. As there is no space available for the dataset, the training is carried out using an external hard disk, so the data transfer can be considered as a factor slowing down the training.

Further information on the proposed method and dataset analyzed during the current study are available from the corresponding author (Oreste.Trematerra@thalesaleniaspace.com) on reasonable request.

In any case, the transfer and positioning of a trained neural network on-board respects the hardware characteristics of the satellite and the inference time over new acquisitions is advantageous in relation to the time-delay delivery problem.

4.2 Future Works

The work has shown an important achievement: it is possible to directly use RAW data to train a Neural Network model in order to extract useful information from them. Extending the work on this topic, an attempt to further improve reconstruction quality will be made by training different neural networks models for different scenarios. For this project, a heterogeneous set of data was collected, but it is possible to simplify the problem by organising different datasets according to the type of Earth area observed (e.g., one dataset for mountains, one dataset for coastlines, etc.) and then training different models for each dataset.

An important feature that should not be underestimated, to use this method directly on board of a spacecraft, is that the model can recognize generic areas. So, these kinds of algorithms could be used to maintain only useful information saving memory and downlink band, which, as known, are limited and precious resources on a spacecraft: understanding the content of the image, it is possible to discard areas of no interest (e.g., only sea) and delete them if unnecessary thus avoiding their download and subsequent processing on the Ground Segment.

Considering data transfer, it was seen that once the RAW data has been acquired, it must be compressed on-board before downloading takes place. From this perspective, a possible solution could be to design and train a suitable Autoencoder to use its latent space as a compressed version of the RAW data. If it learns to reconstruct the RAW data with sufficient accuracy, the FDBAQ compression could be replaced by using the encoder part on-board satellite to compress the data and the decoder part on the Ground Segment to recover it.

The last consideration concerns the use of a Neural Network model for phase reconstruction of the SLC data (e.g., Interferometry applications). Interferometric Synthetic Aperture Radar (InSAR) exploits the phase difference between two complex radar SAR observations taken from slightly different sensor positions to extract information about the Earth's surface. A SAR signal contains amplitude and phase information and the phase is a fundamental information since it is determined primarily by the distance between the satellite antenna and the ground targets. By combining the phase of these two images after coregistration, an interferogram can be generated and its phase will be highly correlated to the terrain topography.

Stating the above and having understood the potential application of these algorithms, research activities on the topic are on-going at Thales Alenia Space in Italy aimed at proving the capability and efficiency of the algorithm and finally to develop it to be embarked and used on-board of future Observation Satellites and Constellations.

References

1. https://eoportal.org/documents/163813/238965/History.pdf
2. Moreira, A., Prats-Iraola, P., Younis, M., Krieger, G., Hajnsek, I., Papathanassiou, K.P.: A tutorial on synthetic aperture radar. IEEE Geosci. Remote. Sens. Mag. **1**(1), 6–43 (2013). https://doi.org/10.1109/MGRS.2013.2248301
3. Ziaja, M., Bosowski, P., Myller, M., Gajoch, G., Gumiela, M., Protich, J., Borda, K., Jayaraman, D., Dividino, R., Nalepa, J.: Benchmarking Deep Learning for On-Board Space Applications
4. https://scihub.copernicus.eu/twiki/do/view/SciHubWebPortal/APIHubDescription
5. Consultative Committee for Space Data Systems. https://public.ccsds.org/Publications/default.aspx
6. Friedt, J.-M.: Sentinel 1 Raw IQ Stream Processing Beyond Synthetic Aperture RADAR Applications
7. https://pytorch.org/docs/stable/index.html
8. Tian, C., Xu, Y., Zuo, W., Du, B., Lin, C.W., Zhang, D.: Designing and Training of A Dual CNN for Image Denoising
9. Pan, J., Liu, S., Sun, D., Zhang, J., Liu, Y., Ren, J., Li, Z., Tang, J., Lu, H., Tai, Y.W., Yang, M.-H.: Learning Dual Convolutional Neural Networks for Low-Level Vision
10. Rittenbach, A., Walters, P.: RDAnet: A Deep Learning Based Approach for Synthetic Aperture Radar Image Formation
11. Simonyan, K., Zisserman, A.: Very Deep Convolutional Networks for Large-Scale Image Recognition (2015). https://arxiv.org/abs/1409.1556 [cs.CV]

12. Tompson, J., Goroshin, R., Jain, A., LeCun, A., Bregler, C.: Efficient Object Localization Using Convolutional Networks (2015). https://arxiv.org/abs/1411.4280 [cs.CV]
13. Hao, S., Li, S.: A Weighted Mean Absolute Error Metric for Image Quality Assessment
14. Kingma, D., Ba, J.A.: A Method for Stochastic Optimization

Canopy Fire Effects Estimation Using Sentinel-2 Imagery and Deep Learning Approach. A Case Study on the Aspromonte National Park

Giandomenico De Luca and Giuseppe Modica

Abstract The accurate estimation of fire severity, in terms of physical effects that occurred on the tree's canopies, as well as the accurate mapping of its spatial distribution, is necessary information to optimally quantify and qualify the damage caused by the fire to ecosystems and address the most correct remedial procedures. The development of even more accurate learning algorithms and higher resolution satellite multispectral data have become essential resources in this framework. This study proposes a deep learning approach, exploiting remotely sensed satellite data, to produce an accurate severity map of the effects caused by the devasting fires affecting the Aspromonte National Park forests during the 2021 fire season. Two multispectral Sentinel-2 data, acquired before and after the fires, were classified using an artificial neural network-based model. All the multispectral fire-sensitive bands (visible, near-infra-red, short-infrared) and the respective temporal difference (post-fire—pre-fire) were involved, while the selection of the training pixels was based on field-based observations. Despite the preliminary nature of this study, the map accuracy reached high values (>95%) of $F\text{-}score_M$ (representing the overall accuracy) already since the first test of this configuration, confirming the validity of this approach. The quanti-qualification of the fire effects reported that 35.26 km^2 of forest cover was affected, of which: 41.03% and the 26.04% of tree's canopies were low and moderately affected respectively; the canopies killed but structurally preserved were the 12.88%; the destroyed trees (very-high severity), instead, were the 20.05%.

Keywords Fire severity · Remote sensing · Machine learning

G. De Luca (✉) · G. Modica
Department of Agriculture, University Mediterranea of Reggio Calabria, Località Feo Di Vito, 89122 Reggio Calabria, Italy
e-mail: giandomenico.deluca@unirc.it

G. Modica
e-mail: giuseppe.modica@unirc.it

C. Ieracitano et al. (eds.), *The Use of Artificial Intelligence for Space Applications*, Studies in Computational Intelligence 1088,
https://doi.org/10.1007/978-3-031-25755-1_27

1 Introduction

Forest fires are among the principal factors affecting the Mediterranean environment, both from an ecological and socio-economic aspect [1]. The spatial distribution of burned area, the frequency of the fires and the amount of structural and physical-chemical modifications that they cause to the affected habitats activate subsequent ecological transformations and alterations of the vegetation cover and of the organic-mineral components of the soil. These factors that impact, at different temporal and spatial scales, the physiologic dynamics, the microclimate and the water balance of ecosystems might be either degradative or positive [2–4]. Indeed, if on the one hand the fire can induce decomposition of biomass into inorganic carbon, desertification, sterility of ecosystems and their loss of resilience, up to the destruction of entire biotic communities, on the other hand, and at the right doses of frequency and severity, the fire can stimulate the interaction and competitiveness of some Mediterranean species, activating regeneration processes and enriching biodiversity [1, 5–8]. The accurate quantitative estimation and qualitative categorization of the short-term effects induced by fire on forest vegetation are therefore fundamental analysis to predict and understand their ecological and socio-economic evolution over the time and therefore to be able to plan suitable post-fire management policies.

The application of remote sensing data and techniques, especially optical satellite imagery, have been providing increasingly efficient outcomes in the characterization and mapping of the burn consequences (such as fire severity) on ecosystems (e.g. [2, 9–12]). The availability of free optical imaging systems with high spatial and temporal resolution, such as the multispectral Sentinel-2 satellites of the Copernicus mission managed by the European space agency (ESA) [13], have encouraged the advancements in this framework. Sentinel-2 platforms (two different satellites, A and B) provide the wavelengths mainly sensitive to the fire effects, namely visible, near infrared (NIR), red-edge and short-wave infrared (SWIR) [14], at a native pixel resolution of 10 × 10 m or 20 × 20 m and with a revisit time of 2–3 days at mid-latitudes.

Simultaneously, the development of open-source and user-friendly libraries and software performing complex machine learning algorithms (e.g. OTB, Scikit-learn, TensorFlow + Keras, etc.) increased the opportunities of draw up efficient prediction and classification workflows [15, 16]. Classic supervised machine learning models (principally random forest and support vector machine) rendered performing results in terms of accuracy (between 70 and 90%) when tested on high resolution optical satellite data for supervised classification of fire effects severity (e.g. [16–19]). Most of these works were based on the associating spectral information and field based measurements carried out using well-established protocols (e.g. CBI [20]) for fire severity gradient estimation. In recent years, some studies went as far as to test advanced deep learning models based on artificial neural networks (ANN) in order to improve burned areas detection [21, 22], overcoming the 90% of accuracy in most of the cases. However, to the best of our knowledge, few experiments attempted to exploit deep learning models for estimating fire effects on vegetation [23, 24].

The aim of this study was to contribute to the lacking state of the art by proponing a deep learning-based approach to categorize the level of fire effects severity hat have affected the forest canopies of the Aspromonte national Park (2021 fire season) using Sentinel-2 multispectral imagery. In particular, taking as training information four field-based visual physical effects that can affect canopy forest (low affected, moderately affected, highly affected and very-highly canopy), the level of fire severity was classified by performing an artificial neural network (ANN) model formed by a sequential of Dense hidden layers implemented in Keras library [25]. The contribution of each single image layer band to the learning process was estimated by operating a feature importance based on a weighted linear regression. The single-class $F\text{-}score_i$ accuracy metric was used to compute the accuracy performed by the classification model for each fire effect severity category. The overall accuracy was instead carried out using the multi-class $F\text{-}score_M$. The analysis continued by comparing the fire effects severity result with a vegetation cover map of the Aspromonte Park [26] so that it was possible to investigate their distribution among the different affected forest types.

2 Study Area

The fire events considered in this study involved the Aspromonte National Park [27], located in the southernmost continental Province of Italy (Reggio Calabria, 38° 16'N; 15° 84'E), affecting a total area of about 70 km^2 and occurring between July and August 2021 (Fig. 1). The examined events were part of a larger (>160 km^2) ensemble of fire events striking the entire Province territory during the assessed fire season. The Aspromonte Park, extending for 641.53 km^2 NE-SW and constituting the last continental extension of the Apennine Mountain range, is characterized by a wide phytoclimatic range (from *Lauretum* to *Fagetum*) and high heterogeneity of flora, resulting from the combination of different environmental and topographic factors (latitude, altitude range, proximity to the sea, slope exposition, etc.). Moreover, it contains 21 special habitat sites afferent to the Natura2000 network (https://ec.europa.eu/environment/nature/natura2000/index_en.htm), pointing out its ecological and socio-economic substantial role for the territory. An exhaustive description of the Aspromonte vegetation could be found in Spampinato [28]. In this work, only the fire affecting forest vegetation was taken into consideration (35.26 km^2), discriminated using the Aspromonte Vegetation Cover Map [26].

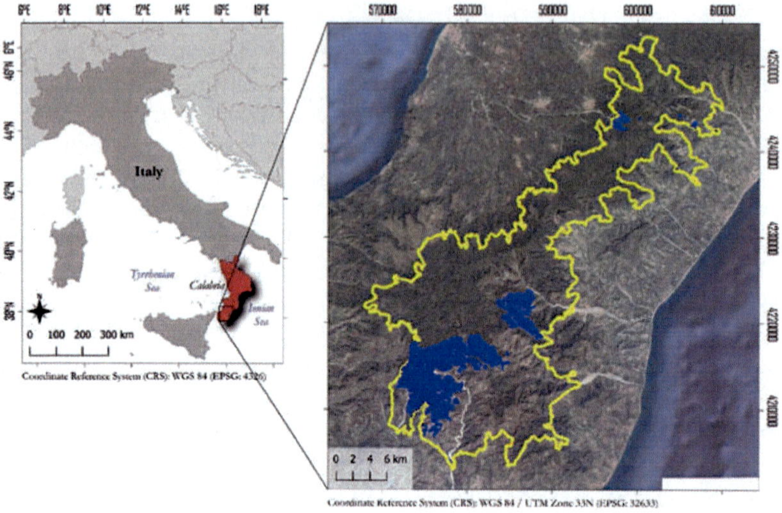

Fig. 1 Location of the study area (left). On the right, the yellow line delimits the Aspromonte national Park perimeter, while the total burned area falling inside the Park are represented in blue

3 Materials and Methods

3.1 Dataset and Pre-Processing

The multispectral Sentinel-2 dataset was composed of two Level 2A (Bottom of Atmosphere reflectance) [13] cloud-free images acquired before (28/07/2021) and after all the fires occurred (16/09/2022) respectively, downloaded through the Copernicus Open Access Hub [29], the official ESA platform for the distribution of Sentinel satellite data.

All the image bands at native resolution of 10 $\times$ 10 m (blue, B2; green, B3; red, B4; NIR, B8) were directly involved to construct the final dataset, while the image bands at native resolution of 20 $\times$ 20 m (red-edge$_{704}$, B5; red-edge$_{739}$, B6; red-edge$_{780}$, B7; NIR$_{864}$, B8A; SWIR$_{1610}$, B11; SWIR$_{2186}$, B12) were resampled to 10 $\times$ 10 m pixels spacing using the red band (B4) as pixel spacing reference and the bilinear interpolation as resampling method.

For each image band, the respective difference between post and pre-fire (delta, Δ) images was carried out. All the post-fire and Δ image bands formed the final dataset employed in subsequent analysis. This according to what was done in a previous work [17] by the same research group in which a dataset formed by post-fire and delta Sentinel-2 image bands was modelled through random forest algorithm to estimate the fire severity in a Mediterranean site.

3.2 Field Measurements and Sampling Points Collection

Several geo-referenced photos and descriptive notes were taken during the field sampling campaign in the period immediately after the fire events (September 2021). Supported by Google Satellite high resolution images [30], the selection of suitable sampling points (one point = one pixel) was subsequently carried out on the basis of those field data. The method of fire effects severity estimation was inspired by the CBI protocol, based on the quantification of structural and physical alterations occurred on several vegetation layers [20]. However, in this study, only the dominant and co-dominant trees cover was assessed, directly categorizing the observed alterations into one of the four severity classes: low, moderate, high, and very-high. In total, 1,000 sampling points for each severity class were retrieved. Moreover, additional 250 samplings points were reserved to trace the bare soil, since it might be present between the canopy ground projection edges, leading to commission errors.

3.3 Artificial Neural Network Construction and Image Classification

ANN architecture and hyperparameters optimization. The ANN structure was characterized by a sequential model of simple Dense hidden layers, in which each neuron is densely connected with each neuron of the previous/next layer. The *relu* (rectified linear unit) function was applied to activate all the hidden layers, except for the last one which was activated with the *softmax* function in order to retrieve the probability distribution of the output data. Each pixel of the dataset is labeled with the class that had reached the highest probability coefficient for that pixel. The weight regularization was implemented using a *L2 kernel regularizer* (based on the square of the value of the weight coefficients) to reduce overfitting. The *RMSProp*, *sparse_categorical_crossentropy* and *accuracy* finally compiled the model as optimizer, loss and metric functions respectively.

The most suitable model hyperparameters values (dimensionality of the output tensor of each hidden layer, epoch, number of batches and kernel regularizer) were determined by testing several possible combinations of pre-choose values and assessing, for each test, the training and test accuracy metric resulted. The best combination of hyperparameters, defined by the reaching of the highest validation accuracy, was kept as final configuration for subsequently prediction step. The tested hyperparameters values were: dimensionality of the layers output tensor (units for all the hidden layers) [50, 100, 250, 500, 1000]; epoch [5, 10, 20, 30, 40]; number of batches (50, 100, 150, 200, 250); kernel regularizer weight coefficient [0.001, 0.002, 0.003, 0.004]. The number of hidden layers was also tested in arrange between 2 and 5 after having set the hyperparameters.

Model training and feature importance calculation. A part (1/2) of the sampling points were randomly choose, equally for each fire effect class, and used to train the ANN model. The remaining part (1/2) of sampling points were used as validation set.

The importance of each image band was estimated by performing a weighted linear regression approach on a smaller sample (20%) of relative training and validation points. The pre-built model *KernelExplainer*, implemented in SHAP library [31], was used for this purpose, in which the returned coefficients and values from game theory determine the crude indicators of importance.

3.4 Accuracy Assessment

Beside the accuracy metric provided by the ANN model, a confusion matrix was constructed using the same validation sampling points (1/2 of the total number of sampling points). For each severity class, the $producer's_i$ and $user's_i$ (i = single severity class) accuracy metrics were retrieved from the confusion matrix, and their respective single-class ($F\text{-}score_i$, Eq. 1) and multi-class ($F\text{-}score_M$, Eq. 2) harmonic means were calculated.

$$F - score_i = (producer's_i \cdot user's_i)/(producer's_i + user's_i) \quad (1)$$

$$F - score_M = (producer's_M \cdot user's_M)/(producer's_M + user's_M) \quad (2)$$

where $user's_M$ (Eq. 3) and $producer's_M$ (Eq. 4) metrics are computed as follows:

$$user's_M = \left(\sum_{i=1} user's_i\right)/n \quad (3)$$

$$producer's_M = \left(\sum_{i=1} producer's_i\right)/n \quad (4)$$

4 Results

4.1 Final Structure of the ANN

The ANN optimal hyperparameters used in the classification process, retrieved by implementing a comparison the accuracy and loss metrics values restituted after having tested the different parameters hyperparameters combinations, are reported in Table 1.

Table 1 Artificial Neural Network optimal hyperparameters used in the classification process, retrieved by comparing the accuracy and loss metrics of various values combinations tested

Hyperparameter name	Optimal value set
Number of hidden layers	6 Dense layers (5 activated with *relu* + 1 activated with *softmax*)
Units (for each Dense hidden layer)	500 (*relu* layers) 5 (*softmax* layer)
Kernel regularizer	0.0025
Epochs	40
Batch size	250

The final structure of the ANN was formed by a concatenation of layers ordered as follows: an input layer (shape: number of input image bands); five Dense hidden layer (units: 500; activation: *relu*) each of which regularized by a *Kernel regularizer* function (weight coefficient: 0.0025); a final Dense hidden layer (units: fire effect categories; activation: *softmax*).

4.2 Classified Fire Effects Map

Figure 2 shows the spatial distribution of the fire effect categories resulted from the classification process (low, green; moderate, yellow; high, orange; very-high, red).

The distribution of the four fire effects among the affected forest cover types (forest cover type labels legend according to Spampinato et al. [26]), inside the Aspromonte Park, is illustrated in Fig. 3.

On 35.26 km^2 of total forest cover burned, the fire had a lower and moderate impact on 14.47 km^2 (41.03%) and 9.18 km^2 (26.04%) of forest vegetation, respectively. A surface equal to 4.54 km^2 (12.88%) and the 7.07 km^2 (20.05%) were instead affected by high and very-high effects severity. Observing the apportionment of the fire effects categories among the forest types (burned surface > 2 km^2), it is noticeable as the most affected were the areas naturally or artificially covered by coniferous, in particular by *Pinus nigra* spp. *Laricio*: "Natural pine *Laricio* forest" (Low, 6.06 km^2; Moderate, 2.92 km^2; High, 1.24 km^2; Very-High, 1.80 km^2), the "Mountain artificial (reforestation) coniferous forest with a prevalence of pine *Laricio*" (Low, 2.23 km^2; Moderate, 1.43 km^2; High, 1.02 km^2; Very-High, 2.35 km^2), the "Degraded natural pine *Laricio* forest" (Low, 2.20 km^2; Moderate, 1.53 km^2; High, 0.67 km^2; Very-High, 1.02 km^2) and the "Hilly artificial (reforestation) coniferous forest with a prevalence of pine *Laricio*" (Low, 0.42 km^2; Moderate, 0.43 km^2; High, 0.52 km^2; Very-High, 0.88 km^2).

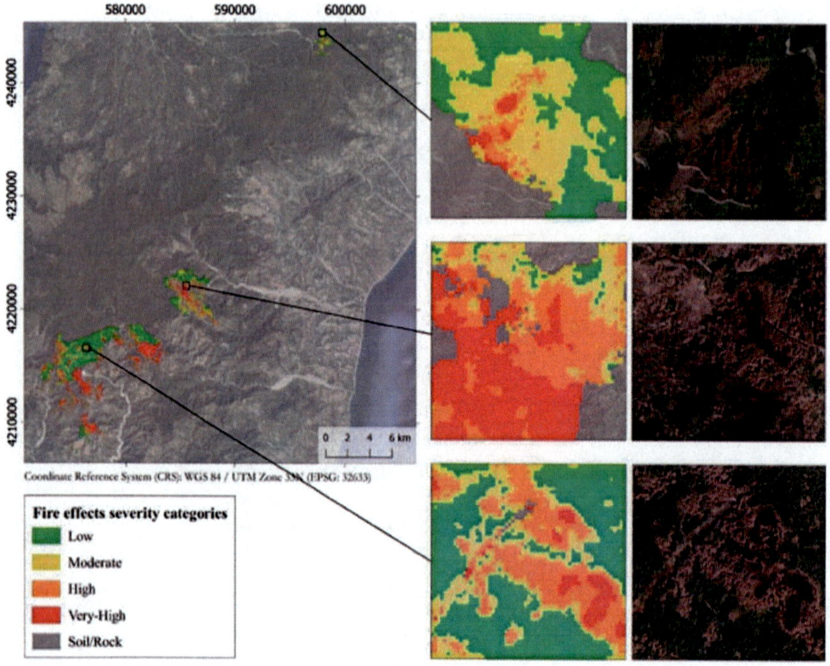

Fig. 2 Fire effects severity map resulted from the classification process showing the spatial distribution of the four fire effects severity categories (Low, Moderate, High and Very-High) on the forest vegetation of the Aspromonte national Park

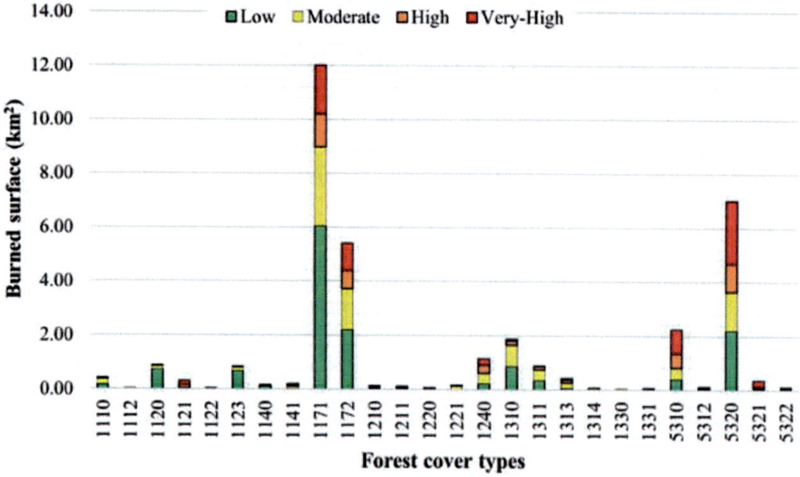

Fig. 3 Surface distribution of the four fire effects severity (Low, Moderate, High and Very-High) among the affected forest types (the legend of forest cover type is according to Spampinato et al. [26])

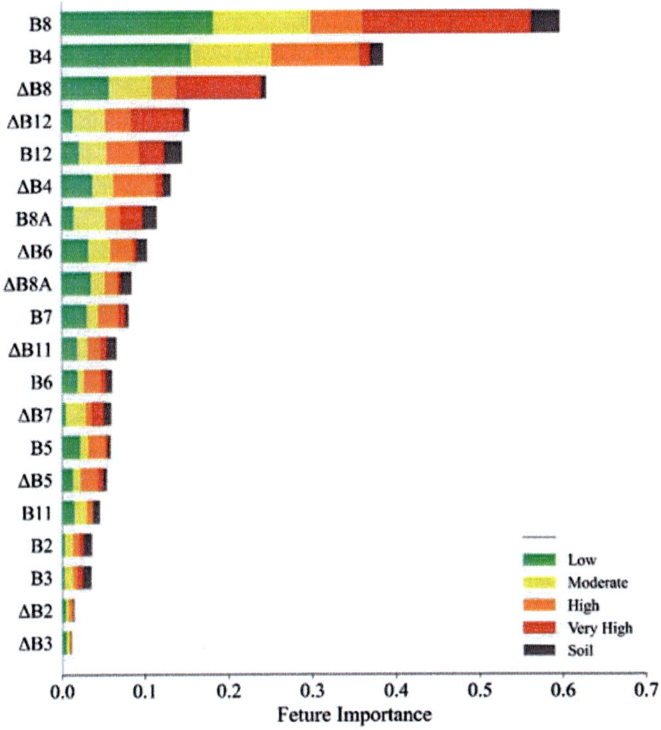

Fig. 4 Feature importance (SHAP *Kernel Explainer*) exerted by each image band layer

4.3 Feature Importance

Figure 4 reports the importance that each image input layer exerted during the learning processing, for each of the four fire effect categories, resulted from the weighted linear regression approach.

The highest importance (>0.6) was achieved by the post-fire NIR band, followed by red band (0.43), the ΔNIR (0.28), the $\Delta SWIR_{2186}$ (0.18) and the $SWIR_{2186}$ (0.17). On the other hand, the green and blue image bands, and the respective Δ, resulted to render lowest influence (<0.05).

4.4 Map Accuracy

Figure 5 report the training and validation loss (upper) and accuracy (bottom) metrics calculated directly by the ANN model.

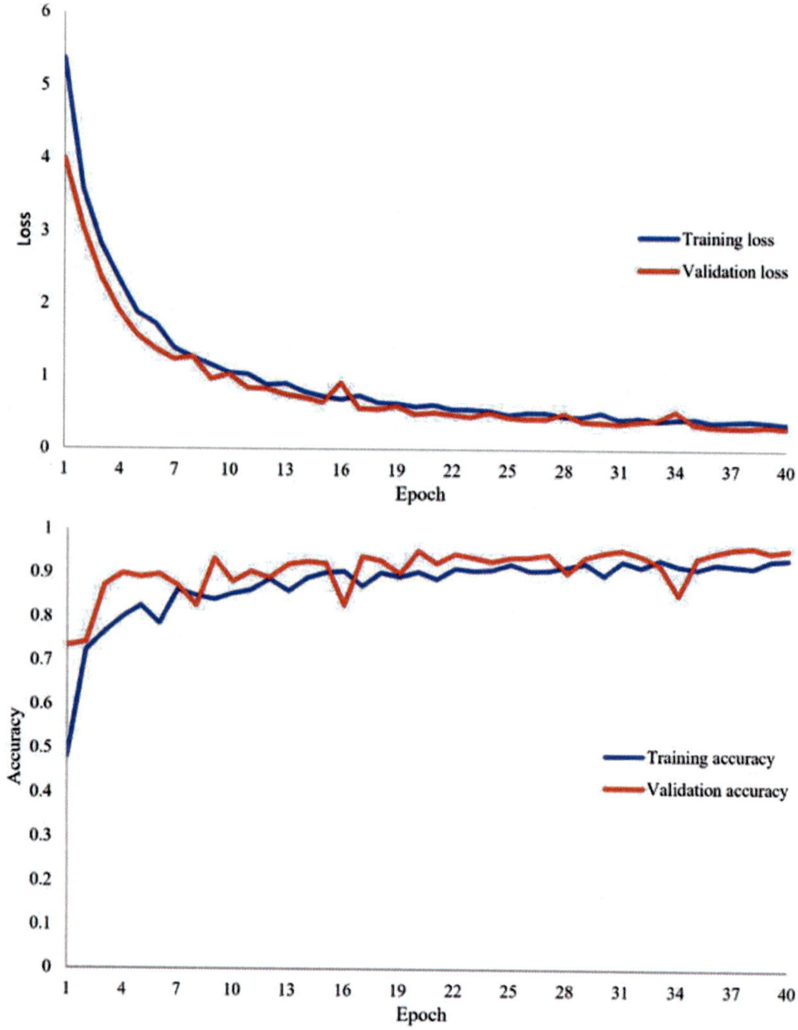

Fig. 5 Plots displaying the values of the loss (top) and the accuracy (bottom) of training and validation processes calculated at each epoch of the classification process

The confusion matrix retrieved in order to assess the accuracy of the map is shown in Fig. 6. From the confusion matrix, the single-class $F\text{-}score_i$ accuracy metric values computed for each fire effect category were 97.52% (Low), 93.01% (Moderate), 95.33% (High), 99.18% (Very-High) and 89.92% (Soil). The multi-class $F\text{-}score_M$, representing the overall accuracy of the map, was 95.12%.

	Low	Mod	High	Very-High	Soil	*User's*$_i$	*F-score*$_i$
Low	491	16	0	0	0	96.84%	97.52%
Mod	9	466	17	0	10	92.83%	93.01%
High	0	17	480	2	9	94.49%	95.33%
Very-High	0	0	1	486	5	98.78%	99.18%
Soil	0	1	1	0	116	98.31%	89.92%
Producer's$_i$	98.20%	93.20%	96.19%	99.59%	82.86%		

Producer's$_M$	94.01%
User's$_M$	96.25%
F-score$_M$	95.12%

Fig. 6 Confusion matrix reporting the accuracy of the classified map. The single-class *user's*$_i$, *producer's*$_i$, *F-score*$_i$ and the multi-class *user's*$_M$, *producer's*$_M$, *F-score*$_M$ accuracy metrics are reported

5 Discussions

This study qualitatively and quantitatively estimated the fire effects severity that affected the forest vegetation of Aspromonte Park during the 2021 fire season. The analysis showed as conifers (*Laricio* pine) forests were the most affected, both in terms of surface and the severity. Although the high susceptibility to fire widespread is common for these forest type [32, 33], other unknowns may have played a relevant role in this case: the location of the ignition points (arson) and the arrangement of the firefighting teams.

The only canopy damages were considered because, thought the sub-foliage layers are important from an ecological point of view (as also provided in the CBI protocol), this is not adaptable to the use of optical radiometric information due to their inability to penetrate the foliage.

The methodology for fire effects severity estimation proposed in this study based on a simple ANN model resulted to be reliable in terms of accuracy. Considering the spatial resolution of Sentinel-2, the accuracy resulting from the confusion matrix is very high and no particular case of overfitting can be found in the curves of Fig. 5. The classified map spatial distribution of the four fire effects severity categories confirms what emerged from accuracy metrics. Minor exceptions are observable for soil class that was confused with moderately affected forest cover, as also proved by the confusion matrix outcomes. This could be due to the lower number of sampling points used for this class. However, the discriminations of bare soil and rocks in burned areas needs more accurate stratagems, such as the use of a longer observations (time series) [34]. In general, spectral confusion is ever expected at moderate severity gradients, caused by the concomitant presence of affected and unburned covers

[17, 35–37] and/or by the presence char and ash posed on the underlying slight exposed layers or on the foliage [20, 38, 39]. Further investigations involving spectral mixture analysis for mixed pixels accounting [35], as well as the use of higher resolution optical images (although not free-available at the moment) could be tested to deal with the above mentioned issues.

Accounting the performance of the feature importance, the red, NIR and SWIR bands (and their respective temporal difference) are the most influent wavelengths for fire effects severity estimation, corroborating what has been widely proved in literature since earlier applications [39, 40]. These are indeed the spectral bands used to construct two of the main vegetation indices already used in this context, the normalized burn ratio (NBR) and the normalized difference vegetation index (NDVI). Observing the Fig. 5, the red band seems more related to the Moderate and High fires, at the expense of Very-High category. The latter appear to have been better managed by NIR and (followed by) SWIR, two chlorophyll and moisture-related wavelengths. The red-edge bands (at 739 and 780 nm) also expressed a medium importance. Other studies pointed out the relevance of red-edge for post-fire effect investigations [17, 41, 42]. In De Luca et al. [17], using the Gini coefficient to calculate the feature importance of Sentinel-2 bands for fire severity estimation, the B6 and B7 bands resulted decisive for the classification process. Filipponi et al. [42] proposed a new spectral indices for burned areas detection based on Sentinel-2 red-edge wavelengths. In this work, the blue and green resulted to be the less influent image bands.

6 Conclusions

This study presented an effective deep learning-based procedure to accurately estimate and map the fire effects severity that affected the Mediterranean forest vegetation of a national park. The use of these advanced artificial intelligence algorithms has been not as investigated in literature as the classic machine learning models (e.g. random forest, support vector machine, etc.). For these reasons, the proposed approach, based on a very simple ANN, was a preliminary trial which was set to test its reliability for the purpose. This study confirmed as the use of a simple Dense ANN achieves very good results and might be taken into consideration in future applications. This represents an optimal compromise between accuracy level and ease of implementation. The use of the python-based open-source and user-friendly Keras interface (on the TensorFlow architecture), in fact, enabled to exploit the calculation operations of the deep learning algorithms without requiring special programming skills. The outcomes were very promising, and next experimentation might involve more robust models for incrementing precision, such as the convolutional neural networks widely employed for image processing.

References

1. Moreira, F., Ascoli, D., Safford, H., Adams, M.A., Moreno, J.M., Pereira, J.M.C., Catry, F.X., Armesto, J., Bond, W., González, M.E., Curt, T., Koutsias, N., McCaw, L., Price, O., Pausas, J.G., Rigolot, E., Stephens, S., Tavsanoglu, C., Vallejo, V.R., Van Wilgen, B.W., Xanthopoulos, G., Fernandes, P.M.: Wildfire management in Mediterranean-type regions: paradigm change needed. Environ. Res. Lett. **15**, 11001 (2020). https://doi.org/10.1088/1748-9326/ab541e
2. Morresi, D., Vitali, A., Urbinati, C., Garbarino, M.: Forest spectral recovery and regeneration dynamics in stand-replacing wildfires of central Apennines derived from Landsat time series. Remote Sens. **11**, (2019). https://doi.org/10.3390/rs11030308
3. Semeraro, T., Vacchiano, G., Aretano, R., Ascoli, D.: Application of vegetation index time series to value fire effect on primary production in a Southern European rare wetland. Ecol. Eng. **134**, 9–17 (2019). https://doi.org/10.1016/j.ecoleng.2019.04.004
4. Rosa, I.M.D., Pereira, J.M.C., Tarantola, S.: Atmospheric emissions from vegetation fires in Portugal (1990–2008): estimates, uncertainty analysis, and sensitivity analysis. Atmos. Chem. Phys. **11**, 2625–2640 (2011). https://doi.org/10.5194/acp-11-2625-2011
5. De Luis, M., Raventós, J., González-Hidalgo, J.C.: Post-fire vegetation succession in Mediterranean gorse shrublands. Acta Oecol. **30**, 54–61 (2006). https://doi.org/10.1016/j.actao.2006.01.005
6. Mitchell, R.J., Simonson, W., Flegg, L.A., Santos, P., Hall, J.: A comparison of the resilience of four habitats to fire, and the implications of changes in community composition for conservation: a case study from the Serra de Monchique, Portugal. Plant Ecol. Divers. **2**, 45–56 (2009). https://doi.org/10.1080/17550870902752421
7. Montès, N., Ballini, C., Bonin, G., Faures, J.: A comparative study of aboveground biomass of three Mediterranean species in a post-fire succession. Acta Oecol. **25**, 1–6 (2004). https://doi.org/10.1016/j.actao.2003.10.002
8. Riaño, D., Chuvieco, E., Ustin, S., Zomer, R., Dennison, P., Roberts, D., Salas, J.: Assessment of vegetation regeneration after fire through multitemporal analysis of AVIRIS images in the Santa Monica Mountains. Remote Sens. Environ. **79**, 60–71 (2002). https://doi.org/10.1016/S0034-4257(01)00239-5
9. Meng, R., Dennison, P.E., Huang, C., Moritz, M.A., D'Antonio, C.: Effects of fire severity and post-fire climate on short-term vegetation recovery of mixed-conifer and red fir forests in the Sierra Nevada Mountains of California. Remote Sens. Environ. **171**, 311–325 (2015). https://doi.org/10.1016/j.rse.2015.10.024
10. Morresi, D., Marzano, R., Lingua, E., Motta, R., Garbarino, M.: Mapping burn severity in the western Italian Alps through phenologically coherent reflectance composites derived from Sentinel-2 imagery. Remote Sens. Environ. **269**, 112800 (2022). https://doi.org/10.1016/j.rse.2021.112800
11. Han, A., Qing, S., Bao, Y., Na, L., Bao, Y., Liu, X., Zhang, J., Wang, C.: Short-term effects of fire severity on vegetation based on Sentinel-2 satellite data. Sustainability **13**, 1–22 (2021). https://doi.org/10.3390/su13010432
12. Saulino, L., Rita, A., Migliozzi, A., Maffei, C., Allevato, E., Garonna, A. Pietro, Saracino, A.: Detecting burn severity across Mediterranean forest types by coupling medium-spatial resolution satellite imagery and field data. Remote Sens. **12**, 1–21 (2020). https://doi.org/10.3390/rs12040741
13. ESA: ESA Sentinel-2 Homepage. https://sentinels.copernicus.eu/web/sentinel/missions/sentinel-2
14. Pereira, J.M.C., Sá, A.C.L., Sousa, A.M.O., Silva, J.M.N., Santos, T.N., Carreiras, J.M.B.: Spectral characterisation and discrimination of burnt areas. In: Remote Sensing of Large Wildfires (1999). https://doi.org/10.1007/978-3-642-60164-4_7
15. Modica, G., De Luca, G., Messina, G., Praticò, S.: Comparison and assessment of different object-based classifications using machine learning algorithms and UAVs multispectral imagery in the framework of precision agriculture. Eur. J. Remote Sens. **54**, 431–460 (2021). https://doi.org/inpress

16. Bot, K., Borges, J.G.: A Systematic Review of Applications of Machine Learning Techniques for Wildfire Management Decision Support (2022). https://doi.org/10.3390/inventions7010015
17. De Luca, G., Silva, J.M.N., Oom, D., Modica, G.: Combined use of Sentinel-1 and Sentinel-2 for burn severity mapping in a mediterranean region. In: Computational Science and Its Applications–ICCSA 2021. Lecture Notes in Computer Science, pp. 139–154 (2021). https://doi.org/10.1007/978-3-030-87007-2_11
18. Sali, M., Piaser, E., Boschetti, M., Brivio, P.A., Sona, G., Bordogna, G., Stroppiana, D.: A burned area mapping algorithm for sentinel-2 data based on approximate reasoning and region growing. Remote Sens. **13**, (2021). https://doi.org/10.3390/rs13112214
19. Amos, C., Petropoulos, G.P., Ferentinos, K.P.: Determining the use of Sentinel-2A MSI for wildfire burning & severity detection. Int. J. Remote Sens. **40**, 905–930 (2019). https://doi.org/10.1080/01431161.2018.1519284
20. Key, C.H., Benson, N.C.: Landscape assessment (LA) sampling and analysis methods. In: FIREMON: Fire Effects Monitoring and Inventory System (2006)
21. Knopp, L., Wieland, M., Rättich, M., Martinis, S.: A deep learning approach for burned area segmentation with Sentinel-2 data. Remote Sens. **12**, (2020). https://doi.org/10.3390/RS12152422
22. Hu, X., Ban, Y., Nascetti, A.: Uni-Temporal Multispectral Imagery for Burned Area Mapping with Deep Learning (2021). https://doi.org/10.3390/rs13081509
23. Monaco, S., Pasini, A., Apiletti, D., Colomba, L., Garza, P., Baralis, E.: Improving wildfire severity classification of deep learning U-nets from satellite images. In: 2020 IEEE International Conference on Big Data (Big Data), pp. 5786–5788 (2020). https://doi.org/10.1109/BigData50022.2020.9377867
24. Farasin, A., Colomba, L., Garza, P.: Double-Step U-Net: A Deep Learning-Based Approach for the Estimation of Wildfire Damage Severity through Sentinel-2 Satellite Data (2020). https://doi.org/10.3390/app10124332
25. Keras Homepage: Keras Homepage. https://keras.io/
26. Spampinato, G., Cameriere, P., Caridi, D., Crisafulli, A.: Carta della vegetazione reale del Parco Nazionale dell'Aspromonte (2002)
27. Aspromonte Park: Aspromonte Park. http://www.parconazionaleaspromonte.it/pagina.php?id=41
28. Spampinato, G.: Guida alla flora dell'Aspromonte. Laruffa Editore (2014)
29. Copernicus Access Hub: Copernicus Access Hub. https://scihub.copernicus.eu/
30. Google Earth: Google Earth. https://earth.google.com/web/
31. SHAP: SHAP KernelExplainer Doc. https://shap-lrjball.readthedocs.io/en/latest/generated/shap.KernelExplainer.html
32. Moreno, J.M., Morales-Molino, C., Torres, I., Arianoutsou, M.: Fire in Mediterranean Pine Forests: Past, Present and Future BT-Pines and Their Mixed Forest Ecosystems in the Mediterranean Basin (2021). https://doi.org/10.1007/978-3-030-63625-8_21
33. Fernández-Manso, A., Quintano, C., Suarez-Seoane, S., Marcos, E., Calvo, L.: Remote Sensing Techniques For Monitoring Fire Damage And Recovery of Mediterranean Pine Forests: Pinus Pinaster and Pinus Halepensis as Case Studies BT-Pines and Their Mixed Forest Ecosystems in the Mediterranean Basin (2021). https://doi.org/10.1007/978-3-030-63625-8_27
34. Pereira-Pires, J.E., Aubard, V., Ribeiro, R.A., Fonseca, J.M., Silva, J.M.N., Mora, A.: Semi-Automatic Methodology for Fire Break Maintenance Operations Detection with Sentinel-2 Imagery and Artificial Neural Network (2020). https://doi.org/10.3390/rs12060909
35. Quintano, C., Fernández-Manso, A., Roberts, D.A.: Multiple endmember spectral mixture analysis (MESMA) to map burn severity levels from Landsat images in Mediterranean countries. Remote Sens. Environ. **136**, 76–88 (2013). https://doi.org/10.1016/j.rse.2013.04.017
36. Smith, A.M.S., Lentile, L.B., Hudak, A.T., Morgan, P.: Evaluation of linear spectral unmixing and ΔNBR for predicting post-fire recovery in a North American ponderosa pine forest. Int. J. Remote Sens. **28**, 5159–5166 (2007). https://doi.org/10.1080/01431160701395161
37. Fernandez-Manso, A., Quintano, C., Roberts, D.A.: Burn severity influence on post-fire vegetation cover resilience from Landsat MESMA fraction images time series in Mediterranean

forest ecosystems. Remote Sens. Environ. **184**, 112–123 (2016). https://doi.org/10.1016/j.rse.2016.06.015
38. Rogan, J., Franklin, J.: Mapping wildfire burn severity in southern california forests and shrublands using enhanced thematic mapper imagery. Geocarto Int. **16**, 91–106 (2001). https://doi.org/10.1080/10106040108542218
39. De Santis, A., Chuvieco, E.: Burn severity estimation from remotely sensed data: performance of simulation versus empirical models. Remote Sens. Environ. **108**, 422–435 (2007). https://doi.org/10.1016/j.rse.2006.11.022
40. García-Llamas, P., Suárez-Seoane, S., Fernández-Guisuraga, J.M., Fernández-García, V., Fernández-Manso, A., Quintano, C., Taboada, A., Marcos, E., Calvo, L.: Evaluation and comparison of Landsat 8, Sentinel-2 and Deimos-1 remote sensing indices for assessing burn severity in Mediterranean fire-prone ecosystems. Int. J. Appl. Earth Obs. Geoinf. **80**, 137–144 (2019). https://doi.org/10.1016/j.jag.2019.04.006
41. Fernández-Manso, A., Fernández-Manso, O., Quintano, C.: SENTINEL-2A red-edge spectral indices suitability for discriminating burn severity. Int. J. Appl. Earth Obs. Geoinf. (2016). https://doi.org/10.1016/j.jag.2016.03.005
42. Filipponi, F.: BAIS2: burned area index for Sentinel-2. In: Proceedings, p. 5177 (2018). https://doi.org/10.3390/ecrs-2-05177

A Supervised Learning-Based Approach to Maneuver Detection Through TLE Data Mining

Riccardo Cipollone, Nugraha Setya Ardi, and Pierluigi Di Lizia

Abstract In the last decades, the Near-Earth environment has grown more and more crowded due to an exponential increase of launches and Space missions. As a direct consequence of that, fragmentations, impacts and debris have caused one of the most important issues of the commercial use of Space to rise: Space Traffic Management. In this framework, Space Surveillance and Tracking (SST) tries to address the problem by tracking and cataloguing Resident Space Objects through ground and space-based sensor networks. The collected data are processed and structured as catalogues in which new measurements are used to obtain orbit estimates. This kind of cataloguing activity implies the generation of a great deal of structured data. This setting is ideal to develop data-driven techniques and infer on the target and its pattern of life. The present work is a first step in this direction, specifically focusing on maneuver detection through Two-Line Element (TLE) time-series by means of a segmentation-aimed neural network. Starting with the extraction of meaningful orbital parameters from the TLE history of a target, the developed technique is capable of spotting maneuvering epochs across a given orbit history and even perform a first classification of general maneuver categories. This achievement is significant, since the available (and publicly accessible) data on maneuvering events are limited to few objects. Training with hybrid data and tests with real ones have been performed to asses accuracy and generalization capability. The results are promising, showing satisfactory performance on real data and proving how versatile this network architecture can be.

Keywords Maneuver detection · Orbital data mining · Machine learning

---

R. Cipollone (✉) · N. Setya Ardi · P. Di Lizia
Department of Aerospace Science and Technology, Politecnico di Milano, Via La Masa 34, 20156 Milano, MI, Italy
e-mail: riccardo.cipollone@polimi.it

N. Setya Ardi
e-mail: nugrahasetya@mail.polimi.it

P. Di Lizia
e-mail: pierluigi.dilizia@polimi.it

C. Ieracitano et al. (eds.), *The Use of Artificial Intelligence for Space Applications*, Studies in Computational Intelligence 1088,
https://doi.org/10.1007/978-3-031-25755-1_28

1 Introduction

Resident Space Object (RSOs) catalogues and their maintenance are key to efficiently monitoring and modelling the progressively overcrowded Near Earth environment, targeted by more and more scientific and, above all, commercial launches. In this context, sensor networks measurements are exploited to generate orbit estimates associated to the observed objects that are continuously stored and updated as part of a large data structure. This framework allows applications of statistical analysis to understand targets behaviour and characterize them as much as possible. A first aspect to be investigated is the detection of maneuvers or, more generally, events that differ from the coastal natural motion that the target should follow. This kind of study is usually carried out to deepen operators' knowledge about RSOs pattern of life and, consequently, to help them tailor the following processing to this additional information. In this way, most of SST-related activities could get a significant improvement in robustness: the Orbit Determination process could take a control input into account to better fit measurements, Correlation could integrate an expense-based metric to understand how likely the candidates can reach the observation data and Sensors Tasking algorithms could include this information to optimally plan activities.

Most of the traditional approaches to this task rely on the most common orbit data file format, i.e. the Two-Line Element (TLE) due to its wide availability. They usually scan TLE time sequences, extracting meaningful orbital parameters from them, and exploit standard statistical analysis to define backgrounds and bounds and perform thresholding. Hence, the peaks trespassing some given limits are flagged as anomalies (Maneuvers or Space Events) and the corresponding epoch is labeled for further investigation. The technique by Patera [1] proposes orbital energy and its derivative as candidate parameters, firstly applying a moving window curve fit via least squares, with a predefined curve order with a 1-TLE stride along the timeseries. This process aims at the computation of a local deviation, filtering data noise but preserving local features at the same time (depending on the window size). As for the threshold to identify anomalies, it is computed as standard deviation of the previously defined local ones and multiplied by a factor defining how strict the threshold is. A similar solution to the problem is reported by Kelecy et al. [2], performing a moving window fit of semi-major axis (detecting shape-changing event and equivalent to orbital energy) but employing a lower order for the curve (respectively offset and linear) on two adjacent windows to analyse the statistical difference at the interface, computed as an extrapolated mid-point time. Events are once again detected by a user defined threshold accounting for noise and outliers. These approaches are characterized by a main limitation, namely, some fundamental parameters are not automatically determined but chosen by the user (window size, threshold) implying ad-hoc selection for every specific case. A first attempt to address this shortcoming is the method by Lemmens et al. [3]. In this case an algorithm is devised to replace the user-defined threshold based on sample variance of unknown distributions with an adaptive, percentile-based one. The process aims at searching the input time series for harmonics and estimating the noise of the ones without harmonics or events to obtain a reliable background. This is done by processing semi-major axis sequences and applying a periodogram to extract peaks in the frequency domain. The low-frequency peaks are associated to maneuvers. By exploiting the interquartile range of the processed time series, a threshold initial value is defined. The actual detection algorithm starts with a running moving window of a given size, which changes across the scan. At the same time the standard threshold can be replaced during the process with a higher local one according to the noise local level. If this threshold is exceeded for a fixed amount of time, an event is reported.

A different branch of methods are instead conceived to leverage the great deal of structured data available from SST databases, by means of data-driven techniques, unsupervised clustering and supervised learning. An example of data mining method is given by X. Bai et al. [4] assessing the performance of unsupervised classification methods for the detection of orbital maneuvers. The available TLE history is firstly divided in categories according to the related maneuver magnitude, therefore it is pre-processed through standardization, filtering and propagation to obtain variations in the selected parameters. In the following step, K-means-based contour maps are computed to identify characteristic variables and optimal clusters number. The actual clustering is then carried out with several methods, to assess and compare performance. As for Supervised learning it focuses on Probability Density Function (PDF) estimation as function of time and classification algorithms to distinguish maneuvering epochs from non-maneuvering ones exploiting different model architectures (Shabarekh et al. [5], Thomas G. Roberts et al. [6]).

The current work leverages a peculiar combination of neural networks to perform an effective feature extraction through the layers of a residual Convolutional Neural Network (CNN) giving them as input to U-Net, an encoder-decoder network usually used to perform segmentation tasks on images. In this way not only are the inputs processed inside the network backbone, but the same training process optimize their values to best fit the mask generation process performed by its head. In this case due to the very nature of the data, being time sequences, the first iteration of the algorithm consists in a 2D-CNN where the first dimension depends on the number of parameters selected while the second one is determined by a fixed time window dimension.

The network architecture employed to build the method implies several differences with respect to both the traditional and the data-driven branches of the state of the art. In particular, its main contributions to it consist in:

- overcoming the need for a threshold to be tailored to the specific case at hand by detecting patterns that are autonomously extracted by the data distribution.
- achieving detection and classification of maneuvers so to enrich a possible pattern of life to be extracted from the data.
- extending the method to multiple orbital parameters analysis to increase detection and characterization robustness.

2 Fundamentals

This section deals with the main theoretical concepts behind the development of the method, from TLE data format to a brief explanation of the orbital parameters involved and what kind of maneuver they can be linked to. Besides, some deep learning notions are reported as well, specifically focused on residual CNNs and segmentation.

2.1 Two Line Element File and Orbital Parameters

The Two-Line Elemenet (TLE) is a data format consisting of orbital elements and information about a RSO at a given epoch. It usually consists of two 69-character lines of data which can be properly processed to extract the orbital state (position and velocity) of the target. An example of TLE is shown in Fig. 1.

Column 1 indicates the line number and hence the format for that line. The next field on each line (columns 3–7) indicates the North American Aerospace Defence and Command (NORAD) Catalog Number of the object. It is a unique identification code assigned by the NORAD for each Earth-orbiting artificial satellite in their catalogue. For a valid Two Line Element, the two fields must be identical. Column 8 indicates the security classification of the data. The 'U' in the example indicates unclassified data. The next three fields (columns 10–17) define the International Designator of the object, which is an additional unique code associated to each object. It indicates the year of launch (columns 10–11 only gives the last two digits), the launch number related to that year (columns 12–14), and the piece of that launch (columns 15–17) for each object. These three fields can be left blank, but all must be present if any of them is available. The next two fields (columns 19–32) define the reference time for the elements set and are jointly referred to as the epoch. Columns 19–20 represent the two digit year and columns 21–32 are the day of that year. The epoch defines the time which all of the parameters in the set are referred to. Columns 34–43 are the first derivative of the mean motion divided by two and columns 45–52 represent the second derivative of the mean motion divided by six. Columns 54–61 define the term called B* (BSTAR) which is an adjusted value of the ballistic coefficient B using the reference value of atmospheric density, ρ_0. Column 63 is the ephemeris type (i.e., orbital model) used to generate the data. Columns 65–68 represent the element set number. This number is usually incremented each time a new element set is generated. The last column on each line (column 69) represents a checksum of the data. To calculate the checksum, simply add the values of all

```
1 36508U 10013A   10098.62668804  .00000003  00000-0  00000+0 0  9997
2 36508 092.0159 286.8207 0021893 257.8601 102.0156 14.54886728    03
```

Fig. 1 TLE format example

the numbers on each line, ignoring all letters, spaces, periods, and plus signs, and assigning a value of 1 to all minus signs. The checksum is the last digit of that sum [7].

Line 2 consists primarily of orbital parameters calculated using the orbital model. Right Ascension, Argument of Perigee and Mean Anomaly are all reported in degrees and can range from 0 up to 360°, while Inclination only goes up to 180°. The Eccentricity (columns 27–33) is a unit-less value with an assumed leading decimal point. The Mean Motion (columns 53–63) is measured in revolutions per day. The final field on line 2, prior to the checksum, is the revolutions number [7].

In order to perform maneuver detection, an effective selection of parameters has to be made, taking into account the ones associated with the highest sensitivity to different kinds of maneuver. A first categorization can be made in terms of in-plane and out-of-plane ones. The former can be identified by looking at:

- Semi-major axis a (associated with the target closed orbit), specific orbital energy E or the specific angular momentum norm h to capture shape variations
- Pericenter anomaly ω (defined as the angle between the apseline and the **i** axis of the Earth Centered Intertial (ECI) frame) to identify in-plane orientation changes.

As for the latter, the parameters to analyze are mainly orbit inclination i defined as the angle between the normal to the orbital plane and the ECI **k** axis and right ascension of the ascending node Ω defined as the angle between the vernal equinox and the line of the nodes or, as an alternative to both of them, specific angular momentum **h** direction.

2.2 *Segmentation-Aimed Network Architecture*

Due to its versatility and its ability to grant high level of performance if used for suitable applications, Machine Learning has become increasingly popular in the last decade. Whenever an adequate quantity of samples is available, data-driven techniques like this one can let datasets speak for themselves performing regression, classification and inference on unknown distributions, with no assumptions of any a priori model. The neural network is a widespread architecture belonging to this algorithm category. It allows, after an optimization process (called training) of a loss function built upon a given dataset, to perform the above-mentioned tasks. The focus of this work is Supervised learning, meaning that the model is trained with labeled data. More specifically, CNN is the basic unit that has been used to build the presented algorithm. It is a deep neural network characterized by Convolution layers, where a given number of kernels (or filters) are multiplied element-wise by different pieces of the input volume. In this way, activation maps can be generated, encoding features that become more and more complex throughout the layers. This kind of architecture is widely used in the field of Computer Vision and image processing, due to its capability to preserve the input data structure.

The depth of a network allows it to encode elaborate features although leading at the same time to degradation in training performance, depending on the system at hand. To tackle this issue, not to lose performance and feature complexity, skip connections are introduced in the architecture. They connect layers apart from each other with an identity step and the deeper end of the link is summed to a residual to get the exact same value of the original structure. This is the rationale that justifies ResNets, as the one employed for the current model. A further exploit to improve a standard CNN is the Squeeze-and-Excitation technique, giving the ability to the network to assign a channel-wise attention score, learnt during the training phase. This is done by means of two operators acting on activation maps.

The Squeeze one performs a global average pooling for every channel generating a 1-by-1-by-C volume (C is the number of channels) on which the Excitation operator grasps dependencies by means of activation functions composition.

The final notion to understand the structure of the developed algorithm concerns the architecture used to perform semantic segmentation, i.e. a classification of every element of the input data as part of a category, enabling to distinguish the diverse targets inside an image and to generate masks to highlight them. U-Net has been chosen as the architecture to perform segmentation of orbital parameters time series. It consists in an encoder-decoder structure, where the input volume features are pooled and encoded to be successively up-sampled and decoded to the corresponding segmented mask. Due to the high amount of layers, skip connections and copy-and-crops are included to help the training process.

3 Method

The first iteration of the developed method consists in a neural network built as a sequence of a 1D-CNN, taking as input a time sequence of the selected orbital parameter and U-Net, used to extract the segmented mask from the original sequence as a one-hot array where the 1's refer to maneuver epochs while the 0's to natural motion. A second version of the algorithm is fed with a 2D input, involving more than one orbital parameter at a time, so that multiple categories of maneuver can be detected.

3.1 Data Structure and Pre-Processing

Data collection and structure is a crucial topic to deal with in the network development process. In this case, the availability of real, labeled data has been the main limitation for the method. The satellites analysed for the purpose of the work are TOPEX/Poseidon, Sentinel-3A, Sentinel-3B, Jason-1, Jason-2, Jason-3, Cryosat-2, Haiyang-2A and Envisat. Their TLEs have been retrieved from Space-Track website and their maneuver history can be found in NASA International Laser Ranging

Service database [10]. Made exception for TOPEX/Poseidon, the maneuver-related ΔV components in radial, along-track, and cross-track direction are also reported in the maneuver histories file.

After a preliminary analysis of both TLEs and maneuver history of every target, three conclusions are drawn:

- Every maneuver changes mean motion. This information is available in all of the collected data.
- Whenever the biggest component of ΔV is in the across-track direction, a significant amount of inclination variation can be noticed as well. This behaviour is reported in Sentinel-3A, Sentinel-3B and Envisat satellites only.
- Correlation between maneuvering events and changes in RAAN and argument of perigee is very low.

Hence, a first approach could be to perform maneuver detection across the time series by analyzing sudden variations in every parameter linked to the mean motion history, such as orbital energy or semi-major axis. Nonetheless most of them, and mean motion itself, cannot capture the changes in eccentricity that are part of every in-plane maneuver, so a more effective way to condense these variations into a single value is to take specific angular momentum into account to detect and characterize this kind of maneuver. Its equation is given below:

$$h = \sqrt{\mu_E a(1 - e^2)} \tag{1}$$

where μ is Earth's standard gravitational parameter, a is the semi-major axis, and e is eccentricity.

As regards the out-of-plane and orientation-changing maneuvers, they can be mostly detected through inclination sudden changes. Analysing the available maneuver history, out-of-plane maneuvers coincide with energy variations too, but characterized by different magnitude. As a consequence to spot them, 2 parameters should be fed to the network. Some examples of inclination and h history plots are reported in Figs. 2 and 3.

Moreover, due to maneuvers diverse duration and aim, their transients in terms of parameter variations can vary as well. For this particular application, they are categorized according to their peak steepness as short or long-term.

In order to cope with the limited availability of labeled data, not to depend too much on propagation and dynamics modelling, the original dataset is augmented acting directly on the parameters sequences. In this way the training phase can rely on a significant amount of realistic data while, for the testing step, part of the original samples can be employed. Two different algorithms have been developed to generate longer sequences, according to the type of variations that the target satellites feature. For the short-term ones the process consists in splitting, shifting and shuffling data between two adjacent maneuvers (i.e., a window) to generate new ones with similar patterns, as shown in Fig. 4. For the long-term, smoother maneuvers instead, a further effort is taken to shape the parameter variations, involving a linear interpolation of randomly generated points between two non-overlapping windows so to link them.

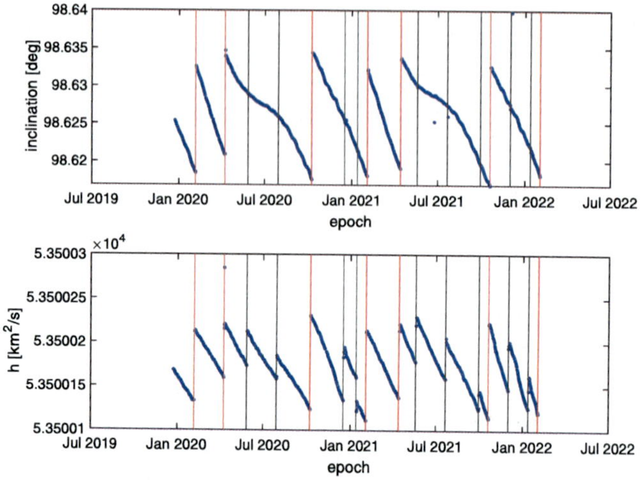

Fig. 2 Sentinel-3B specific angular momentum and inclination history. Red lines denote out-of-plane maneuver epochs while the black ones refer to in-plane

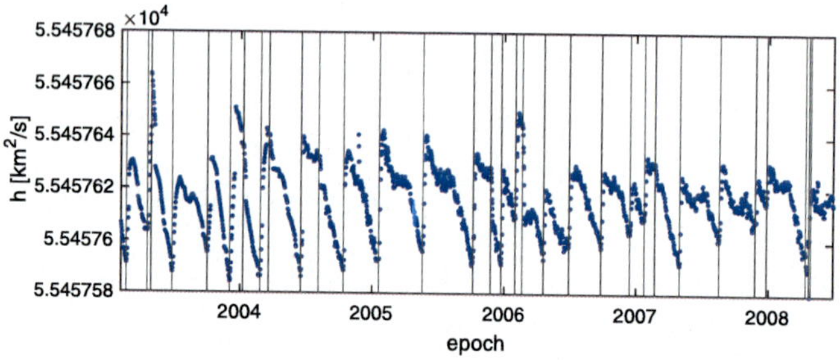

Fig. 3 Jason-1 specific angular momentum history. It is easy to notice how smooth the peaks relative to maneuvers are with respect to Sentinel-3B spikes

This process results in realistic patterns as displayed in Fig. 5 for Jason-1. Both algorithms are included in the workflow charts of Fig. 6. As for the time series scanning method, it consists in a moving window composed of a fixed number of TLE samples n with a given stride. This is done in a way to obtain a constant input dimension and an overlap in time spans. Consequently, for an epoch to be classified as maneuvering, a rule can be set that it shall be detected by a used-defined percentage K of the total number of overlapping windows containing it. For example, if $K = 0.5$, a detection is considered as valid if in 50% of the windows including the epoch of detection a maneuver has been spotted.

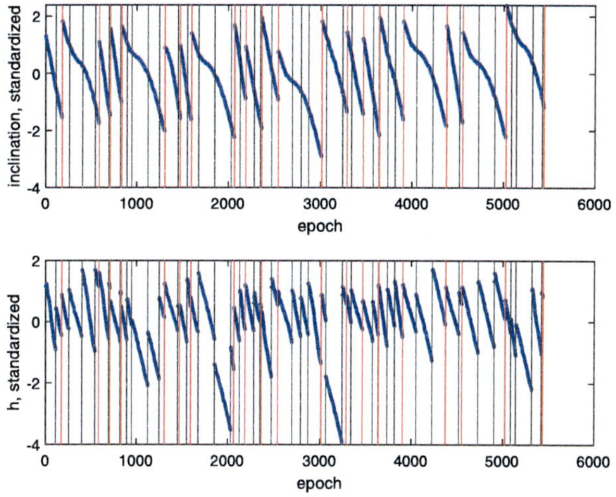

Fig. 4 Augmented angular momentum data sequence from satellite Sentinel-3B

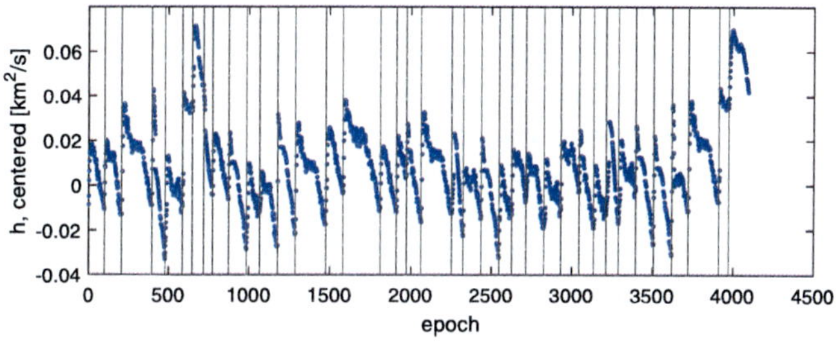

Fig. 5 Augmented angular momentum data sequence from satellite Jason-1

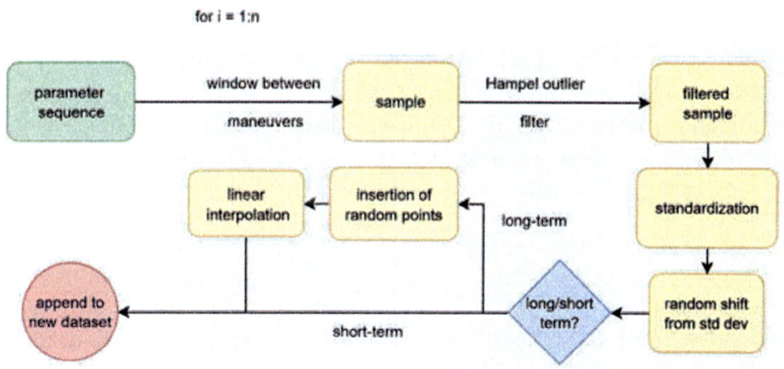

Fig. 6 Data augmentation process scheme

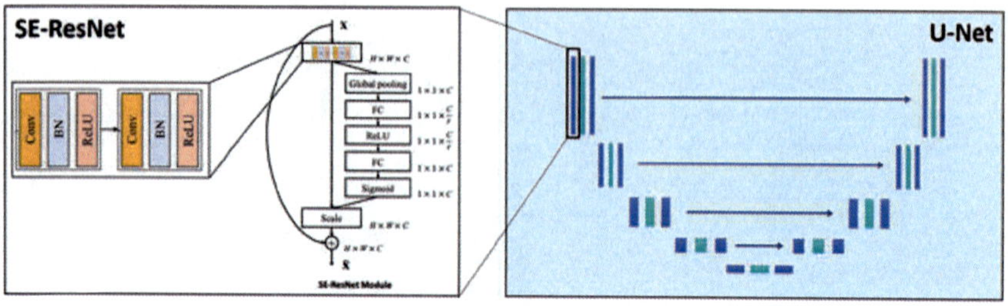

Fig. 7 A scheme representing the network architecture. On the left, the architecture of the convolution blocks composing the encoding segment of U-Net. On the right the U-shape composed of both the encoding and decoding segments of the network

3.2 Network Structure and Training

The network architecture consists in a concatenation of the components described in Sect. 2.2. It uses a Squeeze-and-Excitation residual neural network to encode the given parameter sequence window into suitable features to train the segmentation block with. The latter is based on U-Net architecture and turns the encoded input in an segmented mask, expressed as a one-hot vector. The complete structure is reported in Fig. 7: on the left, the architecture of the convolution blocks composing the encoding segment of U-Net together with pooling layers is displayed; on the right instead, a simplified representation of convolution volumes constitutes the typical "U" shape of the network; this is to highlight how the two halves of the architecture interface through skip connections (represented as blue arrows). The aim is to link the initial layers of the encoding section, storing simple features only but preserving most of the input original structure, to the last layers of the decoding segment, so to help in the generation of the segmented mask.

As for the the training process, it leverages categorical cross-entropy as loss function. It examines each element of the last activation map individually and compares the class predictions (depth-wise element vector) to a one-hot encoded target vector. More specifically, the last layer of U-Net performs classification using a Softmax activation function p_k which generates class scores for each element [9]:

$$p_k(\mathbf{x}) = \frac{e^{a_k(\mathbf{x})}}{\sum_{k'=1}^{K} e^{a_{k'}(\mathbf{x})}} \tag{2}$$

where a_k denotes the k-th kernel channel comprising the final feature map element corresponding to the input one position $\mathbf{x}$, and K represents the total number of classes (a single one in this case). The resulting feature map is then compared with the ground-truth mask using the mentioned loss function type for each output element $\mathbf{x}$:

$$L(\mathbf{x}) = -\sum_{k'=1}^{K} \alpha_{k'}\, t_{k'}(\mathbf{x}) \, \log(p_{k'}(\mathbf{x})) \tag{3}$$

where the distance from the ground-truth pixel mask is considered for each class k' by comparing the true label $t_{k'}$ with $p_{k'}$. $\alpha_{k'}$ is a coefficient assigned to the class k' to weigh every category associated with an event of interest ('maneuver', 'in-plane maneuver', 'out-of-plane maneuver') 200 times the 'no maneuver' class, so to cope with class imbalance.

The employed samples are windows of length 1000 retrieved from the augmented dataset described in Sect. 3.1. They are divided into 80–20% as training-validation split to prevent overfitting. Google's Colaboratory with free GPU and TensorFlow with Keras API has been used to train the network.

Two types of model have been trained with different objectives:

- Network A: a network trained with dataset to specifically fit each satellite, for target-tailored maneuver detection. In this case 9 models aimed at maneuver detection task only (maneuver/no-maneuver) are obtained.
- Network B: a network trained with dataset to specifically fit each satellite, for target-tailored maneuver characterization. In this case, 3 models aimed at maneuver characterization (in-plane, out-of-plane, no-maneuver) are obtained.

3.3 Network Post-Processing and Testing

Concerning post-processing, sometimes the algorithm produces multiple maneuver detections close to each other while there is only one actual maneuver at the specified epoch. To deal with this issue, every clump of positive epochs will be considered as a single maneuver, picking the first detection instant to account for the whole set. Moreover, when comparing predicted and ground-truth maneuvers, an epoch distance threshold is assigned, i.e. a prediction is deemed successful if the epoch difference with respect to actual one is less than a fixed value. This is done because of the complexity in precisely pinpointing an exact maneuver in time with a sparsely-sampled dataset as a TLE time sequence can be. Consequently, the chosen test scores are computed after this post-processing step. Standard metrics have been selected to assess the network performance, namely Precision, Recall, and F1 score, defined as:

$$\begin{aligned} Precision &= \frac{TP}{TP+FP} \\ Recall &= \frac{TP}{TP+FN} \\ F1 &= \frac{2 \times Precision \times Recall}{Precision + Recall} \end{aligned} \tag{4}$$

where TP, FP, FN respectively refer to True Positive, False Positive, and False Negative.

4 Results

The testing phase consists in an assessment of the trained model performance when fed with never-seen-before data. In the case at hand, both network categories (A and B) have been tested with real raw data to assess both network performance and effectiveness of the employed data augmentation technique.

4.1 Network A Testing

For this category, testing performance seem to suit the application in most cases.

The main difference consists in the kind of pattern the object features: for short-term events, generating discontinuous peaks Precision, Recall and F1-score are around 90% testifying that the augmented dataset is close enough to real data. On the contrary, long-term events, characterized by smoother peaks, result in lower scores due to modelling inaccuracies in the augmentation step. Table 1 reports performance for the available satellites, while Fig. 8 shows an example of output for long-term maneuvers hinting at the higher detection complexity with respect to short-term ones.

4.2 Network B Testing

The same patterns can be observed for the maneuver characterization task, for which performance differs again according to the maneuver time scale, as shown in Table 2. As a comparison between Figs. 9 and 10 shows, there is a crucial difference in maneuver pattern and noise levels between the Sentinel satellites and the Envisat, impacting characterization scores significantly.

Table 1 Metric scores of Network A trained and tested with same satellite

Satellite	Precision	Recall	F-1	Pattern type
TOPEX/Pos.	0.917	0.880	0.897	Long-term
Sentinel-3A	0.956	0.956	0.956	Short-term
Sentinel-3B	1.00	1.00	1.00	Short-term
Jason-1	0.844	0.710	0.771	Long-term
Jason-2	0.933	0.700	0.800	Short-term
Jason-3	1.00	1.00	1.00	Short-term
Haiyang-2A	0.872	0.759	0.811	Long-term
Envisat	0.757	0.746	0.751	Long-term
Cryosat-2	0.896	0.941	0.918	Short-term

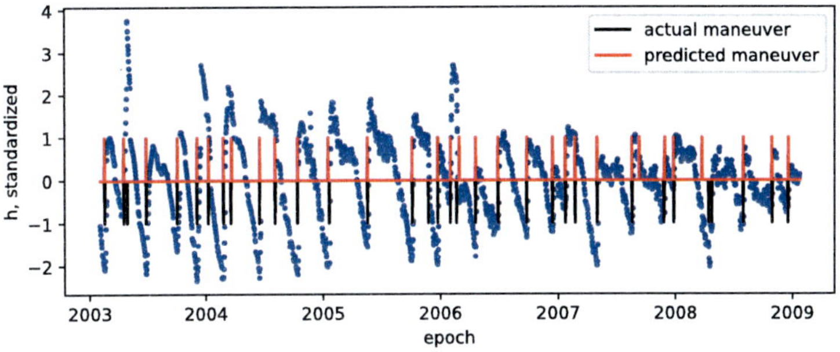

Fig. 8 Network output example of a long-term maneuver patter from Jason-1. Red lines are predicted maneuvers while black ones are the ground-truth ones

Table 2 Metric scores of Network B trained and tested with same satellite

Satellite	Precision	Recall	F-1	Pattern type
Sentinel-3A (in-plane)	1.000	0.920	0.960	Short-term
Sentinel-3A (out-of-plane)	1.000	1.000	0.97	Short-term
Sentinel-3B (in-plane)	1.00	1.00	1.00	Short-term
Sentinel-3B (out-of-plane)	0.89	1.00	0.94	Short-term
Envisat (in-plane)	0.756	0.402	0.525	Long-term
Envisat (out-of-plane)	0.947	0.620	0.75	Long-term

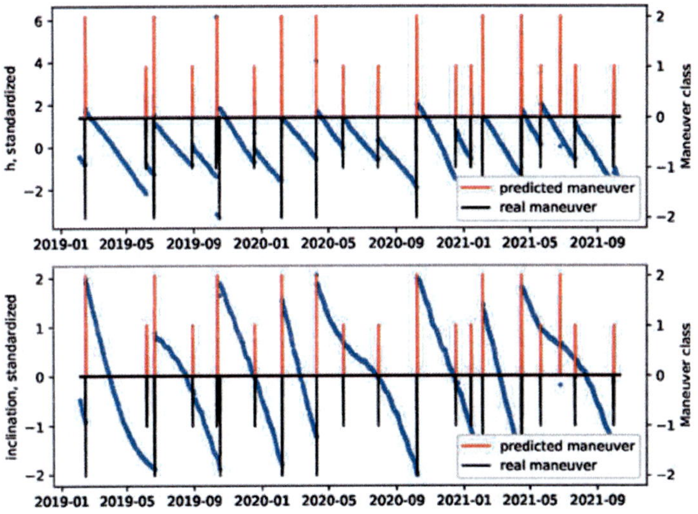

Fig. 9 Network output example of a long-term maneuver patter from Sentinel-3B. Red lines are predicted maneuvers while black ones are the ground truth ones

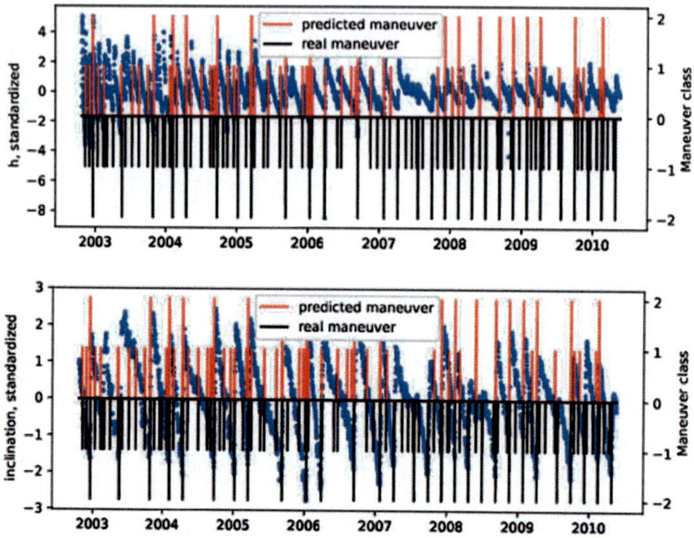

Fig. 10 Network output example of a long-term maneuver patter from Envisat. Red lines are predicted maneuvers while black ones are the ground-truth ones

4.3 Results Comparison

By drawing a comparison with detection models from the current state of the art, some differences can be easily noticed both in results and general aim. The method in [3] is used as a first benchmark. Their results are limited to a few satellites (14 distinct objects) for which at least a maneuver has been detected by a human operator throughout the year 2011. The precision, obtained by some computation, amounts to 88% while the recall reaches 79%, leading to a final F1-score of roughly 83%. This performance level is justified by the limited pattern encoding capability of this kind of thresholding process. While performance is mostly comparable or lower with respect to the one from this work, generalization capability is higher, having no limitations related to a training process or a data-driven cost function. Consequently, the approach can be employed with any target at the expense of complete automation, since user interaction is inevitable.

The algorithm described in [6], on the other hand, offers a more significant bench since it relies on a similar neural network approach aimed at segmentation. Given the higher number of analysed samples, the computed scores are much more representative than the previous case. The achieved precision (retrieved from computations performed on their results and normalized due to strong class imbalance) is 87.7% while recall reaches 96.6% for a final F1-score of about 91%. This level of performance seems comparable to the one achieved for some of the satellites belonging to the short-term category in this paper but the class imbalance of this specific application generated a high number of false positives in [6]. In the current work, this issue has been addressed by introducing category weights in the cost function, so to optimize on different scales according to each class population.

The main remark from this brief examination is the difference in terms of generalization with respect to the state of the art. This could be considered as the main limitation of the developed algorithm but it is not its main objective. Being the focus of the method to detect and characterize maneuvers from orbital history and help labelling new time series, it does not require real-time compliant performance during training nor inference phase. This allows to develop models that are tailored to single satellites or even specific mission phases, so to increase both accuracy and characterization detail. Moreover, diving into an operational scenario, re-training sessions can be planned and adjusted only when enough labelled and certified data from a target have been collected or its mission phase has changed. As a consequence, constraints on processing time are quite loose, leading to a trade-off outcome that tends to prioritize precision and detail.

5 Conclusions

The developed maneuver detection and characterization method has shown promising results despite the scarce availability of data, proving that data-driven techniques are perfectly suited for the SST framework, involving large and structured datasets that allow for inference and pattern of life characterization. Even if the generalization capabilities of the models are currently limited to the same satellite on which training has been performed, there is potential to extend prediction to satellites belonging to the same category, sharing the same mission aim or orbiting in the same regime through encodings to be embedded in the input data, so to provide the network with more context for the analysed parameters sequences. It is important to state that this work is still under development, focusing on encoding structures and maneuver probability prediction, given a reliable event history of a target.

References

1. Patera, R.P.: Space event detection method. J. Spacecraft Rockets **45**(2) (2008)
2. Kelecy, T., Hall, D., Hamada, K., Stocker, D.: Satellite maneuver detection using two-line element (TLE) Data. AMOS (2007)
3. Lemmens, S., Krag, H.: Two-line-elements-based maneuver detection methods for satellites in low earth orbit. J. Guidance Control Dyn. **37**(3) (2014)
4. Bai, X., Liao, C., Pan, X., Xu, M.: Mining two-line element data to detect orbital maneuver for satellite. IEEE Access **7** (2019)
5. Shabarckh, C., Kent-Bryant, J., Keselman, G., Mitidis, A.: A novel method for satellite maneuver prediction. AMOS (2016)
6. Roberts, T.G., Linares, R.: Satellite repositioning maneuver detection in geosynchronous orbit using two-line element (TLE) data. In: 71st International Astronautical Congress (2020)
7. Vallado, D., Cefola, P.: Two-line elements sets-Practice and use. Int. Astronaut, Cong (2012)

8. Varshaneya, V., Balasubramanian, S., Darshan, G.: RES-SE-NET: Boosting performance of resnets by enhancing bridge-connections. Mach. Learn. Aplgorith. Appl. **4** (2021)
9. Ronneberger, O., Fischer, P., Brox, T.: U-Net: Convolutional networks for biomedical image segmentation. In: Medical Image Computing and Computer-Assisted Intervention–MICCAI (2015)
10. NASA International Laser Ranging Service website, Satellite Maneuver Histories. https://ilrs.gsfc.nasa.gov/data_and_products/predictions/maneuver.html (2019)

A Machine Learning Approach for Monitoring of GNSS Signal Quality in Spaceborne Receivers: Evil Waveform and RF Threats

Andrea Emmanuele, Ruggero Colombo, Stefano Zago, and Mirko Salaris

Abstract Artificial Intelligence (AI) is here targeted to GNSS Space applications for the identification of RF threats, in particular "Evil waveform" and its distortions on received signals at GNSS receiver side. These anomalies are generated at SV payload level and, if not removed from the processing, degrade on-board Space GNSS Receiver measurements, positioning navigation and timing (PNT) solution with failures at related services. EGNOS system [1] currently augments GPS users by transmitting dedicated signals to users, including corrections and alerts for anomalous distortions detection, but with an intrinsic processing delay. Interest on AI potential is raised nowadays, for its application in many fields including GNSSs [2], with classification methods developed to distinguish signals conditions, as Line-of-Sight (LoS), non-LoS, Multipath, using Machine Learning (ML) models [3], with attempts focused also in EWF prediction [4]. The proposed work considers a basic ML based algorithm for EWF detection directly on board of the GNSS receiver for Space applications to reduce time to detect distortions for accurate and reliable PNT [5]. Amidst ML techniques, the most promising one is analyzed within Support Vector Machines (SVM) [6], because of lower computational burden [7] and expected feasibility of an embedded implementation in next generation Spaceborne receivers, based on multi-correlator technology and with enhanced memory availability. The receiver signal correlations and link budgets are the observables for the AI detection scheme, collected in different live and simulated conditions. The ML

A. Emmanuele (✉) · R. Colombo · S. Zago
Thales Alenia Space Italia, Gorgonzola, Italy
e-mail: andrea.emmanuele@thalesaleniaspace.com

R. Colombo
e-mail: ruggero.colombo@thalesaleniaspace.com

S. Zago
e-mail: stefano.zago@thalesaleniaspace.com

M. Salaris
DEIB, Politecnico Di Milano, Milan, Italy
e-mail: mirko.salaris@polimi.it

C. Ieracitano et al. (eds.), *The Use of Artificial Intelligence for Space Applications*, Studies in Computational Intelligence 1088,
https://doi.org/10.1007/978-3-031-25755-1_29

algorithm performance are so characterized with specific figures of merit, promising if compared to traditional detection schemes specifications.

Keywords Artificial Intelligence (AI) · Machine-Learning (ML) · Support Vector Machines (SVM) · GNSS spaceborne receiver · RF threats detection · Evil waveform (EWF) · Multipath (MP)

1 Introduction and Motivation for GNSS Space Applications

After the first observed Evil Wave Form (EWF) event in 1993, ICAO Threat Models (TM) and Threat Spaces (TS) [8] were fully standardized and proposed for GPS L1 C/A signal, to characterize these signal anomalies, generated at SV payload level—at analog and/or digital sections—and induced distortions at signal autocorrelation level. They distinguish so payload failures generating the signal irregularities and are used to model the effects at receiver correlation and position/time domain. These distortions at autocorrelation in fact would produce in turn in GNSS receivers more than tenths of meters error at pseudorange measurement level, as function of specific EWF type, receiver RF front-end characteristics and tracking scheme, inducing biases in the Signal-in-Space (SIS) time delay estimation, on top of a relevant degradation at link budget, available at correlators level.

European Geostationary Navigation Overlay Service (EGNOS) system currently augments GPS user applications, with transmitting by dedicated satellites a signal containing information on the reliability and accuracy of the positioning signals sent out by GPS. This information is estimated through a ground located network of Ranging, Integrity and Monitoring Stations (RIMS), which experiencing the SV distortions effects, jointly to central system facilities, determine the user corrections and detect anomalous distortions, which would impair user integrity risk, to be sent at ground segment level. After an intrinsic system delay in the communication and processing, this information is made available at users, which receive EGNOS messages to correct measurements and not use signals which are deemed unhealthy by the system.

The fundamental element of this architecture consists in the Signal Quality Monitoring (SQM) algorithm designed at reference stations level, aimed to detect the anomalies and to protect so GNSS users in safety-critical applications, as civil aviation (e.g. guiding aircraft during landing approaches), from the potential risk of reception of signals affected by these signal anomalies.

Following this logic, a SQM functionality is deemed necessary for Spaceborne GNSS Receivers, where fundamental targets are an accurate and integrity-risk free positioning and velocity trajectory tracking, together with precise time synchronization, to complement, substitute or enhance current existing augmentations systems performance.

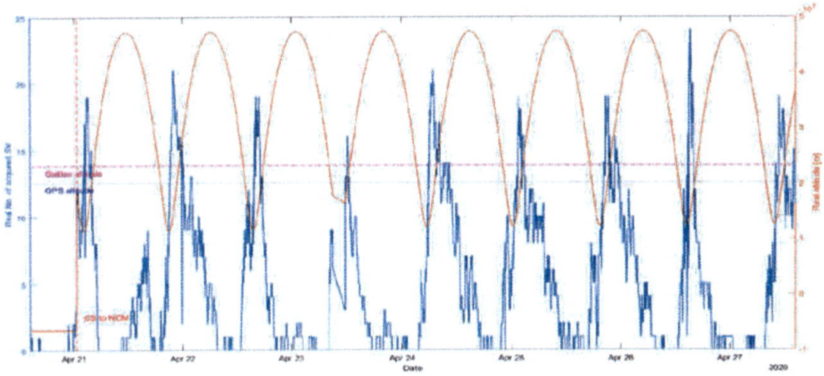

Fig. 1 Measures used in PVT solution versus Altitude in GEO Transfer Orbit

Space applications that consider a GNSS Receiver as a mean of positioning and timing estimation are nowadays a standard, especially in LEO environment. New mission concepts (in which a GNSS receiver follows a High Earth Orbit or, in general, in Geo Transfer Orbit) are becoming of strategic importance. The use of electric propulsion to perform or complement the transfer maneuvers to reach the final GEO orbit is nowadays employed. This imposes long duration orbits that intersect several times the "GPS and Galileo Orbital Spheres", where GNSS Receiver performance are strongly influenced by the availability of measurements [1]. A typical Geo Transfer Orbit, with high apogee and low perigee has periodically very low visibility conditions and long phases without GNSS Signals where the receiver propagates its attitude and trajectory to sustain further acquisition of GNSS signals as soon as they return in view. An example of (GPS and Galileo) satellites in visibility and acquired by receiver is proposed in Fig. 1, in blue lines, where long periods of unavailability of at least 4/5 satellites are shown. In background also the altitude in Geo Transfer Orbit is reported (orange line), to stress links between visibility and specific orbit.

The accuracy of propagated solution is strongly impacted by the quality of signals of the latest measurements available before initiate propagation. In fact, the extreme visibility conditions previously depicted for these applications are also affected by RF impairments at received signals level, linked to Satellite structure and propagation characteristics. Therefore a measurement affected by distortions, to be detected, maintained in PVT solution could deviate the trajectory estimation, on the other hand exclude a valid measure due to a false anomalies detection can initiate the propagation before the effective right time. Then, looking at a signal quality monitoring scheme for distortions detection, to mitigate integrity and continuity degradations, both the miss detection and the false alarm figures of merit become extremely significant in the introduced Space application.

The satellite platforms have several system constraints which push to collocate different equipments on board, with elements which can interact each other, e.g. solar panel and satellite platform, and in particular can obstruct or degrade GNSS received signals and signal processing performance. GNSS antennae will receive in fact, on top of Line-Of-Sight (LOS) signal, also reflected echoes of it, producing so called multipath phenomena. The effect of multipath is to have a nominal distortion of the signal correlation shaping, producing similar even if more limited effects w.r.t. the most degrading EWF distortion models, in the code bias error domain. A two-ray multipath model is so assumed for performance analysis in typical Space platform conditions, impairing in this assessment the target RF threats identification.

In the above introduced scenarios, an innovative SQM scheme for space applications is here proposed considering a basic Artificial Intelligence (AI) based algorithm for detecting EWF threats models, directly on board of the satellite during navigation, in order to reduce related time to detection and mitigation of RF threat distortions' effects, so related integrity risk induced in the application.

The AI family of Machine learning, in particular the Support Vector Machines (SVM) [2] are introduced in the next sections, as firstly matching the signal monitoring scopes and the specific application targets. The Machine learning algorithm performance are finally addressed by considering detection probability and false alarm probability, together with time to detection, as function of the smoothing filter time investigated in this assessment as free parameter to be traded-off.

Considering the SVM classification method under investigation, the inputs provided at the different learning and classification operative stages has been selected to have the highest impact at performance level. In particular, the inputs consisted in the most informative observables available at receiver side, including eleven In-phase signals correlation measurements designed symmetrically around the prompt one, the measured link budget (proportional to the Signal to Noise ratio, SNR), the elevation and azimuth angles. These inputs have been collected in live campaign or in different laboratory controlled scenarios, generating so the fundamental classes presented in the next section used to first classify the observables within the faulty or the faulty-free domains.

2 Observables and Working Hypothesis

2.1 Dataset Description

Typically, for GNSS applications, the received GNSS signal is convolved with an ideal (nominal) signal local replica at received side to obtain the correlation function, basic element for all next signal processing. In particular, a set of minimum three correlation samples is needed—early, late and prompt correlators, sometimes already incorporated in a specific expression to save memory—to ensure tracking of the signal and retrieve navigation information. In a GNSS receiver, this minimum number of correlators are foreseen for each of multiple physical channel, in order allowing the tracking of multiple GNSS signals in parallel and enable processing for positioning and timing.

For monitoring purposes instead, the signal correlation function is sampled within its main lobe width, with a number of samples higher than three, usually constrained by the available receiver resources.

The assessment here proposed uses as sufficient variable for detection test a discrete set of 11 complex signal correlation samples (in-phase and in quadrature-phase, I/Q), with measured link budget value, for each physical channel, stored at 1 Hz rate for the GPS L1CA signal.

Current-generation Spaceborne Receivers operate with 5 complex correlators per channel, mainly due to the limited resources available on hardware space qualified. Next-generation hardware, which is currently under development for usage in future spaceborne GNSS receivers, will support at least 11 complex correlators. This in fact can be considered as a new potential feature to be recommended for any future space missions, to be now enabled by available technologies.

In order to obtain similar monitoring capabilities on legacy HW, dedicated time-division multiplexing on the spacing sets could be as well considered, but through flexible FW/SW configurations in time. This approach could be anyway adopted only if phenomena to be monitored remain stable in time (i.e. EWF presence) w.r.t. time division slots, with observables not varying for propagation effects (narrowband signal without fast variations).

2.2 Live Collection and Synthetic Datasets Generation

The above designed sufficient variable is sampled in different conditions/hypothesis, to cover a priori classifications and the final detection stage, as will be explained in the next sections. It is worth to recall that evil waveforms have a very relevant impact on receiver raw measurements for navigation, but they rarely occur.

Live test campaigns are adopted to collect EWF fault-free datasets, with variable thermal noise and multipath conditions in time. They don't provide data containing evil waveform distortions. For this reason, synthetic datasets of correlators have been foreseen to be generated following the principle of "superposition of effects".

The theoretical autocorrelation (ACF) function is so first synthetized with EWF threats, by adopting the ICAO models for type A, B and C. Examples of ACF distorted by EWF types A, B and C (examples) are reported in Fig. 2 vs reference function without distortion. The functions here reported are already normalized to the prompt, they are sampled at the designed correlator spacing to obtain the distortions taps for analysis, for each EWF threat to be analyzed.

The EWF synthetized set of observables is obtained by multiplying the extracted taps from threat model to the set of measured correlators, collected in live conditions, as described in (1):

$$C_S^i(t) = \frac{C_S^i}{C_N} \cdot C_r^j(t) \tag{1}$$

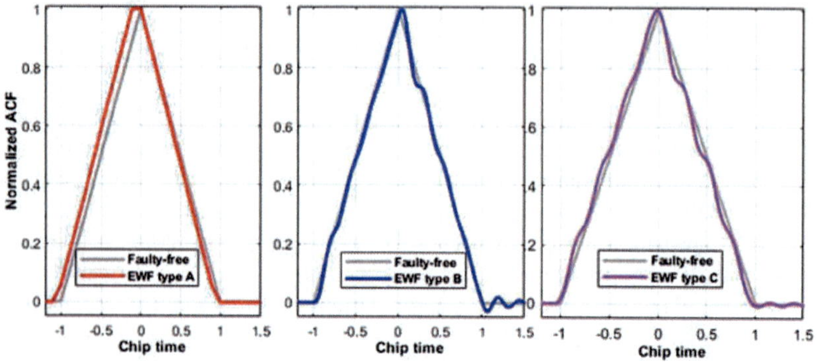

Fig. 2 Normalized ACF distorted by EWF type A, B and C versus faulty-free ACF

where:

i: index of the sampled EWF distortion; j: index of sampled real temporal sequence.

$C_S^i(t)$: synthetic temporal sequence; C_S^i: sampled (ith)EWF distortion.

C_N: reference nominal convolution; $C_r^j(t)$: sampled (jth) real temporal sequence.

This process is considered accurate enough for measurements collected at high C/N0 levels (i.e. high signal w.r.t. noise power), as it is the case of measurements acquired in live ground condition, typical GNSS patch antenna, elevations above 5° mask.

A similar procedure is then also foreseen to synthetize (controlled) MP distortions jointly to EWF threats, applied on the available live collected measurements. Two-rays multipath model with Line-of-sight (LoS) and a single echo is considered, spanning a large sets of amplitudes, phases and delays of the echo w.r.t. the LoS component.

This collection may result pessimistic w.r.t. operative scenario, because live sets already contain multipath contribution, but it is designed to degrade SQM detection in faulty condition (i.e. miss-detection figure in EWF presence), considering "controlled" synthetized levels of multipath.

An approximation to be stressed consists in keeping constant for synthetized datasets the code tracking biases—i.e. prompt delay w.r.t. real one—obtained in live acquisitions, without considering added EWF or MP impact on this parameter. This is linked to the above distortion method, acting only at correlators level and not simulating also tracking process. Again, it can be proven that this model results pessimistic in the frame of detection of distortions and EWF discrimination, because EWF itself would introduce relevant tracking biases, perturbing as well associated correlators (Fig. 3).

Finally hence, for the proposed assessment the overall dataset consists of data from:

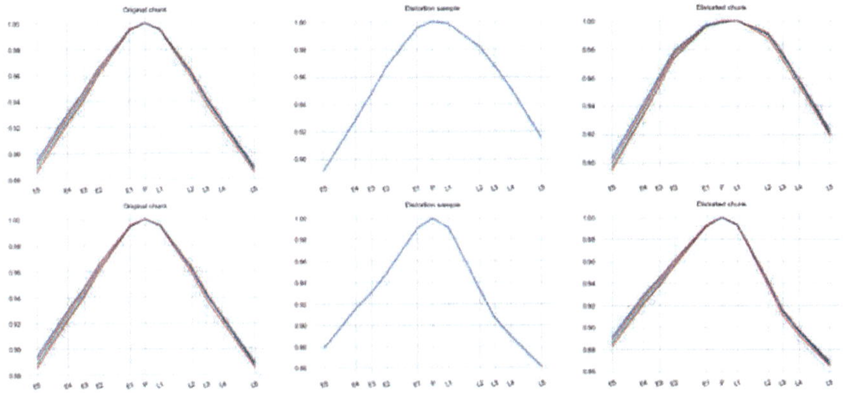

Fig. 3 From the left to the right columns, two examples of live ACF measurements, EWF + MP taps for distortion, distorted ACF via specific EWF + MP

1. Datasets from live ground stations records containing temporal sequences of convolutions, without evil waveform, distorted by multipath and thermal noise, representing "fault-free" condition. These records have been considered as pessimistic w.r.t. on board vehicle ones.
2. Synthetic datasets obtained by post-processing live data,
   - with evil waveforms of type A, B and C, without multipath on top, to generate "faulty" EWF-only condition,
   - with "EWF + MP" models, again to obtain as "faulty" condition, but to degrade AI detection process, with contribution of modelling LoS and a single echo due to satellite platform scattering.

3 Machine Learning: Support Vector Machines Scheme

The aim of this approach is to identify a separator *hyperplane* between the points of each class of modelled signals, within a n-dimensional space of possible combinations of observables.

In particular, a Support Vector Machine (SVM) is able to estimate an optimal *hyperplane* by resolution of an optimization problem which involves maximizing the *margin* between a set of *support vectors*. Support vectors are defined as vectors which comprise the points located on the outer boundaries of each class.

In the specific application of distinguishing undistorted GNSS signals from those affected by RF threats, this process can be seen as the maximization of the distance between "good" and "evil" waveforms, where the latter ones are to be detected and tagged as *not usable* for PVT solution evaluation. An example of this process is reported in Fig. 4, using only 2 dimensional hyperplanes.

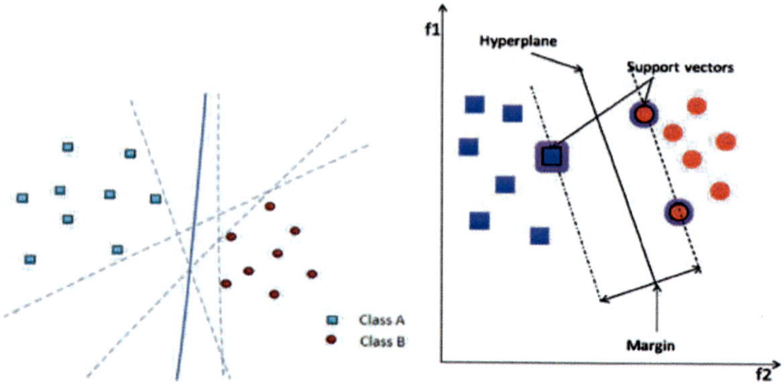

Fig. 4 Hyperplane 2-D representation (left) and distance maximization principle (right)

Mathematically, the hyperplane equation is defined by $w^T x - b = 0$ and the optimization problem is of the form:

$$\begin{aligned} &\min f : {}^1\!/_2 w^2 \\ &s.t.g : Y_i(w \cdot X_i) - b \geq 1 \end{aligned} \tag{2}$$

with "margin" between support vectors and separation region defined by **2/||w||.**

The above theoretical optimization problem is only applicable to classes which are linearly separable in the feature space (determined by the sampled observables). After a preliminary analysis on both live and synthetic datasets, the "good" and "evil" classes involved in the present problem showed not to be linearly separable in the feature space. As a consequence, a non-linear kernel, namely a Radial Basis Function (RBF), has been applied to transform the feature space and obtain an easier to solve problem, albeit at cost of a slightly higher dimensionality and corresponding computational burden.

Training of the chosen SVM classifier is brought out in a *supervised* fashion, using the set of signal samples described in Sect. 2, labeled a-priori to belong to either the *good* or *evil* class. The objective of the training session for the present classifier is not to obtain an optimal solution, but rather to achieve acceptable performance figures while still maintaining an acceptable computational burden for what concerns a real-time embedded implementation of the detection phase.

After the training phase, the SVM produced is ready to be tested in an environment representative of the operational scenario, through new sets of signal samples taken within the faulty or fault-free conditions, different from the training sequences. This allows to quantify in a controlled environment, the model's SQM performance in terms of *miss-detection probability* and *false alarm probability*.

3.1 *The Algorithm: Cross-Validation and Performance Metrics*

The available databases containing the set of data modelling the different classes, are split into 5 subsets of equal size to prepare independent training and test vectors. The model verification approach foresees usage of the method of cross-validation: across 5 parallel training and test runs, each of the 5 data subsets is used exclusively as a test set, with the remaining 4 used for training of the model.

This cross-validation scheme allows to use all the possible combinations between subsets used once for test and the other times for training stage, increasing efficiency of the performance evaluation. The performance figures are finally estimated as the average result from the whole set of cross-validation steps, applied to three independent live available databases.

As typical for decision problems, the test hypothesis (ground truth) and prediction results are reported in a confusion matrix in Fig. 5, which shows all the possible combinations between the set of inputs and corresponding outputs. For the matrix, four combinations are indicated as:

- TP—True Positive: a proper EWF detection for a tested sample distorted by EWF, so in faulty condition;
- FP—False Positive, also known as "False Alarm": a wrong EWF detection on a tested sample not affected by EWF, so in a faulty-free condition;
- TN—True Negative: proper no EWF detection, in a faulty-free condition;
- FN—False Negative, also known as "Miss-detection": a wrong no EWF detection for a tested sample distorted by EWF, so in a faulty condition.

As per above definitions of events in the confusion matrix, the two fundamental performance metrics for a SQM scheme are here defined as:

- *False Positive rate or False Alarm rate*, i.e., the percentage of samples classified by the SVM as *EWF* (or anomaly distorted) which were instead actually *good* signals

$$FPR = \frac{FP}{TN + FP} \tag{3}$$

Fig. 5 Confusion matrix—decision matrix versus hypothesis

		Ground Truth	
		Positive (EWF)	Negative (good)
Prediction	Positive (EWF)	TP	FP (*False Alarm*)
	Negative (good)	FN (*Miss-detection*)	TN

- *False Negative rate or Miss-detection rate*, i.e., the percentage of samples classified by AI as *good* signals, to be used in PVT solution, but which were affected by *EWF* distortions

$$FNR = \frac{FN}{TP + FN} \tag{4}$$

Focusing on the main target of this assessment, both False Alarm and Miss-detection rates represent crucial for spaceborne GNSS receivers.

A high Miss-detection rate would imply a high probability of including a signal affected by EWF in the computation of the PVT solution, leading to potentially harmful errors in the computation of spacecraft Position, Velocity and Time, which could in turn bear impacts on many aspects of spacecraft operations such as Orbit determination and Avionics.

False Alarm rate is considered of relatively lesser importance in the scope of ground receivers: due to the high availability of GNSS satellites nowadays, an EWF false alarm and exclusion of one satellite from PVT computation does not usually cause a big impact on well-equipped GNSS receivers with a high amount of physical channels and multi-constellation support. On the other hand, spaceborne GNSS receivers operate in low-visibility conditions (especially when reaching Geostationary orbit, see Fig. 1), to the point where using or not using a satellite in PVT computation could make decisive difference between even achieving a PVT solution, or having to propagate older information while waiting for more satellites.

PVT availability is usually one of the most important performance metrics for a GNSS receiver; thus the importance of achieving a low False Alarm rate for space applications.

As per above considerations the miss-detection rate and false alarm rates will be so considered in this work in the section with results, as the main figures of merit for performance characterization of the detection algorithm under investigation.

3.2 Considerations on Computational Complexity and Dataset Size

The training phase of an SVM classifier comes with an *exponential time* complexity $\left(O(n^2 * m)\right)$, with n being the size of the training set and m being the number of observables). The detection phase has instead a linear time complexity $(O(n * m)$, with n being the size of the training set and m being the number of observables).

Space complexity of the SVM classifier is also a concern, given that the trained model will need to be stored within the non-volatile memory of the On-Board Computer and loaded in RAM for operation. Namely, the space complexity of an SVM classifier can be defined as linear $(O(n * m)$, with n being the size of the training set and m being the number of observables).

In particular, both the computational cost of the detection phase and the size of the trained model will depend on the amount of Support Vectors, with an upper bound being the whole set of n-dimensional points of the training set.

Given the limited amount of resources available within on-board equipment, a trade-off between detection performance and resource occupation has to be made.

It is worth to be noted that on the specific application of an on-board SVM classifier, the training phase can be reasonably demanded to ground facilities, i.e. in-factory processing, leaving only the runtime detection phase to take into account for an on-board implementation.

Hence, the dataset designed for the training phase shall be properly composed to cover on one side a broad variety of conditions/hypothesis, while keeping anyway narrow its size to still bearing appropriately light computational burden and memory occupation at detection stage for the on-board GNSS processor.

The dataset for the current assessment is composed of 100 k samples, including:

- 25 k EWF-free samples acquired from TAS-I ground reference stations in live conditions, distorted by thermal noise and Multipath, sampled at different C/N_0
- 75 k synthetic samples affected by EWF (25 k affected by EWF type A, 25 k by type B and 25 k by type C), with different amounts of Multipath applied (applied to all samples, with relative amplitudes w.r.t. LoS one sampled from 0, 0.05, 0.1, 0.2, 0.3, 0.5).

According to cross-validation method defined in previous section, performance depicted in next section for the study case are assessed with considering a training dataset of a size of 80 k samples.

4 Study Case and ROC Performance

The application under study can be so finally synthetized as a AI classification of the runtime observed samples, after an in-factory training phase, distinguishing the typical GNSS signals affected by noise and multipath, and those also distorted by EWF threat or EWF plus additional Multipath, assuming as sufficient variable the receiver correlator samples generated in tracking steady-state conditions.

The noise and the multipath are known to introduce nuisance parameters and "unpredictable" distortions on the nominal GNSS autocorrelation functions degrading SQM figures of merit to distinguish good and EWF waveforms, and vice-versa. These phenomena are also variable in time, especially for multipath contribution in a dynamic user application.

To counteract these nuisance contributions, a smoothing pre-processing stage is here considered applied to the correlators in time, to verify any beneficial effects on the detection scheme performance. Hence, the smoothing time or window has been considered as free parameter for the assessment, with the expectation to obtain AI classification performance trends w.r.t this parameter.

In particular, a window of *W consecutive epochs* in time is considered in the study, to analyze classifier performance w.r.t. the signals and multipath time characteristics measured in live conditions. The observables rate for the application is assumed to be 1 Hz, so each epoch length lasts one second. For each window, a set of 11 smoothed correlators and average C/N0 is obtained as sufficient variables for SQM test, simulating so an output of a Low-Pass Finite Impulse Response (FIR) filter which will be implemented on the input raw observables at 1 Hz rate.

The smoothing time window considered in the assessment spans in a range from a minimum of 1 s (detection over a single unfiltered measurement) up to 100 s. The algorithm performance in terms of false alarm and miss-detection rates have been tested, keeping the same number of smoothed samples for training and test, for each analyzed window size, (see cross-validation definition for partition rule), so this means that for increasing window size a proportionally larger (unsmoothed) raw dataset has been processed.

Receiver operating characteristics (ROC) are reported in the chart of Fig. 6, representing both False alarm rate (FPR) and Miss-detection rate (FNR), as the main figures of merit defined in Sect. 3.1, estimated as function of the smoothing time parameter.

The miss-detection rate curve (i.e. FNR, red colored line) shows to decrease as function of the smoothing time window and reaches a plateau at around 35–40 s smoothing time. This result is linked to the smoothing of the noise and *variable multipath* in time, as per original live datasets acquisition. The benefit is anyway limited by the minor contribution of them w.r.t. the EWF + MP injected synthetic distortions, expected to be higher than nominal noise and multipath, kept constant within the smoothing window, so not removed by low-pass filtering process.

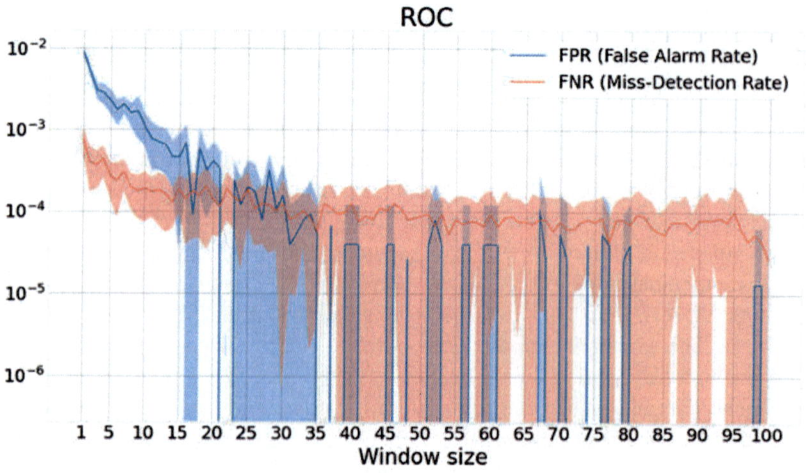

Fig. 6 Receiver Operating Characteristics (ROC): FPR and FNR versus smoothing time

The false alarm rate (FPR, blue colored line) decreases as well with the smoothing time, with a more marked rate and not showing a floor value, at least in the range of analyzed windows. This result is justified by noise and multipath contributions which are filtered out as long as their relevant frequency components in the observables are present and they result higher than low-pass filter cut-off frequency. Some false alarm rates are estimated equal to "zero" in the chart. It is worth to be noted that those inaccurate estimates are linked to the limited number of samples (~100 k) adopted for the assessment and/or an effective rate lower than 1E-5.

It is then worth to be highlighted that, the higher the smoothing time, the lower is the noise and variable multipath contributions, the lower is false alarm rate. On the other hand, a wider smoothing time will produce typically a larger time delay in detection of the EWF threat if present, and in time to alert the user processing SVs raw measurements at 1 Hz rate and potential integrity risk within this delay. So a trade-off between ROC and time to alert shall be performed.

Typical tracking intervals (i.e. visibility) for a SV in space applications can span in LEO orbit between few minutes up to 40 min, whereas in GEO can reach up to few hours, reporting them only as examples.

The levels of miss-detection and false alarm rates reach values around 1E-4 already with a smoothing time of ~35 s. An higher smoothing time seems not useful for miss-detection rate and would be degrading in terms of time to alert, even with benefits in terms of false alarm figure. With such expected rates, the effective number of events in a short visibility window time become acceptable for each specific tracked satellite.

On top of this it is worth to be noted that, these results are also promising if compared to traditional detection schemes specifications. For ground reference stations, in fact, false alarm rates are requested to be of the order 1E-5/1E-6, as function of the visible SV elevation and smoothing time parameter configured with hundreds of seconds, and with miss-detection probabilities below 1E-2.

With proposed detection scheme, performance show to be favorably comparable with above specifications and further assessments can be targeted as function of free parameter and operative conditions.

Finally, to further improve ROC performance and detection delays, aggregates of outputs and decisions, each sampled with a shortened smoothing time window (e.g. 10 s), could be also processed with designing a simple majority voting scheme based on "2 over 3 decisions", to detect or not the threat presence. With this specific decision scheme, the equivalent $FPR_{2/3}$ and $FNR_{2/3}$ can be measured by:

$$FPR_{2/3} = (FPR_{10s})^3 + 3(FPR_{10s})^2(1 - FPR_{10s}) \tag{5}$$

$$FNR_{2/3} = (FNR_{10s})^3 + 3(FNR_{10s})^2(1 - FNR_{10s}) \tag{6}$$

where the false alarm and miss-detection rates $FPR_{10s} = 3\text{E-}4$ and $FNR_{10s} = 1\text{E-}3$, respectively, are extracted at 10 s from the chart of Fig. 6, to be used in (5) and (6).

Considering the majority voting scheme, the decision will be available at most after 30 s, i.e. 3 times the 10 s smoothing time needed for each sub-decisions, with resulting $FPR = 2.7\text{E-}7$ and $FNR = 3\text{E-}6$, so reaching lower probabilities than those obtained with single decision with 35 s smoothing time window. Of course, these results remain valid until the decisions used in the scheme can be assumed independent each other, as it is the case of fast evolving multipath and noise contributions. Otherwise, in case of slow propagating channel, this partitioning can result degrading overall performance, because FPR and FNR would remain equal to those at 10 s (worse than those obtained with detection at 35 s).

4.1 Conclusions and Way Forward

While still requiring fine tuning and further analysis, with increasing the datasets size and modifying sufficient variables involved, the studied approach already at this stage shows to be very promising and the performance very good with respect to both execution speed and metrics.

The next step of the research foresees to include new observables distorted by EWF threats, obtained by using dedicated RF signals generator to feed GNSS receiver, including as well controlled multipath, instead of synthetic observables from live ones, to verify and validate the obtained promising results also in more realistic conditions.

Future works as well foresee to deeply investigate AI schemes and performance, with comparison against traditional models of detection, i.e. correlators ratio metrics based, which are typical for ground reference stations (e.g. EGNOS [2]). In this frame, a possible integration of AI scheme with traditional ones can be also targeted, to increase overall performance.

Furthermore, next generation EGNOS will provide more accuracy, extending augmentation service also to Dual-Frequency multi-constellation receivers, processing all Safety-of-Life (SoL) signals—GPS L1CA, L5, Galileo E1, E5-and protecting users in related applications with redundancy in frequency domain. As well, space application are going to make standard the usage of a dual-frequency PVT solution, to mitigate ionosphere delay errors in navigation performance. For these reasons, a further area of research and interest will cover also SQM assessments for all Safety of Life (SoL) signals and new models of RF threats distortion.

Data Availability Statement: further information on the proposed method and dataset analyzed during the current study are available from the corresponding author (andrea.emmanuele@thalesaleniaspace.com) on reasonable request.

References

1. Zin, A., Zago, S., Ferrario, A., Piccolo, A., Colombo, A.E., Siniscalco, L., Marradi, L., Martella, D., Bidart, P.: Reaching and navigating in geostationary orbit with a GNSS receiver: challenges and in-flight results. In: Proceedings of the 34th International Technical Meeting of the Satellite Division of the Institute of Navigation (ION GNSS+ 2021), pp. 1018–1032. St. Louis (2021)
2. Bauer, F., Greze, G., Haddad, F., Tourtier, A., Rols, B., Urbanska, K.: A study on a New EGNOS V2 release with enhanced system performances. In: Proceedings of the 32nd International Technical Meeting of the Satellite Division of the Institute of Navigation (ION GNSS+ 2019). Miami (2019)
3. Siemuri, A., Kuusniemi, H., Elmusrati, M. S., Välisuo, P., Shamsuzzoha, A.: Machine learning utilization in GNSS—use cases, challenges and future applications. In: 2021 International Conference on Localization and GNSS (ICL-GNSS) (2021)
4. Hsu, L.: GNSS multipath detection using a machine learning approach. In: 2017 IEEE 20th International Conference on Intelligent Transportation Systems (ITSC), pp. 1–6. Yokohama (2017)
5. Louis, A., Raimondi, M.: Neural network based evil waveforms detection. In: Proceedings of the 33rd International Technical Meeting of the Satellite Division of the Institute of Navigation (ION GNSS+ 2020), pp. 1984–1989 (2020)
6. Burges, C.J.C.: A tutorial on Support Vector Machines for pattern recognition. Data Min. Knowl. Discov. **2**, 121–167 (1998)
7. Abdiansah, N.: Time complexity analysis of Support Vector Machines (SVM) in LibSVM. Int. J. Comput. Appl. **128**, (2015). https://doi.org/10.5120/ijca2015906480
8. ICAO. Annex 10-Aeronautical Telecommunications, vol. 1, 6th edn. Radio Navigation Aids (Internet) (2006)